Lecture Notes in Mathematics 1774

Editors:
J.-M. Morel, Cachan
F. Takens, Groningen
B. Teissier, Paris

Springer
Berlin
Heidelberg
New York
Barcelona
Hong Kong
London
Milan
Paris
Tokyo

Volker Runde

Lectures on Amenability

 Springer

Author

Volker Runde
Department of Mathematical and Statistical Sciences
University of Alberta
Edmonton AB
Canada T6G 2G1

E-mail: *vrunde@ualberta.ca*
http://www.math.ualberta.ca/~runde/runde.html

Cataloging-in-Publication Data applied for.

Die Deutsche Bibliothek - CIP-Einheitsaufnahme

Runde, Volker:
Lectures on amenability / Volker Runde. - Berlin ; Heidelberg ; New York ;
Barcelona ; Hong Kong ; London ; Milan ; Paris ; Tokyo : Springer, 2002
  (Lecture notes in mathematics ; 1774)
  ISBN 3-540-42852-6

Mathematics Subject Classification (2000):
46H20 (PRIMARY), 46H25, 46B20, 46L10, 46L05, 46L06, 46L07, 46M18, 46M20, 43A07,
43A20, 22D15, 22D20, 58B99, 47B47, 47B48, 47L25, 46B28

ISSN 0075-8434
ISBN 3-540-42852-6 Springer-Verlag Berlin Heidelberg New York

Springer-Verlag Berlin Heidelberg New York a member of BertelsmannSpringer
Science + Business Media GmbH

http://www.springer.de

© Springer-Verlag Berlin Heidelberg 2002
Printed in Germany

The use of general descriptive names, registered names, trademarks, etc. in this publication does not imply,
even in the absence of a specific statement, that such names are exempt from the relevant protective laws
and regulations and therefore free for general use.

Typesetting: Camera-ready TₑX output by the author

SPIN: 10856657   41/3142-543210 - Printed on acid-free paper

Dedicated to
Annemarie Runde, née Scherer,
1939–1986,
in loving, everlasting memory

Preface

The notion of amenability has its origins in the beginnings of modern measure theory: Does there exist a finitely additive set function which is invariant under a certain group action? Since the 1940s, amenability has become an important concept in abstract harmonic analysis (or rather, more generally, in the theory of semitopological semigroups). In 1972, B. E. Johnson showed that the amenability of a locally compact group G can be characterized in terms of the Hochschild cohomology of its group algebra $L^1(G)$: this initiated the theory of amenable Banach algebras. Since then, amenability has penetrated other branches of mathematics, such as von Neumann algebras, operator spaces, and even differential geometry.

In the summer term of 1999, I taught a course on amenability at the Universität des Saarlandes. My goals were lofty: I wanted to show my students how the concept of amenability orginated from measure theoretic problems — of course, the Banach–Tarski paradox would have to be covered —, how it moved from there to abstract harmonic analysis, how it then ventured into the theory of Banach algebras, and how it impacted areas as diverse as von Neumann algebras and differential geometry. I had also planned to include very recent developments such as C. J. Read's construction of a commutative, radical, amenable Banach algebra or Z.-J. Ruan's notion of operator amenability. On top of all this, I wanted my lectures to be accessible to students who had taken a one-year course in functional analysis (including the basics of Banach and C^*-algebras), but who had not not necessarily any background in operator algebras, homological algebra, or abstract harmonic analysis. Lofty as they were, these goals were unattainable, of course, in a one-semester course...

The present notes are an attempt to resurrect the original plan of my lectures at least in written form. They are a polished (and greatly expanded) version of the notes I actually used in class. In particular, this is *not* a research monograph that exhaustively presents our current state of knowledge on amenability. These notes are intended to introduce second year graduate students to a fascinating area of modern mathematics and lead them, within a reasonable period of time, to a level from where they can go on to read original research papers on the subject. This has, of course, influenced the exposition: The order in which material is presented is pedagogically rather than systematically motivated, and the style is, in general, chatty and informal.

I am a firm believer in learning by doing: Nobody has ever learned anything — especially not mathematics! — just by passively absorbing a teacher's performance. For this reason, there are numerous exercises interspersed in the main text of these notes (I didn't count them). These exercises vary greatly in their degree of difficulty: Some just ask the reader to make a fairly obvious, but nevertheless important observation, others ask him/her to fill in a tedious detail of a proof, and again others challenge him/her to deepen his/her understanding of the material by actively wrestling with it. My frequent interjections in the text serve a similar purpose: Just because something is claimed to be clear/obvious/immediate, that doesn't mean that one can just believe it without really checking.

The background required from a student working through these notes varies somewhat throughout the text:

- **Chapter 0**: This chapter only requires a very modest background in linear and abstract algebra, along with some mathematical maturity.
- **Chapter 1**: Anyone who has taken a course in measure theory and a first course in functional analysis should be able to read this chapter; the necessary background from abstract harmonic analysis is put together in Appendix A.
- **Chapter 2**: If you know the basics of Banach algebra theory, you're fit for this chapter. In addition, some facts about Banach space tensor products are needed: Appendix B contains a crash course on Banach space tensor products that provides the necessary background.
- **Chapter 3**: As for Chapter 2; additional background material from Banach space theory is collected in Appendix C.
- **Chapter 4**: You need a bit more background from Banach algebra theory than in the previous chapters. Also, a few basic facts on von Neumann algebras and one deep C^*-algebraic result are required. References are provided.
- **Chapter 5**: As for the previous three chapters. Some previous exposure to the language of categories and functors helps, of course, but isn't necessary. If you already know about homological algebra, this chapter should be very easy for you.
- **Chapter 6**: This is the most challenging chapter of these notes. Although references to standard texts on C^*- and von Neumann algebras are provided wherever necessary, anyone who wishes to get something out of this chapter, needs a certain degree of fluency in operator algebras.
- **Chapter 7**: As for Chapters 2 to 5; the necessary background from operator spaces can be found in Appendix D.
- **Chapter 8**: This chapter builds on Chapter 2 (and to a much lesser extent on Chapter 4). It contains the necessary background from infinite-dimensional differential geometry.

Each chapter concludes with a section entitled *Notes and comments*: These sections contain references to the original literature, as well as outlines of results that were not included in the main text, along with suggestions for further reading.

Since these notes are not intended to be a monograph, I have perhaps been a little less accurate when it comes to giving credit for results than I would have been otherwise. Hence, the universal disclaimer applies: Just because I didn't explicitly attribute a result to someone else, that doesn't mean that it's due to me. Furthermore, these notes perfectly reflect my preferences and prejudices towards amenability: It is my (highly subjective) belief that Banach algebras are the natural setting in which to deal with amenability. For this reason, most of these notes are devoted to amenable Banach algebras and their next of kin. Amenable groups are only presented to set the stage for amenable Banach algebras; amenable semigroups, (semi)group actions, representations, groupoids, etc., aren't even mentioned. This does not mean that I consider these topics unimportant or uninteresting: I just don't know enough about these things (yet) to be able to teach a course on them, let alone publish my lecture notes.

Finally, I would like to thank my students who attended the actual *Lectures on Amenability* at Saarbrücken and who made the course a joy to teach: Kim Louis, Matthias Neufang, Stefanie Schmidt, and a fourth one whom everyone only knew as "the algebraist". I would also like to thank H. Garth Dales of the University of Leeds for his encouragement when he learned of my project of turning my class notes into some sort of book. Thanks are especially due to Matthias Neufang and Ross Stokke who read preliminary versions of the entire manuscript; both prevented an embarassing number of errors from making it into the final version. Of course, all omissions, inaccuracies, and outright errors that remain are my fault alone.

By the way, unless explicitly stated otherwise, all spaces and algebras in these notes are over $\mathbb{C}$.

Edmonton, August 2001   *Volker Runde*

Table of Contents

0 Paradoxical decompositions

There is a mathematical theorem which implies the following:

> *An orange can be cut into finitely many pieces, and these pieces can be reassembled to yield two oranges of the same size as the original one.*

Admittedly, this sounds more like a version of the feeding of the five thousand than like rigorous mathematics. Nevertheless, it is a perfectly sound application of the Banach–Tarski paradox, whose strongest form is:

> *Let A and B be any two bounded sets in three-dimensional space with non-empty interior. Then there is a partition of A into finitely many sets which can be reassembled to yield B.*

Applications much more bizarre than just making two oranges out of one come to mind immediately:

> *A pea can be split into finitely many pieces which can be recombined to yield a life-sized statue of Stefan Banach (or a solid ball whose diameter is larger than the distance of the Earth from the sun).*

Can a theorem whose consequences so blatantly defy common sense be true? In fact, there is an element of faith to the Banach–Tarski paradox (so that it is not all that removed from the feeding of the five thousand). Its proof rests on two pillars:

– the axiom of choice (and you can believe in that or leave it), and
– the fact that the free group in two generators lacks a property called *amenability*.

In fact, the Banach–Tarski paradox is just one instance of so-called paradoxical decompositions: Some set can be split up and then be recombined to yield two copies of itself. As we shall see in this chapter, paradoxical decompositions (or rather: their non-existence) are at the very heart of the notion of amenability.

0.1 The Banach–Tarski paradox

We begin by making precise what we mean when we say that some set admits a paradoxical decomposition:

Definition 0.1.1 Let G be a group which acts on a (non-empty) set X. Then $E \subset X$ is called *G-paradoxical* if there are pairwise disjoint subsets $A_1, \dots, A_n, B_1, \dots, B_m$ of E along with $g_1, \dots, g_n, h_1, \dots, h_m \in G$ such that $E = \bigcup_{j=1}^{n} g_j \cdot A_j$ and $E = \bigcup_{j=1}^{m} h_j \cdot B_j$.

Exercise 0.1.1 Let G be a group which acts on a set X, let H be a subgroup of G, and let $E \subset X$ be H-paradoxical. Show that E is G-paradoxical.

We shall simply speak of *paradoxical* sets (instead of using the slightly lengthier adjective G-paradoxical) in the following two cases:

- G acts on itself via multiplication from the left, and
- X is a metric space, and G is the group of invertible isometries on X.

Theorem 0.1.2 *The free group in two generators is paradoxical.*

Proof Let a and b be two generators of $\mathbb{F}_2$. For $x \in \{a, b, a^{-1}, b^{-1}\}$, let

$$W(x) := \{w \in \mathbb{F}_2 : w \text{ starts with } x\}.$$

Then

$$\mathbb{F}_2 = \{e_{\mathbb{F}_2}\} \cup W(a) \cup W(b) \cup W(a^{-1}) \cup W(b^{-1}),$$

the union being disjoint. Now observe that, for any $w \in \mathbb{F}_2 \setminus W(a)$, we have $a^{-1}w \in W(a^{-1})$, so that $w = a(a^{-1}w) \in aW(a^{-1})$. It follows that $\mathbb{F}_2 = W(a) \cup aW(a^{-1})$ and, similarly, $\mathbb{F}_2 = W(b) \cup bW(b^{-1})$. $\square$

We shall now see that, given a paradoxical group G that acts on some set X, a mild demand on the group action forces X to G-paradoxical. This is the first time that we encounter the axiom of choice; we indicate the dependence on the axiom of choice by **(AC)**.

Proposition 0.1.3 (AC) *Let G be a paradoxical group which acts on X without non-trivial fixed points, i.e. if there are $g \in G$ and $x \in X$ such that $g \cdot x = x$, then $g = e_G$. Then X is G-paradoxical.*

Proof Let $A_1, \ldots, A_n, B_1, \ldots, B_m \subset G$ and $g_1, \ldots, g_n, h_1, \ldots, h_m$ be as in Definition 0.1.1. Choose a set $M \subset X$ such that M contains exactly one element from every G-orbit. We claim that $\{g \cdot M : g \in G\}$ is a disjoint partition of X. Certainly, $\bigcup_{g \in G} g \cdot M = X$ (since M contains one point from every G-orbit). Suppose now that there are $g, h \in G$ and $x, y \in M$ such that $g \cdot x = h \cdot y$. Then $h^{-1}g \cdot x = y$, so that $x, y \in M$ are in the same G-orbit and thus must be equal. Since G acts on X without non-trivial fixed points, this means $h^{-1}g = e_G$. Let

$$A_j^* = \bigcup \{g \cdot M : g \in A_j\} \qquad (j = 1, \ldots, n)$$

and

$$B_j^* = \bigcup \{g \cdot M : g \in B_j\} \qquad (j = 1, \ldots, m).$$

Clearly, $A_1^*, \ldots, A_n^*, B_1^*, \ldots, B_m^*$ are pairwise disjoint subsets of X such that $X = \bigcup_{j=1}^n g_j \cdot A_j^* = \bigcup_{j=1}^m h_j \cdot B_j^*$. $\square$

Exercise 0.1.2 Show where exactly in the proof of Proposition 0.1.3 the axiom of choice was invoked.

So, suppose that $\mathbb{F}_2$ acts without non-trivial fixed points on some set X. Then Theorem 0.1.2 and Proposition 0.1.3 combined yield that X is $\mathbb{F}_2$-paradoxical. Since — in view of the Banach–Tarski paradox — we want to show that certain subsets of $\mathbb{R}^3$ are paradoxical, we are thus faced with the problem of making $\mathbb{F}_2$ act on $\mathbb{R}^3$ as invertible isometries.

For $n \in \mathbb{N}$, we use the symbol $O(n)$ to denote the real $n \times n$-matrices A such that $A^t A = A A^t = E_n$, where E_n is the identity matrix; $O(n)$ is called the n-dimensional *orthogonal group*. The n-dimensional *special orthogonal group* $SO(n)$ is the subgroup of $O(n)$ consisting of those matrices $A \in O(n)$ such that $\det A = 1$.

Theorem 0.1.4 *There are rotations A and B about lines through the origin in $\mathbb{R}^3$ which generate a subgroup of $SO(3)$ isomorphic to $\mathbb{F}_2$.*

Proof Let

$$A^{\pm} = \begin{bmatrix} \frac{1}{3} & \mp\frac{2\sqrt{2}}{3} & 0 \\ \pm\frac{2\sqrt{2}}{3} & \frac{1}{3} & 0 \\ 0 & 0 & 1 \end{bmatrix} \quad \text{and} \quad B^{\pm} = \begin{bmatrix} 1 & 0 & 0 \\ 0 & \frac{1}{3} & \mp\frac{2\sqrt{2}}{3} \\ 0 & \pm\frac{2\sqrt{2}}{3} & \frac{1}{3} \end{bmatrix}.$$

Let w be a reduced word in A, B, A^{-1}, and B^{-1} which is not the empty word ϵ. We claim that w cannot act as the identity on $\mathbb{R}^3$. To see this, first note that we may assume without loss of generality that w ends in $A^{\pm}$ (otherwise, conjugate with $A^{\pm}$). We claim that

$$w \cdot \begin{bmatrix} 1 \\ 0 \\ 0 \end{bmatrix} = \frac{1}{3^k} \begin{bmatrix} a \\ b\sqrt{2} \\ c \end{bmatrix},$$

where $a, b, c \in \mathbb{Z}$ with $3 \nmid b$ and k is the length of w. This certainly establishes the claim that w does not act as the identity on $\mathbb{R}^3$.

We proceed by induction on k. Suppose first that $k = 1$, i.e. $w = A^{\pm}$, so that

$$w \cdot \begin{bmatrix} 1 \\ 0 \\ 0 \end{bmatrix} = \frac{1}{3} \begin{bmatrix} 1 \\ \pm 2\sqrt{2} \\ 0 \end{bmatrix}.$$

Let $w = A^{\pm}w'$ or $w = B^{\pm}w'$, where

$$w' \cdot \begin{bmatrix} 1 \\ 0 \\ 0 \end{bmatrix} = \frac{1}{3^{k-1}} \begin{bmatrix} a' \\ b'\sqrt{2} \\ c' \end{bmatrix}$$

with $a', b', c' \in \mathbb{Z}$ and $3 \nmid b'$. It follows that

$$w \cdot \begin{bmatrix} 1 \\ 0 \\ 0 \end{bmatrix} = \frac{1}{3^k} \begin{bmatrix} a \\ b\sqrt{2} \\ c \end{bmatrix},$$

where

$$\begin{cases} a = a' \mp 4b', \; b = b' \pm 2a', \quad c = 3c', \text{ in case } w = A^{\pm}w'; \\ a = 3a', \; b = b' \mp 2c', \; c = c' \pm 4b', \text{ in case } w = B^{\pm}w'. \end{cases}$$

It is clear from these identities that $a, b, c \in \mathbb{Z}$. What remains to be shown is that $3 \nmid b$.

Case 1: $w = A^{\pm}B^{\pm}v$ (where possibly $v = \epsilon$).

Then $b = b' \mp 2a'$ with $3|a'$. Since $3 \nmid b'$ by assumption, it follows that $3 \nmid b$.

Case 2: $w = B^{\pm}A^{\pm}v$.

Then $b = b' \mp 2c'$ with $3|c'$. Again, $3 \nmid b'$, so that $3 \nmid b$.

Case 3: $w = A^{\pm}A^{\pm}v$.

By assumption,

$$v \cdot \begin{bmatrix} 1 \\ 0 \\ 0 \end{bmatrix} = \frac{1}{3^{k-2}} \begin{bmatrix} a'' \\ b''\sqrt{2} \\ c'' \end{bmatrix}$$

with $a'', b'', c'' \in \mathbb{Z}$. It follows that

$$b = b' \pm 2a' = b \pm 2(a'' \mp 4b'') = b' + b'' \pm 2a'' - 9b'' = 2b' - 9b'',$$

so that $3 \nmid b$.

Case 4: $w = B^{\pm}B^{\pm}v$.

This is treated like Case 3. □

Exercise 0.1.3 Show that $SO(n)$, for each $n \geq 3$, contains a subgroup isomorphic to $\mathbb{F}_2$.

With $\mathbb{F}_2$ as a subgroup of $SO(3)$ and Proposition 0.1.3 at hand, we can now prove the first classical result on paradoxical decompositions. As is customary, we write S^2 for the unit sphere in $\mathbb{R}^3$:

Theorem 0.1.5 (Hausdorff paradox; AC) *There is a countable subset D of S^2 such that $S^2 \setminus D$ is $SO(3)$-paradoxical.*

Proof Let A and B be as in the proof of Theorem 0.1.4. Then the subgroup G of $SO(3)$ generated by A and B is isomorphic to $\mathbb{F}_2$. Since A and B are rotations about lines through the origin, each $w \in G \setminus \{e_G\}$ has exactly two fixed points in S^2. The set

$$F := \{x \in S^2 : x \text{ is a fixed point for some } w \in G \setminus \{e_G\}\}$$

is thus countable, and so is $D := \bigcup\{w \cdot F : w \in G\}$. Clearly, G acts on $S^2 \setminus D$ without non-trivial fixed points, so that, by Proposition 0.1.3, $S^2 \setminus D$ is G-paradoxical. By Exercise 0.1.1, $S^2 \setminus D$ is $SO(3)$-paradoxical. □

As we shall see very soon, S^2 itself is $SO(3)$-paradoxical.

Definition 0.1.6 Let G be a group which acts on a set X, and let A and B be subsets of X. Then A and B are called G-*equidecomposable* if there are $A_1, \ldots, A_n \subset A$, $B_1, \ldots, B_n \subset B$ and $g_1, \ldots, g_n \in G$ such that:

(i) $A = \bigcup_{j=1}^{n} A_j$ and $B = \bigcup_{j=1}^{n} B_j$;

(ii) $A_j \cap A_k = \varnothing = B_j \cap B_k$ $(j, k \in \{1, \ldots, n\}, j \neq k)$;

(iii) $g_j \cdot A_j = B_j$ $(j = 1, \ldots, n)$.

If A and B are G-equidecomposable, we write $A \sim_G B$. In case G is the group of invertible isometries of a metric space (or if G is obvious), we simply write $A \sim B$.

Exercise 0.1.4 Show that $\sim_G$ is an equivalence relation.

Exercise 0.1.5 Let G be a group that acts on a set X, and let $A, A_1, A_2, B, B_1, B_2 \subset X$. Show that the following hold:

(i) If $A \sim B$, then there is a bijection $\phi \colon A \to B$ with $C \sim \phi(C)$ for all $C \subset A$.

(ii) If $A_1 \cap A_2 = \varnothing = B_1 \cap B_2$ with $A_1 \sim B_1$ and $A_2 \sim B_2$, then $A_1 \cup A_2 \sim B_1 \cup B_2$.

Proposition 0.1.7 *Let $D \subset S^2$ be countable. Then S^2 and $S^2 \setminus D$ are $SO(3)$-equidecomposable.*

Proof Let L be a line in $\mathbb{R}^3$ through the origin such that $L \cap D = \varnothing$, and let W be the collection of all $\theta \in [0, 2\pi)$ with the follow property:

There is $x \in D$ such that $\rho \cdot x \in D$ as well, where — for some $n \in \mathbb{N}$ — ρ is the rotation about L by the angle $n\theta$.

It is clear from the definition that W is countable, so that there is $\theta \in [0, 2\pi) \setminus W$. Let ρ be the rotation about L by the angle θ, then $\rho^n \cdot D \cap D = \varnothing$ for all $n \in \mathbb{N}$. It follows that

$$\rho^n \cdot D \cap \rho^m \cdot D = \varnothing \qquad (n, m \in \mathbb{N},\ n \neq m).$$

Let $\tilde{D} = \bigcup_{n=0}^{\infty} \rho^n \cdot D$. Then

$$S^2 = \tilde{D} \cup (S^2 \setminus \tilde{D}) \sim \rho \cdot \tilde{D} \cup (S^2 \setminus \tilde{D}) = S^2 \setminus D,$$

which establishes the claim. $\square$

The relevance of G-equidecomposability becomes apparent in the next proposition:

Proposition 0.1.8 *Let G be a group that acts on a set X, and let E and E' be subsets of X with $E \sim_G E'$. Then, if E is G-paradoxical, so is E'.*

Proof Let $A_1, \ldots, A_n, B_1, \ldots, B_m \subset E$ be as in in Definition 0.1.1, and let

$$A := \bigcup_{j=1}^{n} A_j \qquad \text{and} \qquad B := \bigcup_{j=1}^{n} B_j,$$

so that $A \sim_G E$ and $B \sim_G E$. From Exercise 0.1.4, it follows that $A \sim_G E'$ and $B \sim_G E'$. This implies that E' is also G-paradoxical. $\square$

Exercise 0.1.6 Verify the last statement in the proof of Proposition 0.1.8.

Together, Theorem 0.1.5, Proposition 0.1.7, and Proposition 0.1.8 yield a stronger version of the Hausdorff paradox:

Corollary 0.1.9 (AC) *S^2 is $SO(3)$-paradoxical.*

We are also now in a position to prove a weak version of the Banach–Tarski paradox (which is already sufficient if all you want is to make two oranges out of one):

Corollary 0.1.10 (Weak Banach–Tarski paradox; AC) *Every closed ball in $\mathbb{R}^3$ is paradoxical.*

Proof It is sufficient to prove that the closed unit ball $B_1[0, \mathbb{R}^3]$ is paradoxical. We first show that $B_1[0, \mathbb{R}^3] \setminus \{0\}$ is paradoxical. Since S^2 is $SO(3)$-paradoxical by Corollary 0.1.9, there are $A_1, \ldots, A_n, B_1, \ldots, B_m \subset S^2 \subset \mathbb{R}^3$ and $g_1, \ldots, g_n, h_1, \ldots, h_m \in SO(3)$ as in Definition 0.1.1. Let

$$A_j^* := \{tx : t \in (0,1], x \in A_j\} \qquad (j = 1, \ldots, n)$$

and

$$B_j^* := \{tx : t \in (0,1], x \in B_j\} \qquad (j = 1, \ldots, m).$$

Then, certainly, $A_1^*, \ldots, A_n^*, B_1^*, \ldots, B_m^* \subset B_1[0,1; \mathbb{R}^3] \setminus \{0\}$ are pairwise disjoint and $B_1[0, \mathbb{R}^3] \setminus \{0\} = \bigcup_{j=1}^n g_j \cdot A_j^* = \bigcup_{j=1}^m h_j \cdot B_j^*$. It follows that $B_1[0, \mathbb{R}^3] \setminus \{0\}$ is indeed paradoxical.

Let $x = \left(0, 0, \frac{1}{2}\right) \in B_1[0, \mathbb{R}^3] \setminus \{0\}$. Let ρ be a rotation of infinite order about a line through x that misses the origin. Let $D := \{\rho^n \cdot 0 : n \in \mathbb{N}_0\}$. Clearly, $\rho \cdot D = D \setminus \{0\}$. Then

$$B_1[0, \mathbb{R}^3] = D \cup B_1[0, \mathbb{R}^3] \setminus D \sim \rho \cdot D \cup B_1[0, \mathbb{R}^3] \setminus D = B_1[0, \mathbb{R}^3] \setminus \{0\},$$

so that $B_1[0, \mathbb{R}^3]$ is paradoxical by Proposition 0.1.8. $\square$

Exercise 0.1.7 Show that $\mathbb{R}^3$ is paradoxical.

Remark 0.1.11 A classical result by S. Mazur and S. Ulam ([M–U]) asserts that a surjective isometry between real Banach spaces is already linear if it maps the origin to the origin. It follows that any surjective isometry between Banach spaces is the composition of a surjective, $\mathbb{R}$-linear isometry with a translation. For the invertible isometries on $\mathbb{R}^3$, this means that every such isometry is the composition of an element of $O(3)$ with a translation.

To establish the Banach–Tarski paradox in its strongest form, we need another definition:

Definition 0.1.12 Let G be a group that acts on the set X, and let A and B be subsets of X. We write $A \preceq_G B$ (or simply: $A \preceq B$) if A and a subset of B are G-equidecomposable.

Exercise 0.1.8 Show that $\preceq_G$ is a reflexive and transitive relation on the equivalence classes of $\mathfrak{P}(X)$ with respect to $\sim_G$.

The following is an analogue of the Cantor–Bernstein theorem for $\preceq$:

Theorem 0.1.13 *Let G be a group that acts on a set X, and let A and B be subsets of X such that $A \preceq_G B$ and $B \preceq_G A$. Then $A \sim_G B$.*

Proof Let $B_1 \subset B$ and $A_1 \subset A$ be such that $A \sim B_1$ and $B \sim A_1$. Let $\phi : A \to B_1$ and $\psi : B \to A_1$ be bijections as in Exercise 0.1.5(i). Let $C_0 := A \setminus A_1$, and define $C_{n+1} := \psi(\phi(C_n))$. Let $C := \bigcup_{n=0}^\infty C_n$. It follows that $\psi^{-1}(A \setminus C) = B \setminus \phi(C)$. By Exercise 0.1.5(i), this means that $A \setminus C \sim B \setminus \phi(C)$. In a similar fashion, we see that $C \sim \phi(C)$. It follows that

$$A = (A \setminus C) \cup C \sim (B \setminus \phi(C)) \cup \phi(C) = B$$

by Exercise 0.1.5(ii). $\square$

We can now prove the Banach–Tarski paradox:

Theorem 0.1.14 (Banach–Tarski paradox; AC) *Let A and B be bounded subsets of $\mathbb{R}^3$, each with non-empty interior. Then $A \sim B$.*

Proof By symmetry and by Theorem 0.1.13, it is sufficient to prove that $A \preceq B$. Since A is bounded, there is $r > 0$ such that $A \subset B_r[0, \mathbb{R}^3]$. Let x be an interior point of B. Then there is $\epsilon > 0$ such that $B_\epsilon[x, \mathbb{R}^3] \subset B$. Since $B_r[0, \mathbb{R}^3]$ is compact, there are invertible isometries $g_1, \dots, g_n \colon \mathbb{R}^3 \to \mathbb{R}^3$ (in fact, translations will do) such that

$$B_r[0, \mathbb{R}^3] \subset g_1 \cdot B_\epsilon[x, \mathbb{R}^3] \cup \cdots \cup g_n \cdot B_\epsilon[x, \mathbb{R}^3].$$

Choose isometries $h_1, \dots, h_n$ on $\mathbb{R}^3$ such that $h_j \cdot B_\epsilon[x, \mathbb{R}^3] \cap h_k \cdot B_\epsilon[x, \mathbb{R}^3] = \varnothing$ for $j \neq k$ (again, translations will do). Let $S := \bigcup_{j=1}^n h_j \cdot B_\epsilon[x, \mathbb{R}^3]$. It follows from the weak Banach–Tarski paradox Corollary 0.1.10, that $S \preceq B_\epsilon[x, \mathbb{R}^3]$ (Why?). Then

$$A \subset B_r[0, \mathbb{R}^3] \subset g_1 \cdot B_\epsilon[x, \mathbb{R}^3] \cup \cdots \cup g_n \cdot B_\epsilon[x, \mathbb{R}^3] \preceq S \preceq B_\epsilon[x, \mathbb{R}^3] \subset B$$

which establishes the claim. □

Exercise 0.1.9 Where exactly in the proof of Theorem 0.1.14 was the axiom of choice used?

0.2 Tarski's theorem

We have mentioned in the introduction of this chapter that one of the key ingredients for the proof of the Banach–Tarski paradox was that $\mathbb{F}_2$ lacks the property of amenability. We have not yet defined what amenability is, but we won't spill any beans by already admitting that for a (discrete) group amenability is the same as not being paradoxical. Nevertheless, the formal definition of amenability, which will be given at the end of this section, looks quite different from Definition 0.1.1.

The reason why these two notions are equivalent is the following deep theorem due to A. Tarski:

Theorem 0.2.1 (Tarski's theorem; AC) *Let G be a group that acts on a set X, and let E be subset of X. Then there is a finitely additive, G-invariant set function $\mu \colon \mathfrak{P}(X) \to [0, \infty]$ with $\mu(E) \in (0, \infty)$ if and only if E is not G-paradoxical.*

Exercise 0.2.1 Let G be a group that acts on a set X, and let $E \subset X$ be G-paradoxical. If $\mu \colon \mathfrak{P}(X) \to [0, \infty]$ is any finitely additive, G-invariant set function, show that $\mu(E) = 0$ or $\mu(E) = \infty$ (and thus prove the easy direction of Tarski's theorem). Do you need the axiom of choice?

The proof of the difficult direction of Tarski's theorem requires some preparation.

We start with some notions from graph theory: A (undirected) *graph* is a triple (V, E, ϕ), where V and E are non-empty sets and ϕ is a map from E into the unordered pairs of elements of V. The elements of V are called *vertices* and the elements of E are called *edges*. If $e \in E$ and $v, w \in V$ with $\phi(e) = (v, w)$, we say that e *joins* v and w or that v and w are the *endpoints* of e; we shall also sometimes be sloppy and simply identify the edge e with its image under ϕ. A *path* in (V, E, ϕ) is a finite sequence $(e_1, \dots, e_n)$ of edges together with a finite sequence $(v_0, \dots, v_n)$ of vertices where v_0 is an endpoint of e_1, v_n is an endpoint of e_n, and v_j is an endpoint of e_j for $j = 1, \dots, n-1$. We say that v_0 and v_n are *joined* by the path. For formal reasons, we will also say that the empty path joins each vertex with itself.

Definition 0.2.2 Let (V, E, ϕ) be a graph, and let $k \in \mathbb{N}$.

(i) If each vertex is the endpoint of exactly k edges, then (V, E, ϕ) is called k-*regular*.

(ii) If $V = X \cup Y$, where X and Y are disjoint, and each edge has one endpoint in X and one in Y, then (V, E, ϕ) is called *bipartite*.

In what follows we shall simply speak of a bipartite graph (X, Y, E, ϕ) when we mean that (V, E, ϕ) is a bipartite graph with V, X and Y as in Definition 0.2.2(ii).

Definition 0.2.3 Let (X, Y, E, ϕ) be a bipartite graph, and let $A \subset X$ and $B \subset Y$. A *perfect matching* of A and B is a subset F of E such that

(i) each element of $A \cup B$ is an endpoint of exactly one $f \in F$, and

(ii) all endpoints of edges in F are in $A \cup B$.

Exercise 0.2.2 Let (X, Y, E, ϕ) be a bipartite graph which is k-regular for some $k \in \mathbb{N}$, and suppose that $|V| < \infty$.

(i) Show that $|E| < \infty$ and that $|X| = |Y|$.

(ii) For any $M \subset V$, let $N(M)$ be the set of those vertices which are joined by an edge with a point of M. Show that $|N(M)| \geq |M|$.

(iii) Let $A \subset X$ and $B \subset Y$ be such that there is a perfect matching F of A and B with $|F|$ maximal. Show that $A = X$. (*Hint:* Assume that there is $x \in X \setminus A$, and consider the set Z of those vertices z which are joined with x by a path $(e_1, \ldots, e_n)$ such that the e_j's lie alternately in F and $E \setminus F$. Use the maximality of $|F|$ to show that $Z \cap Y \subset B$. Then show that $Z \cap Y = N(Z \cap X)$ and use (ii) to arrive at a contradiction.)

(iv) (Marriage theorem) Conclude that there is a perfect matching of X and Y.

The reason why the assertion of Exercise 0.2.2(iv) carries the catchy name "marriage theorem" is the following application: If each girl in town finds exactly k boys attractive and each boy in town finds exactly k girls attractive, then it is possible to arrange marriages between them in such a way that, in each marriage, the partners find one another attractive. How comforting to know ...

We shall now prove König's theorem, which is simply the marriage theorem without any finiteness requirements:

Theorem 0.2.4 (König's theorem; AC) *Let* (X, Y, E, ϕ) *be a bipartite graph which is* k-*regular for some* $k \in \mathbb{N}$. *Then there is a perfect matching of* X *and* Y.

Proof Define an equivalence relation on V: Two vertices are equivalent if they can be joined by a path (remember that each vertex is joined with itself by the empty path). Each equivalence class with respect to this relation is again a k-regular, bipartite graph and has countably many vertices and edges (by k-regularity). If we can find a perfect matching for each such graph, we can put them together to form a matching for X and Y. We may thus suppose that X, Y and E are countably infinite, and that any two points in V can be joined by a path.

Let $\{e_n : n \in \mathbb{N}\}$ be an enumeration of E. Consider the collection of all finite sequences in $\{0, 1\}$; if s and t are such sequences, we write $s \leq t$ if s is an initial segment of t. We call such a sequence s of length n *good* if there are finite sets $V' \subset V$ and $E' \subset E$ such that:

- V' contains all vertices that appear as endpoints of $e_1, \ldots, e_n$;
- $\{e_1, \ldots, e_n\} \subset E'$;
- $(V', E', \phi|_{E'})$ is a k-regular, bipartite graph which has a perfect matching F' for $V' \cap X$ and $V' \cap Y$ such that, for $j < n$, $e_j \in F'$ if and only if $s_j = 1$.

From this definition, it is clear that if $s \leq t$ and if t is good, then so is s.

If $V' \subset V$ and $E' \subset E$ are finite sets such that $(V', E', \phi|_{E'})$ is a graph, then there are finite subsets $V'' \subset V$ and $E'' \subset E$ with $V' \subset V''$ and $E' \subset E''$ such that $(V'', E'', \phi|_{E''})$ is a bipartite, k-regular graph. (How exactly is this accomplished?). By Exercise 0.2.2(iv), there is a perfect matching $F'' \subset E''$ for $V'' \cap X$ and $V'' \cap Y$. Define a good sequence through

$$\mathbb{N} \to \{0, 1\}, \quad j \mapsto \begin{cases} 1, \text{ if } e_j \in F'', \\ 0, \text{ else.} \end{cases}$$

It follows that each good sequence of length n has, for each $m > n$, a good extension of length m.

We can thus inductively define an infinite sequence in $\{0, 1\}$ such that each finite initial segment of s is good and has infinitely many good extensions. Let $F := \{e_n : s_n = 1\}$. Then F is a perfect matching for X and Y. $\square$

Exercise 0.2.3 Once again: Where exactly in the proof of Theorem 0.2.4 was the axiom of choice used?

König's theorem was only the first step on the way to a proof of Tarki's theorem. The next step is to associate, with each group action on some set, an object called the type semigroup of that action. In order to define the type semigroup of a given group action, we first have to enlarge that action:

Definition 0.2.5 Let G be a group that acts on a set X.

(i) Define $X^* := X \times \mathbb{N}_0$ and

$$G^* := \{(g, \pi) : g \in G, \text{ and } \pi \text{ is a permutation of } \mathbb{N}_0\},$$

and let G^* act on X^* via

$$(g, \pi) \cdot (x, n) := (g \cdot x, \pi(n)) \qquad ((g, \pi) \in G^*, (x, n) \in X^*).$$

(ii) If $A \subset X^*$, then those $n \in \mathbb{N}_0$ such that there is an element of A whose second coordinate is n are called the *levels* of A.

Exercise 0.2.4 Let G, X, G^*, and X^* as in Definition 0.2.5. Show that, if $E_1, E_2 \subset X$, then $E_1 \sim_G E_2$ if and only if $E_1 \times \{n\} \sim_{G^*} E_2 \times \{m\}$ for all $n, m \in \mathbb{N}_0$.

Definition 0.2.6 Let G, X, G^*, and X^* be as in Definition 0.2.5.

(i) A set $A \subset X^*$ is called *bounded* if it has only finitely many levels.
(ii) If $A \subset X^*$ is bounded, then the equivalence class of A with respect to $\sim_{G^*}$ is called the *type* of A and is denoted by $[A]$.
(iii) If $E \subset X$, we write $[E] := [E \times \{0\}]$.

(iv) Let $A, B \subset X^*$ be bounded, and let $k \in \mathbb{N}_0$ be such that

$$B' := \{(b, n+k) : (b,n) \in B\} \cap A = \varnothing.$$

Define $[A] + [B] := [A \cup B']$.

(v) Let

$$S := \{[A] : A \subset X^* \text{ is bounded}\}.$$

Then $(S, +)$ is called the *type semigroup* of the action of G on X.

We have just labeled $(S, +)$ the type semigroup, thus implying that it is indeed a semigroup. This requires at least a little verification:

Exercise 0.2.5 Let G be a group acting on a set X, and let $(S, +)$ be the corresponding type semigroup.

(i) Show that $+$ is well defined.

(ii) Show that $(S, +)$ is a commutative semigroup with identity $[\varnothing]$.

If S is any commutative semigroup, define

$$n\alpha := \underbrace{\alpha + \cdots + \alpha}_{n \text{ times}} \qquad (\alpha \in S, \, n \in \mathbb{N}).$$

Also, we write $\alpha \le \beta$ for $\alpha, \beta \in S$ if there is $\gamma \in S$ such that $\alpha + \gamma = \beta$.

Exercise 0.2.6 Let S be a commutative semigroup, and let $\alpha_1, \alpha_2, \beta_1, \beta_2 \in S$ be such that $\alpha_1 \le \beta_1$ and $\alpha_2 \le \beta_2$. Show that $\alpha_1 + \alpha_2 \le \beta_1 + \beta_2$.

Exercise 0.2.7 Let G be a group acting of a set X, and let S be the corresponding type semigroup. Show that:

(i) If $\alpha, \beta \in S$ are such that $\alpha \le \beta$ and $\beta \le \alpha$, then $\alpha = \beta$.

(ii) A set $E \subset X$ is G-paradoxical if and only if $[E] = 2[E]$.

If S is the type semigroup of some group action, we have the following cancellation law:

Theorem 0.2.7 (AC) *Let S be the type semigroup of some group action, and let $\alpha, \beta \in S$ and $n \in \mathbb{N}$ be such that $n\alpha = n\beta$. Then $\alpha = \beta$.*

Proof If $n\alpha = n\beta$, then there are two disjoint, bounded G^*-equidecomposable sets $E, E' \subset X^*$ with pairwise disjoint subsets $A_1, \ldots, A_n \subset E$ and $B_1, \ldots, B_n \subset E'$ such that

- $E = A_1 \cup \cdots \cup A_n$ and $E' = B_1 \cup \cdots \cup B_n$, and
- $[A_j] = \alpha$ and $[B_j] = \beta$ for $j = 1, \ldots, n$.

Let $\chi \colon E \to E'$ and, for $j = 1, \ldots, n$, let $\phi_j \colon A_1 \to A_j$ and $\psi_j \colon B_1 \to B_j$ be bijective maps as in Exercise 0.1.5(i) (choose ϕ_1 and ψ_1 as the identity). For each $a \in A_1$ and $b \in B_1$ let

$$\bar{a} := \{a, \phi_2(a), \ldots, \phi_n(a)\} \qquad \text{and} \qquad \bar{b} := \{b, \phi_2(b), \ldots, \phi_n(b)\}$$

Define a bipartite graph as follows: Let $X := \{\bar{a} : a \in A_1\}$ and $Y := \{\bar{b} : b \in B_1\}$. For each $j = 1, \ldots, n$ form an edge from $\bar{a} \in X$ to $\bar{b} \in Y$ if $\chi(\phi_j(a)) \in \bar{b}$. This graph is n-regular

(Why?). Thus, by König's theorem, it has a perfect matching F. For each $\bar{a} \in X$, there is $\bar{b} \in Y$ and a unique edge $(\bar{a}, \bar{b}) \in F$ such that $\chi(\phi_j(a)) = \psi_k(b)$ for some $j, k \in \{1, \dots, n\}$. For any $j, k \in \{1, \dots, n\}$ define

$$C_{j,k} := \{a \in A_1 : (\bar{a}, \bar{b}) \in F \text{ and } \chi(\phi_j(a)) = \psi_k(b)\}$$

and

$$D_{j,k} := \{b \in B_1 : (\bar{a}, \bar{b}) \in F \text{ and } \chi(\phi_j(a)) = \psi_k(b)\}.$$

Then $\psi_k^{-1} \circ \chi \circ \phi_j$ maps $C_{j,k}$ bijectively onto $D_{j,k}$ as in Exercise 0.1.5(i); in particular $C_{j,k} \sim_{G^*} D_{j,k}$. Since $\{C_{j,k} : j, k = 1, \dots, n\}$ and $\{D_{j,k} : j, k = 1, \dots, n\}$ are disjoint partitions of A_1 and B_1, respectively, it follows from Exercise 0.1.5(ii) that $A_1 \sim_{G^*} B_1$, i.e. $\alpha = \beta$. $\square$

For the proof of Tarski's theorem, we will need the following corollary of the cancellation law:

Corollary 0.2.8 (AC) *Let S be the type semigroup of some group action, and let $\alpha \in S$ and $n \in \mathbb{N}$ be such that $(n+1)\alpha \leq n\alpha$. Then $\alpha = 2\alpha$.*

Proof From the hypothesis, we obtain

$$2\alpha + n\alpha = (n+1)\alpha + \alpha \leq n\alpha + \alpha = (n+1)\alpha \leq n\alpha.$$

Repeating this argument, we obtain $n\alpha \geq n\alpha + n\alpha = 2n\alpha$. Since, trivially, $n\alpha \leq 2n\alpha$, we have $n\alpha = 2n\alpha = n(2\alpha)$ by Exercise 0.2.7(i). From Theorem 0.2.7, $\alpha = 2\alpha$ follows. $\square$

In order to complete the proof of Tarski's theorem, we need one more theorem, for which, in turn, we need the following technical lemma:

Lemma 0.2.9 *Let S be a commutative semigroup, let $S_0 \subset S$ be finite, and let $\epsilon \in S_0$ be such that $(n+1)\epsilon \not\leq n\epsilon$ for $n \in \mathbb{N}$, and such that for each $\alpha \in S$ there is $n \in \mathbb{N}$ with $\alpha \leq n\epsilon$. Then there is a function $\nu \colon S_0 \to [0, \infty]$ with the following properties:*

(i) $\nu(\epsilon) = 1$;

(ii) *if $\alpha_1, \dots, \alpha_n, \beta_1, \dots, \beta_m \in S_0$ are such that $\alpha_1 + \cdots + \alpha_n \leq \beta_1 + \cdots + \beta_m$, then*

$$\sum_{j=1}^{n} \nu(\alpha_j) \leq \sum_{j=1}^{m} \nu(\beta_j).$$

Proof We proceed by induction on the cardinality of $|S_0|$.

If $|S_0| = 1$, then $S_0 = \{\epsilon\}$. In this case $\nu(\epsilon) := 1$ satisfies (i). To verify (ii) let $n, m \in \mathbb{N}$ be such that $n\epsilon \leq m\epsilon$, and assume that $n \geq m + 1$. But then $(m+1)\epsilon \leq n\epsilon \leq m\epsilon$, which contradicts the hypotheses on ϵ.

Suppose now that that there is $\alpha_0 \in S_0 \setminus \{\epsilon\}$. By the induction hypothesis, there is a function $\nu \colon S_0 \setminus \{\alpha_0\} \to [0, \infty]$ satisfying (i) and (ii). Since, for any $\alpha \in S$, there is $n \in \mathbb{N}$ such that $\alpha \leq n\epsilon$, it follows from (ii) that ν attains only finite values. Extend ν to S_0 by letting

$$\nu(\alpha_0) := \inf \left\{ \frac{1}{r} \left(\sum_{j=1}^{p} \nu(\gamma_j) - \sum_{j=1}^{q} \nu(\delta_j) \right) \right\},$$

where the infimum is taken over all $r \in \mathbb{N}$ and $\gamma_1, \ldots, \gamma_p, \delta_1, \ldots, \delta_q \in S_0 \setminus \{\alpha_0\}$ satisfying

$$\delta_1 + \cdots + \delta_q + r\alpha_0 \leq \gamma_1 + \cdots + \gamma_p.$$

It follows that this infimum is taken over a non-empty set (Why?), and that $\nu(\alpha_0) \geq 0$. It remains to be shown that the extended ν still satisfies (ii).

Let $\alpha_1, \ldots, \alpha_n, \beta_1, \ldots, \beta_m \in S_0 \setminus \{\alpha_0\}$ and $s, t \in \mathbb{N}_0$ be such that

$$\alpha_1 + \cdots \alpha_n + s\alpha_0 \leq \beta_1 + \cdots + \beta_m + t\alpha_0. \tag{0.1}$$

If $s = t = 0$, the claim is clear from the induction hypothesis.

Case 1: $s = 0$ and $t > 0$. We have to show that

$$\sum_{j=1}^{n} \nu(\alpha_j) \leq t\nu(\alpha_0) + \sum_{j=1}^{m} \nu(\beta_j).$$

i.e.

$$\nu(\alpha_0) \geq \frac{1}{t} \left(\sum_{j=1}^{n} \nu(\alpha_p) - \sum_{j=1}^{m} \nu(\beta_j) \right) =: w.$$

Let $r \in \mathbb{N}$ and $\gamma_1, \ldots, \gamma_p, \delta_1, \ldots, \delta_q \in S_0 \setminus \{\alpha_0\}$ satisfy

$$\delta_1 + \cdots + \delta_q + r\alpha_0 \leq \gamma_1 + \cdots + \gamma_p. \tag{0.2}$$

It is sufficient to show that

$$\frac{1}{r} \left(\sum_{j=1}^{p} \nu(\gamma_j) - \sum_{j=1}^{q} \nu(\delta_j) \right) \geq w. \tag{0.3}$$

From (0.1) — note that $s = 0$ — we obtain by multiplication with r and adding the same terms on both sides:

$$r\alpha_1 + \cdots + r\alpha_n + t\delta_1 + \cdots + t\delta_q \leq r\beta_1 + \cdots + r\beta_m + rt\alpha_0 + t\delta_1 + \cdots + t\delta_q.$$

Substituting (0.2) yields:

$$r\alpha_1 + \cdots + r\alpha_n + t\delta_1 + \cdots + t\delta_q \leq r\beta_1 + \cdots + r\beta_m + t\gamma_1 + \cdots + t\gamma_p.$$

Applying the induction hypothesis, we obtain

$$r \sum_{j=1}^{n} \nu(\alpha_j) + t \sum_{j=1}^{q} \nu(\delta_j) \leq r \sum_{j=1}^{m} \nu(\beta_j) + t \sum_{j=1}^{n} \nu(\gamma_j),$$

which implies (0.3).

Case 2: Suppose that $s > 0$.

It suffices to show that

$$s\nu(\alpha_0) + \sum_{j=1}^{n} \nu(\alpha_j) \leq z_1 + \cdots + z_t + \sum_{j=1}^{m} \nu(\beta_j),$$

where $z_1, \ldots, z_t$ are any of the numbers whose infimum defines $\nu(\alpha_0)$. Without loss of generality, we may suppose that $z_1 = \cdots = z_t =: z$ (Why?). We thus must prove that

$$s\nu(\alpha_0) + \sum_{j=1}^{n} \nu(\alpha_j) \leq tz + \sum_{j=1}^{m} \nu(\beta_j).$$

Multiplying (0.1) by r and adding the same terms on both sides yields:

$$r\alpha_1 + \cdots + r\alpha_n + rs\alpha_0 + t\delta_1 + \cdots + t\delta_q \leq r\beta_1 + \cdots + r\beta_m + rt\alpha_0 + t\delta_1 + \cdots + t\delta_q. \quad (0.4)$$

Let $\gamma_1, \ldots, \gamma_p, \delta_1, \ldots, \delta_q \in S_0 \setminus \{\alpha_0\}$ satisfy (0.2) with

$$z = \frac{1}{r}\left(\sum_{j=1}^{p} \nu(\gamma_j) - \sum_{j=1}^{q} \nu(\delta_j)\right).$$

Substituting (0.2) into (0.4) yields:

$$r\alpha_1 + \cdots + r\alpha_n + t\delta_1 + \cdots + t\delta_q + rs\alpha_0 \leq r\beta_1 + \cdots + r\beta_m + t\gamma_1 + \cdots + t\gamma_p.$$

From this inequality and from the definition of $\nu(\alpha_0)$, it follows that

$$s\nu(\alpha_0) + \sum_{j=1}^{n} \nu(\alpha_j) \leq \sum_{j=1}^{n} \nu(\alpha_j) + \frac{s}{sr}\left(r\sum_{j=1}^{m} \nu(\beta_j) + t\sum_{j=1}^{p} \nu(\gamma_j) - r\sum_{j=1}^{n} \nu(\alpha_j) - t\sum_{j=1}^{q} \nu(\delta_j)\right)$$

$$= tz + \sum_{j=1}^{m} \nu(\beta_j).$$

This finishes the proof. $\square$

Theorem 0.2.10 (AC) *Let $(S, +)$ be a commutative semigroup with neutral element 0, and let ϵ be an element of S. Then the following are equivalent:*

(i) *$(n+1)\epsilon \not\leq n\epsilon$ for all $n \in \mathbb{N}$.*

(ii) *There is a semigroup homomorphism $\nu\colon (S, +) \to ([0, \infty], +)$ such that $\nu(\epsilon) = 1$.*

Proof The implication (ii) $\Longrightarrow$ (i) is easy (Work it out for yourself!). We shall thus focus on the proof of (i) $\Longrightarrow$ (ii).

Without loss of generality suppose that, for each $\alpha \in S$, there is $n \in \mathbb{N}$ such that $\alpha \leq n\epsilon$ (otherwise, first disregard those elements α lacking this property, and later define $\nu(\alpha) := \infty$ for all such α).

For any finite subset S_0 of S containing ϵ, let M_{S_0} be the set of all $\kappa\colon S \to [0, \infty]$ such that

$- \kappa(\epsilon) = 1$, and

$- \kappa(\alpha + \beta) = \kappa(\alpha) + \kappa(\beta)$ for $\alpha, \beta, \alpha + \beta \in S_0$.

It follows from Lemma 0.2.9 that $M_{\mathcal{S}_0} \neq \varnothing$ (Why?).

Let $[0, \infty]^{\mathcal{S}}$ be equipped with the product topology. Then it is compact by Tychonoff's theorem. It is easily seen (but nevertheless check it yourself) that, for each finite subset $\mathcal{S}_0$ of $\mathcal{S}$, the set $M_{\mathcal{S}_0}$ is closed in $[0, \infty]^{\mathcal{S}}$. It is equally easily seen that the family

$$\{M_{\mathcal{S}_0} : \mathcal{S}_0 \subset \mathcal{S} \text{ is finite}\}$$

has the finite intersection property. Hence, $\bigcap\{M_{\mathcal{S}_0} : \mathcal{S}_0 \subset \mathcal{S} \text{ is finite}\}$ contains at least one map $\nu : \mathcal{S} \to [0, \infty]$. It is clear that $\nu(\epsilon) = 1$, and if $\alpha, \beta \in \mathcal{S}$, then $\nu(\alpha + \beta) = \nu(\alpha) + \nu(\beta)$ because $\nu \in M_{\{\alpha, \beta, \alpha + \beta\}}$. $\square$

Exercise 0.2.8 One last time: Where in the proof of Theorem 0.2.10 was the axiom of choice used?

We are finally able to complete the proof of Tarski's theorem:

Proof of (the difficult direction of) Tarski's theorem Let $\mathcal{S}$ be the type semigroup of the action of G on X. Suppose that E is not G-paradoxical. By Exercise 0.2.7(ii), this means that $[E] \neq 2[E]$, and from Corollary 0.2.8 it follows that $(n + 1)[E] \not\leq n[E]$ for all $n \in \mathbb{N}$. Hence, Theorem 0.2.10 implies that there is an additive map $\nu : \mathcal{S} \to [0, \infty]$ such that $\nu([E]) = 1$. Then

$$\mu : \mathfrak{P}(X) \to [0, \infty], \quad A \mapsto \nu([A])$$

is the desired set function. $\square$ If X is any set, and $\mu : \mathfrak{P}(X) \to [0, \infty]$ is a finitely additive set function with $\mu(X) < \infty$, then we may define $m \in \ell^{\infty}(X)^{*}$ through

$$\langle \phi, m \rangle := \int_{X} \phi(x) \, d\mu(x) \qquad (\phi \in \ell^{\infty}(X))$$

(for a detailed exposition on integration with respect to not necessarily countably additive set functions, see [D–S, Chapter III]). The following is thus an easy consequence of Tarski's theorem:

Corollary 0.2.11 (AC) *For a group G the following are equivalent:*

(i) G *is not paradoxical.*
(ii) *There is a finitely additive set function* $\mu : \mathfrak{P}(G) \to [0, \infty)$ *such that* $\mu(G) = 1$ *and* $\mu(gE) = \mu(E)$ *for all* $g \in G$ *and* $E \subset G$.
(iii) *There is* $m \in \ell^{\infty}(G)^{*}$ *with:*
(a) $\langle 1, m \rangle = \|m\| = 1$;
(b) $\langle \delta_g * \phi, m \rangle = \langle \phi, m \rangle \qquad (g \in G, \phi \in \ell^{\infty}(G))$.

Groups satisfying Corollary 0.2.11(iii) are called *amenable* ...

0.3 Notes and comments

As you might have guessed, the interest in paradoxical decompositions does not go as far back as to biblical times. Nevertheless, their origins can be traced back to the beginnings of modern measure theory.

As we all know, n-dimensional Lebesgue measure λ^n is (a) σ-additive, (b) invariant under invertible isometries (Why?), and (c) normalized in the sense that a certain set, $[0,1]^n$ say, is mapped to 1. As we all know as well, λ^n cannot be extended to all of $\mathfrak{P}(\mathbb{R}^n)$ in such a way that (a), (b), and (c) remain valid. In [Hau], F. Hausdorff raised the question of whether there was a (a)' *finitely additive* set function on $\mathfrak{P}(\mathbb{R}^n)$ that still satisfied (b) and (c). With the help of his paradox, Hausdorff was able to answer this question in the negative for $n \geq 3$. Interestingly, for $n = 1, 2$, there is a set function on all of $\mathfrak{P}(\mathbb{R}^n)$ for which (a)', (b), and (c) still hold ([Ban]). Further investigations in this direction led S. Banach and A. Tarski to the paradox that now carries their name ([B–T]). Tarski's theorem first saw the light of day in [Tar].

Our exposition is based on [Wag], which is a monograph devoted entirely to paradoxical decompositions; another modern account of the Banach–Tarski paradox is [Stro]. Popular expositions of the Banach–Tarski paradox are [Fre] and [Run 2]. The article [Fre] attempts to be humorous and contains three pictures showing its author (i) with one orange, then (ii) cutting up the orange with a knife into finitely many pieces, and eventually (iii) with *two* oranges. Why is any attempt to implement the Banach–Tarski paradox using a knife bound to fail?

Finally, if you had problems solving Exercise 0.2.2, look it up in [B–M] or [Pat 1].

1 Amenable, locally compact groups

In view of Corollary 0.2.11, there are three possible ways of extending the notion of a-menability from discrete groups to general locally compact groups:

– Require that the group G be not paradoxical with some Borel overtones added to Definition 0.1.1, i.e. we demand that the sets $A_1, \ldots, A_n$ and $B_1, \ldots, B_m$ in Definition 0.1.1 be Borel sets.

– Require the existence of a finitely additive set function μ on the Borel subsets of G with $\mu(G) = 1$ such that $\mu(gE) = \mu(E)$ for all $g \in G$ and all Borel subsets E of G.

– Require the existence of a linear functional m as in Corollary 0.2.11(iii) on some space that can replace $\ell^\infty(G)$ in the general locally compact situation, such as $L^\infty(G)$ or $\mathcal{C}_b(G)$.

It is this third definition of amenability which has turned out to be the "right" one: The class of amenable, locally compact groups is large enough to encompass lots of interesting examples — all compact and all abelian groups are amenable —, but, on the other hand, is small enough to allow for the development of a strong theory.

Since the objects of this chapter are locally compact groups, some background in abstract harmonic analysis is indispensable for understanding it. For the reader's convenience, we have collected the necessary concepts and facts from abstract harmonic analysis in Appendix A. We would like to point out here that our use of the convolution symbol $*$ is a bit unorthodox: For the convolution of a measure from the right with an L^∞-function, it differs from the usual definition (compare Definitions A.1.6 and A.1.14).

By the way, since we define amenable, locally compact groups in terms of bounded, linear functionals, we are now entering the realm of functional analysis, where virtually nothing works without the axiom of choice; therefore, from now on, we shall no longer indicate whether a results depends on the axiom of choice or not, but simply suppose that this axiom holds.

1.1 Invariant means on locally compact groups

We start with the definition of a mean:

Definition 1.1.1 Let G be a locally compact group, and let E be a subspace of $L^\infty(G)$ containing the constant functions. A *mean* on E is a functional $m \in E^*$ such that $\langle 1, m \rangle = \|m\| = 1$.

For the subspaces of $L^\infty(G)$ we are interested in — $L^\infty(G)$ itself, $\mathcal{C}_b(G)$, $LUC(G)$, $RUC(G)$, $UC(G)$ (see Appendix A for the definitions of these spaces) — the following characterization of means holds:

Proposition 1.1.2 *Let G be a locally compact group, and let E be a subspace of $L^\infty(G)$ containing the constant functions and closed under complex conjugation. Then, for a linear functional $m\colon E \to \mathbb{C}$ with $\langle 1, m\rangle = 1$, the following are equivalent:*

(i) *m is a mean on E.*
(ii) *m is positive, i.e.*

$$\langle \phi, m\rangle \geq 0 \qquad (\phi \in E,\ \phi \geq 0).$$

Proof (i) $\Longrightarrow$ (ii): We first prove that $\langle \phi, m\rangle \in \mathbb{R}$ for all $\phi \in E$ with $\phi(G) \subset \mathbb{R}$. Let $\phi \in E$ be real-valued, and suppose without loss of generality that $\|\phi\|_\infty \leq 1$. Let $\alpha, \beta \in \mathbb{R}$ be such that $\langle \phi, m\rangle = \alpha + \beta i$. For any $t \in \mathbb{R}$, we then have:

$$(\beta + t)^2 \leq |\alpha + i(\beta + t)|^2 = |\langle \phi + it1, m\rangle|^2 \leq \|\phi + it1\|_\infty^2 \leq 1 + t^2$$

(Where did we use the fact that $\phi(G) \subset \mathbb{R}$?). It follows that

$$2\beta t \leq 1 - \beta^2 \qquad (t \in \mathbb{R}),$$

so that $\beta = 0$.

Now, let $\phi \in E$ be such that $\phi \geq 0$. Without loss of generality, suppose that $0 \leq \phi \leq 1$. Let $\psi := 2\phi - 1$, so that $\psi(G) \subset \mathbb{R}$ and $\|\psi\|_\infty \leq 1$. Since m is a mean, we have $|\langle \psi, m\rangle| \leq \|\psi\|_\infty \leq 1$, so that

$$\langle \phi, m\rangle = \frac{1}{2}\langle 1 + \psi, m\rangle = \frac{1}{2}(1 + \langle \psi, m\rangle) \geq 0.$$

(ii) $\Longrightarrow$ (i): Let $\phi \in E$ be real-valued. Then $\psi := \|\phi\|_\infty 1 - \phi \geq 0$, so that $\langle \psi, m\rangle \geq 0$. It follows that $\langle \phi, m\rangle \in \mathbb{R}$ and $\langle \phi, m\rangle \leq \|\phi\|_\infty$. Replacing ϕ with $-\phi$, we obtain $-\langle \phi, m\rangle \leq \|\phi\|_\infty$, so that, as a whole, $|\langle \phi, m\rangle| \leq \|\phi\|_\infty$.

For arbitrary $\phi \in E$, choose $\lambda \in \mathbb{C}$ such that $\langle \lambda\phi, m\rangle = |\langle \phi, m\rangle|$. Since E is closed under complex conjugation, there are real-valued $\phi_1, \phi_2 \in E$ such that $\lambda\phi = \phi_1 + i\phi_2$ (Why exactly?). Since $\langle \phi_1, m\rangle + i\langle \phi_2, m\rangle = \langle \lambda\phi, m\rangle \geq 0$ and since $\langle \phi_1, m\rangle, \langle \phi_2, m\rangle \in \mathbb{R}$, it follows that $\langle \phi_2, m\rangle = 0$. We thus have

$$|\langle \phi, m\rangle| = \langle \phi_1, m\rangle \leq \|\phi_1\|_\infty \leq \|\lambda\phi\|_\infty = \|\phi\|_\infty,$$

which establishes (i). $\quad\square$

Exercise 1.1.1 Let G be a locally compact group, and let E be a subspace of $L^\infty(G)$ containing the constant functions and closed under complex conjugation. Show that the set of means on E is w^*-compact in E^*.

We now define what we mean by a (left) invariant mean:

Definition 1.1.3 Let G be a locally compact group, and let E be a subspace of $L^\infty(G)$ containing the constant functions and closed under complex conjugation..

(i) E is called *left invariant* if $\delta_g * \phi \in E$ for all $\phi \in E$ and $g \in G$.

(ii) If E is left invariant, then a mean m on E is called *left invariant* if

$$\langle \delta_g * \phi, m \rangle = \langle \phi, m \rangle \qquad (g \in G, \, \phi \in E).$$

Definition 1.1.4 A locally compact group G is *amenable* if there is a left invariant mean on $L^\infty(G)$.

By Corollary 0.2.11, for a discrete group, being amenable is the same as not being paradoxical.

We now give a few examples of amenable (and non-amenable) groups:

Examples 1.1.5   (a) By Theorem 0.1.2 and Corollary 0.2.11, $\mathbb{F}_2$ is not amenable (since this conclusion only depends on the easy direction of Tarski's theorem, it is independent of the axiom of choice).

(b) Let G be a compact group, so that $L^\infty(G) \subset L^1(G)$. Let m_G denote normalized (left) Haar measure on G, and define

$$\langle \phi, m \rangle := \int_G \phi(g) \, dm_G(g) \qquad (\phi \in L^\infty(G)).$$

It is obvious, that m is a left invariant mean on $L^\infty(G)$, so that G is amenable.

(c) Let G be a locally compact, abelian group. We claim that G is amenable. To see this, let K be the set of all means on $L^\infty(G)$. It is easy to see that K is convex, and by Exercise 1.1.1, it is compact. For each $g \in G$ define a linear map $T_g \colon L^\infty(G)^* \to L^\infty(G)^*$ through

$$\langle T_g n, \phi \rangle := \langle \delta_g * \phi, n \rangle \qquad (\phi \in L^\infty(G), \, n \in L^\infty(G)^*).$$

It is easy to see (but you should nevertheless check it), that each map T_g is w^*-continuous and leaves K invariant. Moreover, it is immediate that $T_{gh} = T_g T_h$ for $g, h \in G$. By the Markov–Kakutani fixed point theorem ([D–S, Theorem V.10.6]), there is $m \in K$ such that $T_g m = m$ for all $g \in G$. This means that m is a left invariant mean on $L^\infty(G)$.

The argument used in Examples 1.1.5(c) to establish the amenability of locally compact, abelian groups is rather non-constructive: We get the existence of a left invariant mean, but we can't lay our hands on it. For concrete abelian groups, however, it is sometimes possible to give a more explicit description of a left invariant mean:

Exercise 1.1.2 Let $\mathcal{U}$ be a a free ultrafilter on $\mathbb{N}$, and define $m \colon \ell^\infty(\mathbb{Z}) \to \mathbb{C}$ through

$$\langle \phi, m \rangle := \lim_{\mathcal{U}} \frac{1}{2n+1} \sum_{k=-n}^{n} \phi(k) \qquad (\phi \in \ell^\infty(\mathbb{Z})).$$

Show that m is a left invariant mean on $\ell^\infty(\mathbb{Z})$. Do you have any idea of how a left invariant mean on $L^\infty(\mathbb{R})$ might look?

We have defined amenable, locally compact groups G in terms of left invariant means on $L^\infty(G)$. This definition is not always convenient to work with. For example, if we have a closed subgroup H of G, then Haar measure on H need not be the restriction to H of Haar measure on G, so that there is, in general, no relation between $L^\infty(H)$ and $L^\infty(G)$. This is particularly awkward if we want to investigate the hereditary properties of amenability. We shall now characterize amenable, locally compact groups in terms of other (left invariant) subspaces of $L^\infty(G)$.

We require the following definition for technical reasons:

Definition 1.1.6 Let G be a locally compact group.

(i) Let

$$P(G) := \{f \in L^1(G) : f \geq 0, \|f\|_1 = 1\}.$$

(ii) Let E be one of the following spaces: $L^\infty(G)$, $C_b(G)$, $LUC(G)$, $RUC(G)$, or $UC(G)$. Then a mean $m \in E^*$ is called *topologically left invariant* if

$$\langle f * \phi, m \rangle = \langle \phi, m \rangle \qquad (\phi \in E, \, f \in P(G)).$$

Exercise 1.1.3 Show that $P(G)$ is a subsemigroup of $(L^1(G), *)$.

Exercise 1.1.4 Show that a topologically invariant mean in the sense of Definition 1.1.6(ii) is well defined, i.e. if E is $L^\infty(G)$, $C_b(G)$, $LUC(G)$, $RUC(G)$, or $UC(G)$, and if $\phi \in E$ and $f \in P(G)$, then $f * \phi \in E$ as well.

We first clarify the relationship between left invariant means and topologically invariant means:

Lemma 1.1.7 *Let G be a locally compact group, and let E be $L^\infty(G)$, $C_b(G)$, $LUC(G)$, $RUC(G)$, or $UC(G)$. Then every topologically left invariant mean on E is left invariant.*

Proof Let $m \in E^*$ be a topologically left invariant mean. Let $\phi \in E$, and let $g \in G$. Fix $f \in P(G)$. Then we have

$$\langle \delta_g * \phi, m \rangle = \langle \underbrace{f * \delta_g}_{\in P(G)} *\phi, m \rangle = \langle \phi, m \rangle,$$

so that m is left invariant. $\square$

Lemma 1.1.8 *Let G be a locally compact group, and let $m \in UC(G)^*$ be a left invariant mean. Then m is topologically left invariant.*

Proof Let $\phi \in UC(G)$, and let $f \in L^1(G)$. Since m is left invariant, we obtain

$$\langle \delta_g * f * \phi, m \rangle = \langle f * \phi, m \rangle \qquad (g \in G). \tag{1.1}$$

For each $\phi \in UC(G)$, the functional $L^1(G) \ni f \mapsto \langle f * \phi, m \rangle$ is continuous, so that there is $\psi \in L^\infty(G)$ such that

$$\langle f * \phi, m \rangle = \int_G f(g)\psi(g)\, dm_G(g) \qquad (f \in L^1(G)).$$

From (1.1), it follows that $\delta_g * \psi = \psi$ for all $g \in G$, so that ψ is in fact constant. Thus, for any $\phi \in UC(G)$, there is $C(\phi) \in \mathbb{C}$ such that

$$\langle f * \phi, m \rangle = C(\phi) \int_G f(g) \, dm_G(g) \qquad (f \in L^1(G)).$$

For $f \in P(G)$, it follows that

$$\langle f * \phi, m \rangle = C(\phi) \qquad (\phi \in UC(G)).$$

By Theorem A.1.8, there is a bounded approximate identity $(e_\alpha)_\alpha$ for $L^1(G)$ contained in $P(G)$. Then $(e_\alpha)_\alpha$ is bounded left approximate identity for the left Banach $L^1(G)$-module $UC(G)$ (by Proposition A.2.5). For $f \in P(G)$ and $\phi \in UC(G)$, we thus have

$$\langle f * \phi, m \rangle = \lim_\alpha \langle f * e_\alpha * \phi, m \rangle = C(\phi) = \lim_\alpha \langle e_\alpha * \phi, m \rangle = \langle \phi, m \rangle,$$

i.e. m is topologically left invariant. $\square$

Exercise 1.1.5 Let G be a locally compact group. For which of the following subspaces of $L^\infty(G)$ does the proof of Lemma 1.1.8 still work: $L^\infty(G)$, $\mathcal{C}_b(G)$, $LUC(G)$, $RUC(G)$?

Theorem 1.1.9 *For a locally compact group G, the following are equivalent:*

(i) *G is amenable.*
(ii) *There is a left invariant mean on $\mathcal{C}_b(G)$.*
(iii) *There is a left invariant mean on $LUC(G)$.*
(iv) *There is a left invariant mean on $RUC(G)$.*
(v) *There is a left invariant mean on $UC(G)$.*

Proof It is obvious that only (v) $\Longrightarrow$ (i) needs proof.

Let m be a left invariant mean on $UC(G)$. By Lemma 1.1.8, m is also topologically left invariant. Let $(e_\alpha)_\alpha$ be a bounded approximate identity for $L^1(G)$ in $P(G)$, and choose an ultrafilter $\mathcal{U}$ on the index set of $(e_\alpha)_\alpha$ that dominates the order filter. Define

$$\tilde{m} \colon L^\infty(G) \to \mathbb{C}, \quad \phi \mapsto \lim_{\mathcal{U}} \langle e_\alpha * \phi * e_\alpha, m \rangle.$$

By Proposition A.2.5(iii), this is well defined. It is easily seen that $\tilde{m}$ is a mean on $L^\infty(G)$ (but check for yourself). For $\phi \in L^\infty(G)$ and $f \in P(G)$, we then have:

$$\begin{aligned}
\langle f * \phi, \tilde{m} \rangle &= \lim_{\mathcal{U}} \langle e_\alpha * f * \phi * e_\alpha, m \rangle \\
&= \lim_{\mathcal{U}} \langle f * e_\alpha * \phi * e_\alpha, m \rangle \\
&= \lim_{\mathcal{U}} \langle e_\alpha * \phi * e_\alpha, m \rangle, \qquad \text{since } m \text{ is topologically left invariant,} \\
&= \langle \phi, \tilde{m} \rangle,
\end{aligned}$$

i.e. $\tilde{m}$ is topologically left invariant. By Lemma 1.1.7, $\tilde{m}$ is left invariant. $\square$

Corollary 1.1.10 *For a locally compact group G, let G_d be the same group equipped with the discrete topology. Then, if G_d is amenable, so is G.*

Proof If G_d is amenable, then there is a left invariant mean m on $\ell^\infty(G)$. Since $\mathcal{C}_b(G) \subset \ell^\infty(G)$, $m|_{\mathcal{C}_b(G)}$ is a left invariant mean on $\mathcal{C}_b(G)$, so that G is amenable. $\square$

As we shall see in the next section, the converse of Corollary 1.1.10 is false.

An apparent drawback of Definition 1.1.4 is that it is not symmetric: We have defined amenability through left invariant means, so that groups satisfying that definition should rather be called left amenable. We could as well have defined amenability via right invariant means (it is so obvious what that is that we won't give a formal definition) or (two-sided) invariant means (it is also clear what that is supposed to mean). Fortunately, all these variants of Definition 1.1.4 characterize the same class of locally compact groups, as we shall now see.

We leave some preparations to the reader:

Exercise 1.1.6 Let G be a locally compact group, and identify $L^1(G)$ with its canonical image in $L^\infty(G)^*$.

(i) Show that $P(G)$ is w^*-dense in the set of all means on $L^\infty(G)$.
(ii) Show that G is amenable if and only if there is a net $(f_\alpha)_\alpha$ in $P(G)$ such that $\lim_\alpha \|\delta_g * f_\alpha - f_\alpha\|_1 = 0$ for all $g \in G$. (*Hint*: In any normed space, the norm closure and the weak closure of a convex set are the same.)

Theorem 1.1.11 *For a locally compact group G the following are equivalent:*

(i) *G is amenable.*
(ii) *There is a right invariant mean on $L^\infty(G)$.*
(iii) *There is an invariant mean on $L^\infty(G)$.*

Proof (i) $\implies$ (ii): Let m be a left invariant mean on $L^\infty(G)$. For $\phi \in L^\infty(G)$ define $\check\phi \in L^\infty(G)$ by letting $\check\phi(g) := \phi(g^{-1})$ for $g \in G$. Define

$$\check m \colon L^\infty(G) \to \mathbb{C}, \quad \phi \mapsto \langle \check\phi, m \rangle.$$

Then $\check m$ is a right invariant mean on $L^\infty(G)$.

(ii) $\implies$ (i) is proved in exactly the same fashion.

To prove (i) $\implies$ (iii), let $(f_\alpha)_\alpha$ be a net in $P(G)$ as specified in Exercise 1.1.6(ii). For $f \in L^1(G)$, let $\tilde f \in L^1(G)$ be defined as in Exercise A.1.7. By Exercise A.1.7(ii), $(\tilde f_\alpha)_\alpha$ is a net in $P(G)$ such that $\lim_\alpha \|\tilde f_\alpha * \delta_g - \tilde f_\alpha\|_1 = 0$ for all $g \in G$ (Why does that latter property hold?). It follows that $(f_\alpha * \tilde f_\alpha)_\alpha$ is a net in $P(G)$ satisfying

$$\lim_\alpha \|\delta_g * f_\alpha * \tilde f_\alpha - f_\alpha * \tilde f_\alpha\|_1 = \lim_\alpha \|f_\alpha * \tilde f_\alpha * \delta_g - f_\alpha * \tilde f_\alpha\|_1 = 0 \qquad (g \in G). \qquad (1.2)$$

Let m be any w^*-accumulation point of $(f_\alpha * \tilde f_\alpha)_\alpha$ in $L^\infty(G)^*$. Then clearly m is a mean on $L^\infty(G)$, and (1.2) implies that m is invariant.

Finally, (iii) $\implies$ (i) holds trivially. $\square$

In view of Exercise 1.1.6(ii), one might ask if, for a locally compact group G, a left invariant mean on $L^\infty(G)$ can be chosen in $P(G)$ itself. We leave it to the reader to verify that this is possible only in the obvious, i.e. compact, case:

Exercise 1.1.7 Let G be a locally compact group. Show that there is a left invariant mean on $L^\infty(G)$ belonging to $P(G)$ if and only if G is compact.

1.2 Hereditary properties

Next, we investigate how amenability behaves under the standard operations on locally compact groups, such as forming closed subgroups, quotients, or short, exact sequences. As a byproduct of this investigation, we will increase our stock of examples of amenable (and non-amenable) groups.

We start with quotients, but will in fact prove the following, more general result:

Proposition 1.2.1 *Let G be an amenable, locally compact group, let H be another locally compact group, and let $\theta \colon G \to H$ be a continuous homomorphism with dense range. Then H is amenable as well.*

Proof Define a continous homomorphism of Banach algebras

$$\theta^* \colon C_b(H) \to C_b(G), \quad \phi \mapsto \phi \circ \theta.$$

Let $\phi \in LUC(H)$, and let $(g_\alpha)_\alpha$ be a net in G converging to some $g \in G$. Then

$$\lim_\alpha \delta_{g_\alpha} * \theta^* \phi = \lim_\alpha \theta^*(\delta_{\theta(g_\alpha)} * \phi) = \theta^*(\delta_{\theta(g)} * \phi) = \delta_g * \theta^* \phi,$$

so that $\theta^* \phi \in LUC(G)$.

Let m be a left invariant mean on $LUC(G)$, and define $\tilde{m} \in LUC(H)^*$ through

$$\langle \phi, \tilde{m} \rangle := \langle \theta^* \phi, m \rangle \qquad (\phi \in LUC(H)).$$

It is immediate that $\tilde{m}$ is a mean on $LUC(H)$. Let $g \in G$. Then

$$\langle \delta_{\theta(g)} * \phi, \tilde{m} \rangle = \langle \theta^*(\delta_{\theta(g)} * \phi), m \rangle = \langle \delta_g * \theta^* \phi, m \rangle = \langle \theta^* \phi, m \rangle = \langle \phi, \tilde{m} \rangle. \tag{1.3}$$

Let $h \in H$ be arbitrary. Since $\theta(G)$ is dense in H, there is a net $(g_\alpha)_\alpha$ in G such that $\lim_\alpha \theta(g_\alpha) = h$. For $\phi \in LUC(H)$, we then have $\lim_\alpha \delta_{\theta(g_\alpha)} * \phi = \delta_h * \phi$ by the definition of $LUC(H)$. Together with (1.3), this yields

$$\langle \delta_h * \phi, \tilde{m} \rangle = \lim_\alpha \langle \delta_{\theta(g_\alpha)} * \phi, \tilde{m} \rangle = \langle \phi, \tilde{m} \rangle,$$

so that $\tilde{m}$ is left invariant. $\square$

Corollary 1.2.2 *Let G be an amenable, locally compact group, and let N be a closed, normal subgroup of G. Then G/N is amenable.*

The fact that amenability is inherited by closed subgroups is surprisingly hard to obtain. We need the following definition:

Definition 1.2.3 Let G be a locally compact group, and let H be a closed subgroup of G. A *Bruhat function* for H is a continuous, positive function β on G with the following properties:

(i) For each compact subset K of G, we have that $\mathrm{supp}(\beta|_{KH})$ is compact.
(ii) We have

$$\int_H \beta(gh) \, dm_H(h) = 1 \qquad (g \in G).$$

We shall see that Bruhat functions always exist. This requires a little preparation:

Lemma 1.2.4 *Let G be a locally compact group, let H be a closed subgroup, and let U be an open, symmetric, i.e. $U = U^{-1}$, relatively compact neighborhood of e_G. Then there is a subset S of G with the following properties:*

(i) *For each $g \in G$, there is $s \in S$ such that $gH \cap Us \neq \varnothing$.*

(ii) *If $K \subset G$ is compact, then $\{s \in S : KH \cap \overline{U}s \neq \varnothing\}$ is finite.*

Proof Use Zorn's lemma to obtain a subset S of G with the property that $s \notin UtH$ for all $s, t \in S$ with $s \neq t$ which is maximal with respect to set inclusion (Do you always believe it when someone says or writes that something exists by Zorn's lemma?).

For (i), let $g \in G$. If $gH \cap Us = \varnothing$, we have $g \notin UsH$ for all $s \in S$. From the maximality of S, it follows that $g \in S$ (Why? It is here where the symmetry of U comes in ...). Since $g \in gH \cap Ug$, this is a contradiction.

In order to prove (ii), assume that there is a compact subset K of G such that $\{s \in S : KH \cap \overline{U}s \neq \varnothing\}$ is infinite. We may thus find sequences $(s_n)_{n=1}^{\infty}$ in S and $(h_n)_{n=1}^{\infty}$ in H such that $s_n \neq s_m$ whenever $n \neq m$ and $s_n h_n \in \overline{U}K$ for all $n \in \mathbb{N}$. Since $\overline{U}K$ is compact, the sequence $(s_n h_n)_{n=1}^{\infty}$ has a cluster point in $g \in \overline{U}K$. Choose a symmetric neighborhood V of e_G such that $VV \subset U$ (Why does one exist?). We may thus find $n, m \in \mathbb{N}$ with $n \neq m$ such that $s_n h_n, s_m h_m \in Vg$ and thus $s_n h_n \in Us_m h_m$. But this means that $s_n \in Us_m H$, so that $s_n = s_m$ by the definition of S. This is a contradiction. □

Lemma 1.2.5 *Let G be a locally compact group, and let H be a closed subgroup. Then there is a continuous function $f: G \to [0, \infty)$ with the following properties:*

(i) *For each $g \in G$, we have $\{h \in G : f(h) > 0\} \cap gH \neq \varnothing$.*

(ii) *If $K \subset G$ is compact, $\mathrm{supp}(f|_{KH})$ is compact.*

Proof Choose a positive function $\phi \in \mathcal{C}_{00}(G)$ with

$$\phi(g) = \phi(g^{-1}) \quad (g \in G) \quad \text{and} \quad \phi(e_G) = 1,$$

and let $U := \{g \in G : \phi(g) > 0\}$. Then U is an open, symmetric, relatively compact neighborhood of e_G. Let S be as in Lemma 1.2.4, and define $f: G \to [0, \infty)$ through

$$f(g) := \sum_{s \in S} \phi(gs^{-1}) \quad (g \in G). \tag{1.4}$$

By Lemma 1.2.4(ii), the sum in (1.4) is finite whenver g ranges through an arbitrary compact subset of G. This establishes at once that f is well-defined and continuous.

Then (i) is an immediate consequence of Lemma 1.2.4 and the definition of U.

Let $K \subset G$ be compact. By Lemma 1.2.4(ii), there are $s_1, \ldots, s_n$ such that

$$\{g \in KH : f(g) > 0\} = \bigcup_{s \in S} Us \cap KH \subset Us_1 \cup \cdots \cup Us_n.$$

Since $Us_1 \cup \cdots \cup Us_n$ is relatively compact, it follows that $\mathrm{supp}(f|_{KH})$ is compact. □

Exercise 1.2.1 Let G be a locally compact group, and let H be a closed subgroup. Let $f: G \to \mathbb{C}$ be continuous such that $\mathrm{supp}(f|_{KH})$ is compact for every compact subset K of G. Show that, for any $\phi \in L^\infty(H)$, the function

$$G \to \mathbb{C}, \quad g \mapsto \int_H \phi(h) f(g^{-1}h) \, dm_H(h)$$

is continuous. (*Hint*: Use the fact that $\mathcal{C}_{00}(G) \subset LUC(G)$ by Remark A.2.4.)

Proposition 1.2.6 *Let G be a locally compact group, and let H be a closed subgroup of G. Then there is a Bruhat function for H.*

Proof Let f be as in Lemma 1.2.5, and define

$$\alpha: G \to \mathbb{C}, \quad g \mapsto \int_G f(gh) \, dm_H(h).$$

It follows from Lemma 1.2.5(ii) that α is well defined, and Lemma 1.2.5(i) and Exercise 1.2.1 establish that α is continuous. Also, Lemma 1.2.5 and Exercise A.1.4 yield that $\alpha(g) > 0$ for all $g \in G$. Define

$$\beta: G \to \mathbb{C}, \quad g \mapsto \frac{f(g)}{\alpha(g)}.$$

It is immediate from Lemma 1.2.5(ii) that β is continuous and satisfies Definition 1.2.3(i). Since m_H is left invariant, α is constant on the left cosets of H. Hence, β satisfies Definition 1.2.3(ii) as well. $\square$

Theorem 1.2.7 *Let G be an amenable, locally compact group, and let H be a closed subgroup of G. Then H is amenable.*

Proof Let $\beta: G \to \mathbb{C}$ be a Bruhat function for H, and define $T: \mathcal{C}_b(H) \to \ell^\infty(G)$ through

$$(T\phi)(g) := \int_H \phi(h) \beta(g^{-1}h) \, dm_H(h) \qquad (g \in G).$$

It is immediate that T is a contraction such that $T1 = 1$. By Exercise 1.2.1, we have $T\mathcal{C}_b(H) \subset \mathcal{C}_b(G)$. Let m be a left invariant mean on $\mathcal{C}_b(G)$, and define $\tilde{m} := T^* m$. It is immediate that $\tilde{m}$ is a mean on $\mathcal{C}_b(H)$.

Let $h \in H$ and $\phi \in \mathcal{C}_b(H)$; we have:

$$
\begin{aligned}
T(\delta_h * \phi)(g) &= \int_H \phi(h^{-1}k) \beta(g^{-1}k) \, dm_H(k) \\
&= \int_H \phi(k) \beta((g^{-1}h)k) \, dm_H(k) \\
&= (T\phi)(h^{-1}g) \\
&= (\delta_h * T\phi)(g) \qquad (g \in G).
\end{aligned}
$$

It follows that $\tilde{m}$ is left invariant. $\square$

Exercise 1.2.2 What's wrong with the following simple "proof" of Theorem 1.2.7?

We define a map $T: \mathcal{C}_b(H) \to L^\infty(G)$ as follows. Let $(g_\alpha)_\alpha$ be a family of representatives of the right cosets of H. Let $g \in G$. Since G is the disjoint union of $(Hg_\alpha)_\alpha$, there is a unique g_α such that $g \in Hg_\alpha$, so that $gg_\alpha^{-1} \in H$. Define

$$(T\phi)(g) := \phi(gg_\alpha^{-1}) \qquad (\phi \in \mathcal{C}_b(H)).$$

It follows immediately from that definition that $T\phi|_{Hg_\alpha} \in \mathcal{C}_b(Hg_\alpha)$ for all $\phi \in \mathcal{C}_b(H)$ and any g_α. Consequently, $T\phi \in L^\infty(G)$ for all $\phi \in \mathcal{C}_b(H)$. The following properties of the map $T: \mathcal{C}_b(H) \to L^\infty(G)$ are easily verified:

– T is a linear isometry;
– $T1 = 1$;
– $T(\delta_h * \phi) = \delta_h * T\phi \qquad (h \in H, \phi \in \mathcal{C}_b(H))$

Let m be a left invariant mean on $L^\infty(G)$, and define $\tilde{m} \in \mathcal{C}_b(H)^*$ through

$$\langle \phi, \tilde{m} \rangle := \langle T\phi, m \rangle \qquad (\phi \in \mathcal{C}_b(H)).$$

Then $\tilde{m}$ is a left invariant mean on $\mathcal{C}_b(H)$.

Show that this argument does indeed work if H is open or G_d is amenable.

Together, Theorem 1.2.7 and Examples 1.1.5(a) yield:

Corollary 1.2.8 *Let G be a locally compact group that contains a closed subgroup isomorphic to $\mathbb{F}_2$. Then G is not amenable.*

As a consequence, the converse of Corollary 1.1.10 does not hold:

Example 1.2.9 For $n \geq 3$, the group $SO(n)$ is compact and thus amenable. However, by Exercise 0.1.3, $SO(n)$ has a subgroup isomorphic to $\mathbb{F}_2$. Since in the discrete topology every subgroup is closed, $SO(n)_d$ is not amenable.

Next, we consider short, exact sequences:

Theorem 1.2.10 *Let G be a locally compact group, and let N be a closed, normal subgroup such that both N and G/N are amenable. Then G is amenable.*

Proof Let $\tilde{m}$ be a left invariant mean on $\mathcal{C}_b(N)$. For $\phi \in LUC(G)$, define $T\phi: G \to \mathbb{C}$ through

$$(T\phi)(g) := \langle (\delta_g * \phi)|_N, \tilde{m} \rangle.$$

Since $\phi \in LUC(G)$, we have $T\phi \in \mathcal{C}_b(G)$ (Why exactly?). Obviously, $T: LUC(G) \to \mathcal{C}_b(G)$ is a bounded, linear map. Let $g_1, g_2 \in G$ belong to the same coset of G, i.e. there is $n \in N$ such that $g_1 = ng_2$. For $\phi \in LUC(G)$, this means

$$(T\phi)(g_1) = \langle (\delta_{g_1} * \phi)|_N, \tilde{m} \rangle = \langle \delta_n * ((\delta_{g_2} * \phi)|_N), \tilde{m} \rangle = \langle (\delta_{g_2} * \phi)|_N, \tilde{m} \rangle = (T\phi)(g_2),$$

i.e. the value of $T\phi$ at a point $g \in G$ depends only on the coset of g. Hence, $T\phi$ drops to a function in $\mathcal{C}_b(G/N)$, so that T in induces a bounded linear map $\tilde{T}: LUC(G) \to \mathcal{C}_b(G/N)$. It is immediate that $\tilde{T}1 = 1$ and that $T\phi \geq 0$ for $\phi \geq 0$.

Let $\bar{m}$ be a right invariant mean on $\mathcal{C}_b(G/N)$. Define $m \in LUC(G)^*$ through

$$\langle \phi, m \rangle := \langle \tilde{T}\phi, \bar{m} \rangle \qquad (\phi \in LUC(G)).$$

It follows that $\langle 1, m \rangle = 1$ and $\langle \phi, m \rangle \geq 0$ for $\phi \in LUC(G)$ with $\phi \geq 0$, so that, by Proposition 1.1.2, m is a mean on $LUC(G)$. Let $g \in G$, and let $\phi \in LUC(G)$. Then we have

$$\langle \delta_g * \phi, m \rangle = \langle \tilde{T}(\delta_g * \phi), \bar{m} \rangle = \langle (\tilde{T}\phi) * \delta_{g^{-1}N}, \bar{m} \rangle = \langle \tilde{T}\phi, \bar{m} \rangle = \langle \phi, m \rangle. \tag{1.5}$$

Therefore m is left invariant. $\square$

Exercise 1.2.3 Justify in detail why the second equality in (1.5) holds.

Example 1.2.11 A group G is called *solvable*, if there are normal subgroups $N_0, \ldots, N_n$ of G with

$$\begin{cases} \{e_G\} = N_0 \subset \cdots \subset N_n = G; \\ N_j/N_{j-1} \text{ is abelian for } j = 1, \ldots, n. \end{cases}$$

Since abelian groups are amenable by Examples 1.1.5(c), it follows from successively applying Theorem 1.2.10 that all solvable groups are amenable as discrete groups. If a locally compact group G is solvable, G_d is amenable (since solvability does not depend on the topology), and by Corollary 1.1.10, G is amenable.

Exercise 1.2.4 Show that the *Heisenberg group*

$$\left\{ \begin{bmatrix} 1 & x & y \\ 0 & 1 & z \\ 0 & 0 & 1 \end{bmatrix} : x, y, z \in \mathbb{R} \right\}$$

is solvable and thus amenable.

Finally, we consider directed unions:

Proposition 1.2.12 *Let G be a locally compact group, and let $(H_\alpha)_\alpha$ be a directed family of closed sugroups of G such that $\bigcup_\alpha H_\alpha$ is dense in G and each group H_α is amenable. Then G is amenable.*

Proof For each H_α, let m_α be a translation invariant mean on $C_b(H_\alpha)$. Define $\tilde{m}_\alpha \in C_b(G)^*$ through

$$\langle \phi, \tilde{m}_\alpha \rangle := \langle \phi|_{H_\alpha}, m_\alpha \rangle \qquad (\phi \in C_b(G)).$$

Let m be a w^*-accumulation point of $(\tilde{m}_\alpha)_\alpha$ in $C_b(G)^*$. Then, certainly, m is a mean on $C_b(G)$. Let $\phi \in C_b(G)$, and let $g \in \bigcup_\alpha H_\alpha$. Since the family $(H_\alpha)_\alpha$ is directed, it follows that $\langle \delta_g * \phi, m \rangle = \langle \phi, m \rangle$. As in the proof of Proposition 1.2.1, we see that

$$\langle \delta_g * \phi, m \rangle = \langle \phi, m \rangle \qquad (g \in G, \phi \in LUC(G)),$$

i.e. $m|_{LUG(G)}$ is a left invariant mean on $LUC(G)$. $\square$

Example 1.2.13 A group G is called *locally finite* if every finite set of elements of G generates a finite subgroup of G. Let $\mathcal{F}$ be the collection of all finite subsets of G. For each $F \in \mathcal{F}$, let $\langle F \rangle$ denote the group generated by F. If G is locally finite, then $(\langle F \rangle)_{F \in \mathcal{F}}$ is a directed family of subgroups of G such that $\bigcup_{F \in \mathcal{F}} \langle F \rangle = G$. Since finite groups are amenable (Why?), every locally finite group is amenable.

Exercise 1.2.5 Show that

$$\{\pi : \mathbb{N} \to \mathbb{N} : \pi \text{ is bijective, and there is } N \in \mathbb{N} \text{ such that } \pi(n) = n \text{ for } n \geq N\}$$

is amenable.

For $n \in \mathbb{N}$, let

$$GL(n, \mathbb{R}) := \{A \in \mathbb{M}_n(\mathbb{R}) : A \text{ is invertible}\}$$

and

$$SL(n, \mathbb{R}) := \{A \in \mathbb{M}_n(\mathbb{R}) : \det A = 1\};$$

analoguously, we define $GL(n, \mathbb{C})$ and $SL(n, \mathbb{C})$. We conclude this section with an exercise, whose goal is to establish the non-amenability of these matrix groups:

Exercise 1.2.6 Recall from complex variables the notion of a linear fractional transformation ([Conw 1, Definition 3.5]). We use $\mathcal{G}$ to denote the group of all fractional linear transformations. For each matrix $\begin{bmatrix} a & b \\ c & d \end{bmatrix} \in GL(2, \mathbb{C})$, there is an element $h_A \in \mathcal{G}$ associated with it:

$$h_A(z) := \frac{az + b}{cz + d}.$$

(i) Show that the assignment

$$GL(2, \mathbb{C}) \to \mathcal{G}, \quad A \mapsto h_A \tag{1.6}$$

is a group homomorphism.

(ii) Let $PSL(2, \mathbb{R})$ be the image of $SL(2, \mathbb{R})$ under (1.6), and let it be equipped with the quotient topology. Show that $PSL(2, \mathbb{R}) \cong SL(2, \mathbb{R})/\{-E_2, E_2\}$.

(iii) Let

$$K_1 := \{z \in \mathbb{C} : |z + 2| = 1\},$$
$$\tilde{K}_1 := \{z \in \mathbb{C} : |z - 2| = 1\},$$
$$K_2 := \{z \in \mathbb{C} : |z + 5| = 1\},$$

and

$$\tilde{K}_2 := \{z \in \mathbb{C} : |z - 5| = 1\}.$$

Further, define

$$h_1(z) := \frac{2z + 3}{z + 2} \quad \text{and} \quad h_2(z) := \frac{5z + 24}{z + 5}$$

(so that $h_1, h_2 \in PSL(2, \mathbb{R})$). Let $j \in \{1, 2\}$, and let z be exterior to both circles K_j and $\tilde{K}_j$. Show that then $h_j(z)$ is in the interior of $\tilde{K}_j$ and $h_j^{-1}(z)$ is in the interior of K_j. (This and the next part of the problem require some knowledge about the geometric properties of fractional linear transformations; see, e.g., [Conw 1, §III.3].)

(iv) Let K_1, $\tilde{K}_1$, K_2, and $\tilde{K}_2$ and h_1 and h_2 be as in (iii). Again, let $j \in \{1, 2\}$, and let z be exterior to both K_j and $\tilde{K}_j$. Show that $h_j(z)$ and $h_j^{-1}(z)$ are exterior to both K_k and $\tilde{K}_k$ for $k \neq j$.

(v) Let z be exterior to all of $K_1 \cup \tilde{K}_1 \cup K_2 \cup \tilde{K}_2$, and let d be the distance of z to $K_1 \cup \tilde{K}_1 \cup K_2 \cup \tilde{K}_2$. Show that $|h(z) - z| > d$ for any $h \in PSL(2, \mathbb{R})$ that corresponds to a reduced word $w \neq \epsilon$ in h_1, h_2, h_1^{-1}, and h_2^{-1}. (*Hint*: Show by induction on the length of w that, depending on the element of $\{h_1, h_2, h_1^{-1}, h_2^{-1}\}$ with which w starts, $h(z)$ belongs to the interior of one of the circles K_1, $\tilde{K}_1$, K_2, and $\tilde{K}_2$.)

(vi) Show that the subgroup of $PSL(2, \mathbb{R})$ generated by h_1 and h_2 is closed and isomorphic to $\mathbb{F}_2$ (so that, consequently, $PSL(2, \mathbb{R})$ is not amenable).

(vii) Conclude that $SL(2, \mathbb{R})$ is not amenable.

(viii) Show that, for $n \geq 2$, the groups $SL(n, \mathbb{R})$, $GL(n, \mathbb{R})$, $SL(n, \mathbb{C})$, and $GL(n, \mathbb{C})$ are not amenable.

1.3 Day's fixed point theorem

In Examples 1.1.5(c), we used the Markov–Kakutani fixed point theorem to show that abelian groups are amenable. In this section, we shall give a characterization of amenable, locally compact groups — in principle due to M. M. Day — in terms of a fixed point property.

We leave a little bit of preparation to the reader:

Exercise 1.3.1 Let G be a locally compact group. Show that the set of means of the form

$$\sum_{j=1}^{n} t_j \delta_{g_j} \qquad (n \in \mathbb{N}, \ g_1, \ldots, g_n \in G, \ t_1, \ldots, t_n \geq 0, \ t_1 + \cdots + t_n = 1) \qquad (1.7)$$

is w^*-dense in the set of all means on $\mathcal{C}_b(G)$.

Theorem 1.3.1 (Day's fixed point theorem) *For a locally compact group G, the following are equivalent:*

(i) *G is amenable.*

(ii) *If G acts affinely on a compact, convex subset K of a locally convex vector space E, i.e.*

$$g \cdot (tx + (1-t)y) = t(g \cdot x) + (1-t)(g \cdot y) \qquad (g \in G, \ x, y \in K, \ t \in [0, 1]),$$

such that

$$G \times K \to K, \quad (g, x) \mapsto g \cdot x \qquad (1.8)$$

is separately continuous, then there is $x \in K$ such that $g \cdot x = x$ for all $g \in G$.

Proof (of (i) $\Longrightarrow$ (ii)) Fix $x_0 \in K$ once and for all. Let $A(K)$ denote the set of all continuous, affine functions on K; note that $\{\xi|_K : \xi \in E^*\} \subset A(K)$. For each $\psi \in A(K)$, define

$$\phi_\psi : G \to \mathbb{C}, \quad g \mapsto \psi(g \cdot x_0).$$

Since (1.8) is continuous in the first variable, it follows that $\phi_\psi \in \mathcal{C}_b(G)$. Let m be a left invariant mean on $\mathcal{C}_b(G)$. We claim that

– there is $x \in K$ such that

$$\langle \phi_\psi, m \rangle = \psi(x) \qquad (\psi \in A(K)),$$

and that

– this x is the desired fixed point.

Let $(m_\alpha)_\alpha$ be a net of means of the form (1.7) on $\mathcal{C}_b(G)$ such that $m = w^*\text{-}\lim_\alpha m_\alpha$. It is immediate (Really?) from (1.7) that, for each m_α, there is $x_\alpha \in K$ such that $\langle \phi_\psi, m_\alpha \rangle = \psi(x_\alpha)$ for all $\psi \in A(K)$. Without loss of generality, suppose that $(x_\alpha)_\alpha$ converges to some $x \in K$. Then we have, for any $\psi \in A(K)$,

$$\psi(x) = \lim_\alpha \psi(x_\alpha) = \lim_\alpha \langle \phi_\psi, m_\alpha \rangle = \langle \phi_\psi, m \rangle,$$

as claimed.

For $g \in G$ and $\xi \in E^*$, the function

$$\psi_{g,\xi} \colon K \to \mathbb{C}, \quad x \mapsto \langle g \cdot x, \xi \rangle$$

belongs to $A(K)$ (here, the continuity of (1.8) in the second variable is used). Note that

$$\phi_{\psi_{g,\xi}}(h) = \psi_{g,\xi}(h \cdot x_0) = \langle gh \cdot x_0, \xi \rangle = (\delta_{g^{-1}} * \phi_{\psi_{e_G,\xi}})(h) \qquad (g, h \in G, \; \xi \in E^*) \quad (1.9)$$

Then we have:

$$\begin{aligned}
\langle g \cdot x, \xi \rangle &= \psi_{g,\xi}(x) \\
&= \langle \phi_{\psi_{g,\xi}}, m \rangle \\
&= \langle \delta_{g^{-1}} * \phi_{\psi_{e_G,\xi}}, m \rangle, \qquad \text{by (1.9)}, \\
&= \langle \phi_{\psi_{e_G,\xi}}, m \rangle \\
&= \psi_{e_G,\xi}(x) \\
&= \langle x, \xi \rangle \qquad (g \in G, \; \xi \in E^*).
\end{aligned}$$

Since E^* separates the points of E, it follows that x is indeed a fixed point. $\quad\square$

Exercise 1.3.2 Prove (ii) $\Longrightarrow$ (i) of Theorem 1.3.1.

1.4 Representations on Hilbert space

To further illustrate the many nice properties of amenable, locally compact groups, we present a classical application of invariant means to representation theory.

Definition 1.4.1 Let G be a locally compact group, and let E be a Banach space. A *representation* of G on E is a group homomorphism π from G into the invertible bounded operators on E, which is continuous with respect to the given topology on G and the weak operator topology on $\mathcal{L}(E)$.

For those of you who have forgotten: The *weak operator topology* on $\mathcal{L}(E)$ is given by the family of seminorms $(p_{x,\phi})_{x \in E, \phi \in E^*}$, where

$$p_{x,\phi}(T) := \langle Tx, \phi \rangle \qquad (x \in E, \, \phi \in E^*, \, T \in \mathcal{L}(E)).$$

Similarly, the *strong operator topology* on $\mathcal{L}(E)$ is given through the family of seminorms $(p_x)_{x \in E}$, where

$$p_x(T) := \|Tx\| \qquad (x \in E, \, T \in \mathcal{L}(E)).$$

It is obvious, that the weak operator topology on $\mathcal{L}(E)$ is coarser than the strong operator topology.

Exercise 1.4.1 Let E be a Banach space.

(i) Show that the following are equivalent for a linear functional $\phi \colon \mathcal{L}(E) \to \mathbb{C}$:
(a) ϕ is continuous with respect to the strong operator topology;
(b) ϕ is continuous with resepct to the weak operator topology;
(c) there are $x_1, \dots, x_n \in E$ and $\psi_1, \dots, \psi_n \in E^*$ such that

$$\langle T, \phi \rangle = \sum_{j=1}^{n} \langle Tx_j, \psi_j \rangle \qquad (T \in \mathcal{L}(E)).$$

(ii) Show that the strong and the weak operator topology coincide if and only if $\dim E < \infty$.

Despite Exercise 1.4.1(ii), we have:

Lemma 1.4.2 *Let G be a locally compact group, let E be a Banach space, and let $\pi \colon G \to \mathcal{L}(E)$ be a representation of G on E. Then π is continuous with respect to the given topology on G and the strong operator topology.*

Proof Choose a compact neighborhood K of e_G in G. Then $\{\pi(g) : g \in K\}$ is compact in the weak operator topology, so that, $\{\pi(g)x : g \in K\}$ is compact in the weak operator topology for each $x \in E$. By the uniform boundedness principle, there is $C > 0$ with $\sup_{g \in K} \|\pi(g)\| \le C$.

Let U be another neighborhood of e_G such that $U = U^{-1}$ and $U^2 \subset K$. By Theorem A.1.8, there is a bounded approximate identity let $(e_\alpha)_\alpha$ for $L^1(G)$ consisting of continuous functions with support in U. Let $x \in E$. Since $\{\pi(g) : g \in K\}$ is weakly compact, there is, for each $h \in U$, a unique element $y \in E$ such that

$$\langle y, \phi \rangle = \int_G e_\alpha(h^{-1}g) \langle \pi(g)x, \phi \rangle \, dm_G(g) \qquad (\phi \in E^*) \tag{1.10}$$

(Why exactly?); we simply denote this y by $\int_G e_\alpha(h^{-1}g)\pi(g)x \, dm_G(g)$. One verifies (One should indeed do this using (1.10)!) that

$$\pi(h) \left(\int_G e_\alpha(g)\pi(g)x \, dm_G(g) \right) = \int_G e_\alpha(g)\pi(hg)x \, dm_G(g)$$

$$= \int_G e_\alpha(h^{-1}g)\pi(g)x \, dm_G(g) \qquad (h \in U). \tag{1.11}$$

Define a net $(x_\alpha)_\alpha$ in E through

$$x_\alpha := \int_G e_\alpha(g)\pi(g)x\, dm_G(g).$$

Then, for $h \in U$, we have

$$\|\pi(h)x_\alpha - x_\alpha\| = \left\| \int_G e_\alpha(h^{-1}g) - e_\alpha(g))\pi(g)x\, dm_G(g) \right\|, \qquad \text{by (1.11)},$$
$$\leq \|\delta_h * e_\alpha - e_\alpha\|_1 C \|x\|$$

It follows that the map

$$G \to E, \quad g \mapsto \pi(g)x_\alpha$$

is continuous with respect to the norm topology on E (it is clearly sufficient to have continuity at e_G).

Let

$$F := \{z \in E : G \ni g \mapsto \pi(g)z \text{ is continuous with respect to the norm topology on } E\}.$$

Clearly, F is a linear subspace of E containing $(x_\alpha)_\alpha$. Let $(z_n)_{n=1}^\infty$ be a sequence in F converging to $z \in E$, and let $(g_\beta)_\beta$ be a net in K converging to e_G. Then we have

$$\|\pi(g_\beta)z - z\| = \|\pi(g_\beta)z - \pi(g_\beta)z_n\| + \|\pi(g_\beta)z_n - z_n\| + \|z_n - z\|$$
$$\leq (C+1)\|z_n - z\| + \|\pi(g_\beta)z_n - z_n\|$$

Let $\epsilon > 0$. Choose $n_0 \in \mathbb{N}$ such that $\|z_n - z\| < \frac{\epsilon}{2(C+1)}$ for $n \geq n_0$, and choose β_0 such that $\|\pi(g_\beta)z_{n_0} - z_{n_0}\| < \frac{\epsilon}{2}$ for $\beta \geq \beta_0$. It follows that $\|\pi(g_\beta)z - z\| < \epsilon$ for $\beta \geq \beta_0$. Therfore, $z \in F$ as well. Consequently, F is a closed subspace of E and thus weakly closed.

From (1.10) and the definition of $(x_\alpha)_\alpha$, it follows that x is the weak limit of $(x_\alpha)_\alpha$. Since $(x_\alpha)_\alpha$ is contained in F, this means $x \in F$. Since $x \in E$ was arbitrarily chosen, we have $F = E$. This completes the proof. $\square$

The following is related, but much easier:

Exercise 1.4.2 Let $\mathfrak{H}$ be a Hilbert space. Then the weak and the strong operator topology coincide on the group of unitary operators on $\mathfrak{H}$.

Definition 1.4.3 Let G be a locally compact group, and let E be a Banach space.

(i) Two representations π_1 and π_2 of G on E are *similar* if there is an invertible operator $T \in \mathcal{L}(E)$ such that

$$\pi_1(g) = T\pi_2(g)T^{-1} \qquad (g \in G).$$

(ii) A representation π of G on E is *uniformly bounded* if $\sup_{g \in G} \|\pi(g)\| < \infty$.

(iii) A representation π of G on a Hilbert space is *unitary* if $\pi(G)$ consists of unitary operators.

Exercise 1.4.3 Give an example of a representation of $\mathbb{R}$ on a Hilbert space which is not uniformly bounded.

Exercise 1.4.4 Let G be a locally compact group, let G be a Banach space, and let $\pi \colon G \to \mathcal{L}(E)$ be a uniformly bounded representation. Show that there is an equivalent norm on E such that, with respect to this new norm, $\pi(G)$ consists of isometries.

In the representation theory of locally compact groups, the unitary representations play the predominant rôle because they establish a link between locally compact groups and C^*-algebras (see, e.g., [Dixm 2, Chapter 13]). So, when is an arbitrary representation of a locally compact group similar to a unitary representation? A necessary requirement is, obviously, that the representation be uniformly bounded. But is this necessary condition also sufficient? This is indeed the case for amenable groups:

Theorem 1.4.4 *Let G be an amenable locally compact group, let $\mathfrak{H}$ be a Hilbert space, and let $\pi \colon G \to \mathcal{L}(\mathfrak{H})$ be a uniformly bounded representation. Then π is similar to a unitary representation.*

Proof For $\xi, \eta \in \mathfrak{H}$, define

$$\phi_{\xi,\eta} \colon G \to \mathbb{C}, \quad g \mapsto \langle \pi(g^{-1})\xi, \pi(g^{-1})\eta \rangle.$$

Let $C := \sup_{g \in G} \|\pi(g)\|$. Since, by Lemma 1.4.2, π is continuous with respect to the strong operator topology on $\mathcal{L}(\mathfrak{H})$, it follows from

$$
\begin{aligned}
&|\phi_{\xi,\eta}(g) - \phi_{\xi,\eta}(h)| \\
&= |\langle \pi(g^{-1})\xi, \pi(g^{-1})\eta \rangle - \langle \pi(h^{-1})\xi, \pi(h^{-1})\eta \rangle| \\
&\leq |\langle \pi(g^{-1})\xi, \pi(g^{-1})\eta \rangle - \langle \pi(g^{-1})\xi, \pi(h^{-1})\eta \rangle| \\
&\quad + |\langle \pi(g^{-1})\xi, \pi(h^{-1})\eta \rangle - \langle \pi(h^{-1})\xi, \pi(h^{-1})\eta \rangle| \\
&\leq C\|\xi\| \|\pi(g^{-1})\eta - \pi(h^{-1})\eta\| + \|\pi(g^{-1})\xi - \pi(h^{-1})\xi\| C \|\eta\| \qquad (g, h \in G)
\end{aligned}
$$

that $\phi_{\xi,\eta} \in \mathcal{C}_b(G)$.

Let m be a left invariant mean on $\mathcal{C}_b(G)$, and define

$$[\xi, \eta] := \langle \phi_{\xi,\eta}, m \rangle \qquad (\xi, \eta \in \mathfrak{H}).$$

It is easy to see that $[\cdot, \cdot]$ is a positive semidefinite, sesquilinear form on $\mathfrak{H}$. Let

$$\|\|\xi\|\| := [\xi, \xi]^{\frac{1}{2}} \qquad (\xi \in \mathfrak{H}).$$

It is immediate that $\|\|\xi\|\| \leq C\|\xi\|$ for $\xi \in \mathfrak{H}$. Also, since

$$\|\xi\|^2 \leq \|\pi(g)\|^2 \phi_{\xi,\xi}(g) \leq C^2 \phi_{\xi,\xi}(g) \qquad (\xi \in \mathfrak{H}, g \in G),$$

$\|\xi\| \leq C\|\|\xi\|\|$ for $\xi \in \mathfrak{H}$. Consequently, $\|\cdot\|$ and $\|\|\cdot\|\|$ are equivalent. By elementary Hilbert space theory, there is $S \in \mathcal{L}(\mathfrak{H})$ such that

$$\langle S\xi, \eta \rangle = [\xi, \eta] \qquad (\xi, \eta \in \mathfrak{H}).$$

Since $[\cdot, \cdot]$ is positive definite, S is a positive operator and, in particular, self-adjoint. Letting $T := S^{\frac{1}{2}}$, we thus have

$$\langle T\xi, T\eta \rangle = [\xi, \eta] \qquad (\xi, \eta \in \mathfrak{H}).$$

Since $\| \cdot \|$ and $\|| \cdot \||$ are equivalent, it follows that T is invertible.

Noting that

$$\phi_{\pi(g)\xi, \pi(g)\eta} = \delta_g * \phi_{\xi, \eta} \qquad (g \in G, \ \xi, \eta \in \mathfrak{H}), \tag{1.12}$$

we then have for $\xi, \eta \in \mathfrak{H}$ and $g \in G$:

$$\begin{aligned}
\langle T\pi(g)T^{-1}\xi, T\pi(g)T^{-1}\eta \rangle &= [\pi(g)T^{-1}\xi, \pi(g)T^{-1}\eta] \\
&= \langle \phi_{\pi(g)T^{-1}\xi, \pi(g)T^{-1}\eta}, m \rangle \\
&= \langle \delta_g * \phi_{T^{-1}\xi, T^{-1}\eta}, m \rangle, \qquad \text{by (1.12),} \\
&= \langle \phi_{T^{-1}\xi, T^{-1}\eta}, m \rangle \\
&= [T^{-1}\xi, T^{-1}\eta] \\
&= \langle \xi, \eta \rangle.
\end{aligned}$$

This means that the representation $T\pi(\cdot)T^{-1}$ is indeed unitary. $\quad\square$

Exercise 1.4.5 An *unconditional basis* for a Banach space E is a sequence $(x_n)_{n=1}^{\infty}$ in E with the property that, for each $x \in E$, there is a unique sequence $(\lambda_n)_{n=1}^{\infty}$ in $\mathbb{C}$ such that $x = \sum_{n=1}^{\infty} \lambda_n x_n$, where the series converges unconditionally. Use Theorem 1.4.4 to prove the Köthe–Lorch theorem: For any unconditional basis $(e_n)_{n=1}^{\infty}$ of a Hilbert space $\mathfrak{H}$, there is an invertible operator $T \in \mathcal{L}(\mathfrak{H})$ such that $(Te_n)_{n=1}^{\infty}$ is an orthonormal basis. (*Hint*: Let $G = \{-1, 1\}^{\mathbb{N}}$, and define a representation $\pi \colon G \to \mathcal{L}(\mathfrak{H})$ by letting $\pi(g)\xi := \sum_{n=1}^{\infty} \lambda_n \epsilon_n e_n$ for $g = (\epsilon_n)_{n=1}^{\infty} \in G$ and $\xi = \sum_{n=1}^{\infty} \lambda_n e_n \in \mathfrak{H}$. Then apply Theorem 1.4.4 to this representation.)

1.5 Notes and comments

Amenable (discrete) groups were first considered by J. von Neumann, albeit under a different name ([Neu]). The first to use the adjective "amenable" was M. M. Day in [Day], apparently with a pun in mind: These groups G are called a*men*able because they have an invariant mean on $L^{\infty}(G)$, but also since they are particularly pleasant to deal with and thus are truly *amenable* — just in the sense of that adjective in colloquial English. The introductions of [Pie] and [Pat 1] give sketches of the historical developement of the subject (including more references to early papers).

Our exposition borrows heavily from the monographs [Gre], [Pie], and [Pat 1]. An alternative (but not easier) proof of Theorem 1.2.7, which avoids Bruhat functions, is given in [Gre]; the underlying idea is essentially the same as in the false "proof" in Exercise 1.2.2. The following question was open for quite some time: Is the class of amenable (discrete) groups the smallest class of groups which contains all abelian and all finite groups and which is closed under taking (a) quotients, (b) subgroups, (c) extensions, and (d) directed unions? A counterexample was constructed in 1983 by R. I. Grigorchuk ([Gri]).

Let G be a locally compact group which has a paradoxical decomposition such that the sets $A_1, \ldots, A_n$ and $B_1, \ldots, B_m$ in Definition 0.1.1 are Borel sets. Then G is not amenable: This can either be seen directly, as in Exercise 0.2.1, or be infered from Corollaries 0.2.11 and

1.1.10. The question of whether the non-existence of such Borel paradoxical decompositions characterizes the amenable, locally compact groups seems to be open ([Wag]).

Theorem 1.4.4 is due to J. Dixmier ([Dixm 1]); the special case $G = \mathbb{R}$ was established earlier by B. Sz.-Nagy ([SzN]). Lemma 1.4.2 is [G–L, Theorem 2.8]. The idea to use Theorem 1.4.4 to prove the Köthe–Lorch theorem is from [Wil]. It is a still open problem, posed by J. Dixmier, whether a converse to Theorem 1.4.4 holds: Is a locally compact group such that every uniformly bounded representation on a Hilbert space is similar to a unitary representation already amenable? For each $n = 2, \ldots, \infty$, the free group in n generators has a uniformly bounded representation that is *not* similar to a unitary representation ([Pis 2, Corollary 2.3 and Lemma 2.7(iii)]). Other partial results towards a converse of Theorem 1.4.4 in the discrete case are given in [Pis 4].

Of course, our treatment of amenable, locally compact groups is far from exhaustive. In fact, we have only scratched at the surface. For two reasons, we don't feel too bad about it:

− There are at least three monographs devoted to the subject, namely the aforementioned [Gre], [Pie], and [Pat 1].
− We are primarily interested in the Banach algebraic aspects of amenability, and we have been dealing with amenable, locally compact groups mainly to set the stage for Johnson's theorem in the next chapter.

Nevertheless, in order to give the reader a rough impression of why Day's name stuck, we would like to mention two strikingly beautiful characterizations of amenability for locally compact groups (if you want more, read [Gre], [Pie], and [Pat 1]):

Følner type conditions

A locally compact group G is said to satisfy the *Følner condition* if, for each compact subset K of G and for each $\epsilon > 0$, there is a Borel subset E of G with $0 < m_G(E) < \infty$ such that

$$\frac{m_G(gE \Delta E)}{m_G(E)} < \epsilon \qquad (g \in K),$$

where Δ denotes the symmetric difference of two sets. This condition is named in the honor of E. Følner, who studied it for the first time in [Føl] (in the discrete case only). There are variants of the Følner condition — the weak Følner condition, the strong Følner condition, the Leptin condtion —, some of which are formally weaker, some formally stronger. They all turn out to be equivalent to G being amenable.

Weak containment of unitary representations

If π is a unitary representation of a locally compact group G on some Hilbert space $\mathfrak{H}$, it induces a *-representation $\tilde{\pi}$ of $L^1(G)$ on $\mathfrak{H}$ via

$$\langle \tilde{\pi}(f)\xi, \eta \rangle := \int_G f(g) \langle \pi(g)\xi, \eta \rangle \, dm_G(g) \qquad (f \in L^1(G), \, \xi, \eta \in \mathfrak{H}),$$

which, in turn, extends to a *-representation $\tilde{\pi}$ of the group C^*-algebra $C^*(G)$ (the definition of $C^*(G)$ is sketched on page 240; see [Dixm 2, Chapter 13] for further details). If π_1 and

π_2 are two unitary representations of G — not necessarily on the same Hilbert space —, we say that π_1 is *weakly contained* in π_2 if $\ker \bar{\pi}_2 \subset \ker \bar{\pi}_1$. Let λ_2 denote the left regular represention of G on $L^2(G)$ (Definition A.3.1(i)). The following are equivalent:

(i) G is amenable.

(ii) Every (topologically) irreducible, unitary representation of G is weakly contained in λ_2.

(iii) The trivial represenation of G on $\mathbb{C}$ (the one that maps every element of G to 1) is weakly contained in λ_2.

The equivalence of (i) and (ii) originates with A. Hulanicki ([Hul]) whereas the equivalence of (ii) and (iii) is due to R. Godement ([God]).

Let $C^*_{\lambda_2}(G)$ denote the *reduced group C^*-algebra*, i.e. the closure of $\tilde{\lambda}_2(L^1(G))$ with respect to the operator norm. Then (i), (ii), and (iii) are equivalent to $C^*(G) \cong C^*_{\lambda_2}(G)$. Since $C^*_{\lambda_2}(G)$ is, in general, much more tractable than $C^*(G)$, this is yet another nice feature of amenable groups.

2 Amenable Banach algebras

We now switch from groups to Banach algebras.

There are various Banach spaces and algebras associated with a locally compact group G (see Appendix A). Perhaps the most important one is the convolution algebra $L^1(G)$. It is a complete invariant for G: If G_1 and G_2 are locally compact groups such that $L^1(G_1)$ and $L^1(G_2)$ are isometrically isomorphic, then G_1 and G_2 are topologically isomorphic. Hence, all information about G is already encoded in $L^1(G)$.

In 1972, B. E. Johnson proved that the amenable locally compact groups G can be characterized by the vanishing of certain cohomology groups of $L^1(G)$. This is an important result for two reasons:

– Homological algebra is an extremely powerful toolkit in mathematics. Knowing that a certain property can be characterized in terms of cohomology is thus good in itself.
– The cohomological triviality condition used by Johnson to characterize the amenable, locally compact groups makes sense for arbitrary Banach algebras, which may have nothing to do with locally compact groups. We can thus meaningfully speak of *amenable* Banach algebras.

As in the case of amenable, locally compact groups, the notion of an amenable Banach algebra has turned out to be an extremely fruitful one: We have a variety of interesting examples — $L^1(G)$ for amenable, locally compact groups G, nuclear C^*-algebras, algebras $\mathcal{K}(E)$ for Banach spaces E with certain approximation properties, and even radical Banach algebras — whose only common feature seems to be their amenability. On the other hand, there is still an abundance of substantial theorems for general amenable Banach algebras.

We suppose in this chapter that the reader is familiar with the basics of Banach algebra theory. Some of the background we require is not usually treated in an introductory course on the subject; in particular, we need the basics on bounded approximate identities ([B–D, §11]). We will also have to deal extensively with projective tensor products; a reader not familiar with them should first consult Appendix B.

2.1 Johnson's theorem

Johnson's theorem asserts that amenable, locally compact groups G can be characterized in terms of the *Hochschild cohomology* of their convolution algebras $L^1(G)$. We begin by introducing the necessary concepts.

Definition 2.1.1 Let $\mathfrak{A}$ be a Banach algebra. A Banach space E which is also a left $\mathfrak{A}$-module is called a *left Banach $\mathfrak{A}$-module* if there is $\kappa > 0$ such that

$$\|a \cdot x\| \leq \kappa \|a\| \|x\| \qquad (a \in \mathfrak{A},\, x \in E).$$

You can now guess the respective definitions of a right Banach $\mathfrak{A}$-module and of a Banach $\mathfrak{A}$-bimodule ...

Let $\mathfrak{A}$ be a Banach algebra, and let E be a Banach $\mathfrak{A}$-bimodule. A bounded linear map $D : \mathfrak{A} \to E$ is called a *derivation* if

$$D(ab) = a \cdot Db + (Da) \cdot b \qquad (a, b \in \mathfrak{A}).$$

Let $x \in E$. We define

$$\mathrm{ad}_x : \mathfrak{A} \to E, \quad a \mapsto a \cdot x - x \cdot a$$

It is easily seen (but should nevertheless be checked) that ad_x is a derivation. Derivations of this form are called *inner derivations*. For reasons that will become apparent later, the set of all derivations from $\mathfrak{A}$ into E is denoted by $\mathcal{Z}^1(\mathfrak{A}, E)$, and the set of all inner derivations is denoted by $\mathcal{B}^1(\mathfrak{A}, E)$. Clearly, $\mathcal{Z}^1(\mathfrak{A}, E)$ is a closed subspace of $\mathcal{L}(\mathfrak{A}, E)$, and $\mathcal{B}^1(\mathfrak{A}, E)$ is a (not necessarily closed) subspace of $\mathcal{Z}^1(\mathfrak{A}, E)$.

Definition 2.1.2 Let $\mathfrak{A}$ be a Banach algebra, and let E be a Banach $\mathfrak{A}$-bimodule. Then

$$\mathcal{H}^1(\mathfrak{A}, E) := \mathcal{Z}^1(\mathfrak{A}, E) / \mathcal{B}^1(\mathfrak{A}, E)$$

is the *first Hochschild cohomology group* of $\mathfrak{A}$ with coefficients in E.

Since $\mathcal{Z}^1(\mathfrak{A}, E)$ and $\mathcal{B}^1(\mathfrak{A}, E)$ are linear spaces, $\mathcal{H}^1(\mathfrak{A}, E)$ should rather be called a cohomology *space*, but it is customary to speak of cohomology groups.

We have now defined cohomology groups — in fact only first order cohomology groups —, but what can we do with them? Is there any relevant information on $\mathfrak{A}$ and E encoded in $\mathcal{H}^1(\mathfrak{A}, E)$?

Exercise 2.1.1 Let $\mathfrak{A}$ be a Banach algebra. Verify the following:

(i) If E is a left Banach $\mathfrak{A}$-module, then E^* becomes a right Banach $\mathfrak{A}$-module through

$$\langle x, \phi \cdot a \rangle := \langle a \cdot x, \phi \rangle \qquad (a \in \mathfrak{A},\, x \in E,\, \phi \in E^*).$$

(ii) If E is a right Banach $\mathfrak{A}$-module, then E^* becomes a left Banach $\mathfrak{A}$-module through

$$\langle x, a \cdot \phi \rangle := \langle x \cdot a, \phi \rangle \qquad (a \in \mathfrak{A},\, x \in E,\, \phi \in E^*).$$

(iii) If E is a Banach $\mathfrak{A}$-bimodule, then E^* equipped with the left and right module actions of $\mathfrak{A}$ from (ii) and (i), respectively, is a Banach $\mathfrak{A}$-bimodule.

Dual spaces of Banach modules (left, right, or bi-) equipped with the module operations pointed out in Exercise 2.1.1 are called *dual Banach modules*.

Exercise 2.1.2 Let $\mathfrak{A}$ be a Banach algebra, and let E be a Banach $\mathfrak{A}$-module (left, right, or bi-), and let F be a w^*-closed submodule of E^*. Then F is a dual Banach module.

One significant advantage of dual Banach modules is that we have the compact w^*-topology on their closed unit balls:

Proposition 2.1.3 *Let $\mathfrak{A}$ be a Banach algebra with a bounded right approximate identity, and let E be a Banach $\mathfrak{A}$-bimodule such that $\mathfrak{A} \cdot E = \{0\}$. Then $\mathcal{H}^1(\mathfrak{A}, E^*) = \{0\}$.*

Proof It is clear that $E^* \cdot \mathfrak{A} = \{0\}$. Let $D \in \mathcal{Z}^1(\mathfrak{A}, E^*)$. Then it follows that

$$D(ab) = a \cdot Db \qquad (a, b \in \mathfrak{A}).$$

Let $(e_\alpha)_\alpha$ be a bounded right approximate identity for $\mathfrak{A}$, and let $\phi \in E^*$ be a w^*-accumulation point of $(De_\alpha)_\alpha$. Without loss of generality, we may suppose that $\phi = w^*\text{-}\lim_\alpha De_\alpha$. It follows that

$$Da = \lim_\alpha D(ae_\alpha) = \lim_\alpha a \cdot De_\alpha = a \cdot \phi \qquad (a \in \mathfrak{A}),$$

so that $D = \mathrm{ad}_\phi$. $\square$

Of course, there is an analogue of Proposition 2.1.3 for Banach algebras with bounded left approximate identities where the module action from the right is trivial.

Exercise 2.1.3 Wendel's theorem asserts: If G is a locally compact group, and if $T \in \mathcal{L}(L^1(G))$ is such that $T(f * g) = f * Tg$ for all $f, g \in L^1(G)$, then there is $\mu \in M(G)$ such that $Tf = f * \mu$ for all $f \in L^1(G)$. Derive Wendel's theorem from Proposition 2.1.3.

If $\mathfrak{A}$ is a unital Banach algebra, a Banach $\mathfrak{A}$-bimodule E is called *unital* if

$$e_\mathfrak{A} \cdot x = x \cdot e_\mathfrak{A} = x \qquad (x \in E).$$

For non-unital Banach algebras $\mathfrak{A}$ (such as $L^1(G)$ for non-discrete G), the following class of Banach $\mathfrak{A}$-bimodules plays a similar rôle:

Definition 2.1.4 Let $\mathfrak{A}$ be a Banach algebra. A Banach $\mathfrak{A}$-bimodule E is called *pseudo-unital* if

$$E = \{a \cdot x \cdot b : a, b \in \mathfrak{A}, x \in E\}.$$

Similarly, one defines pseudo-unital left and right Banach modules.

Exercise 2.1.4 Let $\mathfrak{A}$ be a Banach algebra, and let E be a Banach $\mathfrak{A}$-bimodule. We say that E is *essential* if the linear hull of $\{a \cdot x \cdot b : a, b \in \mathfrak{A}, x \in\}$ is dense in E. Use Cohen's factorization theorem [B–D, Theorem 11.10] to show: If $\mathfrak{A}$ has a bounded approximate identity and if E is essential, then E is pseudo unital.

The following proposition is often useful:

Proposition 2.1.5 *For a Banach algebra $\mathfrak{A}$ with a bounded approximate identity the following are equivalent:*

(i) $\mathcal{H}^1(\mathfrak{A}, E^*) = \{0\}$ *for each Banach $\mathfrak{A}$-bimodule E.*
(ii) $\mathcal{H}^1(\mathfrak{A}, E^*) = \{0\}$ *for each pseudo-unital Banach $\mathfrak{A}$-bimodule E.*

Proof Obviously, (i) $\Longrightarrow$ (ii) holds.

For the converse, let E be any Banach $\mathfrak{A}$-bimodule, and let $D \in \mathcal{Z}^1(\mathfrak{A}, E^*)$. Let

$$E_0 := \{a \cdot x \cdot b : a, b \in \mathfrak{A}, x \in E\}.$$

By Cohen's factorization theorem [B–D, Theorem 11.10], E_0 is a closed submodule of E. Let $\pi : E^* \to E_0^*$ be the restriction map. It is routinely checked that π is a module homomorphism, so that $\pi \circ D \in \mathcal{Z}^1(\mathfrak{A}, E_0^*)$. Since $\mathcal{H}^1(\mathfrak{A}, E_0^*) = \{0\}$, there is $\phi_0 \in E_0^*$ such that $\pi \circ D = \mathrm{ad}_{\phi_0}$. Choose $\phi \in E^*$ such that $\phi|_{E_0} = \phi_0$. It follows that $\tilde{D} := D - \mathrm{ad}_\phi \in \mathcal{Z}^1(\mathfrak{A}, E_0^\perp)$. We have $E_0^\perp \cong (E/E_0)^*$ (as Banach $\mathfrak{A}$-bimodules). From the definiton of E_0, it follows that $\mathfrak{A} \cdot (E/E_0) = \{0\}$, so that $\mathcal{H}^1(\mathfrak{A}, E_0^\perp) = \{0\}$ by Proposition 2.1.3. Hence, there is $\psi \in E_0^\perp$ such that $\tilde{D} = \mathrm{ad}_\psi$, i.e. $D = \mathrm{ad}_{\phi - \psi}$. $\square$

It is often convenient to extend a derivation to a larger algebra. If a Banach algebra $\mathfrak{A}$ is contained as a closed ideal in another Banach algebra $\mathfrak{B}$ then the *strict topology* on $\mathfrak{B}$ with respect to $\mathfrak{A}$ is defined through the family of seminorms $(p_a)_{a \in \mathfrak{A}}$, where

$$p_a(b) := \|ba\| + \|ab\| \qquad (b \in \mathfrak{B}).$$

Note that the strict topology is Hausdorff only if $\{b \in \mathfrak{B} : b \cdot \mathfrak{A} = \mathfrak{A} \cdot b = \{0\}\} = \{0\}$.

Proposition 2.1.6 *Let $\mathfrak{A}$ be a Banach algebra with a bounded approximate identity which is contained as closed ideal in a Banach algebra $\mathfrak{B}$, let E be a pseudo-unital Banach $\mathfrak{A}$-bimodule, and let $D \in \mathcal{Z}^1(\mathfrak{A}, E^*)$. Then E is a Banach $\mathfrak{B}$-bimodule in a canonical fashion, and there is a unique $\tilde{D} \in \mathcal{Z}^1(\mathfrak{B}, E^*)$ such that:*

(i) *$\tilde{D}|_{\mathfrak{A}} = D$;*

(ii) *$\tilde{D}$ is continuous with respect to the strict topology on $\mathfrak{B}$ and the w^*-topology on E^*.*

Proof For $x \in E$, let $a \in \mathfrak{A}$ and $y \in E$ be such that $x = a \cdot y$. For $b \in \mathfrak{B}$, define $b \cdot x := ba \cdot y$. We claim that $b \cdot x$ is well defined, i.e. independent of the choices for a and y. Let $a' \in \mathfrak{A}$ and $y' \in E$ be such that $x = a' \cdot y'$, and let $(e_\alpha)_\alpha$ be a bounded approximate identity for $\mathfrak{A}$. Then

$$ba \cdot y = \lim_\alpha be_\alpha a \cdot y = \lim_\alpha be_\alpha a' \cdot y' = ba' \cdot y' \qquad (b \in \mathfrak{B}).$$

It is obvious that this operation of $\mathfrak{B}$ on E turns E into a left Banach $\mathfrak{B}$-module. Similarly, one defines a right Banach $\mathfrak{B}$-module structure on E, so that, eventually, E becomes a Banach $\mathfrak{B}$-bimodule.

To extend D, let

$$\tilde{D} : \mathfrak{B} \to E^*, \qquad b \mapsto w^*\text{-}\lim_\alpha (D(be_\alpha) - b \cdot De_\alpha). \tag{2.1}$$

We claim that $\tilde{D}$ is well-defined, i.e. the limit in (2.1) does exist. Let $x \in E$, and let $a \in \mathfrak{A}$ and $y \in E$ be such that $x = y \cdot a$. Then

$$
\begin{aligned}
\langle x, D(be_\alpha) - b \cdot De_\alpha \rangle &= \langle y \cdot a, D(be_\alpha) - b \cdot De_\alpha \rangle \\
&= \langle y, a \cdot D(be_\alpha) - ab \cdot De_\alpha \rangle \\
&= \langle y, D(abe_\alpha) - (Da) \cdot be_\alpha - D(abe_\alpha) + D(ab) \cdot e_\alpha \rangle \\
&= \langle e_\alpha \cdot y, D(ab) \rangle - \langle be_\alpha \cdot y, Da \rangle \\
&\to \langle y, D(ab) \rangle - \langle b \cdot y, Da \rangle \qquad (b \in \mathfrak{B}),
\end{aligned}
$$

so that the limit in (2.1) exists. Moreover, for $a \in \mathfrak{A}$,

$$\tilde{D}a = w^*\text{-}\lim_{\alpha}(D(ae_\alpha) - a \cdot De_\alpha) = w^*\text{-}\lim_{\alpha}(Da) \cdot e_\alpha = Da,$$

so that $\tilde{D}$ extends D. Furthermore, we have for $b \in \mathfrak{B}$ and $a \in \mathfrak{A}$ that

$$\begin{aligned}
(\tilde{D}b) \cdot a &= w^*\text{-}\lim_{\alpha}(D(be_\alpha) \cdot a - b \cdot (De_\alpha) \cdot a) \\
&= w^*\text{-}\lim_{\alpha}(D(be_\alpha a) - be_\alpha \cdot Da - b \cdot D(e_\alpha a) + be_\alpha \cdot Da) \\
&= D(ba) - b \cdot Da.
\end{aligned}$$

It follows (and you should provide the details) that $\tilde{D}$ is continuous with respect to the strict topology on $\mathfrak{B}$ and the w^*-topology on E^*.

It remains to be shown that $\tilde{D}$ is a derivation. From the definition of the strict topology it is clear that $be_\alpha \to b$ in this topology for all $b \in \mathfrak{B}$. Let $b, c \in \mathfrak{B}$. Then

$$\begin{aligned}
\tilde{D}(bc) &= w^*\text{-}\lim_{\alpha} w^*\text{-}\lim_{\beta} D((be_\alpha)(ce_\beta)) \\
&= w^*\text{-}\lim_{\alpha} w^*\text{-}\lim_{\beta}(be_\alpha \cdot D(ce_\beta) + D(be_\alpha) \cdot ce_\beta) \\
&= b \cdot \tilde{D}c + (\tilde{D}b) \cdot c,
\end{aligned}$$

i.e. $\tilde{D} \in \mathcal{Z}^1(\mathfrak{B}, E^*)$. $\square$

For the case of $\mathfrak{A} = L^1(G)$ and $\mathfrak{B} = M(G)$, where G is a locally compact group, Proposition 2.1.6 yields:

Corollary 2.1.7 *Let G be a locally compact group, let E be a pseudo-unital Banach $L^1(G)$-module, and let $D \in \mathcal{Z}^1(L^1(G), E^*)$. Then E is a Banach $M(G)$-bimodule in a canonical fashion, and there is a unique $\tilde{D} \in \mathcal{Z}^1(M(G), E^*)$ that extends D and is continuous with respect to the strict topology on $M(G)$ and the w^*-topology on E^*. In particular, $\tilde{D}$ is uniquely determined by its values on $\{\delta_g : g \in G\}$.*

Proof Everything follows immediately from Proposition 2.1.6, except the last clause. However, by Exercise A.2.4(iii), $\ell^1(G)$ is strictly dense in $M(G)$, which shows that $\tilde{D}$ is uniquely determined through its values on $\ell^1(G)$ and hence on $\{\delta_g : g \in G\}$. $\square$

We can now prove the main result of this section:

Theorem 2.1.8 (Johnson's theorem) *For a locally compact group G, the following are equivalent:*

(i) *G is amenable.*
(ii) *$\mathcal{H}^1(L^1(G), E^*) = \{0\}$ for each Banach $L^1(G)$-bimodule E.*

Proof (i) $\Longrightarrow$ (ii): Let E be a Banach $L^1(G)$-bimodule. By Proposition 2.1.5 there is no loss of generality if we suppose that E is pseudo-unital. Let $D \in \mathcal{Z}^1(L^1(G), E^*)$, and let $\tilde{D} \in \mathcal{Z}^1(M(G), E^*)$ be the extension of D according to Corollary 2.1.7.

We shall see that $\tilde{D}$ is inner using Theorem 1.3.1. Let K be the w^*-closed convex hull of the set $\{(\tilde{D}\delta_g) \cdot \delta_{g^{-1}} : g \in G\}$. Then K is w^*-compact and convex. Define an action of G on E^* through

$$g \cdot \phi := \delta_g \cdot \phi \cdot \delta_{g^{-1}} + (\tilde{D}\delta_g) \cdot \delta_{g^{-1}} \qquad (g \in G, \, \phi \in E^*).$$

We claim that this definition actually defines a group action:

$$\begin{aligned} gh \cdot \phi &= \delta_{gh} \cdot \phi \cdot \delta_{(gh)^{-1}} + (\tilde{D}\delta_{gh}) \cdot \delta_{(gh)^{-1}} \\ &= \delta_g \cdot (\delta_h \cdot \phi \cdot \delta_{h^{-1}}) \cdot \delta_{g^{-1}} + (\delta_g \cdot \tilde{D}(\delta_h) + \tilde{D}(\delta_g) \cdot \delta_h) \cdot \delta_{h^{-1}} * \delta_{g^{-1}} \\ &= \delta_g \cdot (\delta_h \cdot \phi \cdot \delta_{h^{-1}} + \tilde{D}(\delta_h) \cdot \delta_{h^{-1}}) \cdot \delta_{g^{-1}} + \tilde{D}(\delta_g) \cdot \delta_{g^{-1}} \\ &= g \cdot (h \cdot \phi) \qquad (g, h \in G, \, \phi \in E^*). \end{aligned} \qquad (2.2)$$

It is clear that, with this definiton, G acts affinely on E^* It is also easy to see that this group action is w^*-continuous in the second variable. It is a bit harder to see that it is also continuous in the first variable with respect to the given topology on G and the w^*-topology on E^* (Try it yourself! Use the fact that, if $(g_\alpha)_\alpha$ is net in G converging to some $g \in G$ in the given topology, then $\delta_{g_\alpha} \to \delta_g$ in the strict topology). It also follows from (2.2) that K is invariant under the group action. By Theorem 1.3.1, there is thus $\phi \in K$ such that

$$g \cdot \phi = \delta_g \cdot \phi \cdot \delta_{g^{-1}} + (\tilde{D}\delta_g) \cdot \delta_{g^{-1}} = \phi \qquad (g \in G)$$

or, equivalently,

$$\tilde{D}\delta_g = \phi \cdot \delta_g - \delta_g \cdot \phi \qquad (g \in G),$$

i.e. $\tilde{D} = \mathrm{ad}_{-\phi}$.

(ii) $\implies$ (i): Define an $L^1(G)$-bimodule action on $L^\infty(G)$ through

$$f \cdot \phi := f * \phi \quad \text{and} \quad \phi \cdot f := \left(\int_G f(g) \, dm_G(g) \right) \phi \qquad (f \in L^1(G), \, \phi \in L^\infty(G)).$$

Choose $n \in L^\infty(G)^*$ with $\langle 1, n \rangle = 1$, and define

$$D: L^1(G) \to L^\infty(G)^*, \quad f \mapsto f \cdot n - n \cdot f.$$

Then

$$\langle 1, Df \rangle = \left\langle \int_G f(g) \, dm_G(g) - f * 1, n \right\rangle = 0 \qquad (f \in L^1(G)) \qquad (2.3)$$

because $f * 1 = \int_G f(g) \, dm_G(g)$ for all $f \in L^1(G)$. It is easy to see that $\mathbb{C}1$ is a submodule of $L^\infty(G)$. Let $E := L^\infty(G)/\mathbb{C}1$. By (2.3), $D(L^1(G)) \subset E^*$. Since $\mathcal{H}^1(L^1(G), E^*) = \{0\}$, there is $\tilde{n} \in E^*$ such that $D = \mathrm{ad}_{\tilde{n}}$. Let $\tilde{m} := n - \tilde{n}$; then $\langle 1, \tilde{m} \rangle = 1$. Further, we have for $\phi \in L^\infty(G)$ and $f \in P(G)$:

$$\langle f * \phi, \tilde{m} \rangle = \langle f * \phi, n - \tilde{n} \rangle = \langle f \cdot \phi, n - \tilde{n} \rangle = \langle \phi \cdot f, n - \tilde{n} \rangle = \langle \phi, \tilde{m} \rangle.$$

As in the proof of Lemma 1.1.7, we conclude that

$$\langle \delta_g * \phi, \tilde{m} \rangle = \langle \phi, \tilde{m} \rangle \qquad (g \in G, \, \phi \in L^\infty(G)).$$

Finally, view $\tilde{m}$ as a measure on the character space of the commutative C^*-algebra $L^\infty(G)$ (by the Riesz representation theorem). Then $|\tilde{m}| \neq 0$ is left invariant (Why exactly?). Letting $m := \langle 1, |\tilde{m}| \rangle^{-1} |\tilde{m}|$, we obtain a left invariant mean on $L^\infty(G)$. $\quad \square$

Exercise 2.1.5 Use the Ryll-Nardzewski fixed point theorem ([Gre, Theorem A.2.2]) to prove: $\mathcal{H}^1(L^1(G), E) = \{0\}$ for every locally compact group G and for every Banach $L^1(G)$-bimodule E whose underlying Banach space is reflexive.

In light of Theorem 2.1.8, the use of the word amenable in the following definition makes sense:

Definition 2.1.9 A Banach algebra $\mathfrak{A}$ is called *amenable* if $\mathcal{H}^1(\mathfrak{A}, E^*) = \{0\}$ for every Banach $\mathfrak{A}$-bimodule E.

Exercise 2.1.6 Let $\mathfrak{A}$ be a Banach algebra, let E be a Banach $\mathfrak{A}$-bimodule. A derivation $D: \mathfrak{A} \to E$ is called *approximately inner* if there is a bounded net $(x_\alpha)_\alpha$ in E such that $Da = \lim_\alpha \mathrm{ad}_{x_\alpha} a$ for all $a \in \mathfrak{A}$. Show that a Banach algebra $\mathfrak{A}$ is amenable if and only if every derivation from $\mathfrak{A}$ into a Banach $\mathfrak{A}$-bimodule is approximately inner. (*Hint*: Same as for Exercise 1.1.6(ii).)

2.2 Virtual and approximate diagonals

In this section, we give an intrinsic characterization of amenable Banach algebras. We begin with a preliminary result:

Proposition 2.2.1 *Let $\mathfrak{A}$ be an amenable Banach algebra. Then $\mathfrak{A}$ has a bounded approximate identity.*

Proof Let A be the Banach $\mathfrak{A}$-bimodule whose underlying space is $\mathfrak{A}$, but on which $\mathfrak{A}$ acts via

$$a \cdot x := ax \quad \text{and} \quad x \cdot a := 0 \qquad (a \in \mathfrak{A}, \, x \in A).$$

Let $D : \mathfrak{A} \to A^{**}$ be the canonical embedding of $\mathfrak{A}$ into its second dual. It is routinely checked that $D \in \mathcal{Z}^1(\mathfrak{A}, A^{**}) = \mathcal{B}^1(\mathfrak{A}, A^{**})$. This means that there is $E \in A^{**}$ such that $a = a \cdot E$ for all $a \in \mathfrak{A}$. Let $(e_\alpha)_\alpha$ be a bounded net in $\mathfrak{A}$ such that $E = w^*\text{-}\lim_\alpha e_\alpha$. Then

$$a = w\text{-}\lim_\alpha ae_\alpha \qquad (a \in \mathfrak{A}).$$

Passing to convex combinations, we can suppose (How exactly? Think of how you — hopefully — solved Exercises 1.1.6(ii) and 2.1.6.) that

$$a = \lim_\alpha ae_\alpha \qquad (a \in \mathfrak{A}),$$

i.e. (e_α) is a bounded right approximate identity for $\mathfrak{A}$.

In an analoguous fashion, we obtain a bounded left approximate identity $(f_\beta)_\beta$ for $\mathfrak{A}$. Define $\tilde{e}_{\alpha,\beta} := e_\alpha + f_\beta - e_\alpha f_\beta$. Then, for $a \in \mathfrak{A}$, we have:

$$\|a\tilde{e}_{\alpha,\beta} - a\| \leq \|ae_\alpha - a\| + \|af_\beta - ae_\alpha f_\beta\| \leq \|ae_\alpha - a\| + \|a - ae_\alpha\| \|f_\beta\| \to 0.$$

Similarly, one sees that $a = \lim_{\alpha,\beta} \tilde{e}_{\alpha,\beta} a$ for $a \in \mathfrak{A}$. $\square$

If $\mathfrak{A}$ is a Banach algebra, then the corresponding *diagonal operator* is defined through

$$\Delta_{\mathfrak{A}} : \mathfrak{A} \hat{\otimes} \mathfrak{A} \to \mathfrak{A}, \quad a \otimes b \to ab;$$

if it is clear to which algebra $\mathfrak{A}$ we refer, we simply write Δ. By Exercise B.2.13, $\mathfrak{A}\hat{\otimes}\mathfrak{A}$ becomes a Banach $\mathfrak{A}$-bimodule through

$$a \cdot (b \otimes c) := ab \otimes c \quad \text{and} \quad (b \otimes c) \cdot a := b \otimes ca \qquad (a, b, c \in \mathfrak{A}). \tag{2.4}$$

It is clear that $\Delta_{\mathfrak{A}}$ is a bimodule homomorphism with respect to this module structure on $\mathfrak{A}\hat{\otimes}\mathfrak{A}$.

Definition 2.2.2 Let $\mathfrak{A}$ be a Banach algebra.

(i) An element $\mathbf{M} \in (\mathfrak{A}\hat{\otimes}\mathfrak{A})^{**}$ is called a *virtual diagonal* for $\mathfrak{A}$ if

$$a \cdot \mathbf{M} = \mathbf{M} \cdot a \quad \text{and} \quad a \cdot \Delta_{\mathfrak{A}}^{**}\mathbf{M} = a \qquad (a \in \mathfrak{A}).$$

(ii) A bounded net $(\mathbf{m}_{\alpha})_{\alpha}$ in $\mathfrak{A}\hat{\otimes}\mathfrak{A}$ is called an *approximate diagonal* for $\mathfrak{A}$ if

$$a \cdot \mathbf{m}_{\alpha} - \mathbf{m}_{\alpha} \cdot a \to 0 \quad \text{and} \quad a\Delta_{\mathfrak{A}}\mathbf{m}_{\alpha} \to a \qquad (a \in \mathfrak{A}).$$

So, when does a Banach algebra $\mathfrak{A}$ have a(n) virtual/approximate diagonal? We begin with an easy example:

Example 2.2.3 Let $\mathfrak{A} := \mathbb{M}_n$, the (Banach) algebra of complex $n \times n$ matrices. Let $\{e_{j,k} : j, k = 1, \dots, n\}$ be the set of canonical matrix units of $\mathfrak{A}$, and let

$$\mathbf{m} := \frac{1}{n} \sum_{j,k=1}^{n} e_{j,k} \otimes e_{k,j}.$$

Then

$$\Delta\mathbf{m} = \frac{1}{n} \sum_{j=1}^{n} n e_{j,j} = E_n,$$

so that

$$a\Delta\mathbf{m} = a \qquad (a \in \mathfrak{A}) \tag{2.5}$$

and, for $\ell, m \in \{1, \dots, n\}$, we have

$$e_{\ell,m} \cdot \mathbf{m} = \frac{1}{n} \sum_{j,k=1}^{n} e_{\ell,m}e_{j,k} \otimes e_{k,j} = \frac{1}{n} \sum_{j=1}^{n} e_{\ell,j} \otimes e_{j,\ell} = \frac{1}{n} \sum_{j,k=1}^{n} e_{j,k} \otimes e_{k,j}e_{\ell,m} = \mathbf{m} \cdot e_{\ell,m}.$$

It follows that

$$a \cdot \mathbf{m} - \mathbf{m} \cdot a = 0 \qquad (a \in \mathfrak{A}). \tag{2.6}$$

Exercise 2.2.1 For a Banach algebra $\mathfrak{A}$, let $\mathfrak{A}^{\mathrm{op}}$ denote its *opposite algebra*, i.e. the algebra whose underlying linear space is $\mathfrak{A}$, but whose multiplication is the multiplication in $\mathfrak{A}$ reversed.

(i) How does the module operation (2.4) of $\mathfrak{A}^{\mathrm{op}}$ on $\mathfrak{A}^{\mathrm{op}}\hat{\otimes}\mathfrak{A}^{\mathrm{op}}$ look like in terms of $\mathfrak{A}$ and $\mathfrak{A}\hat{\otimes}\mathfrak{A}$?

We call any element $\mathbf{m} \in \mathfrak{A}\hat{\otimes}\mathfrak{A}$ satisfying both (2.5) and (2.6) a *diagonal* for $\mathfrak{A}$. For the remainder of this exercise, let $\mathfrak{A} := \mathbb{M}_n$, and let $\mathbf{m}$ be as in Example 2.2.3. Note that, in Example 2.2.3, we showed that $\mathbf{m}$ is a diagonal for $\mathfrak{A}$.

(i) Show that **m** is also a diagonal for $\mathfrak{A}^{\mathrm{op}}$.

(ii) Let $\mathbf{m}' \in \mathfrak{A}\hat{\otimes}\mathfrak{A}$ be a diagonal both for $\mathfrak{A}$ and $\mathfrak{A}^{\mathrm{op}}$. Show that $\mathbf{m}' = \mathbf{m}$.

(iii) Let G be a finite, *irreducible* $n \times n$ *matrix group*, i.e. a subgroup of $GL(n, \mathbb{C})$ whose linear span in $\mathbb{M}_n$ is all of $\mathbb{M}_n$. Show that

$$\mathbf{m} = \frac{1}{|G|} \sum_{g \in G} g \otimes g^{-1}.$$

Exercise 2.2.2 Let G be an amenable (discrete) group with left invariant mean m on $\ell^\infty(G)$. For $\phi \in \ell^\infty(G \times G)$, define $\tilde{\phi} \in \ell^\infty(G)$ through

$$\tilde{\phi}(g) := \phi(g, g^{-1}) \qquad (g \in G)$$

Define $\mathbf{M} \in \ell^\infty(G \times G)^* \cong (\ell^1(G)\hat{\otimes}\ell^1(G))^{**}$ through

$$\langle \phi, \mathbf{M} \rangle := \langle \tilde{\phi}, m \rangle \qquad (\phi \in \ell^\infty(G \times G)).$$

Show that $\mathbf{M}$ is a virtual diagonal for $\ell^1(G)$.

This last exercise suggests some connection between amenability and the existence of virtual diagonals, and, indeed, there is more than just some connection:

Theorem 2.2.4 *For a Banach algebra $\mathfrak{A}$ the following are equivalent:*

(i) $\mathfrak{A}$ *is amenable.*

(ii) *There is an approximate diagonal for $\mathfrak{A}$.*

(iii) *There is a virtual diagonal for $\mathfrak{A}$.*

Proof (i) $\implies$ (iii): By Proposition 2.2.1, $\mathfrak{A}$ has a bounded approximate identity $(e_\alpha)_\alpha$. Let $\mathbf{E} \in (\mathfrak{A}\hat{\otimes}\mathfrak{A})^{**}$ be a w^*-accumulation point of $(e_\alpha \otimes e_\alpha)$. Then, for $a \in \mathfrak{A}$, we have

$$\Delta^{**}(a \cdot \mathbf{E} - \mathbf{E} \cdot a) = w^*\text{-}\lim_\alpha \Delta(ae_\alpha \otimes e_\alpha - e_\alpha \otimes e_\alpha a) = \lim_\alpha (ae_\alpha^2 - e_\alpha^2 a) = 0,$$

so that $\mathrm{ad}_{\mathbf{E}}(\mathfrak{A}) \subset \ker \Delta^{**}$. Since Δ is a bimodule homomorphism, so is Δ^{**} (little, very easy exercise), and consequently, $\ker \Delta^{**}$ is a Banach $\mathfrak{A}$-bimodule. Also, since $\mathfrak{A}$ has a bounded approximate identity, Cohen's factorization theorem ([B–D, Theorem 11.10]) implies that Δ is surjective and thus open. Consequently, $\ker \Delta^{**} \cong (\ker \Delta)^{**}$, so that $\ker \Delta^{**}$ is in fact a dual Banach $\mathfrak{A}$-bimodule. Since $\mathfrak{A}$ is amenable, there is $\mathbf{N} \in \ker \Delta^{**}$ such that $\mathrm{ad}_{\mathbf{E}} = \mathrm{ad}_{\mathbf{N}}$. Letting $\mathbf{M} := \mathbf{E} - \mathbf{N}$, we obtain

$$\left.\begin{cases} a \cdot \Delta^{**}\mathbf{M} = a \cdot \Delta^{**}\mathbf{E} = \lim_\alpha ae_\alpha^2 = a \\ a \cdot \mathbf{M} - \mathbf{M} \cdot a = \mathrm{ad}_{\mathbf{M}}a = \mathrm{ad}_{\mathbf{E}-\mathbf{N}}a = \mathrm{ad}_{\mathbf{E}}a - \mathrm{ad}_{\mathbf{N}}a = 0 \end{cases}\right\} \qquad (a \in \mathfrak{A}),$$

so that $\mathbf{M}$ is a virtual diagonal for $\mathfrak{A}$.

(iii) $\implies$ (ii): Let $\mathbf{M}$ be a virtual diagonal for $\mathfrak{A}$, and let $(\mathbf{m}_\alpha)_\alpha$ be a bounded net in $\mathfrak{A}\hat{\otimes}\mathfrak{A}$ with $\mathbf{M} = w^*\text{-}\lim_\alpha m_\alpha$. It follows that

$$w\text{-}\lim_\alpha(a \cdot \mathbf{m}_\alpha - \mathbf{m}_\alpha \cdot a) = 0 \quad \text{and} \quad w\text{-}\lim_\alpha a\Delta\mathbf{m}_\alpha = a \qquad (a \in \mathfrak{A}). \tag{2.7}$$

Passing to convex combinations (you should know by now how that goes), we can replace the weak limits in (2.7) by norm limits and thus obtain an approximate diagonal.

(ii) $\implies$ (i): Let $(\mathbf{m}_\alpha)_\alpha$ be an approximate diagonal for $\mathfrak{A}$. Then $(\Delta\mathbf{m}_\alpha)_\alpha$ is a bounded approximate identity for $\mathfrak{A}$. Let E be a Banach $\mathfrak{A}$-bimodule. We want to show that $\mathcal{H}^1(\mathfrak{A}, E^*) = \{0\}$. By Proposition 2.1.5, there is no loss of generality if we suppose that E is pseudo-unital. Let $D \in \mathcal{Z}^1(\mathfrak{A}, E^*)$, and let

$$\mathbf{m}_\alpha = \sum_{n=1}^\infty a_n^{(\alpha)} \otimes b_n^{(\alpha)}$$

be with $\sum_{n=1}^\infty \|a_n^{(\alpha)}\|\|b_n^{(\alpha)}\| < \infty$. Then $\left(\sum_{n=1}^\infty a_n^{(\alpha)} \cdot Db_n^{(\alpha)}\right)_\alpha$ is a bounded net in E^*, which has a w^*-accumulation point, say $\phi \in E^*$; without loss of generality, we may suppose that ϕ is the w^*-limit of $\left(\sum_{n=1}^\infty a_n^{(\alpha)} \cdot Db_n^{(\alpha)}\right)_\alpha$ Then, for $a \in \mathfrak{A}$ and $x \in E$, we have:

$$\langle x, a \cdot \phi \rangle = \lim_\alpha \left\langle x, \sum_{n=1}^\infty a a_n^{(\alpha)} \cdot Db_n^{(\alpha)} \right\rangle$$

$$= \lim_\alpha \left\langle x, \sum_{n=1}^\infty a_n^{(\alpha)} \cdot D(b_n^{(\alpha)} a) \right\rangle \qquad \text{(Why?)}$$

$$= \lim_\alpha \left\langle x, \sum_{n=1}^\infty (a_n^{(\alpha)} b_n^{(\alpha)} \cdot Da + a_n^{(\alpha)} \cdot Db_n^{(\alpha)} \cdot a) \right\rangle$$

$$= \lim_\alpha \left\langle x \cdot \sum_{n=1}^\infty a_n^{(\alpha)} b_n^{(\alpha)}, Da \right\rangle + \langle x, \phi \cdot a \rangle$$

$$= \langle x, Da \rangle + \langle x, \phi \cdot a \rangle.$$

It follows that $D = \mathrm{ad}_\phi$. $\square$

Exercise 2.2.3 Let $\mathfrak{A}$ be a unital, amenable Banach algebra. Show that $\mathfrak{A}$ has an approximate diagonal $(m_\alpha)_\alpha$ such that $\Delta m_\alpha = e_{\mathfrak{A}}$ for all m_α.

2.3 Hereditary properties

As in our discussion of amenable, locally compact groups, we now study the hereditary properties of amenability for Banach algebras.

Proposition 2.3.1 *Let $\mathfrak{A}$ be an amenable Banach algebra, let $\mathfrak{B}$ be a Banach algebra, and let $\theta \colon \mathfrak{A} \to \mathfrak{B}$ be a (continuous) homomorphism with dense range. Then $\mathfrak{B}$ is amenable.*

Proof Let $(\mathbf{m}_\alpha)_\alpha$ be an approximate diagonal for $\mathfrak{A}$. Then $((\theta \otimes \theta)\mathbf{m}_\alpha)_\alpha$ is a bounded approximate diagonal for $\mathfrak{B}$. By Theorem 2.2.4, this means that $\mathfrak{B}$ is also amenable. $\square$

Exercise 2.3.1 Give an alternative proof of Proposition 2.3.1 using only Definition 2.1.9.

Corollary 2.3.2 *If $\mathfrak{A}$ is an amenable Banach algebra and if I is a closed ideal in $\mathfrak{A}$, then $\mathfrak{A}/I$ is amenable as well.*

What about passing to substructures such as closed subalgebras and closed ideals? There is virtually nothing that can be said about closed subalgebras. By Proposition 2.2.1, a necessary condition for a closed ideal of an amenable Banach algebra to be amenable is that is has a bounded approximate identity. We shall see that this condition is also necessary:

Proposition 2.3.3 *Let $\mathfrak{A}$ be an amenable Banach algebra, and let I be a closed ideal of $\mathfrak{A}$. The the following are equivalent:*

(i) *I is amenable.*

(ii) *I has a bounded approximate identity.*

Proof Only (ii) $\Longrightarrow$ (i) needs proof.

Let E be a Banach I-bimodule. By Proposition 2.1.5, we may suppose that E is pseudo-unital. By Proposition 2.1.6, the module action of I on E extends to $\mathfrak{A}$ in a canonical fashion. Let $D \in \mathcal{Z}^1(\mathfrak{A}, E^*)$. Then, again by Proposition 2.1.6, D has an extension $\tilde{D} \in \mathcal{Z}^1(\mathfrak{A}, E^*)$. Since $\mathfrak{A}$ is amenable, we have $\tilde{D} \in \mathcal{B}^1(\mathfrak{A}, E^*)$, so that $D = \tilde{D}|_I \in \mathcal{B}^1(I, E^*)$. $\square$

Exercise 2.3.2 Prove Proposition 2.3.3 using Theorem 2.2.4 instead of Definition 2.1.9.

Example 2.3.4 Let $\mathfrak{A}$ be a commutative C^*-algebra. First, consider the case where $\mathfrak{A}$ is unital. Then $\mathfrak{A} \cong \mathcal{C}(\Omega)$ for some compact Hausdorff space Ω. Let

$$G := \{\exp(if) : f \in \mathcal{C}(\Omega), \, f(\Omega) \subset \mathbb{R}\}.$$

Then G is a (necessarily abelian) subgroup of $\operatorname{Inv} \mathcal{C}(\Omega)$, so that $\ell^1(G)$ is amenable. Define

$$\theta \colon \ell^1(G) \to \mathcal{C}(\Omega), \quad \sum_{g \in G} \lambda_g \delta_g \mapsto \sum_{g \in G} \lambda_g g.$$

Then θ is a homomorphism whose range is dense by the Stone–Weierstraß theorem. By Proposition 2.3.1, $\mathcal{C}(\Omega)$ is amenable. Suppose now that $\mathfrak{A}$ is not unital. Then its unitization $\mathfrak{A}^{\#}$ is a unital C^*-algebra and thus amenable by the foregoing. Since $\mathfrak{A}$ is a closed ideal in $\mathfrak{A}^{\#}$ and, like any C^*-algebra, has a bounded approximate identity, $\mathfrak{A}$ itself is amenable by Proposition 2.3.3.

There is another characterization of the closed ideals of an amenable Banach algebras which are amenable themselves.

Definition 2.3.5 Let E be a Banach space. Then a closed subspace F of E is called *weakly complemented* in E if

$$F^{\perp} = \{\phi \in E^* : \langle x, \phi \rangle = 0 \text{ for all } x \in F\}$$

is complemented in E^*.

It is easy to see that every complemented subspace is weakly complemented. The converse, however, is not true:

Exercise 2.3.3 Show that c_0 is weakly complemented in ℓ^∞, but not complemented:

(i) Show that there is an uncountable family $(A_\alpha)_\alpha$ of infinite subsets of $\mathbb{N}$ such that $A_\alpha \cap A_\beta$ is finite for all $\alpha \neq \beta$. (*Hint:* Replace $\mathbb{N}$ by $\mathbb{Q}$ (you can do that because they have the same cardinality), use $\mathbb{R}$ as your index set, und utilize the fact that every real number is the limit of a sequence in $\mathbb{Q}$.)

(ii) There is no countable subset of $(\ell^\infty/c_0)^*$ that separates the points of ℓ^∞/c_0. (*Hint:* Choose $(A_\alpha)_\alpha$ as in (i) and consider the family $(\phi_\alpha)_\alpha$ of the cosets of the indicator functions of the sets A_α; show that, for fixed $m \in (\ell^\infty/c_0)^*$, the set $\{\phi_\alpha : \langle \phi_\alpha, m \rangle \neq 0\}$ is at most countable.)

(iii) Conclude from (i) and (ii) that c_0 is not complemented in ℓ^∞.

(iv) Show that c_0 is weakly complemented in ℓ^∞.

Lemma 2.3.6 *Let $\mathfrak{A}$ be an amenable Banach algebra, and let L be a closed left ideal of $\mathfrak{A}$. Then the following are equivalent:*

(i) *L has a bounded right approximate identity.*

(ii) *L is weakly compelemeted in $\mathfrak{A}$.*

Proof (i) $\Longrightarrow$ (ii): Let $(e_\alpha)_\alpha$ be a bounded right approximate identity for L, and let $\mathcal{U}$ be an ultrafilter on the index set of $(e_\alpha)_\alpha$ that dominates the order filter. Define

$$P: \mathfrak{A}^* \to \mathfrak{A}^*, \quad \phi \mapsto w^*\text{-}\lim_{\mathcal{U}}(\phi - e_\alpha \cdot \phi).$$

Since

$$\langle a, P\phi \rangle = \lim_{\mathcal{U}}\langle a, \phi - e_\alpha \cdot \phi \rangle = \lim_{\mathcal{U}}\langle a - ae_\alpha, \phi \rangle = 0 \qquad (a \in L, \phi \in \mathfrak{A}^*),$$

it follows that $P\mathfrak{A}^* \subset L^\perp$. If $\phi \in L^\perp$ and $a \in \mathfrak{A}$, we have

$$\langle a, P\phi \rangle = \lim_{\mathcal{U}}(\langle a, \phi \rangle - \langle \underbrace{ae_\alpha}_{\in L}, \phi \rangle) = \langle a, \phi \rangle,$$

so that $P^2 = P$. (Have we used at all in this part of the proof that $\mathfrak{A}$ was amenable?)

(ii) $\Longrightarrow$ (i) : Let $(\mathbf{m}_\alpha)_\alpha$ be an approximate diagonal for $\mathfrak{A}$, where

$$\mathbf{m}_\alpha = \sum_{n=1}^{\infty} a_n^{(\alpha)} \otimes b_n^{(\alpha)}$$

with $\sum_{n=1}^{\infty} \|a_n^{(\alpha)}\| \| b_n^{(\alpha)}\| < \infty$. Let $P: \mathfrak{A}^* \to \mathfrak{A}^*$ be a projection onto $L^\perp$. Let $Q := \mathrm{id}_{\mathfrak{A}^{**}} - P^{**}$. Then $Q: \mathfrak{A}^{**} \to \mathfrak{A}^{**}$ is a projection onto $L^{\perp\perp} \cong L^{**}$. Define

$$E_\alpha := \sum_{n=1}^{\infty} a_n^{(\alpha)} \cdot Q b_n^{(\alpha)}.$$

Then, for $a \in L$, we have:

$$\lim_{\alpha} a \cdot E_\alpha = \lim_{\alpha} \sum_{n=1}^{\infty} a a_n^{(\alpha)} \cdot Q b_n^{(\alpha)}$$

$$= \lim_{\alpha} \sum_{n=1}^{\infty} a_n^{(\alpha)} \cdot Q(\underbrace{b_n^{(\alpha)} a}_{\in L})$$

$$= \lim_{\alpha} \sum_{n=1}^{\infty} a_n^{(\alpha)} b_n^{(\alpha)} a$$

$$= a.$$

If E is a w^*-accumulation point of $(E_\alpha)_\alpha$, this means

$$a \cdot E = a \qquad (a \in L).$$

As in the proof of Proposition 2.2.1, we obtain a bounded right approximate identity for L. $\square$

For closed ideals (two-sided), we finally obtain:

Theorem 2.3.7 *Let $\mathfrak{A}$ be an amenable Banach algebra, and let I be a closed ideal of $\mathfrak{A}$. Then the following are equivalent:*

(i) *I is amenable.*

(ii) *I has a bounded approximate identity.*

(iii) *I is weakly complemented.*

Proof Most of the theorem has already been proved (Proposition 2.3.3 and Lemma 2.3.6). Only (iii) $\Longrightarrow$ (i) still needs some consideration: By Lemma 2.3.6, I has a bounded right approximate identity. Passing to $\mathfrak{A}^{\mathrm{op}}$ and applying Lemma 2.3.6 once again yields a bounded left approximate identity for I. As in the proof of Proposition 2.2.1, we form a bounded (two-sided) approximate identity for I out of these two one-sided bounded approximate identities. $\square$

Since ideals with finite dimension or codimension are trivially complemented, we have:

Corollary 2.3.8 *Let $\mathfrak{A}$ be an amenable Banach algebra, and let I be a closed ideal of $\mathfrak{A}$ with finite dimension or codimension. Then I is amenable.*

This corollary applies, in particular, to the maximal modular ideals of amenable, commutative Banach algebras.

Exercise 2.3.4 Let $\mathfrak{A}$ be an amenable Banach algebra, and let I be a finite-dimensional ideal of $\mathfrak{A}$. Show that I has an identity. Conclude that an amenable Banach algebra $\mathfrak{A}$ whose Jacobson radical $\mathrm{rad}(\mathfrak{A})$ is finite-dimensional must be semisimple.

For group algebras, Theorem 2.3.7 allows for further improvement:

Theorem 2.3.9 *For a locally compact group G, the following are equivalent:*

(i) *G is amenable.*

(ii) *The augmentation ideal*

$$L_0^1(G) := \left\{ f \in L^1(G) : \int_G f(g)\, dm_G(g) = 0 \right\}$$

of $L^1(G)$ has a bounded right approximate identity.

(iii) *Every closed ideal of $L^1(G)$ that is weakly complemented has a bounded approximate identity.*

Proof It is easily checked (Exercise 2.3.5 below) that $L_0^1(G)$ is a closed ideal of $L^1(G)$ with codimension one. Hence, only (ii) $\Longrightarrow$ (i) still needs proof.

Let $(e_\alpha)_\alpha$ be a bounded right approximate identity for $L_0^1(G)$. Without loss of generality, suppose that $(e_\alpha)_\alpha$ is w^*-convergent in $LUC(G)^*$ (with the canonical identification of $L^1(G)$ with a subspace of $LUC(G)^*$). Let $f \in L_0^1(G)$, and let $\phi \in L^\infty(G)$. Then $f * \phi \in LUC(G)$ by Proposition A.2.3 and

$$\lim_\alpha \langle f * \phi, e_\alpha \rangle = \lim_\alpha \langle \phi, \tilde{f} * e_\alpha \rangle = \langle \phi, \tilde{f} \rangle = (f * \phi)(e_G), \tag{2.8}$$

with $\tilde{f}$ defined as in Exercise A.1.7. Define

$$\tilde{m} : LUC(G) \to \mathbb{C}, \quad \phi \mapsto \phi(e_G) - \lim_\alpha \langle \phi, e_\alpha \rangle.$$

Then, certainly, $\langle 1, \tilde{m} \rangle = 1$. Let $g \in G$, and let $\phi \in LUC(G)$. By Exercise A.2.3(i), there are $f \in L^1(G)$ and $\psi \in LUC(G)$ such that $\phi = f * \psi$. Note that $\delta_g * f - f \in L^1_0(G)$. It then follows that

$$
\begin{aligned}
\langle \delta_g * \phi - \phi, \tilde{m} \rangle &= \phi(g^{-1}) - \phi(e_G) - \lim_\alpha \langle (\delta_g * f - f) * \psi, e_\alpha \rangle \\
&= \phi(g^{-1}) - \phi(e_G) - ((\delta_g * f - f) * \psi)(e_G), \qquad \text{by (2.8)}, \\
&= \phi(g^{-1}) - \phi(e_G) - (\phi(g^{-1}) - \phi(e_G)) \\
&= 0,
\end{aligned}
$$

i.e.

$$
\langle \delta_g * \phi, \tilde{m} \rangle = \langle \phi, \tilde{m} \rangle \qquad (g \in G, \ \phi \in LUC(G)).
$$

Although $\tilde{m}$ need not be positive, we obtain a left invariant mean on $LUC(G)$ as in the proof of Theorem 2.1.8: Consider $\tilde{m}$ as a measure on the character space of $LUC(G)$, and normalize $|\tilde{m}|$. $\square$

Exercise 2.3.5 Let G be a locally compact group. Show that $L^1_0(G)$ is a closed ideal of $L^1(G)$ with codimension one.

Next, we consider short, exact sequences:

Theorem 2.3.10 *Let $\mathfrak{A}$ be a Banach algebra, and let I be a closed ideal of $\mathfrak{A}$ such that both I and $\mathfrak{A}/I$ are amenable. Then $\mathfrak{A}$ is amenable.*

Proof Let E be a Banach $\mathfrak{A}$-bimodule, and let $D \in \mathcal{Z}^1(\mathfrak{A}, E^*)$. Then $D|_I \in \mathcal{Z}^1(I, E^*)$. Since I is amenable, there is $\phi_1 \in E^*$ such that

$$
Da = a \cdot \phi_1 - \phi_1 \cdot a \qquad (a \in I).
$$

Let $\tilde{D} = D - \mathrm{ad}_{\phi_1}$. Then $\tilde{D}|_I = 0$ and thus induces a map from $\mathfrak{A}/I$ into E^*, which we denote likewise by $\tilde{D}$. Let

$$
F := \{ \psi \in E^* : a \cdot \psi = \psi \cdot a = 0 \text{ for all } a \in I \}
$$

and

$$
E_0 := \mathrm{lin}\{ a \cdot x + y \cdot b : a, b \in I, \ x, y \in E \}.
$$

Then $F \cong (E/E_0)^*$ is a dual Banach $\mathfrak{A}/I$-bimodule.

Let $a \in I$ and $b \in \mathfrak{A}$. Then $a \cdot \tilde{D}b = \tilde{D}(ab) - (\tilde{D}a) \cdot b = 0$ because $\tilde{D}$ vanishes on I; similarly, $\tilde{D}b \cdot a = 0$. It follows that $\tilde{D}(\mathfrak{A}/I) \subset F$. Since $\mathfrak{A}/I$ is amenable, there is $\phi_2 \in F$ such that $\tilde{D} = \mathrm{ad}_{\phi_2}$. Consequently, $D = \mathrm{ad}_{\phi_1 + \phi_2}$. $\square$

Corollary 2.3.11 *A Banach algebra $\mathfrak{A}$ is amenable if and only if its unitization $\mathfrak{A}^\#$ is amenable.*

We now take a first look at splitting properties of exact sequences of Banach modules over amenable Banach algebras:

Definition 2.3.12 Let $\mathfrak{A}$ be a Banach algebra, let E be a Banach $\mathfrak{A}$-module (left, right, or bi-), and let F be a closed submodule of E. Then the short, exact sequence

$$\{0\} \to F \to E \overset{\pi}{\to} E/F \to \{0\} \tag{2.9}$$

(i) is *admissible* if $\pi\colon E \to E/F$ has a bounded right inverse, and

(ii) *splits* if π has a bounded right inverse that is also a module homomorphism (left, right, or bi-).

Obviously, (2.9) is admissible if and only if F is complemented in E, and (2.9) splits if and only there is a projection onto F that is also a module homomorphism. Certainly, (2.9) is admissible whenever

– F or E/F is finite-dimensional, or
– the underlying Banach space of E is a Hilbert space.

Clearly, every short exact sequence of Banach modules that splits is admissible. For modules over amenable Banach algebras we have a partial converse:

Theorem 2.3.13 *Let $\mathfrak{A}$ be an amenable Banach algebra. Then every admissible, short, exact sequence (2.9) of left Banach $\mathfrak{A}$-modules splits provided that F is a dual module.*

Proof Without loss of generality, we may suppose that $\mathfrak{A}$ is unital, and that E is a unital left Banach $\mathfrak{A}$-module, i.e. $e_{\mathfrak{A}} \cdot x = x$ for all $x \in E$. Otherwise, replace $\mathfrak{A}$ by its unitization, which is also amenable by Corollary 2.3.11, and extend the module action by letting

$$(a + \lambda e) \cdot x := a \cdot x + \lambda x \qquad (a \in \mathfrak{A}, \, \lambda \in \mathbb{C}, \, x \in E).$$

Let $(\mathbf{m}_\alpha)_\alpha$ be an approximate diagonal for $\mathfrak{A}$, where

$$\mathbf{m}_\alpha = \sum_{n=1}^{\infty} a_n^{(\alpha)} \otimes b_n^{(\alpha)}$$

with $\sum_{n=1}^{\infty} \|a_n^{(\alpha)}\| \|b_n^{(\alpha)}\| < \infty$. By Exercise 2.2.3, we may suppose that

$$\sum_{n=1}^{\infty} a_n^{(\alpha)} b_n^{(\alpha)} = e_{\mathfrak{A}}.$$

Let $Q\colon E \to E$ be a projection onto F. Let $\mathcal{U}$ be an ultrafilter on the index set of $(\mathbf{m}_\alpha)_\alpha$ that dominates the order filter. Define

$$P\colon E \to E, \quad x \mapsto w^*\text{-}\lim_{\mathcal{U}} \sum_{n=1}^{\infty} a_n^{(\alpha)} \cdot Q(b_n^{(\alpha)} \cdot x);$$

since the expressions whose limit is taken always belong to F, and since F is a dual space, this definition makes sense.

For $a \in \mathfrak{A}$ and $x \in E$, we have

$$P(a \cdot x) = w^*\text{-}\lim_{\mathcal{U}} \sum_{n=1}^{\infty} a_n^{(\alpha)} \cdot Q(b_n^{(\alpha)} a \cdot x)$$

$$= w^*\text{-}\lim_{\mathcal{U}} \sum_{n=1}^{\infty} a a_n^{(\alpha)} \cdot Q(b_n^{(\alpha)} \cdot x)$$

$$= a \cdot Px,$$

so that P is a module homomorphism. Furthermore, for $x \in F$,

$$Px = w^*\text{-}\lim_{\mathcal{U}} \sum_{n=1}^{\infty} a_n^{(\alpha)} \cdot Q(\underbrace{b_n^{(\alpha)} \cdot x}_{\in F}) = w^*\text{-}\lim_{\mathcal{U}} \sum_{n=1}^{\infty} a_n^{(\alpha)} b_n^{(\alpha)} \cdot x = x.$$

Since $PE \subset F$ by definition, it follows that P is a projection onto F. $\square$

Clearly, there is also a right module variant of Theorem 2.3.13. We leave it to the reader to prove a bimodule version of Theorem 2.3.13.

Exercise 2.3.6 Let $\mathfrak{A}$ be a Banach algebra.

(i) Suppose that $\mathfrak{A}$ is amenable, and that $\mathfrak{B}$ is also an amenable Banach algebra. Show that $\mathfrak{A} \hat{\otimes} \mathfrak{B}$ is amenable. (*Hint*: Construct an approximate diagonal for $\mathfrak{A} \hat{\otimes} \mathfrak{B}$ from approximate diagonals for $\mathfrak{A}$ and for $\mathfrak{B}$.)

Let Ω be a locally compact Hausdorff space, and let $\mathcal{C}_0(\Omega) \check{\otimes} \mathfrak{A}$ be equipped with the unique extension of $\bullet$ to $\mathcal{C}_0(\Omega) \check{\otimes} \mathfrak{A}$ (Exercise B.2.6).

(ii) Suppose that $\mathfrak{A}$ is amenable. Show that $\mathcal{C}_0(\Omega) \check{\otimes} \mathfrak{A}$ is amenable as well.
(iii) Let $\mathfrak{A}$ be a C^*-algebra. Show that $\mathcal{C}_0(\Omega) \check{\otimes} \mathfrak{A}$ is a C^*-algebra in a canonical fashion.

Exercise 2.3.7 Prove Theorem 2.3.13 for short, exact sequences of bimodules. (*Hint*: Note that, for a Banach algebra $\mathfrak{A}$, every Banach $\mathfrak{A}$-bimodule E becomes a left Banach $\mathfrak{A} \hat{\otimes} \mathfrak{A}^{\mathrm{op}}$-module via

$$(a \otimes b) \cdot x := a \cdot x \cdot b \qquad (a, b \in \mathfrak{A}, \, x \in E).$$

Then use Exercise 2.3.6(i) and Theorem 2.3.13 for left modules.)

In fact, amenable Banach algebras can be characterized by the splitting of particular short, exact sequences:

Exercise 2.3.8 Let $\mathfrak{A}$ be a Banach algebra with a bounded left or right approximate identity.

(i) Show that the short, exact sequence

$$\{0\} \to \mathfrak{A}^* \xrightarrow{\Delta^*} (\mathfrak{A} \hat{\otimes} \mathfrak{A})^* \to (\ker \Delta)^* \to \{0\} \tag{2.10}$$

of Banach $\mathfrak{A}$-bimodules is admissible.
(ii) Show that the following are equivalent:
(a) $\mathfrak{A}$ is amenable.
(b) (2.10) splits.
(c) The short, exact sequence

$$\{0\} \to (\ker \Delta)^{**} \to (\mathfrak{A} \hat{\otimes} \mathfrak{A})^{**} \xrightarrow{\Delta^{**}} \mathfrak{A}^{**} \to \{0\}$$

of Banach $\mathfrak{A}$-bimodules splits.

An application of Theorem 2.3.13 is the following:

Theorem 2.3.14 *Let $\mathfrak{A}$ be an amenable, uniform Banach algebra, i.e. a closed subalgebra of a commutative C^*-algebra. Then $\mathfrak{A}$ is a commutative C^*-algebra.*

Proof Without loss of generality, let $\mathfrak{A}$ be unital with compact character space Ω. Assume towards a contradiction that $\mathfrak{A} \subsetneq C(\Omega)$. By the Hahn–Banach theorem and the Riesz representation theorem, there is $\mu \in M(\Omega) \setminus \{0\}$ such that

$$\int_\Omega f \, d\mu = 0 \qquad (f \in \mathfrak{A}).$$

Let $\mathfrak{H} := L^2(|\mu|)$, and let $\mathfrak{K}$ be the closure of $\mathfrak{A}$ in $\mathfrak{H}$. Then both $\mathfrak{H}$ and $\mathfrak{K}$ are (left) Banach $\mathfrak{A}$-modules via left multiplication. Clearly, the short, exact sequence

$$\{0\} \to \mathfrak{K} \to \mathfrak{H} \to \mathfrak{H}/\mathfrak{K} \to \{0\}$$

is admissible and thus splits by Theorem 2.3.13. Let $P : \mathfrak{H} \to \mathfrak{H}$ be a projection onto $\mathfrak{K}$ which is also a left $\mathfrak{A}$-module homomorphism. For $f \in C(\Omega)$, let $L_f : \mathfrak{H} \to \mathfrak{H}$ be the corresponding multiplication operator; clearly, L_f is normal with $L_f^* = L_{\bar{f}}$. Since P is an $\mathfrak{A}$-module homomorphism, we have

$$L_f P = P L_f \qquad (f \in \mathfrak{A}). \tag{2.11}$$

The Fuglede–Putnam theorem ([Conw 2, IX.6.7]), implies that also

$$L_{\bar{f}} P = P L_{\bar{f}} \qquad (f \in \mathfrak{A}). \tag{2.12}$$

Clearly, $\{f \in C(\Omega) : L_f P = P L_f\}$ is a closed subalgebra of $C(\Omega)$ which, by (2.11), (2.12), and the Stone–Weierstraß theorem must be all of $C(\Omega)$. It follows that

$$f = L_f 1 = L_f P 1 = P(L_f 1) = P f \in \mathfrak{K} \qquad (f \in C(\Omega)),$$

so that $\mathfrak{H} = \mathfrak{K}$.

Let $f \in C(\Omega)$. Then there is a sequence $(f_n)_{n=1}^\infty$ in $\mathfrak{A}$ such that $\|f - f_n\|_2 \to 0$ in $\mathfrak{H}$. Then,

$$|\langle f, \mu \rangle|^2 = |\langle f - f_n, \mu \rangle|^2 \leq \int_\Omega |f - f_n|^2 \, d|\mu| = \|f - f_n\|_2^2 \to 0,$$

contradicting that $\mu \neq 0$. $\square$

We wish to conclude this section with an analogue of Proposition 1.2.12. We require another definition:

Definition 2.3.15 Let $C \geq 1$. A Banach algebra $\mathfrak{A}$ is called C-amenable if it has an approximate diagonal bounded by C.

Clearly, every amenable Banach algebra is C-amenable for some C, and it doesn't make much sense to define C-amenability for constants $C < 1$ (Why?).

Exercise 2.3.9 Let $C \geq 1$ and let $\mathfrak{A}$ be a Banach algebra. Show that $\mathfrak{A}$ is C-amenable if and only if $\mathfrak{A}$ has a virtual diagonal with norm not exceeding C.

Exercise 2.3.10 Let G be an amenable (discrete) group. Show that $\ell^1(G)$ is 1-amenable.

Exercise 2.3.11 Show that every unital, commutative C^*-algebra is 1-amenable.

Example 2.3.16 For $n \in \mathbb{N}$ and $p \in [1, \infty]$, let ℓ_n^p be $\mathbb{C}^n$ equipped with the corresponding ℓ^p-norm, i.e.

$$\|(\lambda_1, \ldots, \lambda_n)\|_p := \left(\sum_{j=1}^{n} |\lambda_j|^p \right)^{\frac{1}{p}} \quad (\lambda_1, \ldots, \lambda_n \in \mathbb{C})$$

in the case where $p < \infty$ and

$$\|(\lambda_1, \ldots, \lambda_n)\|_\infty := \max\{|\lambda_1|, \ldots, |\lambda_n|\} \quad (\lambda_1, \ldots, \lambda_n \in \mathbb{C})$$

We can algebraically identify $\mathcal{L}(\ell_n^p)$ with $\mathbb{M}_n$. Let $\mathbf{m} \in \mathbb{M}_n \otimes \mathbb{M}_n$ be the diagonal defined in Example 2.2.3. We wish to estimate $\|\mathbf{m}\|$ in $\mathcal{L}(\ell_n^p) \hat{\otimes} \mathcal{L}(\ell_n^p)$. Fix $j \in \{1, \ldots, n\}$. We shall use the description of $\mathbf{m}$ in Exercise 2.2.1(iv) for an appropriate choice of G. Let $\mathfrak{S}_n$ be the group of permutations of $\{1, \ldots, n\}$; for every $\sigma \in \mathfrak{S}_n$, let A_σ denote the corresponding permutation matrix, and, for every $\epsilon \in \{-1, 1\}^n$, let D_ϵ be the corresponding diagonal matrix. Let

$$G := \{D_\epsilon A_\sigma : \epsilon \in \{-1, 1\}^n, \sigma \in \mathfrak{S}_n\}.$$

Then G is a finite, irreducible $n \times n$ matrix group (check it), and, certainly, $\|g\|_p = 1$ for all $g \in G$. It follows that

$$\|\mathbf{m}\| \leq \frac{1}{|G|} \sum_{g \in G} \|g\| \|g^{-1}\| = 1,$$

so that $\mathcal{L}(\ell_n^p)$ is 1-amenable.

Proposition 2.3.17 *Let $C \geq 1$, let $\mathfrak{A}$ be a Banach algebra, and let $(\mathfrak{A}_\alpha)_\alpha$ be a directed family of closed, C-amenable subalgebras of $\mathfrak{A}$ such that $\bigcup_\alpha \mathfrak{A}_\alpha$ is dense in $\mathfrak{A}$. Then $\mathfrak{A}$ is C-amenable.*

Proof Let $\mathcal{F}$ be the family of all finite subsets of $\bigcup_\alpha \mathfrak{A}_\alpha$. Let $F \in \mathcal{F}$, and let $\epsilon > 0$. Choose $\mathfrak{A}_\alpha$ with $F \subset \mathfrak{A}_\alpha$. Since $\mathfrak{A}_\alpha$ is C-amenable, there is $\mathbf{m}_{F,\epsilon} \in \mathfrak{A}_\alpha \hat{\otimes} \mathfrak{A}_\alpha$ with $\|m_{F,\epsilon}\| \leq C$ such that

$$\|a \cdot \mathbf{m}_{F,\epsilon} - \mathbf{m}_{F,\epsilon} \cdot a\| < \epsilon \quad \text{and} \quad \|a \Delta \mathbf{m}_{F,\epsilon} - a\| < \epsilon \quad (a \in F).$$

It follows that $(\mathbf{m}_{F,\epsilon})_{F \in \mathcal{F}, \epsilon > 0}$ is an approximate diagonal for $\mathfrak{A}$ bounded by C. $\square$

Exercise 2.3.12 Show that ℓ^2 equipped with pointwise multiplication does not have a bounded approximate identity. Use this example to refute the following "improvement" of Proposition 2.3.17: If $\mathfrak{A}$ is a Banach algebra and if $(\mathfrak{A}_\alpha)_\alpha$ is a directed family of closed, amenable subalgebras of $\mathfrak{A}$ such that $\bigcup_\alpha \mathfrak{A}_\alpha$ is dense in $\mathfrak{A}$, then $\mathfrak{A}$ is amenable.

Example 2.3.18 Let $E = c_0$ or $E = \ell^p$ with $p \in (1, \infty)$. For $n \in \mathbb{N}$, let $P_n : E \to E$ be the projection onto the first n coordinates, and let $\mathfrak{A}_n := P_n \mathcal{K}(E) P_n$. It follows that $\overline{\bigcup_{n=1}^\infty \mathfrak{A}_n} = \mathcal{K}(E)$. Since $\mathfrak{A}_n \cong \mathcal{L}(\ell_n^p)$ for $E = \ell^p$ with $p \in (1, \infty)$ and $\mathfrak{A}_n \cong \mathcal{L}(\ell_n^\infty)$ for $E = c_0$, all the algebras $\mathfrak{A}_n$ are 1-amenable by Example 2.3.16. By Proposition 2.3.17, $\mathcal{K}(E)$ is 1-amenable.

Exercise 2.3.13 Does the argument in Example 2.3.18 also establish the amenability of $\mathcal{K}(\ell^1)$?

It is not hard to see (Really?) that, for C-amenable Banach algebras $\mathfrak{A}_1, \ldots, \mathfrak{A}_n$, their ℓ^∞-direct sum $\mathfrak{A}_1 \oplus \cdots \oplus \mathfrak{A}_n$ is also C-amenable. Together with Proposition 2.3.17 this yields:

Corollary 2.3.19 *Let $C \geq 1$, and let $(\mathfrak{A}_\alpha)_\alpha$ be a family of C-amenable Banach algebras. Then their c_0-direct sum $c_0\text{-}\bigoplus_\alpha \mathfrak{A}_\alpha$ is C-amenable.*

2.4 Hochschild cohomology

At the beginning of this chapter, we introduced, for a Banach algebra $\mathfrak{A}$ and a Banach $\mathfrak{A}$-bimodule E, the spaces $\mathcal{Z}^1(\mathfrak{A}, E)$, $\mathcal{B}^1(\mathfrak{A}, E)$, and $\mathcal{H}^1(\mathfrak{A}, E)$, and we called $\mathcal{H}^1(\mathfrak{A}, E)$ the first Hochschild cohomology group of $\mathfrak{A}$ with coefficients in E. This suggest that there should be second and third, etc., cohomology groups, and indeed there are.

We begin by introducing the basic notions of Hochschild cohomology:

Definition 2.4.1 Let $\mathfrak{A}$ be a Banach algebra, and let E be a Banach $\mathfrak{A}$-bimodule.

(i) Let $\mathcal{L}^0(\mathfrak{A}, E) := E$, and, for $n \in \mathbb{N}$, let

$$\mathcal{L}^n(\mathfrak{A}, E) := \{T : \mathfrak{A}^n \to E : T \text{ is bounded and } n\text{-linear}\}.$$

The elements of $\mathcal{L}^n(\mathfrak{A}, E)$ are called *n-cochains*.

(ii) For $n \in \mathbb{N}_0$, define $\delta^n : \mathcal{L}^n(\mathfrak{A}, E) \to \mathcal{L}^{n+1}(\mathfrak{A}, E)$ through

$$
\begin{aligned}
(\delta^n T)(a_1, \ldots, a_{n+1}) := {} & a_1 \cdot T(a_2, \ldots, a_{n+1}) \\
& + \sum_{k=1}^{n} (-1)^k T(a_1, \ldots, a_k a_{k+1}, \ldots, a_{n+1}) \\
& + (-1)^{n+1} T(a_1, \ldots, a_n) \cdot a_{n+1} \\
& (T \in \mathcal{L}^n(\mathfrak{A}, E), \, a_1, \ldots, a_{n+1} \in \mathfrak{A}).
\end{aligned}
$$

The map δ^n is called the *n-coboundary operator*.

(iii) Let $\mathcal{B}^0(\mathfrak{A}, E) := \{0\}$, and, for $n \in \mathbb{N}$, let $\mathcal{B}^n(\mathfrak{A}, E) := \operatorname{ran} \delta^{n-1}$. The elements of $\mathcal{B}^n(\mathfrak{A}, E)$ are called *n-coboundaries*.

(iv) For $n \in \mathbb{N}_0$ let $\mathcal{Z}^n(\mathfrak{A}, E) := \ker \delta^n$. The elements of $\mathcal{Z}^n(\mathfrak{A}, E)$ are called *n-cocycles*.

The sequence

$$\{0\} \to E \xrightarrow{\delta^0} \mathcal{L}(\mathfrak{A}, E) \xrightarrow{\delta^1} \mathcal{L}^2(\mathfrak{A}, E) \xrightarrow{\delta^2} \cdots \xrightarrow{\delta^{n-1}} \mathcal{L}^n(\mathfrak{A}, E) \xrightarrow{\delta^n} \mathcal{L}^{n+1}(\mathfrak{A}, E) \xrightarrow{\delta^{n+1}} \cdots \qquad (2.13)$$

is called the *Hochschild cochain complex*.

Exercise 2.4.1 Let $\mathfrak{A}$ be a Banach algebra, and let E be a Banach $\mathfrak{A}$-bimodule. Describe $\mathcal{Z}^0(\mathfrak{A}, E)$, and show that the definitions of $\mathcal{Z}^1(\mathfrak{A}, E)$ and $\mathcal{B}^1(\mathfrak{A}, E)$ from Section 2.1 and Definition 2.4.1 coincide.

The next lemma is crucial for the definition of Hochschild cohomology:

Lemma 2.4.2 *Let $\mathfrak{A}$ be a Banach algebra, and let E be a Banach $\mathfrak{A}$-bimodule. Then $\mathcal{B}^n(\mathfrak{A}, E) \subset \mathcal{Z}^n(\mathfrak{A}, E)$ for all $n \in \mathbb{N}_0$.*

For $n = 0$, the conclusion of Lemma 2.4.2 is trivial, and for $n = 1$, we have already checked it. The proof of the general case is lengthy and extremely tedious — so tedious that we omit it altogether. If you want to get an impression of just *how* tedious the proof of Lemma 2.4.2 is, try to prove it only for $n = 2$. Anyway, with Lemma 2.4.2, the next definition makes sense:

Definition 2.4.3 Let $\mathfrak{A}$ be a Banach algebra, and let E be a Banach $\mathfrak{A}$-bimodule. For $n \in \mathbb{N}_0$, let

$$\mathcal{H}^n(\mathfrak{A}, E) := \mathcal{Z}^n(\mathfrak{A}, E)/\mathcal{B}^n(\mathfrak{A}, E)$$

be the *n-th Hochschild cohomology group* of $\mathfrak{A}$ with coefficients in E.

So, for a Banach algebra $\mathfrak{A}$, a Banach $\mathfrak{A}$-bimodule E, and $n \in \mathbb{N}$, we have defined $\mathcal{H}^n(\mathfrak{A}, E)$, and two questions come (at least, should come) to mind immediately:

- What is $\mathcal{H}^n(\mathfrak{A}, E)$ good for, i.e. what relevant information is encoded in $\mathcal{H}^n(\mathfrak{A}, E)$? Thanks to Theorem 2.1.8, we have seen that there is some use for $\mathcal{H}^1(\mathfrak{A}, E)$ (in case E is dual). But what about $n \geq 2$?
- How do we lay our hands on, i.e. calculate, $\mathcal{H}^n(\mathfrak{A}, E)$?

We don't have the space here to deal with these questions in any depth, and will only indicate possible answers. We deal with the second question first.

Lemma 2.4.4 *Let $\mathfrak{A}$ be a Banach algebra, and let E be a Banach $\mathfrak{A}$-bimodule. Then, for $k \in \mathbb{N}$, the space $\mathcal{L}^k(\mathfrak{A}, E)$ becomes a Banach $\mathfrak{A}$-bimodule through*

$$(a \cdot T)(a_1, \ldots, a_k) := a \cdot T(a_1, \ldots, a_k) \qquad (a \in \mathfrak{A}, \, T \in \mathcal{L}^k(\mathfrak{A}, E), \, a_1, \ldots, a_k \in \mathfrak{A})$$

and

$$\begin{aligned}
(T \cdot a)(a_1, \ldots, a_k) := \; & T(aa_1, \ldots, a_k) \\
& + \sum_{j=1}^{k-1} (-1)^j T(a, a_1, \ldots, a_j a_{j+1}, \ldots, a_k) \\
& + (-1)^k T(a, a_1, \ldots, a_{k-1}) \cdot a_k \\
& (a \in \mathfrak{A}, \, T \in \mathcal{L}^k(\mathfrak{A}, E), \, a_1, \ldots, a_k \in \mathfrak{A}).
\end{aligned}$$

Let $\mathfrak{A}$ be a Banach algebra, let E be a Banach $\mathfrak{A}$-bimodule, and let $k \in \mathbb{N}$. If the space $\mathcal{L}^k(\mathfrak{A}, E)$ is turned into a Banach $\mathfrak{A}$-bimodule as in Lemma 2.4.4, we let

$$\delta_k^n \colon \mathcal{L}^n(\mathfrak{A}, \mathcal{L}^k(\mathfrak{A}, E)) \to \mathcal{L}^{n+1}(\mathfrak{A}, \mathcal{L}^k(\mathfrak{A}, E)) \qquad (n \in \mathbb{N}_0)$$

denote the corresponding n-coboundary operator, whereas the coboundary operators of (2.13) are still denoted by δ^n for $n \in \mathbb{N}_0$.

Lemma 2.4.5 *Let $\mathfrak{A}$ be a Banach algebra, let E be a Banach $\mathfrak{A}$-bimodule, and let $k \in \mathbb{N}$. Then, for $n \in \mathbb{N}_0$, the map*

$$\tau^n \colon \mathcal{L}^{n+k}(\mathfrak{A}, E) \to \mathcal{L}^n(\mathfrak{A}, \mathcal{L}^k(\mathfrak{A}, E)),$$

defined by

$$((\tau^n T)(a_1, \dots, a_n))(a_{n+1}, \dots, a_{n+k})$$
$$:= T(a_1, \dots, a_n, a_{n+1}, \dots, a_{n+k})$$
$$(T \in \mathcal{L}^{n+k}(\mathfrak{A}, E), \, a_1, \dots, a_n, a_{n+1}, \dots, a_{n+k} \in \mathfrak{A})$$

is an isometric isomorphism such that

$$\delta_k^n \circ \tau^n = \tau^{n+1} \circ \delta^{n+k}.$$

Exercise 2.4.2 Prove Lemmas 2.4.4 and 2.4.5

Theorem 2.4.6 (reduction of dimension) *Let* $\mathfrak{A}$ *be a Banach algebra, let* E *be a Banach* $\mathfrak{A}$-*bimodule, and let* $k \in \mathbb{N}_0$. *Then we have*

$$\mathcal{H}^{n+k}(\mathfrak{A}, E) \cong \mathcal{H}^n(\mathfrak{A}, \mathcal{L}^k(\mathfrak{A}, E)) \qquad (n \in \mathbb{N}_0).$$

Proof For $k = 0$, the claim is trivially true. Therefore, let $k \in \mathbb{N}$. For $n \in \mathbb{N}$, let τ^n : $\mathcal{L}^{n+k}(\mathfrak{A}, E) \to \mathcal{L}^n(A, \mathcal{L}^k(\mathfrak{A}, E))$ be as in Lemma 2.4.5. We want to show that τ^n induces an isomorphism $\mathcal{H}^{n+k}(\mathfrak{A}, E) \cong \mathcal{H}^n(\mathfrak{A}, \mathcal{L}^k(\mathfrak{A}, E))$.

By Lemma 2.4.5,

$$\begin{aligned}
T \in \mathcal{Z}^{n+k}(\mathfrak{A}, E) &\iff \delta^{n+k} T = 0 \\
&\iff (\tau^{n+1} \circ \delta^{n+k}) T = 0 \\
&\iff (\delta_k^n \circ \tau^n) T = 0 \\
&\iff \tau^n T \in \mathcal{Z}^n(\mathfrak{A}, \mathcal{L}^k(\mathfrak{A}, E))
\end{aligned}$$

holds.

Let $T \in \mathcal{B}^{n+k}(\mathfrak{A}, E)$. By definition, there is $S \in \mathcal{L}^{(n-1)+k}(\mathfrak{A}, E)$ such that $\delta^{(n-1)+k} S = T$. Again by Lemma 2.4.5, we have

$$\tau^n T = (\tau^n \circ \delta^{(n-1)+k}) S = \delta_k^{n-1}(\tau^{n-1} S) \in \mathcal{B}^n(\mathfrak{A}, \mathcal{L}^k(\mathfrak{A}, E)).$$

Conversely, let $T \in \mathcal{L}^{n+k}(\mathfrak{A}, E)$ with $\tau^n T \in \mathcal{B}^n(\mathfrak{A}, \mathcal{L}^k(\mathfrak{A}, E))$. Again by definition, there is $S \in \mathcal{L}^{n-1}(\mathfrak{A}, \mathcal{L}^k(\mathfrak{A}, E))$ such that $\tau^n T = \delta_k^{n-1} S$. Since τ^n is an isomorphism, there is $\tilde{S} \in \mathcal{L}^{(n-1)-k}(\mathfrak{A}, E)$ such that $\tau^{n-1} \tilde{S} = S$. Hence, we have:

$$\tau^n T = (\delta_k^{n-1} \circ \tau^{n-1}) \tilde{S} = (\tau^n \circ \delta^{(n-1)+k}) \tilde{S} = \tau^n(\delta^{(n-1)+k} \tilde{S}).$$

Since τ^n is injective, this means that $T = \delta^{(n-1)+k} \tilde{S} \in \mathcal{B}^{n+k}(\mathfrak{A}, E)$.

This establishes the claim for $n \in \mathbb{N}$. $\square$

Exercise 2.4.3 Prove Theorem 2.4.6 for $n = 0$.

Roughly speaking, Theorem 2.4.6 asserts that every Hochschild cohomology group is already a first Hochschild cohomology group. This does *not* mean that higher Hochschild cohomology groups are uninteresting: In order to reduce the order of the cohomology group as in Theorem 2.4.6, we have to trade the (perhaps quite simple) coefficient module E for the much more complicated module $\mathcal{L}^k(\mathfrak{A}, E)$.

An application of Theorem 2.4.6 is the following:

Theorem 2.4.7 *For a Banach algebra $\mathfrak{A}$ the following are equivalent:*

(i) $\mathfrak{A}$ *is amenable.*

(ii) $\mathcal{H}^n(\mathfrak{A}, E^*) = \{0\}$ *for each Banach $\mathfrak{A}$-bimodule E and for all $n \in \mathbb{N}$.*

Proof Of course, only (i) $\implies$ (ii) needs proof.

For $n \in \mathbb{N}$, let

$$F := \underbrace{\mathfrak{A} \hat{\otimes} \cdots \hat{\otimes} \mathfrak{A}}_{(n-1)\text{-times}} \hat{\otimes} E.$$

Define a bimodule action of $\mathfrak{A}$ on F by letting

$$(a_1 \otimes \cdots \otimes a_{n-1} \otimes x) \cdot a := a_1 \otimes \cdots \otimes a_{n-1} \otimes x \cdot a \qquad (a, a_1, \ldots, a_{n-1} \in \mathfrak{A}, x \in E)$$

and

$$a \cdot (a_1 \otimes \cdots \otimes a_{n-1} \otimes x) := aa_1 \otimes \cdots \otimes a_{n-1}$$
$$+ \sum_{j=1}^{n-2} (-1)^j a \otimes a_1 \otimes \cdots \otimes a_j a_{j+1} \otimes \cdots \otimes a_{n-1} \otimes x$$
$$+ (-1)^{n-1} a \otimes a_1 \otimes \cdots \otimes a_{n-2} \otimes a_{n-1} \cdot x$$
$$(a, a_1, \ldots, a_{n-1} \in \mathfrak{A}, x \in E).$$

Identifying F^* with $\mathcal{L}^{n-1}(\mathfrak{A}, E^*)$ as in Exercise B.2.10, we see that the dual module action of $\mathfrak{A}$ on F^* is the module action of $\mathfrak{A}$ on $\mathcal{L}^{n-1}(\mathfrak{A}, E^*)$ (Check this!). By Theorem 2.4.6, we then have

$$\mathcal{H}^n(\mathfrak{A}, E^*) \cong \mathcal{H}^1(\mathfrak{A}, \mathcal{L}^{n-1}(\mathfrak{A}, E^*)) \cong \mathcal{H}^1(\mathfrak{A}, F^*) = \{0\}.$$

This establishes (ii). $\square$

We want to conclude this section with an application of second Hochschild cohomology groups.

If $\mathfrak{A}$ is a finite-dimensional algebra with (Jacobson) radical R, then the classical Wedderburn decomposition theorem asserts that there is a subalgebra $\mathfrak{B}$ of $\mathfrak{A}$ such that $\mathfrak{A} = \mathfrak{B} \oplus R$ ($\oplus$ denoting a vector space direct sum). If an arbitrary algebra has such a decomposition, we say that $\mathfrak{A}$ has a *Wedderburn decomposition*. Not every algebra has a Wedderburn decomposition.

For Banach algebras, the following variant is often more appropriate:

Definition 2.4.8 Let $\mathfrak{A}$ be a Banach algebra with radical R. We say that $\mathfrak{A}$ has a *strong Wedderburn decomposition* if there is a closed subalgebra $\mathfrak{B}$ of $\mathfrak{A}$ such that $\mathfrak{A} = \mathfrak{B} \oplus R$.

The question that we wish to now briefly touch upon is which Banach algebras have a strong Wedderburn decomposition.

Lemma 2.4.9 *Let $\mathfrak{A}$ be a Banach algebra whose radical R is complemented and satisfies $R^2 = \{0\}$. Then, if $\rho : \mathfrak{A}/R \to \mathfrak{A}$ is a bounded right inverse of the quotient map $\pi : \mathfrak{A} \to \mathfrak{A}/R$, we turn R into a Banach $\mathfrak{A}/R$-bimodule by letting*

$$(a + R) \cdot r := \rho(a + R)r \quad and \quad r \cdot (a + R) := r\rho(a + R) \qquad (a \in \mathfrak{A}, r \in R).$$

This module action is independent of the choice of ρ.

Proof Let $b_1, b_2 \in \mathfrak{A}/R$. Obviously,

$$\pi(\rho(b_1)\rho(b_2) - \rho(b_1 b_2)) = \pi(\rho(b_1))\pi(\rho(b_2)) - \pi(\rho(b_1 b_2)) = 0$$

holds, so that $\rho(b_1)\rho(b_2) - \rho(b_1 b_2) \in R$. Since $R^2 = \{0\}$, we have

$$b_1 \cdot (b_2 \cdot r) - (b_1 b_2) \cdot r = [\rho(b_1)\rho(b_2) - \rho(b_1 b_2)]r = 0 \qquad (r \in R)$$

meaning that R is a left Banach $\mathfrak{A}/R$-module. Analoguously, one sees that R is also a right Banach $\mathfrak{A}/R$-module. It is easy to see that the two module actions are compatible in such a way that R becomes a Banach $\mathfrak{A}/R$-bimodule.

Let $\tilde{\rho}$ be another continuous right inverse of π. Then, we have $\rho(b) - \tilde{\rho}(b) \in R$ for all $b \in \mathfrak{A}/R$. Again due to the fact that $R^2 = \{0\}$, we obtain

$$\rho(b) \cdot r = \tilde{\rho}(b) \cdot r \qquad (b \in \mathfrak{A}/R, r \in R).$$

Analoguously, one shows that the right module action is independent of the choice of ρ. $\square$

In the next theorem, we refer to the module action defined in Lemma 2.4.9.

Theorem 2.4.10 *Let $\mathfrak{A}$ be a Banach algebra whose radical R is complemented and satisfies $R^2 = \{0\}$. Suppose that $\mathcal{H}^2(\mathfrak{A}/R, R) = \{0\}$. Then $\mathfrak{A}$ has a strong Wedderburn decomposition.*

Proof Once again, let $\rho \colon \mathfrak{A}/R \to \mathfrak{A}$ be a bounded right inverse of the quotient map $\pi \colon \mathfrak{A} \to \mathfrak{A}/R$. Define $T \in \mathcal{L}^2(\mathfrak{A}/R, R)$ through

$$T(b_1, b_2) = \rho(b_1)\rho(b_2) - \rho(b_1 b_2) \qquad (b_1, b_2 \in \mathfrak{A}/R).$$

For $b_1, b_2, b_3 \in \mathfrak{A}/R$ we have

$$
\begin{aligned}
(\delta^2 T)(b_1, b_2, b_3) &= b_1 \cdot T(b_2, b_3) - T(b_1 b_2, b_3) + T(b_1, b_2 b_3) - T(b_1, b_2) \cdot b_3 \\
&= \rho(b_1)[\rho(b_2)\rho(b_3) - \rho(b_2 b_3)] - \rho(b_1 b_2)\rho(b_3) + \rho(b_1 b_2 b_3) \\
&\quad + \rho(b_1)\rho(b_2 b_3) - \rho(b_1 b_2 b_3) - [\rho(b_1)\rho(b_2) - \rho(b_1 b_2)]\rho(b_3) \\
&= \rho(b_1)\rho(b_2)\rho(b_3) - \rho(b_1)\rho(b_2 b_3) - \rho(b_1 b_2)\rho(b_3) + \rho(b_1 b_2 b_3) \\
&\quad + \rho(b_1)\rho(b_2 b_3) - \rho(b_1 b_2 b_3) - \rho(b_1)\rho(b_2)\rho(b_3) + \rho(b_1 b_2)\rho(b_3) \\
&= 0,
\end{aligned}
$$

meaning that $T \in \mathcal{Z}^2(\mathfrak{A}/R, R)$. Since $\mathcal{H}^2(\mathfrak{A}/R, R) = \{0\}$, there is $S \in \mathcal{L}^1(\mathfrak{A}/R, R)$ with $T = \delta^1 S$. Let $\theta := \rho - S$. Then we have for $b_1, b_2 \in \mathfrak{A}/R$:

$$
\begin{aligned}
\theta(b_1)\theta(b_2) - \theta(b_1 b_2) &= (\rho(b_1) - Sb_1)(\rho(b_2) - Sb_2) - \rho(b_1 b_2) + S(b_1 b_2) \\
&= \rho(b_1)\rho(b_2) - \rho(b_1 b_2) - \rho(b_1)Sb_2 + S(b_1 b_2) - (Sb_1)\rho(b_2) \\
&= T(b_1, b_2) - b_1 \cdot Sb_2 + S(b_1 b_2) - (Sb_1) \cdot b_2 \\
&= (T - \delta^1 S)(b_1, b_2) \\
&= 0.
\end{aligned}
$$

Hence, $\theta \colon \mathfrak{A}/R \to \mathfrak{A}$ is a bounded homomorphism of Banach algebras such that

$$(\pi \circ \theta)(b) = (\pi \circ \rho)(b) - (\pi \circ S)(b) = (\pi \circ \rho)(b) = b \qquad (b \in \mathfrak{A}/R). \tag{2.14}$$

Let $\mathfrak{B} := \theta(\mathfrak{A}/R)$. Clearly, $\mathfrak{B}$ is a subalgebra of $\mathfrak{A}$. It follows from (2.14) that $\mathfrak{B} \cap R = \{0\}$. For $a \in \mathfrak{A}$, let $b := (\theta \circ \pi)(a)$ and $r := a - (\theta \circ \pi)(a)$, so that $a = b + r$. By (2.14),

$$\pi(r) = \pi(a - (\theta \circ \pi)(a)) = \pi(a) - \pi(a) = 0$$

holds, so that $r \in R$. All in all, we have $\mathfrak{A} = \mathfrak{B} \oplus R$. Let $(b_n)_{n=1}^{\infty}$ be a sequence in $\mathfrak{B}$ converging in $\mathfrak{A}$ to some element a. It is immediate that $\pi(a) = \lim_{n\to\infty} \pi(b_n)$. Again by (2.14), $\theta \circ \pi = \mathrm{id}_{\mathfrak{B}}$; hence,

$$a = \lim_{n\to\infty} b_n = \lim_{n\to\infty} (\theta \circ \pi)(b_n) = \theta(\pi(a)) \in \mathfrak{B}.$$

holds. Consequently, $\mathfrak{B}$ is closed, and $\mathfrak{A}$ has a strong Wedderburn decomposition. $\square$

Corollary 2.4.11 *Let $\mathfrak{A}$ be a Banach algebra with radical R such that $\dim R < \infty$ and $R^2 = \{0\}$, and suppose that $\mathfrak{A}/R$ is amenable. Then $\mathfrak{A}$ has a strong Wedderburn decomposition.*

Proof Since R is finite-dimensional, R is complemeted in $\mathfrak{A}$ and trivially reflexive. Since $\mathfrak{A}/R$ is amenable, Theorem 2.4.7 yields:

$$\mathcal{H}^2(\mathfrak{A}/R, R) \cong \mathcal{H}^2(\mathfrak{A}/R, R^{**}) = \{0\}.$$

By Theorem 2.4.10, this yields the claim. $\square$

2.5 Notes and comments

B. E. Johnson's memoir [Joh 1] is the big bang of amenable Banach algebras. It contains Theorem 2.1.8 along with most of Section 2.1. Our pseudo-unital Banach modules are called neo-unital in [Joh 1] (and in most of the existing literature). The proof of (i) $\Longrightarrow$ (ii) of Theorem 2.1.8 we give (with Day's fixed point theorem) is different from the one in [Joh 1], and was suggested by R. Stokke. This approach not only shows that any $D \in \mathcal{Z}^1(L^1(G), E^*)$ is inner, but also that the implementing element can be chosen from the w^*-closed, convex hull of the set $\{-(\tilde{D}\delta_g) \cdot \delta_{g^{-1}} : g \in G\}$.

Theorem 2.2.4 is from [Joh 2]. If $\mathfrak{A}$ is a Banach algebra, then an element of $\mathfrak{A}\hat{\otimes}\mathfrak{A}$ is called *symmetric* if it is a fixed point of the flip map $\mathfrak{A}\hat{\otimes}\mathfrak{A} \ni a \otimes b \mapsto b \otimes a$. If $\mathfrak{A}$ has an approximate diagonal consisting of symmetric tensors, it is called *symmetrically amenable*. The symmetrically amenable Banach algebras were introduced in [Joh 8]. They have hereditary properties similar to those of amenable Banach algebras. If G is a locally compact group, then $L^1(G)$ is already symmetrically amenable.

The hereditary properties of amenable Banach algebras as we present them go back to [Joh 1]. Theorem 2.3.13 was first obtained by the Russian school of Banach algebraists led by A. Ya. Helemskiĭ ([Hel 3]). Our proof is from [C–L]. Theorem 2.3.14 is due to M. V. Sheĭnberg ([Sheĭ]).

Examples of amenable Banach algebras which are not 1-amenable can be found in [Joh 7] and [L–L–W].

Hochschild cohomology is named in honor of G. Hochschild who introduced it in [Hoch 1] and [Hoch 2]; one of his motivations was to generalize Wedderburn's decomposition theorem. In [Hoch 1] a proof of Lemma 2.4.2 is given; it is the only reference for a proof I know. Hochschild's original definitions were for arbitrary associative algebras, i.e. without any topology. The functional analytic overtones required to adapt the theory for Banach algebras were added by H. Kamowitz ([Kam]; see also [Gui]). For this reason, the Banach algebraic Hochschild cohomology we are discussing is sometimes called Kamowitz–Hochschild cohomology (another name by which it goes is Johnson–Hochschild cohomology).

Theorem 2.4.7 motivates the following definition: Let $n \in \mathbb{N}$. Then a Banach algebra $\mathfrak{A}$ is called *n-amenable* if $\mathcal{H}^m(\mathfrak{A}, E^*) = \{0\}$ for all $m \geq n$ ([Pat 2]). There is a characterization of n-amenable Banach algebras similar to Theorem 2.2.4; see [Pat 2] for details.

Already Kamowitz used Hochschild cohomology to show that certain Banach algebras have strong Wedderburn decompositions. Our Theorem 2.4.10 is essentially due to him (Kamowitz only considers commutative Banach algebras, but all the ideas are already given in [Kam]). In fact, Theorem 2.4.10 is rather feeble: The hypotheses concerning the radical can be weakened substantially. Strong Wedderburn decompositions are a just one instance of what are called strong splittings of extensions of Banach algebras. A recent account of splittings of extensions of Banach algebras is [B–D–L].

Third Hochschild cohomology groups occur naturally in the study of perturbations of Banach algebras ([Joh 3], [R–T]). I don't know of any Banach algebraic interpretation of n-th Hochschild cohomology groups for $n \geq 4$.

Of course, one can also consider, for a Banach algebra $\mathfrak{A}$ and a Banach $\mathfrak{A}$-bimodule E, the purely algebraic cohomology groups $H^n(\mathfrak{A}, E)$ as originally defined by Hochschild. For each $n \in \mathbb{N}_0$, there is a canonical group homomorphism $c^n : \mathcal{H}^n(\mathfrak{A}, E) \to H^n(\mathfrak{A}, E)$, the *cohomology comparison map*. Clearly, c^0 is an isomorphism, and c^1 is injective. Already for c^2, the situation becomes more complicated ([Sol] and [D–V]). In [Wod], M. Wodzicki showed that, if $\mathfrak{A}$ is amenable and has the cardinality $\aleph_n$, then $c^m \equiv 0$ for all $m \geq n + 3$. If we assume the continuum hypothesis, this means that for amenable Banach algebras with the cardinality of the continuum, $c^m = 0$ for all $m \geq 4$. Hence, algebraic and "our" Hochschild cohomology can be surprisingly unrelated. Nothing seems to be known about c^3.

The introduction to Hochschild cohomology we have presented in this chapter is fairly straightforward. A more sohpisticated and more powerful approach is propagated by Helemskiĭ's Russian school. We give an introduction to that approach in Chapter 5.

3 Examples of amenable Banach algebras

What are the amenable Banach algebras we know so far? First, there are, of course, the group algebras of amenable groups. Then we have commutative C^*-algebras, and the algebras of compact operators on certain classical sequence spaces. And we can use constructions like quotients, tensor products, c_0-direct sums, etc., to get further examples from the old ones. If you think that's not impressive, you're probably right.

In this chapter, we shall thus look out for more amenable Banach algebras that aren't just obtained by applying standard constructions to algebras we already know.

3.1 Banach algebras of compact operators

In Example 2.3.18, we saw that $\mathcal{K}(E) = \mathcal{A}(E)$ is amenable whenever $E = c_0$ or $E = \ell^p$ for $p \in (1, \infty)$. A natural questions is thus to ask for which Banach spaces E precisely the Banach algebra $\mathcal{K}(E)$ — or rather: $\mathcal{A}(E)$ — is amenable. For every E, perhaps?

Since amenable Banach algebra have bounded approximate identities by Proposition 2.2.1, we first look at the question of when $\mathcal{K}(E)$ — or rather: $\mathcal{A}(E)$ — has a (bounded) approximate identity. The approximation property and related notions are discussed in Section C.1.

Proposition 3.1.1 *Let E be a Banach space with the approximation property. Then $\mathcal{K}(E)$ has a left approximate identity belonging to $\mathcal{F}(E)$. In particular, $\mathcal{K}(E) = \mathcal{A}(E)$ holds.*

Proof For each compact subset K of E and for each $\epsilon > 0$, there is $S_{K,\epsilon} \in \mathcal{F}(E)$ such that

$$\sup\{\|x - S_{K,\epsilon}x\| : x \in K\} < \epsilon.$$

Since, for any $T \in \mathcal{K}(E)$, the the image of the closed unit ball of E under T is relatively compact, it follows that $\lim_{(K,\epsilon)} \|S_{K,\epsilon}T - T\| = 0$, so that $(S_{K,\epsilon})_{K,\epsilon}$ is a left approximate identity for $\mathcal{K}(E)$. $\square$

We don't know if the converse of Proposition 3.1.1 holds, i.e. if E necessarily has the approximation property when $\mathcal{K}(E)$ has a left approximate identity belonging to $\mathcal{F}(E)$. For the bounded approximation property, the situation is better understood:

Theorem 3.1.2 *Let E be a Banach space, and let $C \geq 1$. Then the following are equivalent:*

  (i) *E has the C-approximation property.*
  (ii) *$\mathcal{K}(E)$ has a left approximate identity bounded by C and belonging to $\mathcal{F}(E)$.*
  (iii) *$\mathcal{A}(E)$ has a left approximate identity bounded by C.*

Proof (i) $\Longrightarrow$ (ii) is proved like Proposition 3.1.1, and (ii) $\Longrightarrow$ (iii) is clear.

For the proof of (iii) $\Longrightarrow$ (i), let E be such that $\mathcal{A}(E)$ has a left approximate identity $(S_\alpha)_\alpha$ bounded by C. We may suppose that $(S_\alpha)_\alpha$ lies in $\mathcal{F}(E)$. It follows that $S_\alpha \to \mathrm{id}_E$ in the strong operator topology, and thus on all finite subsets of E. Since $(S_\alpha)_\alpha$ is bounded, it follows that $S_\alpha \to \mathrm{id}_E$ uniformly on all compact subsets of E. $\square$

Corollary 3.1.3 *The following are equivalent for a Banach space E:*

(i) *E has the bounded approximation property.*

(ii) *$\mathcal{A}(E)$ has a bounded left approximate identity.*

In particular, if we are looking for Banach spaces E such that $\mathcal{A}(E)$ is amenable, spaces without the bounded approximation property are out. But, of course, being amenable is much stronger a demand than just having a bounded left approximate identity. So, what does the existence of a bounded right approximate identity or even a bounded (two-sided) approximate identity for $\mathcal{A}(E)$ imply for E?

Theorem 3.1.4 *For a Banach space E and a constant $C \geq 1$, the following are equivalent:*

(i) *E^* has the C-approximation property.*

(ii) *There is a net $(S_\alpha)_\alpha$ in $\mathcal{F}(E)$ bounded by C such that $\lim_\alpha S_\alpha^* = \mathrm{id}_{E^*}$ uniformly on the compact subsets of E^*.*

(iii) *$\mathcal{A}(E^*)$ has a left approximate identity bounded by C.*

(iv) *$\mathcal{A}(E)$ has a right approximate identity bounded by C.*

Proof (i) $\Longrightarrow$ (ii): Let $\phi_1, \dots, \phi_n \in E^*$, and let $\epsilon > 0$. By (i), there is $S \in \mathcal{F}(E^*)$ with $\|S\| < C$ such that $\|\phi_j - S\phi_j\| < \epsilon$ for $j = 1, \dots, n$. Let $\psi_1, \dots, \psi_m \in E^*$ and $X_1, \dots, X_m \in E^{**}$ be such that $S = \sum_{j=1}^m \psi_j \odot X_j$. Let F be the linear span of $X_1, \dots, X_m$ in E^{**}. By the local reflexivity principle Theorem C.3.3, there is $\tau \colon F \to E$ such that $\|\tau\| < 1 + \epsilon$ and

$$\langle \tau(X), \phi_j \rangle = \langle \phi_j, X \rangle \qquad (X \in F, j = 1, \dots, n). \tag{3.1}$$

Let $\tilde{S} = \sum_{j=1}^m \psi_j \odot \tau(X_j) \in \mathcal{F}(E^*)$. Then, clearly, $\tilde{S}$ is the adjoint of an operator in $\mathcal{F}(E)$. Moreover, we have

$$\|\phi_j - \tilde{S}\phi_j\| = \|\phi_j - S\phi_j\| < \epsilon \qquad (j = 1, \dots, n)$$

by (3.1). Identifying $\mathcal{A}(E^*)$ with $E^* \check\otimes E^{**}$ as in Proposition B.2.7 and invoking Exercise B.2.3, we see that $\|\tilde{S}\| < C(1 + \epsilon)$. This implies that a net as required in (ii) exists (Why?).

(ii) $\Longrightarrow$ (iv): As in the proof of Proposition 3.1.1, we see that $\lim_\alpha S_\alpha^* T = T$ for all $T \in \mathcal{A}(E^*)$. In particular, we have $\lim_\alpha S_\alpha^* T^* = T^*$ for all $T \in \mathcal{A}(E)$. Consequently, $\lim_\alpha T S_\alpha = T$ for all $T \in \mathcal{A}(E)$.

(iv) $\Longrightarrow$ (ii): Let $(S_\alpha)_\alpha$ be a right approximate identity for $\mathcal{A}(E)$ bounded by C. We may suppose that $(S_\alpha)_\alpha$ is contained in $\mathcal{F}(E)$. Then $\lim_\alpha S_\alpha^* T^* = T^*$ for all $T \in \mathcal{A}(E)$. As in the proof of Theorem 3.1.2, one sees (How exactly?) that $\lim_\alpha S_\alpha^* = \mathrm{id}_{E^*}$ uniformly on the compact subsets of E^*.

(ii) $\Longrightarrow$ (i) is trivial, and (i) $\Longleftrightarrow$ (iii) is Theorem 3.1.2. $\square$

Since a Banach space E such that E^* has the C-approximation property for some $C \geq 1$ itself has the C-approximation property (Exercise C.3.1), we obtain:

Corollary 3.1.5 *The following are equivalent for a Banach space E:*

(i) E^* *has the bounded approximation property.*
(ii) $\mathcal{A}(E)$ *has a bounded right approximate identity.*
(iii) $\mathcal{A}(E)$ *has bounded approximate identity.*

Proof Suppose that E^* has the bounded approximation property. By Theorem 3.1.4, $\mathcal{A}(E)$ has a bounded right approximate identity. Since E also has the bounded approximation property, $\mathcal{A}(E)$ has a bounded left approximate identity by Theorem 3.1.2. As in the proof of Proposition 2.2.1, we can combine these two one-sided approximate identities to obtain a bounded approximate identity. $\square$

Hence, for $\mathcal{A}(E)$ to be amenable we have to require — at least! — that E^* have the bounded approximation property.

Example 3.1.6 Let $\mathfrak{H}$ be a infinite-dimensional Hilbert space. Then $(\mathfrak{H} \hat{\otimes} \mathfrak{H})^* \cong \mathcal{L}(\mathfrak{H})$ lacks the approximation property ([Sza]). Consequently, $\mathcal{A}(\mathfrak{H} \hat{\otimes} \mathfrak{H})$ does not have a bounded approximate identity and thus is not amenable. Note that, however, $\mathcal{A}(\mathfrak{H} \hat{\otimes} \mathfrak{H})$ has a bounded left approximate identity by Exercise C.1.2 and Theorem 3.1.2.

We now introduce a rather strong approximation property for a Banach space E which forces $\mathcal{A}(E)$ to be amenable.

Definition 3.1.7 Let E be a Banach space. A *finite, biorthogonal system* for E is a set

$$\{(x_j, \phi_k) : j, k = 1, \ldots, n\} \tag{3.2}$$

with $x_1, \ldots, x_n \in E$ and $\phi_1, \ldots, \phi_n \in E^*$ such that

$$\langle x_j, \phi_k \rangle = \delta_{j,k} \qquad (j, k = 1, \ldots, n).$$

Exercise 3.1.1 Let E be a Banach space, and let (3.2) be a finite, biorthogonal system. Then the map $\theta \colon \mathbb{M}_n \to \mathcal{F}(E)$ given by

$$\theta(A) := \sum_{j,k=1}^{n} a_{j,k} x_j \odot \phi_k \qquad \left(A = [a_{j,k}]_{\substack{j=1,\ldots,n \\ k=1,\ldots,n}} \in \mathbb{M}_n \right)$$

is a homomorphism.

Definition 3.1.8 A Banach space E has *property* $(\mathbb{A})$ if there is a net of finite, biorthogonal systems

$$\left\{ \left(x_j^{(\alpha)}, \phi_k^{(\alpha)} \right) : j, k = 1, \ldots, n_\alpha \right\}$$

for E with corresponding homomorphisms $\theta_\alpha \colon \mathbb{M}_{n_\alpha} \to \mathcal{F}(E)$ such that:

(i) $\lim_\alpha \theta_\alpha(E_{n_\alpha}) = \mathrm{id}_E$ uniformly on the compact subsets of E.
(ii) $\lim_\alpha \theta_\alpha(E_{n_\alpha})^* = \mathrm{id}_{E^*}$ uniformly on the compact subsets of E^*.
(iii) For each index α, there is a finite, irreducible $n_\alpha \times n_\alpha$ matrix group G_α such that

$$\sup_\alpha \sup_{g \in G_\alpha} \|\theta_\alpha(g)\| < \infty.$$

It is immediate from this definition that each Banach space with property (A) has the bounded approximation property (Why?).

Exercise 3.1.2 Show that (i) and (ii) in Definition 3.1.8 are equivalent to $(\theta_\alpha(E_{n_\alpha}))_\alpha$ being an approximate identity for $\mathcal{A}(E)$.

Theorem 3.1.9 *Let E be a Banach space with property (A), and let C denote the supremum in Definition 3.1.8(iii). Then $\mathcal{A}(E)$ is C^2-amenable.*

Proof With notation as in Definition 3.1.8, let

$$\mathbf{m}_\alpha := \frac{1}{|G_\alpha|} \sum_{g \in G_\alpha} \theta_\alpha(g) \otimes \theta_\alpha(g^{-1}),$$

so that $\|\mathbf{m}_\alpha\| \le C^2$. We claim that $(\mathbf{m}_\alpha)_\alpha$ is an approximate diagonal for $\mathcal{A}(E)$.

Let $P_\alpha := \theta(E_{n_\alpha}) = \Delta_{\mathcal{A}(E)}\mathbf{m}_\alpha$. Then $(P_\alpha)_\alpha$ is a bounded approximate identity for $\mathcal{A}(E)$, so that, in particular,

$$T\Delta_{\mathcal{A}(E)}\mathbf{m}_\alpha \to T \qquad (T \in \mathcal{A}(E)).$$

Let $T \in \mathcal{A}(E)$ be arbitrary. Then

$$
\begin{aligned}
T \cdot \mathbf{m}_\alpha - \mathbf{m}_\alpha \cdot T &= (T - P_\alpha T P_\alpha) \cdot \mathbf{m}_\alpha - \mathbf{m}_\alpha \cdot (T - P_\alpha T P_\alpha) + P_\alpha T P_\alpha \cdot \mathbf{m}_\alpha - \mathbf{m}_\alpha \cdot P_\alpha T P_\alpha \\
&= (T - P_\alpha T P_\alpha) \cdot \mathbf{m}_\alpha - \mathbf{m}_\alpha \cdot (T - P_\alpha T P_\alpha) \qquad \text{(Why?)} \\
&\to 0,
\end{aligned}
$$

since $(P_\alpha)_\alpha$ is a bounded approximate identity for $\mathcal{A}(E)$.   □

Exercise 3.1.3 Let $\mathfrak{A}$ be a Banach algebra. An element of $\mathfrak{A}\hat{\otimes}\mathfrak{A}$ is called *symmetric* if it is invariant under the flip map $\mathfrak{A} \otimes \mathfrak{A} \ni a \otimes b \mapsto b \otimes a$. If $\mathfrak{A}$ has an approximate diagonal consisting of symmetric tensors, it is called *symmetrically amenable* (see the notes and comments at the end of the previous chapter). Show that, if E is a Banach space with property (A), then $\mathcal{A}(E)$ is symmetrically amenable.

It is easy to see that what we did in Example 2.3.18 to show that $\mathcal{K}(E) = \mathcal{A}(E)$ is amenable for $E = c_0$ or $E = \ell^p$ with $p \in (1, \infty)$ was in fact a demonstration that these spaces have property (A). Before we exhibit further examples of spaces with property (A), we first take a look at the hereditary properties of (A):

Theorem 3.1.10 *Let E be a Banach space such that E^* has property (A). Then E has property (A) as well.*

Proof Let

$$\left\{ \left(\phi_j^{(\alpha)}, X_k^{(\alpha)} \right) : j, k = 1, \ldots, n_\alpha \right\} \tag{3.3}$$

be a net of finite, biorthogonal systems for E^* as in the definition of property (A).

Let $\mathcal{F}$ and $\mathcal{F}^*$ be the collections of all finite subsets of E and E^*, respectively. For each index α, for each $F \in \mathcal{F}$, and for each $\Phi \in \mathcal{F}^*$, the principle of local reflexivity yields a linear map

$$\tau_{\alpha, F, \Phi} \colon \text{lin}\left(\left\{ X_j^{(\alpha)} : j = 1, \ldots, n_\alpha \right\} \cup F \right) \to E$$

such that $\|\tau_{\alpha,F,\Phi}\| \leq 2$, $\tau_{\alpha,F,\Phi}|_F = \mathrm{id}_F$, and

$$\left\langle \tau_{\alpha,F,\Phi}\left(X_j^{(\alpha)}\right), \phi \right\rangle = \left\langle \phi, X_j^{(\alpha)} \right\rangle \quad \left(j = 1, \ldots, n_\alpha, \ \phi \in \mathrm{lin}\left(\left\{\phi_j^{(\alpha)} : j = 1, \ldots, n_\alpha\right\} \cup \Phi\right)\right).$$

It is then immediate that

$$\left\{\left(\tau_{\alpha,F,\Phi}(X_j), \phi_j^{(\alpha)}\right) : j, k = 1, \ldots, n_\alpha\right\} \tag{3.4}$$

is a net of finite, biorthogonal systems for E. It is also clear that (with $G_{\alpha,F,\Phi} := G_\alpha$) Definition 3.1.8(iii) is satisfied (the supremum increases at most by a factor 2).

Let $\theta_\alpha : \mathbb{M}_{n_\alpha} \to \mathcal{F}(E^*)$ and $\theta_{\alpha,F,\Phi} : \mathbb{M}_{n_\alpha} \to \mathcal{F}(E)$ be the homomorphisms corresponding to (3.3) and (3.4), respectively, and let $P_\alpha := \theta_\alpha(E_{n_\alpha})$ and $P_{\alpha,F,\Phi} := \theta_{\alpha,F,\Phi}(E_{n_\alpha})$. The net $(P_{\alpha,F,\Phi})_{\alpha,F,\Phi}$ is clearly bounded, and for any $F \in \mathcal{F}$ and $\Phi \in \mathcal{F}^*$, we have

$$\|P_{\alpha,F,\Phi}x - x\| = \|\tau_{\alpha,F,\Phi}(P_\alpha^* x) - x\| = \|\tau_{\alpha,F,\Phi}(P_\alpha^* x - x)\| \leq 2\|P_\alpha^* x - x\| \quad (x \in F).$$

It follows that Definition 3.1.8(i) holds (Why exactly?). Similarly,

$$P_{\alpha,F,\Phi}^*\phi = \sum_{j=1}^{n_\alpha} \left\langle \tau_{\alpha,F,\Phi}\left(X_j^{(\alpha)}\right), \phi \right\rangle X_j^{(\alpha)} = \sum_{j=1}^{n_\alpha} \left\langle \phi, X_j^{(\alpha)} \right\rangle X_j^{(\alpha)} = P_\alpha^*\phi \quad (\phi \in \Phi)$$

holds, so that Definition 3.1.8(ii) is equally satisfied. $\quad\square$

We shall soon encounter a Banach space enjoying (A) whose dual lacks that property. For a discussion of the Radon–Nikodým property, see Section C.2.

Theorem 3.1.11 *Let E and F be Banach spaces with property* (A) *such that*

(i) *E^* or F^* has the approximation property, and*

(ii) *E^* or F^* has the Radon–Nikodým property.*

Then $E \check{\otimes} F$ has property (A) *as well.*

Proof Let

$$\left\{\left(x_j^{(\alpha)}, \phi_k^{(\alpha)}\right) : j, k = 1, \ldots, n_\alpha\right\} \quad \text{and} \quad \left\{\left(y_j^{(\beta)}, \psi_k^{(\beta)}\right) : j, k = 1, \ldots, n_\beta\right\}$$

be the nets of finite, biorthogonal systems for E and F, respectively, as required by Definition 3.1.8. We then define a net of finite, biorthogonal systems for $E \check{\otimes} F$:

$$\left\{\left(x_j^{(\alpha)} \otimes y_\nu^{(\beta)}, \phi_k^{(\alpha)} \otimes \psi_\mu^{(\beta)}\right) : j, k = 1, \ldots, n_\alpha, \ \nu, \mu = 1, \ldots, n_\beta\right\}.$$

Identifying $\mathbb{M}_{n_\alpha} \otimes \mathbb{M}_{n_\beta}$ with $\mathbb{M}_{n_\alpha n_\beta}$ (Exercise B.1.7(iii)), we see that Definition 3.1.8(i) is satisfied. By Theorem C.2.3, $(E \check{\otimes} F)^* \cong E^* \check{\otimes} F^*$ holds. In particular, $E^* \otimes F^*$ is dense in $(E \check{\otimes} F)^*$; it follows that Definition 3.1.8(ii) is satisfied as well. Finally, if $G_\alpha \subset GL(n_\alpha, \mathbb{C})$ and $H_\alpha \subset GL(n_\beta, \mathbb{C})$ are finite, irreducible matrix groups, then $G_\alpha \otimes H_\beta \subset GL(n_\alpha n_\beta, \mathbb{C})$ is also a finite, irreducible matrix group. Hence, Definition 3.1.8(iii) is satisfied. $\quad\square$

Theorem 3.1.11 is no longer true if we replace the injective tensor product by the projective tensor product, as we will soon see.

Examples 3.1.12   (a) Let $(\Omega, \mathcal{S}, \mu)$ be a measure space, and let $p \in (1, \infty)$. Consider the collection of all families $\mathcal{T}$ consisting of finitely many, pairwise disjoint sets in $\mathcal{S}$ such that $\mu(A) \in (0, \infty)$ for each $A \in \mathcal{T}$. For two such families $\mathcal{T}_1$ and $\mathcal{T}_2$ define $\mathcal{T}_1 \prec \mathcal{T}_2$ if each member of $\mathcal{T}_1$ is the union of a subfamily of $\mathcal{T}_2$. For each $\mathcal{T}$, we have a corresponding finite, biorthogonal system:

$$\left\{ \left(\frac{1}{\mu(A)^{\frac{1}{p}}} \chi_A, \frac{1}{\mu(B)^{\frac{1}{q}}} \chi_B \right) : A, B \in \mathcal{T} \right\},$$

where $q \in (1, \infty)$ is dual to p, i.e. $\frac{1}{p} + \frac{1}{q} = 1$. Let $\theta_{\mathcal{T}} : \mathbb{M}_{n_{\mathcal{T}}} \to \mathcal{F}(L^p(\Omega, \mathcal{S}, \mu))$ the corresponding homomorphism. It is immediate that

$$\theta_{\mathcal{T}}(E_{n_{\mathcal{T}}}) \chi_A = \chi_A \quad \text{and} \quad \theta_{\mathcal{T}}(E_{n_{\mathcal{T}}})^* \chi_A = \chi_A \qquad (A \in \mathcal{S}, \{A\} \prec \mathcal{T}).$$

This implies that Definition 3.1.8(i) and (ii) are satisfied (Why?). Let $G_{\mathcal{T}}$ be the finite, irreducible matrix group introduced in Example 2.3.16. For $\sigma \in \mathfrak{S}_{n_{\mathcal{T}}}$ and $\epsilon \in \{-1, 1\}^{n_{\mathcal{T}}}$, and $\mathcal{T} = \{A_1, \ldots, A_{n_{\mathcal{T}}}\}$, we have:

$$
\begin{aligned}
\|\theta_{\mathcal{T}}(D_\epsilon A_\sigma)f\|_p^p &= \left\| \sum_{j=1}^{n_{\mathcal{T}}} \left(\frac{\epsilon_j}{\mu(A_j)^{\frac{1}{p}}} \int_{A_j} f \, d\mu \right) \frac{1}{\mu(A_{\sigma(j)})^{\frac{1}{q}}} \chi_{A_{\sigma(j)}} \right\|_p^p \\
&= \sum_{j=1}^{n_{\mathcal{T}}} \frac{1}{\mu(A_j)^{\frac{p}{q}}} \left| \int_{A_j} f \, d\mu \right|^p \\
&\leq \sum_{j=1}^{n_{\mathcal{T}}} \frac{1}{\mu(A_j)^{\frac{p}{q}}} \mu(A_j)^{\frac{p}{q}} \int_{A_j} |f|^p \, d\mu, \qquad \text{by Hölder's inequality,} \\
&= \sum_{j=1}^{n_{\mathcal{T}}} \int_{A_j} |f|^p \, d\mu \\
&= \|f\|_p^p \qquad (f \in L^p(\Omega, \mathcal{S}, \mu)).
\end{aligned}
$$

Consequently, $L^p(\Omega, \mathcal{S}, \mu)$ has property $(\mathbb{A})$, and, by Theorem 3.1.9, $\mathcal{A}(L^p(\Omega, \mathcal{S}, \mu))$ is amenable (in fact, 1-amenable).

(b) Similarly, $L^1(\Omega, \mathcal{S}, \mu)$ has property $(\mathbb{A})$, and $\mathcal{A}(L^1(\Omega, \mathcal{S}, \mu))$ is 1-amenable (Exercise 3.1.4).

(c) Let Ω be a locally compact Hausdorff space. Then $\mathcal{C}_0(\Omega)^* \cong M(\Omega)$ is an L^1-space for an approriate measure space: This follows from the Kakutani–Bohnenblust–Nakano theorem ([Al–B, Theorem 12.26]). Hence, $\mathcal{C}_0(\Omega)^*$ has property $(\mathbb{A})$ by (b) and so has $\mathcal{C}_0(\Omega)$ by Theorem 3.1.10.

(d) For any measure space $(\Omega, \mathcal{S}, \mu)$, the space $L^\infty(\Omega, \mathcal{S}, \mu)$ equipped with pointwise multiplication is a commutative C^*-algebra. By the Gelfand–Naimark theorem, there is a compact Hausdorff space K such that $L^\infty(\Omega, \mathcal{S}, \mu) \cong \mathcal{C}(K)$. It thus follows from (c) that $L^\infty(\Omega, \mathcal{S}, \mu)$ has property $(\mathbb{A})$.

(e) Let Ω be a locally compact Hausdorff space, and let E be a Banach space with property $(\mathbb{A})$ such that E^* has the Radon–Nikodým property. It follows from Theorem B.2.5 and Theorem 3.1.11 that $\mathcal{C}_0(\Omega, E)$ has property $(\mathbb{A})$.

Exercise 3.1.4 Verify Example 3.1.12(b). (*Hint*: First, treat the case of a finite measure space, and proceed as in Examples 3.1.12(a). Then "approximate" an arbitrary measure space with finite measure spaces.)

Remark 3.1.13 Let $\mathfrak{H}$ be an infinite-dimensional Hilbert space. By Examples 3.1.12(a), $\mathfrak{H}$ has property (A), and so has $\mathfrak{H} \check{\otimes} \mathfrak{H}$ by Theorem 3.1.11. However, as we have seen in Example 3.1.6, $\mathcal{A}(\mathfrak{H} \check{\otimes} \mathfrak{H})$ is not amenable, so that $(\mathfrak{H} \check{\otimes} \mathfrak{H})^* \cong \mathfrak{H} \hat{\otimes} \mathfrak{H}$ cannot have property (A). This shows at the same time that, in Theorem 3.1.10, the rôles of E and E^* cannot be interchanged and that, in Theorem 3.1.11, the injective tensor product cannot be replaced by the projective one.

Does every Banach space E such that $\mathcal{A}(E)$ is amenable have property (A)? It seems doubtful: By Exercise 3.1.3, this would mean that any amenable $\mathcal{A}(E)$ is already symmetrically amenable, which seems too good to be true.

Concluding this section, we want to give an example of a (otherwise very nice) Banach space E such that $\mathcal{A}(E)$ is not amenable. In fact, $E = \ell^p \oplus \ell^q$ will do if $p \in (2, \infty)$ and $q \in (1, \infty)$ is such that $\frac{1}{p} + \frac{1}{q} = 1$.

We first need some purely Banach algebraic prerequisites:

Definition 3.1.14 A Banach algebra $\mathfrak{A}$ has *trivial virtual center* if, for each $X \in \mathfrak{A}^{**}$ that commutes with every $a \in \mathfrak{A}$, there is $\lambda \in \mathbb{C}$ such that

$$a \cdot X = \lambda a = X \cdot a \qquad (a \in \mathfrak{A}).$$

For the notion of the *multiplier* (or: *double centralizer*) *algebra* $\mathcal{M}(\mathfrak{A})$ of a Banach algebra $\mathfrak{A}$, we refer to [Pal].

Proposition 3.1.15 *Let $\mathfrak{A}$ be an amenable Banach algebra, let $P_1 \in \mathcal{M}(\mathfrak{A})$ be a projection, and let $P_2 := \mathrm{id}_{\mathfrak{A}} - P_1$. Then, if both $P_1\mathfrak{A}P_1$ and $P_2\mathfrak{A}P_2$ have trivial virtual center and if not both $P_1\mathfrak{A}P_2$ and $P_2\mathfrak{A}P_1$ are zero, one of the following holds:*

(i) $\Delta_{\mathfrak{A}}(P_1\mathfrak{A}P_2 \hat{\otimes} P_2\mathfrak{A}P_1) = P_1\mathfrak{A}P_1$;
(ii) $\Delta_{\mathfrak{A}}(P_2\mathfrak{A}P_1 \hat{\otimes} P_1\mathfrak{A}P_2) = P_2\mathfrak{A}P_2$.

Proof For notational simplicity, let $\mathfrak{A}_{j,k} := P_j\mathfrak{A}P_k$ for $j, k = 1, 2$. For $j, k = 1, 2$ with $k \neq j$, let $\mathfrak{A}_j^\circ := \Delta_{\mathfrak{A}}(\mathfrak{A}_{j,k} \hat{\otimes} \mathfrak{A}_{k,j})$ be equipped with the quotient norm $\| \cdot \|_j^\circ$. Note that $\mathfrak{A}_j^\circ$ is an ideal in $\mathfrak{A}_{j,j}$ for $j = 1, 2$. Let

$$E := \{a \in \mathfrak{A} : P_j a P_j \in \mathfrak{A}_j^\circ, \, j = 1, 2\}.$$

For $x \in E$, let

$$|||x||| = \max\{\|P_1 x P_1\|_1^\circ, \|P_1 x P_2\|, \|P_2 x P_1\|, \|P_2 x P_2\|_2^\circ\}.$$

It is routinely checked (Do it!) that $(E, ||| \cdot |||)$ is a Banach $\mathfrak{A}$-bimodule. Define

$$D : \mathfrak{A} \to E, \quad a \mapsto P_1 a P_2 - P_2 a P_1 (= P_1 a - a P_1).$$

Then D is a derivation, and from the amenability of $\mathfrak{A}$, it follows that there is $X \in E^{**}$ such that $D = \mathrm{ad}_X$.

For $j, k = 1, 2$, let $X_{j,k} = P_j X P_k \in P_j E^{**} P_k$. Note that $X_{j,j} \in (\mathfrak{A}_j^\circ)^{**}$. If $a \in \mathfrak{A}_{j,j}$ for some $j \in \{1, 2\}$, we have $Da = 0$, i.e. $a \cdot X - X \cdot a = 0$ and thus $a \cdot X_{j,j} - X_{j,j} \cdot a = 0$. Since the bitranspose of the inclusion $\mathfrak{A}_j^\circ \hookrightarrow \mathfrak{A}_{j,j}$ embeds $(\mathfrak{A}_j^\circ)^{**}$ into $\mathfrak{A}_{j,j}^{**}$, and since $\mathfrak{A}_{j,j}$ has trivial virtual center, there is $\lambda_j \in \mathbb{C}$ such that

$$a \cdot X_{j,j} = \lambda_j a = X_{j,j} \cdot a \qquad (a \in \mathfrak{A}_{j,j}).$$

Without loss of generality suppose that $\mathfrak{A}_{1,2} \neq \{0\}$. Choose $a \in \mathfrak{A}_{1,2} \setminus \{0\}$. It is not hard to show (Exercise 3.1.5 below) that $\mathfrak{A}_{1,2}$ is a pseudo-unital left Banach $\mathfrak{A}_{1,1}$-module as well as a pseudo-unital right Banach $\mathfrak{A}_{2,2}$-module. This means that there are $a' \in \mathfrak{A}_{1,2}$, $b \in \mathfrak{A}_{1,1}$ and $c \in \mathfrak{A}_{2,2}$ such that $a = ba'c$. Since

$$a = P_1 a P_2 - P_2 a P_1 = a \cdot X - X \cdot a = a \cdot X_{2,2} - X_{1,1} \cdot a,$$

this yields

$$a = ba'c \cdot X_{2,2} - X_{1,1} \cdot ba'c = \lambda_2 ba'c - \lambda_1 ba'c = (\lambda_2 - \lambda_1)a,$$

so that $\lambda_2 - \lambda_1 = 1$.

Without loss of generality suppose that $\lambda_1 \neq 0$. Then $a \cdot \lambda_1^{-1} X_{1,1} = a$ for all $a \in \mathfrak{A}_{1,1}$. With essentially the same argument as in the proof of Proposition 2.2.1 (Work it out if you have doubts!), we obtain a bounded (with respect to $\| \cdot \|_1^\circ$) net $(e_\alpha)_\alpha$ in $\mathfrak{A}_1^\circ$ which is a right approximate identity for $\mathfrak{A}_{1,1}$. Let $C := \sup_\alpha \|e_\alpha\|_1^\circ$. Then, for each $a \in \mathfrak{A}_{1,1}$ and for each $\epsilon > 0$, there is $b \mathfrak{A}_1^\circ$ with $\|a - b\| < \epsilon$ and $\|b\|_1^\circ \leq 2C\|a\|$ — just choose $b := ae_\alpha$ for an appropriate index α. Inductively, (How precisely?) we thus obtain, for each $a \in \mathfrak{A}_{1,1}$, a sequence $(b_n)_{n=1}^\infty$ in $\mathfrak{A}_1^\circ$ with $\sum_{n=1}^\infty \|b_n\|_1^\circ < \infty$ and $\sum_{n=1}^\infty b_n = a$. This means that $\mathfrak{A}_1^\circ = \mathfrak{A}_{1,1}$ as claimed.   □

Exercise 3.1.5 Verify that $\mathfrak{A}_{1,2}$ is a pseudo-unital left Banach $\mathfrak{A}_{1,1}$ module and a pseudo-unital right Banach $\mathfrak{A}_{2,2}$-module. (*Hint:* Given a bounded approximate identity $(e_\alpha)_\alpha$ for $\mathfrak{A}$, show that $(P_1 e_\alpha P_1)_\alpha$ is a bounded left approximate identity for $\mathfrak{A}_{1,2}$, and apply Cohen's factorization theorem ([B–D, Theorem 11.10]). Proceed similarly with $(P_2 e_\alpha P_2)_\alpha$.)

The situation we want to apply Proposition 3.1.15 to is the following: Given two Banach spaces E and F, let $\mathfrak{A} := \mathcal{A}(E \oplus F)$, and let $P_1 \in \mathcal{L}(E \oplus F)$ be the projection onto E corresponding to the direct sum decomposition. Of course, in order to be able to apply Proposition 3.1.15, we require that both $\mathcal{A}(E)$ and $\mathcal{A}(F)$ have trivial virtual center:

Exercise 3.1.6 Let E be a Banach space.

(i) Show that $\mathcal{A}(E)$ is *topologically simple*, i.e. it has no closed (two-sided) ideals other than $\{0\}$ and $\mathcal{A}(E)$. (*Hint:* Show first that $\mathcal{F}(E)$ is (algebraically) simple, i.e. it has no ideals other than $\{0\}$ and itself.)

(ii) Show that $\mathcal{A}(E)$ has trivial virtual center. (*Hint:* Let $X \in \mathcal{A}(E)^{**}$ be such that $T \cdot X = X \cdot T$ for all $T \in \mathcal{A}(E)$. Pick a rank one projection P on E and show that there is $\lambda \in \mathbb{C}$ such that $P \cdot X = \lambda P$. Then use (a) to conclude that $T \cdot X = \lambda T = X \cdot T$ for all $T \in \mathcal{A}(E)$.)

Corollary 3.1.16 *Let E and F be Banach spaces such that $\mathcal{A}(E \oplus F)$ is amenable. Then one of the canonical maps*

$$\mathcal{A}(F, E) \hat{\otimes} \mathcal{A}(E, F) \to \mathcal{A}(E) \qquad and \qquad \mathcal{A}(E, F) \hat{\otimes} \mathcal{A}(F, E) \to \mathcal{A}(F)$$

is surjective.

Example 3.1.17 Let $p, q \in (1, \infty)$ such that $\mathcal{A}(\ell^p \oplus \ell^q)$ is amenable. Then Corollary 3.1.16 implies that one of the maps

$$\mathcal{A}(\ell^q, \ell^p) \hat{\otimes} \mathcal{A}(\ell^p, \ell^q) \to \mathcal{A}(\ell^p) \tag{3.5}$$

and

$$\mathcal{A}(\ell^p, \ell^q) \hat{\otimes} \mathcal{A}(\ell^q, \ell^p) \to \mathcal{A}(\ell^q) \tag{3.6}$$

is surjective.

Assume first that (3.5) is surjective. Then, by the open mapping, theorem, there is $C > 0$ such that, for each $T \in \mathcal{A}(\ell^p)$ there are sequences $(S_n)_{n=1}^{\infty}$ in $\mathcal{A}(\ell^q, \ell^p)$ and $(R_n)_{n=1}^{\infty}$ in $\mathcal{A}(\ell^p, \ell^q)$ with $\sum_{n=1}^{\infty} \|S_n\| \|R_n\| \leq C \|T\|$ and $T = \sum_{n=1}^{\infty} S_n R_n$. Since $\ell^p \cong \ell^p\text{-}\bigoplus_{n=1}^{\infty} \ell^p$ (and similarly for ℓ^q), it follows (How exactly?) that, for each $T \in \mathcal{A}(\ell^p)$ there are $S \in \mathcal{A}(\ell^q, \ell^p)$ and $R \in \mathcal{A}(\ell^p, \ell^q)$ with $\|S\| \|R\| \leq C \|T\|$ and $T = SR$.

For $n \in \mathbb{N}$, let $P_n \in \mathcal{A}(\ell^p)$ be the projection onto the first n coordinates. Let $S_n \in \mathcal{A}(\ell^q, \ell^p)$ and $R_n \in \mathcal{A}(\ell^p, \ell^q)$ with $\|S_n\| \|R_n\| \leq C$ and $P_n = S_n R_n$. Let $Q_n := R_n S_n$; certainly, $Q_n \in \mathcal{A}(\ell^q)$ is a projection with $\|Q_n\| \leq C$. Define $\tau_n := Q_n R_n P_n$ and $\sigma_n := P_n S_n Q_n$. Then τ_n is an isomorphism from $P_n \ell^p$ onto $Q_n \ell^q$ whose inverse is $\sigma_n|_{Q \ell^q}$. It is immediate from the definitions of τ_n and σ_n that $\|\tau_n\| \|\tau_n^{-1}|_{Q_n \ell^q}\| \leq C^3$. Since $\bigcup \{P_n \ell^p : n \in \mathbb{N}\}$ is dense in ℓ^p, it follows from Theorem C.3.7 that ℓ^p is finitely representable (Definition C.3.2) in ℓ^q. By Corollary C.3.10, this is impossible for $p > 2$ and $q \leq 2$, so that (3.5) cannot be surjective in this case.

Assume now that (3.6) is surjective for $p > 2$ and q such that $\frac{1}{p} + \frac{1}{q} = 1$; in particular, $q < 2$. Then, for each $T \in \mathcal{A}(\ell^p)$, there are $(S_n)_{n=1}^{\infty}$ in $\mathcal{A}(\ell^p, \ell^q)$ and $(R_n)_{n=1}^{\infty}$ in $\mathcal{A}(\ell^q, \ell^p)$ with $\sum_{n=1}^{\infty} \|S_n\| \|R_n\| < \infty$ and $T^* = \sum_{n=1}^{\infty} S_n R_n$. Taking the transpose, we obtain $T = \sum_{n=1}^{\infty} R_n^* S_n^*$, so that (3.5) is a surjection. But as we have just seen, this is impossible.

All in all, if $p > 2$, and if $q \in (1, \infty)$ is such that $\frac{1}{p} + \frac{1}{q} = 1$, then neither (3.5) nor (3.6) is surjective, so that $\mathcal{A}(\ell^p \oplus \ell^q)$ cannot be amenable.

Nevertheless, $\mathcal{A}(\ell^p \oplus \ell^q)$ has a bounded approximate identity for all $p, q \in [1, \infty)$.

3.2 A commutative, radical, amenable Banach algebra

So far, all the examples of amenable Banach algebras we have encountered have been semisimple. However, the class of amenable Banach algebras is much richer: We shall now present C. J. Read's construction of a commutative, radical, amenable Banach algebra.

The idea underlying the construction is simple: We construct a sequence of finite-dimensional, nilpotent, commutative Banach algebras such that the inductive limit will have an approximate diagonal (bounded by 1). Nevertheless, the technical difficulties of the construction are considerable.

Definition 3.2.1 Let $\mathfrak{B}$ be a commutative Banach algebra, let $\mathfrak{A}$ be a subalgebra of $\mathfrak{B}$, and let $\delta > 0$. A *metric, δ-approximate unit* for $\mathfrak{A}$ is an element $u \in \mathfrak{B}$ with $\|u\| \leq 1$ such that

$$\|ua - a\| \leq \delta \|a\| \qquad (a \in \mathfrak{A}).$$

For the sake of brevity, we shall call a finite-dimensional, nilpotent, commutative algebra an *FDNC-algebra*. A Banach FDNC-algebra is a Banach algebra which is also an FDNC-algebra, and a *Banach FDNC-extension* of a Banach FDNC-algebra $\mathfrak{A}$ is a Banach FDNC-algebra $\mathfrak{B}$ containing $\mathfrak{A}$ as a subalgebra such that the canonical embedding is an isometry.

Lemma 3.2.2 *Let $\mathfrak{A}$ be a Banach FDNC-algebra, and let $\delta > 0$. Then there is a Banach FDNC-extension of $\mathfrak{A}$ that contains a metric, δ-approximate unit for $\mathfrak{A}$.*

We postpone the proof.

Definition 3.2.3 Let $\mathfrak{A}$ be a commutative Banach algebra, let $a, u \in \mathfrak{A}$ be such that $\|u\| \leq 1$, and let $\zeta, \eta > 0$.

(i) An element $\mathbf{m} \in \mathfrak{A} \hat{\otimes} \mathfrak{A}$ with $\|\mathbf{m}\| \leq 1$ is called a *metric, ζ-approximate commutatant* for a with image u if $\Delta_{\mathfrak{A}} \mathbf{m} = u$ and

$$\|a \cdot \mathbf{m} - \mathbf{m} \cdot a\| \leq \zeta \|a\|.$$

(ii) An element $\mathbf{m} \in \mathfrak{A} \hat{\otimes} \mathfrak{A}$ with $\|\mathbf{m}\| \leq 1$ is called a *weak metric, (η, ζ)-approximate commutant* for a with image u if $\Delta_{\mathfrak{A}} \mathbf{m} = u$, and if there is $x \in \mathfrak{A}$ such that

$$\|x - a\| \leq \eta \|a\| \qquad \text{and} \qquad \|x \cdot \mathbf{m} - \mathbf{m} \cdot x\| \leq \zeta \|a\|.$$

There is an analogue of Lemma 3.2.2 for weak metric, approximate commutants:

Lemma 3.2.4 *Let $\mathfrak{A}$ be a Banach FDNC-algebra, let $u \in \mathfrak{A}$ with $\|u\| \leq 1$, let $\zeta > 0$, and let $\eta \in \left[\frac{9}{10}, 1\right]$. Then for each $a \in \mathfrak{A}$, there is a Banach FDNC-extension $\mathfrak{B}$ of $\mathfrak{A}$ such that $\mathfrak{B} \hat{\otimes} \mathfrak{B}$ contains a weak metric, (η, ζ)-approximate commutant for a with image u.*

Again, we postpone the proof.

Once Lemmas 3.2.2 and 3.2.4 have been established, the remainder of the construction is relatively easy.

Lemma 3.2.5 *Let $\mathfrak{A}$ be a Banach FDNC-algebra, let $u \in \mathfrak{A}$ be such that $\|u\| \leq 1$, let $\zeta > 0$, and let $\eta \in \left[\frac{9}{10}, 1\right]$. Then, for each $a \in \mathfrak{A}$ with $\|a\| \leq 1$ and for each $n \in \mathbb{N}$, there is a Banach FDNC-extension $\mathfrak{B}_n$ of $\mathfrak{A}$ such that $\mathfrak{B}_n \hat{\otimes} \mathfrak{B}_n$ contains a weak metric, $(\eta^n, n\zeta)$-approximate commutant for a with image u^n.*

Proof We proceed by induction on n. For $n = 1$, the claim follows immediately from Lemma 3.2.4.

Suppose that $\mathfrak{B}_n$ as required has already been found. Then there is $x \in \mathfrak{B}_n$ and $\mathbf{m}_n \in \mathfrak{B}_n \hat{\otimes} \mathfrak{B}_n$ with $\Delta_{\mathfrak{B}_n} \mathbf{m}_n = u^n$ such that

$$\|x - a\| \leq \eta^n \qquad \text{and} \qquad \|x \cdot \mathbf{m}_n - \mathbf{m}_n \cdot x\| \leq n\zeta.$$

Let $b := \frac{1}{\eta^n}(x - a)$, so that $\|b\| \leq 1$. Again by Lemma 3.2.4, there is a Banach FDNC-extension $\mathfrak{B}_{n+1}$ of $\mathfrak{B}_n$ containing a weak metric, (η, ζ)-approximate commutant $\mathbf{m} \in \mathfrak{B}_{n+1} \hat{\otimes} \mathfrak{B}_{n+1}$ for b with image u. In particular, there is $y \in \mathfrak{B}_{n+1}$ such that

$$\|y - b\| \leq \eta \qquad \text{and} \qquad \|y \cdot \mathbf{m} - \mathbf{m} \cdot y\| \leq \zeta.$$

Let $\mathbf{m}_{n+1} := \mathbf{m} \bullet \mathbf{m}_n$. Then $\|\mathbf{m}_n\| \le 1$, and $\Delta_{\mathfrak{B}_{n+1}} \mathbf{m}_{n+1} = u^{n+1}$. Let $z := x - \eta^n y$. Then

$$\|z - a\| = \|x - a - \eta^n y\| = \eta^n \|b - y\| \le \eta^{n+1}$$

and

$$
\begin{aligned}
\|z \cdot \mathbf{m}_{n+1} - \mathbf{m}_{n+1} \cdot z\| &\le \|x \cdot \mathbf{m}_{n+1} - \mathbf{m}_{n+1} \cdot x\| + \|\eta^n y \cdot \mathbf{m}_{n+1} - \mathbf{m}_{n+1} \cdot \eta^n y\| \\
&\le \|\mathbf{m}\|\|x \cdot \mathbf{m}_n - \mathbf{m}_n \cdot x\| + \|\mathbf{m}_n\|\|\eta^n y \cdot \mathbf{m} - \mathbf{m} \cdot \eta^n y\| \\
&\le \|x \cdot \mathbf{m}_n - \mathbf{m}_n \cdot x\| + \|\eta^n y \cdot \mathbf{m} - \mathbf{m} \cdot \eta^n y\| \\
&\le n\zeta + \eta^n \zeta \\
&\le (n+1)\zeta,
\end{aligned}
$$

so that $\mathbf{m}_{n+1}$ is a weak metric, $(\eta^{n+1}, (n+1)\zeta)$-approximate commutant for a with image u^{n+1}. $\square$

Lemma 3.2.6 *Let $\mathfrak{A}$ be a Banach FDNC-algebra, and let $\delta > 0$. Then there is a Banach FDNC-extension $\mathfrak{B}$ of $\mathfrak{A}$ such that:*

(i) *$\mathfrak{B}$ contains a metric, δ-approximate unit u for $\mathfrak{A}$.*

(ii) *$\mathfrak{B} \hat{\otimes} \mathfrak{B}$ contains an element $\mathbf{m}$ which is a metric, δ-approximate commutant for each $a \in \mathfrak{A}$ with image u.*

Proof Choose a basis $a_1, \dots, a_n$ of $\mathfrak{A}$, and choose $C \ge 0$ such that, for each $a \in \mathfrak{A}$ with $a = \sum_{j=1}^n \lambda_j a_j$, we have:

$$\sum_{j=1}^n |\lambda_j| \le C\|a\|$$

(Why does such a constant C exist?). Further, choose $\eta \in \left[\frac{9}{10}, 1\right)$ and $n \in \mathbb{N}$ such that $2\eta^n C \le \frac{\delta}{2}$. Then choose $\zeta > 0$ such that $n\zeta C < \frac{\delta}{2}$.

By Lemma 3.2.2, there is a Banach FDNC-extension $\mathfrak{B}_0$ of $\mathfrak{A}$ containing a metric, $\frac{\delta}{n^2}$-approximate unit u_0 for $\mathfrak{A}$. Then use Lemma 3.2.5, to find a Banach FDNC-extension $\mathfrak{B}_1$ of $\mathfrak{B}_0$ such that $\mathfrak{B}_1 \hat{\otimes} \mathfrak{B}_1$ contains a weak metric, $(\eta^n, n\zeta)$-approximate commutant $\mathbf{m}_1$ for a_1 with image u_0^n. Successively we obtain, from Lemma 3.2.5, Banach FDNC-algebras $\mathfrak{B}_2, \dots, \mathfrak{B}_n$ such that, for $j = 2, \dots, n$,

– the algebra $\mathfrak{B}_j$ is a Banach FDNC-extension of $\mathfrak{B}_{j-1}$, and

– $\mathfrak{B}_j \hat{\otimes} \mathfrak{B}_j$ contains a metric, $(\eta^n, n\zeta)$-approximate commutant $\mathbf{m}_j$ for a_j with image u_0^n.

Let $\mathfrak{B} := \mathfrak{B}_{n+1}$, and let $\mathbf{m} := \mathbf{m}_1 \bullet \cdots \bullet \mathbf{m}_n$. Then, clearly, $\|\mathbf{m}\| \le 1$, and for $u := u_0^{n^2} = \Delta_{\mathfrak{B}} \mathbf{m}$, we have

$$\|ua - a\| \le \sum_{j=0}^{n^2 - 1} \|u_0^{j+1} a - u_0^j a\| \le n^2 \frac{\delta}{n^2} \|a\| = \delta\|a\| \qquad (a \in \mathfrak{A}),$$

so that u is a metric, δ-approximate unit for $\mathfrak{A}$.

Let $a = \sum_{j=1}^n \lambda_j a_j \in \mathfrak{A}$, and note that

$$a \cdot \mathbf{m} - \mathbf{m} \cdot a = \sum_{j=1}^{n} \lambda_j (a_j \cdot \mathbf{m}_j - \mathbf{m}_j \cdot a_j) \bullet \prod_{k \neq j} \mathbf{m}_k.$$

It follows from the choice of C that

$$\|a \cdot \mathbf{m} - \mathbf{m} \cdot a\| \leq \sum_{j=1}^{n} |\lambda_j| \|a_j \cdot \mathbf{m}_j - \mathbf{m}_j \cdot a_j\| \leq C\|a\| \max_{j=1,\dots,n} \|a_j \cdot \mathbf{m}_j - \mathbf{m}_j \cdot a_j\|. \quad (3.7)$$

For each $j = 1, \dots, n$, there is $x_j \in \mathfrak{B}_j \subset \mathfrak{B}$ such that

$$\|x_j - a_j\| \leq \eta^n \qquad \text{and} \qquad \|x_j \cdot \mathbf{m}_j - \mathbf{m}_j \cdot x_j\| \leq n\zeta.$$

Since, clearly,

$$\|(x_j - a_j) \cdot \mathbf{m}_j - \mathbf{m}_j \cdot (x_j - a_j)\| \leq 2\|x_j - a_j\| \|\mathbf{m}_j\| \leq 2\eta^n,$$

it follows that

$$\|a_j \cdot \mathbf{m}_j - \mathbf{m}_j \cdot a_j\| \leq 2\eta^n + n\zeta \qquad (j = 1, \dots, n).$$

Together with (3.7), this implies that

$$\|a \cdot \mathbf{m} - \mathbf{m} \cdot a\| \leq C\|a\|(2\eta^n + n\zeta) \leq \delta\|a\|$$

(Remember how η and ζ were chosen?). Since $a \in \mathfrak{A}$ was arbitrary, this means that $\mathbf{m}$ is indeed a metric, δ-approximate commutant for each $a \in \mathfrak{A}$. $\square$

We are now in a position to prove the main theorem of this section:

Theorem 3.2.7 *A Banach FDNC-algebra can be isometrically embedded into a commutative, radical, 1-amenable Banach algebra.*

Proof Let $\mathfrak{A}$ be Banach FDNC-algebra. Choose a sequence $(\delta_n)_{n=1}^{\infty}$ of positive numbers such that $\lim_{n\to\infty} \delta_n = 0$. Applying Lemma 3.2.6, we inductively obtain a sequence $(\mathfrak{A}_n)_{n=1}^{\infty}$ of Banach FDNC-algebras with the following properties:

- $\mathfrak{A}_1 = \mathfrak{A}$.
- For each $n \in \mathbb{N}$, the algebra $\mathfrak{A}_{n+1}$ is a Banach FDNC-extension of $\mathfrak{A}_n$.
- For each $n \in \mathbb{N}$, the tensor product $\mathfrak{A}_{n+1} \hat{\otimes} \mathfrak{A}_{n+1}$ contains an element $\mathbf{m}_n$ which is a metric, δ_n-approximate commutant for each $a \in \mathfrak{A}_n$ and such that $\Delta_{\mathfrak{A}_{n+1}} \mathbf{m}_n$ is a metric, δ_n-approximate unit for $\mathfrak{A}_n$.

Let $\mathfrak{R}$ be the completion of $\bigcup_{n=1}^{\infty} \mathfrak{A}_n$. Let $a \in \bigcup_{n=1}^{\infty} \mathfrak{A}_n$ be arbitrary. Then there is $N \in \mathbb{N}$ such that $a \in \mathfrak{A}_N$. It follows that, for $n \geq N$,

$$\|(\Delta_{\mathfrak{R}} \mathbf{m}_n)a - a\| \leq \delta_n \|a\| \qquad \text{and} \qquad \|a \cdot \mathbf{m}_n - \mathbf{m}_n \cdot a\| \leq \delta_n \|a\|,$$

so that $(\mathbf{m}_n)_{n=1}^{\infty}$ is an approximate diagonal for $\mathfrak{R}$ (bounded by 1). Hence, $\mathfrak{R}$ is amenable.

By construction, the set of nilpotent elements is dense in $\mathfrak{R}$. Since $\mathfrak{R}$ is commutative, this means that every element of $\mathfrak{R}$ is quasi-nilpotent, i.e. $\mathfrak{R}$ is radical. $\square$

Corollary 3.2.8 *There is a non-zero, commutative, radical, amenable Banach algebra.*

What remains to be done, of course, is to prove Lemmas 3.2.2 and 3.2.4. And this is, in fact, where the *real* work for the proof of Theorem 3.2.7 will be done.

We begin with the proof of Lemma 3.2.2 because it is easier and will thus (hopefully) better clarify the idea common to the proofs of both Lemma 3.2.2 and Lemma 3.2.4. As a preparation for both proofs we give an easy exercise:

Exercise 3.2.1 Let $\mathfrak{A}$ be a commutative, unital, normed algebra, and let S be a subsemigroup of the multiplicative semigroup of $\mathfrak{A}$ containing $e_{\mathfrak{A}}$ and bounded by $C \geq 0$.

(i) Show that

$$\|a\|_S := \inf \left\{ \sum_{j=1}^{n} \|a_j\| : n \in \mathbb{N},\ a_1, \ldots, a_n \in \mathfrak{A},\ s_1, \ldots, s_n \in S,\ a = \sum_{j=1}^{n} a_j s_j \right\} \qquad (a \in \mathfrak{A})$$

defines an algebra norm on $\mathfrak{A}$ such that

$$\frac{1}{C}\|a\| \leq \|a\|_S \leq \|a\| \quad (a \in \mathfrak{A}) \qquad \text{and} \qquad \|s\|_S \leq 1 \quad (s \in S).$$

(ii) Show that $\phi \in \mathfrak{A}^*$ has norm at most one with respect to $\|\cdot\|_S$ if and only if $|\langle as, \phi \rangle| \leq 1$ for all $s \in S$ and $a \in \mathfrak{A}$ with $\|a\| = 1$.

Proof of Lemma 3.2.2 Let the unitization $\mathfrak{A}^\#$ of $\mathfrak{A}$ be equipped with the ℓ^1-norm, i.e.

$$\|\lambda e + a\| := |\lambda| + \|a\| \qquad (\lambda \in \mathbb{C},\ a \in \mathfrak{A}). \tag{3.8}$$

Consider the algebra $\mathfrak{A}^\#[X]$ of all polynomials in one variable with coefficients in $\mathfrak{A}^\#$. We use $\mathfrak{A}^\#[X]_0$ to denote the subalgebra consisting of those polynomials whose constant term lies in $\mathfrak{A}$. For $N \in \mathbb{N}$, let I_N be the principal ideal in $\mathfrak{A}^\#[X]$ generated by X^N. Obviously, $I_N \subset \mathfrak{A}^\#[X]_0$ so that $\mathfrak{B} := \mathfrak{A}^\#[X]_0/I_N$ is well defined. It is clear from this definition that $\mathfrak{B}$ is an FDNC-algebra. In fact, if $d \in \mathbb{N}$ is such that $a^d = 0$ for all $a \in \mathfrak{A}$, then $b^{N+d-1} = 0$ for all $b \in \mathfrak{B}$ (check this for yourself). For $a_0, \ldots, a_{N-1} \in \mathfrak{A}^\#$ define

$$\left\| \sum_{j=0}^{N-1} a_j X^j \right\| := \sum_{j=0}^{N-1} \|a_j\|.$$

Clearly, this defines an algebra norm on $\mathfrak{A}^\#[X]/I_N \cong \mathfrak{B}^\#$ that extends the norm on $\mathfrak{A}^\#$. Let

$$S := \left\{ \frac{1}{\delta^k} \prod_{j=1}^{k} (a_j X - a_j) : k \in \mathbb{N}_0,\ a_1 \ldots, a_k \in \mathfrak{A},\ \|a_1\| = \cdots = \|a_k\| = 1 \right\} \subset \mathfrak{B}^\#.$$

Then S is a subsemigroup of the multiplicative semigroup of $\mathfrak{B}^\#$ and contains the identity of $\mathfrak{B}^\#$. Let $n := \dim \mathfrak{A}$. Then $a_1 \cdots a_{nd} = 0$ for all $a_1, \ldots, a_{nd} \in \mathfrak{A}$ (Why?). It follows that

$$\|s\| \leq \left(\frac{2}{\delta} \right)^{nd-1} \qquad (s \in S)$$

(at least if we suppose that $\delta \leq 2$, which we may do). Let $\|\cdot\|_S$ be the algebra norm on $\mathfrak{B}^\#$ defined as in Exercise 3.2.1. Then, for each $a \in \mathfrak{A}$ with $\|a\| = 1$, we have $\frac{1}{\delta}(aX - a) \in S$, so that

$$\|aX - a\|_{\mathcal{S}} \leq \delta \|a\|_{\mathcal{S}} \qquad (a \in \mathfrak{A}).$$

Since $\|X\|_{\mathcal{S}} \leq \|X\| = 1$, it follows that X (or rather its coset in $\mathfrak{B}$) is metric, δ-approximate unit for $(\mathfrak{A}, \| \cdot \|_{\mathcal{S}})$.

We shall now see that, for sufficiently large N, the norm $\| \cdot \|_{\mathcal{S}}$ coincides on $\mathfrak{A}$ with the given norm, so that $\mathfrak{B}$ becomes a Banach FDNC-extension of $\mathfrak{A}$. We do so by dualizing: We show that, for N sufficiently large, each $\phi \in \mathfrak{A}^*$ with $\|\phi\| = 1$ has a norm one extension to $(\mathfrak{B}^\#, \| \cdot \|_{\mathcal{S}})$. First extend ϕ to $\mathfrak{A}^\#$ by letting $\langle e, \phi \rangle = 0$ and then define $\bar{\phi} \colon \mathfrak{B}^\# \to \mathbb{C}$ through

$$\langle aX^k, \bar{\phi} \rangle = \left(1 - \frac{k}{N}\right) \langle a, \phi \rangle \qquad (a \in \mathfrak{A}^\#, \, k = 0, \ldots, N-1)$$

By Exercise 3.2.1(ii), it is sufficient to show that $|\langle bs, \bar{\phi} \rangle| \leq 1$ for all $s \in S$ and $b \in \mathfrak{B}^\#$ with $\|b\| = 1$. Without loss of generality (Really?), we can suppose that b is of the form aX^l with $a \in \mathfrak{A}^\#$ such that $\|a\| = 1$ and $l \in \{0, \ldots, N-1\}$. The element $c := bs$ is then of the form

$$c = \frac{1}{\delta^k} aX^l \prod_{j=1}^{k} (a_j X - a_j), \tag{3.9}$$

where $a_1, \ldots, a_k \in \mathfrak{A}$ are such that $\|a_1\| = \cdots = \|a_n\| = 1$. If $k = 0$, it is easy to see that $|\langle c, \bar{\phi} \rangle| \leq 1$ (How?). Thus, let $k \in \{0, \ldots, dn\}$.

Case 1: $k + l < N$. In this case, the highest power of X occurring in c is $N-1$. Hence, with $\tilde{a} = a \prod_{j=1}^{k} a_j$, we have

$$\langle c, \bar{\phi} \rangle = \frac{1}{\delta^k} \langle \tilde{a}, \phi \rangle \sum_{\nu=0}^{k} (-1)^\nu \binom{k}{\nu} \left(1 - \frac{\nu + l}{N}\right). \tag{3.10}$$

For $k > 1$, we have

$$\sum_{\nu=0}^{k} \binom{k}{\nu} (-1)^\nu = \sum_{\nu=0}^{k} \nu \binom{k}{\nu} (-1)^\nu = 0,$$

so that (3.10) becomes

$$\langle c, \bar{\phi} \rangle = \begin{cases} \dfrac{1}{\delta N} \langle \tilde{a}, \phi \rangle, & k = 1, \\ 0, & k > 1. \end{cases}$$

For $N > \frac{1}{\delta}$, we thus have $|\langle c, \bar{\phi} \rangle| \leq 1$.

Case 2: $k + l \geq N$. Since $k < dn$, for each $j = 0, \ldots, k$, we have either $\tilde{a} X^{j+l} \in I_N$ (if $j + l \geq N$) or, in case $j + l < N$,

$$|\langle \tilde{a} X^{j+l}, \bar{\phi} \rangle| = \left(1 - \frac{j+l}{N}\right) |\langle \tilde{a}, \phi \rangle| \leq \frac{dn}{N}.$$

Using (3.9) to compute $\langle c, \bar{\phi} \rangle$, we thus obtain

$$|\langle c, \bar{\phi} \rangle| \leq \frac{1}{\delta^k} \sum_{\nu=0}^{k} \binom{k}{\nu} \frac{dn}{N} = \left(\frac{2}{\delta}\right)^k \frac{dn}{N} \leq \left(\frac{2}{\delta}\right)^{dn} \frac{dn}{N},$$

where the last inequality holds due to the fact that without loss of generality $\delta \leq 2$. Choosing $N > \left(\frac{2}{\delta}\right)^{dn} dn$, we obtain again that $|\langle c, \bar{\phi} \rangle| \leq 1$.   □ If you now think that the proof of Lemma 3.2.2 was hard, you haven't seen yet what lies ahead ...

Proof of Lemma 3.2.4   First of all, there is no loss of generality if we suppose that $\|a\| = 1$. The idea of the proof is similar to that of the proof of Lemma 3.2.2: Equip $\mathfrak{A}^\#$ with the norm defined in (3.8), and consider the algebra $\mathfrak{A}^\#[X_1, \ldots, X_N]$ of polynomials in N variables with coefficients in $\mathfrak{A}^\#$. We write $\mathfrak{A}^\#[X_1, \ldots, X_N]_0$ for the subalgebra of $\mathfrak{A}^\#[X_1, \ldots, X_N]$ consisiting of those polynomials whose constant term lies in $\mathfrak{A}$. We use the customary multiindex notation, i.e. for $\alpha = (\alpha_1, \ldots, \alpha_N) \in \mathbb{N}_0^N$, we let $X^\alpha := X_1^{\alpha_1} \cdots X_N^{\alpha_N}$ and write $|\alpha|$ for $\alpha_1 + \cdots + \alpha_N$; one particular multiindex is $\mathbf{1} = (1, \ldots, 1)$. Our Banach FDNC-extension $\mathfrak{B}$ of $\mathfrak{A}$ will be a quotient of $\mathfrak{A}^\#[X_1, \ldots, X_N]_0$.

Let I_0 be the ideal of $\mathfrak{A}^\#[X_1, \ldots, X_N]$ generated by the set $\{X^\alpha : |\alpha| \geq N^2 d\}$ (where $d \in \mathbb{N}$ is again such that $\tilde{a}^d = 0$ for each $\tilde{a} \in \mathfrak{A}$). Furthermore, let I_1 be the principal ideal of $\mathfrak{A}^\#[X_1, \ldots, X_N]$ generated by $X^{N\mathbf{1}} - u$. Then $I := I_0 + I_1$ is contained in $\mathfrak{A}^\#[X_1, \ldots, X_N]_0$, so that $\mathfrak{B} := \mathfrak{A}^\#[X_1, \ldots, X_N]_0/I$ is well defined with $\mathfrak{B}^\# \cong \mathfrak{A}^\#[X_1, \ldots, X_N]/I$. Clearly, (Once again: Why?) $\mathfrak{B}$ is an FDNC-algebra.

We first extend the norm on $\mathfrak{A}^\#$ to $\mathfrak{A}^\#[X_1, \ldots, X_N]/I_0$ by letting

$$\left\| \sum_{|\alpha| < N^2 d} a_\alpha X^\alpha \right\| := \sum_{|\alpha| < N^2 d} \|a_\alpha\|$$

for $a_\alpha \in \mathfrak{A}^\#$. This induces a quotient norm on $\mathfrak{B}^\#$, which we will likewise denote by $\| \cdot \|$. Let

$$S := \left\{ \frac{1}{\eta^k} \left(\frac{1}{2N} \sum_{j=1}^N X_j - a \right)^k X^\alpha : k \in \mathbb{N}_0, \alpha \in \mathbb{N}_0^N \right\} \subset \mathfrak{B}^\#.$$

Then S is a finite (Why?) subsemigroup of the multiplicative semigroup of $\mathfrak{B}^\#$ containing the identity, so that we may define again another algebra norm $\| \cdot \|_S$ on $\mathfrak{B}^\#$ as in Exercise 3.2.1. Let $x := \frac{1}{2N} \sum_{j=1}^N X_j$. Then, since $\frac{1}{\eta}(x - a) \in S$, we have $\|x - a\|_S \leq \eta$. For $j = 1, \ldots, N$, let

$$\mathbf{m}_j := \frac{1}{N-1} \sum_{k=1}^{N-1} X_j^k \otimes X_j^{N-k} \in \mathfrak{B} \hat{\otimes} \mathfrak{B}.$$

Clearly, $\|\mathbf{m}_j\|_S \leq 1$ and $\|X_j \cdot \mathbf{m}_j - \mathbf{m}_j \cdot X_j\|_S \leq \frac{2}{N-1}$. Let $\mathbf{m} := \mathbf{m}_1 \bullet \cdots \bullet \mathbf{m}_N$, so that $\|\mathbf{m}\|_S \leq 1$ and

$$\|X_j \cdot \mathbf{m} - \mathbf{m} \cdot X_j\|_S \leq \frac{2}{N-1} \qquad (j = 1, \ldots, N).$$

It follows that $\|x \cdot \mathbf{m} - \mathbf{m} \cdot x\| \leq \frac{1}{N-1}$. For $N > 1 + \frac{1}{\zeta}$, this means that $\|x \cdot \mathbf{m} - \mathbf{m} \cdot x\| \leq \zeta$. Furthermore, it is clear from the definition of I_1 that $\Delta_{\mathfrak{B}^\#} \mathbf{m} = u$. All in all, it is clear that $\mathbf{m}$ is a weak metric, (η, ζ)-approximate commutant for a — provided $\mathfrak{A}$ is equippped with (the restriction of) $\| \cdot \|_S$. Hence, all we have to show is that for sufficiently large N the norm $\| \cdot \|_S$ coincides with the given norm on $\mathfrak{A}$. And this is the *really* hard part of the proof ...

As in the proof of Lemma 3.2.2, we proceed by dualizing. Given $\phi \in \mathfrak{A}^*$ with $\|\phi\| = 1$, we first extend again ϕ to $\mathfrak{A}^\#$ through $\langle e, \phi \rangle = 0$. We then define an extension $\bar{\phi} : \mathfrak{A}^\#[X_1, \ldots, X_N] \to \mathbb{C}$ as follows: For $\alpha \in \mathbb{N}_0^N$ and $\tilde{a} \in \mathfrak{A}^\#$, let

$$\langle \tilde{a} X^\alpha, \bar{\phi} \rangle := \begin{cases} \langle u^k a^l \tilde{a}, \phi \rangle, & \text{if } \alpha = kN\mathbf{1} + \beta \text{ with } k \in [0, d) \text{ and } |\beta| = l \in [0, d), \\ 0, & \text{otherwise} \end{cases}$$

If $N^2 \geq d$, this is well defined (You should check this for yourself).

We first claim that $\bar{\phi}$ vanishes on both I_0 and I_1, and thus induces a linear functional on $\mathfrak{B}^\#$.

Let $\tilde{a} \in \mathfrak{A}^\#$, and let $\alpha \in \mathbb{N}_0^N$ be such that $|\alpha| \geq N^2 d$. Assume that $\langle \tilde{a} X^\alpha, \bar{\phi} \rangle \neq 0$. By the definition of $\bar{\phi}$, however, this implies that $|\alpha| \in [kN^2, kN^2 + d)$ with $k \in [0, d)$, so that $|\alpha| \leq (d-1)N^2 + d - 1 < N^2 d$ since $N^2 \geq d$, which is a contradiction. Hence, $\bar{\phi}$ annihilates I_0.

In order to prove that $\bar{\phi}$ vanishes on I_1 as well, it is sufficient to show that we have $\langle \tilde{a}(X^{N\mathbf{1}} - u)X^\alpha, \bar{\phi} \rangle = 0$ for all $\alpha \in \mathbb{N}_0^N$ and all $\tilde{a} \in \mathfrak{A}^\#$. Suppose that $\alpha = kN\mathbf{1} + \beta$ with $k, |\beta| \in [0, d)$. Then, by the definition of $\bar{\phi}$, we have

$$\langle \tilde{a} u X^\alpha, \bar{\phi} \rangle = \langle u^{k+1} a^{|\beta|} \tilde{a}, \phi \rangle.$$

For all other $\alpha \in \mathbb{N}_0^N$, we have $\langle \tilde{a} y X^\alpha, \bar{\phi} \rangle = 0$. Suppose now that $\alpha = kN\mathbf{1} + \beta$ with $|\beta| \in [0, d)$, but $k \in [0, d-1)$. Then, again by the definition of $\bar{\phi}$, we obtain

$$\langle \tilde{a} X^{N\mathbf{1}+\alpha}, \bar{\phi} \rangle = \langle u^{k+1} a^{|\beta|} \tilde{a}, \phi \rangle (= \langle \tilde{a} y X^\alpha, \bar{\phi} \rangle).$$

Finally, let $\alpha \in \mathbb{N}_0^N$ be such that $\alpha = (d-1)N\mathbf{1} + \beta$ with $|\beta| \in [0, d)$. Then, since $u^d = 0$, it follows that

$$\langle \tilde{a} X^{N\mathbf{1}+\alpha}, \bar{\phi} \rangle = 0 (= \langle \tilde{a} u X^\alpha, \bar{\phi} \rangle).$$

Hence, $\bar{\phi}$ does indeed vanish on I_1.

By Exercise 3.2.1(ii), it is sufficient to show that

$$\left| \left\langle \tilde{a} \left(\frac{1}{2N} \sum_{j=1}^N X_j - a \right)^k X^\alpha, \bar{\phi} \right\rangle \right| \leq \eta^k \tag{3.11}$$

for all $k \in \mathbb{N}_0$, $\alpha \in \mathbb{N}_0$ and $\tilde{a} \in \mathfrak{A}^\#$ with $\|\tilde{a}\| = 1$. As in the proof of Lemma 3.2.4, we have to treat several cases separately.

Case 1: $k \geq \frac{d \log 4}{\log 6 - \log 5}$. We have

$$\left(\frac{1}{2N} \sum_{j=1}^N X_j - a \right)^k = \sum_{|\beta| \leq k} \frac{1}{(2N)^{|\beta|}} \binom{k}{\beta_1, \ldots, \beta_N} (-a)^{k-|\beta|} X^\beta. \tag{3.12}$$

Since $a^d = 0$, no coefficient of X^α in the right hand side of (3.12) has a norm larger than 4^d times the norm of the corresponding coefficient in the multinomial expansion of $\left(\frac{1}{2N} \sum_{j=1}^N X_j - \frac{1}{4}a \right)^k$. Since $\|a\| = 1$, the sum of those norms is at most $\left(\frac{3}{4} \right)^k$. Hence, the sum of the norms of the coeffiecents of the right hand side of (3.12) is at most $4^d \left(\frac{3}{4} \right)^k$. Since $\|\tilde{a}\| = 1$, we have

$$\tilde{a}\left(\frac{1}{2N}\sum_{j=1}^{N}X_j - a\right)^k X^\alpha = \sum_\beta b_\beta X^\beta$$

with $\sum_\beta \|b_\beta\| \le 4^d \left(\frac{3}{4}\right)^k$. From the definition of $\bar{\phi}$, it follows that $|\langle b_\beta X^\beta, \bar{\phi}\rangle| \le \|b_\beta\|$ (Why?). The triangle inequality then yields

$$\left|\left\langle \tilde{a}\left(\frac{1}{2N}\sum_{j=1}^{N}X_j - a\right)^k X^\alpha, \bar{\phi}\right\rangle\right| \le 4^d \left(\frac{3}{4}\right)^k$$

Consequently, (3.11) is satisfied provided that $k \ge \frac{d\log 4}{\log(4\eta)-\log 3}$. This, however, does indeed hold since $\eta \ge \frac{9}{10}$ and $k \ge \frac{d\log 4}{\log 6 - \log 5}$.

$Case\ 2$: $k \le \frac{d\log 4}{\log 6 - \log 5}$ and $\alpha = \nu N\mathbf{1} + \gamma$ where $\nu \in [0,d)$ and $\gamma \in \mathbb{N}_0^N$ with $|\gamma| \in [0,d)$. Again, we use a multinomial expansion:

$$\tilde{a}\left(\frac{1}{2N}\sum_{j=1}^{N}X_j - a\right)^k X^\alpha = \sum_{|\beta|\le k}\frac{1}{(2N)^{|\beta|}}\binom{k}{\beta_1,\dots,\beta_N}(-a)^{k-|\beta|}\tilde{a}X^{\alpha+\beta}$$

$$= \sum_{|\beta|\le k}\frac{1}{(2N)^{|\beta|}}\binom{k}{\beta_1,\dots,\beta_N}(-a)^{k-|\beta|}\tilde{a}X^{\nu N\mathbf{1}+\gamma+\beta}. \quad (3.13)$$

In case $|\gamma| + |\beta| < d$, the definition of $\bar{\phi}$ yields

$$\langle a^{k-|\beta|}\tilde{a}X^{\nu N\mathbf{1}+\gamma+\beta}, \bar{\phi}\rangle = \langle u^\nu a^{k+|\gamma|}\tilde{a}, \phi\rangle. \quad (3.14)$$

If $d \le |\gamma| + |\beta| < N^2$, we have $k + |\gamma| \ge |\beta| + |\gamma| \ge d$, so that $a^{k+|\gamma|} = 0$. This means, however, that both sides of (3.14) are zero in this case, so that again the equality (3.14) holds. Provided that N is sufficiently large, i.e. if $N^2 > d\left(1 + \frac{\log 4}{\log 6 - \log 5}\right)$, the case where $|\gamma| + |\beta| \ge N^2$ can never occur. Hence, we obtain from (3.13) that

$$\left\langle \tilde{a}\left(\frac{1}{2N}\sum_{j=1}^{N}X_j - a\right)^k X^\alpha, \bar{\phi}\right\rangle = \sum_{|\beta|\le k}\frac{1}{(2N)^{|\beta|}}\binom{k}{\beta_1,\dots,\beta_N}(-1)^{k-|\beta|}\langle u^\nu a^{k+|\gamma|}\tilde{a}, \phi\rangle$$

$$= \left(\frac{1}{2} - 1\right)^k \langle u^\nu a^{k+|\gamma|}\tilde{a}, \phi\rangle \quad \text{(Why?)},$$

so that

$$\left|\left\langle \tilde{a}\left(\frac{1}{2N}\sum_{j=1}^{N}X_j - a\right)^k X^\alpha, \bar{\phi}\right\rangle\right| \le \left(\frac{1}{2}\right)^k \le \eta^k.$$

This, again, establishes (3.11).

$Case\ 3$: $k \le \frac{d\log 4}{\log 6 - \log 5}$ and all values of α not covered under Case 2. Unless every summand in the multinomial expansion

$$\left\langle \tilde{a}\left(\frac{1}{2N}\sum_{j=1}^{N}X_j - a\right)^k X^\alpha, \bar{\phi}\right\rangle = \sum_{|\beta|\le k}\frac{1}{(2N)^{|\beta|}}\binom{k}{\beta_1,\dots,\beta_N}(-1)^{k-|\beta|}\langle a^{k-|\beta|}\tilde{a}X^{\alpha+\beta}, \phi\rangle$$

is zero — in which case nothing has to be shown —, it follows from the definition of $\bar{\phi}$ that at least for some $\beta \in \mathbb{N}_0^N$ with $|\beta| \leq k$, we have $\alpha + \beta = \nu N \mathbf{1} + \gamma$ with $\nu \in [0, d)$ and $\gamma \in \mathbb{N}_0^N$ such that $|\gamma| \in [0, d)$. If we choose $N^2 > d \left(1 + \frac{\log 4}{\log 6 - \log 5}\right)$, the smallest interval in $\mathbb{N}_0^N$ containing all possible values of $|\alpha| + |\beta|$ must have a width less than $N^2 - d$. Hence, there is only one possible value $\nu \in [0, d)$ such that $\alpha + \beta = \nu N \mathbf{1} + \gamma$ with $\gamma \in \mathbb{N}_0^N$ such that $|\gamma| \in [0, d)$. Let β_0 be the positive part of $\nu N \mathbf{1} - \alpha$; note that $|\beta_0| \geq 1$. Then all other multiindices β for which $\alpha + \beta$ can be expressed in the aforementioned way must satisfy $\beta \geq \beta_0$. Consequently, we obtain:

$$\left| \left\langle \tilde{a} \left(\frac{1}{2N} \sum_{j=1}^{N} X_j - a \right)^k X^\alpha, \bar{\phi} \right\rangle \right| \leq \sum_{\beta \geq \beta_0, |\beta| \leq k} \binom{k}{\beta_1, \ldots, \beta_N} \frac{1}{(2N)^{|\beta|}}$$

$$\leq \sum_{\mu=0}^{k-|\beta_0|} \frac{k!}{(2N)^{\mu + |\beta_0|}} N^\mu \qquad \text{(Why?)}$$

$$\leq \frac{(k+1)!}{2N}. \qquad (3.15)$$

Let $K := \left\lfloor \frac{d \log 4}{\log 6 - \log 5} \right\rfloor$, and choose $N > \frac{1}{\eta^K} \frac{(K+1)!}{2}$. Then (3.11) follows from (3.15). $\quad \square$

3.3 Notes and comments

The results on bounded approximate identities in Banach algebras of compact or rather approximable operators can be found in [B–P], [C–L–M], [Dixo], [G–W], and [Sam]; there seem to be a lot of re- and parallel discoveries in this area.

Our discussion of Banach algebras of compact operators is taken from [G–J–W]. Our Example 3.1.17 is a poor man's version of [G–J–W, Theorem 6.9]: As we have seen in Example 3.1.17, the amenability of $\mathcal{A}(\ell^p \oplus \ell^q)$ for any $p, q \in (1, \infty)$ implies that ℓ^p is finitely representable in ℓ^q (and vice versa). Using more sophisticated Banach space theoretic arguments than we have at our disposal here, N. Grønbæk, B. E. Johnson, and G. A. Willis show that this is impossible whenever $p \neq q$ and $p \neq 2 \neq q$. The big open question in connection with the amenability of algebras of compact or rather approximable operators is: What is the property — described in Banach space theoretic terms — which characterizes those spaces E for which $\mathcal{A}(E)$ (or $\mathcal{K}(E)$) is amenable? It is plausible to expect it to be some strong form of approximation property, but still weaker than property $(\mathbb{A})$.

A related question about which very little is known is whether there is an infinite-dimensional Banach space E such that $\mathcal{L}(E)$ is amenable. We shall see in our chapter on C^*-algebras that $\mathcal{L}(\mathfrak{H})$ for a Hilbert space $\mathfrak{H}$ is amenable only if $\dim \mathfrak{H} < \infty$. Further, if E is a Banach space such that E^* has the approximation property and $\mathcal{L}(E)$ is amenable, then $\mathcal{A}(E)$ must be amenable as well (this is a consequence of Corollary 3.1.5 and Theorem 2.3.7). Consequently, $\mathcal{L}(\ell^p \oplus \ell^q)$ cannot be amenable if $p, q \in (1, \infty) \setminus \{2\}$ with $p \neq q$. It seems to be unknown whether $\mathcal{L}(\ell^p)$ is amenable for $p \in (1, \infty) \setminus \{2\}$. Only very recently, C. J. Read was able to prove that $\mathcal{L}(\ell^1)$ is not amenable ([Rea 2]); his proof involves the theory of random hypergraphs. As an inspection of Read's proof shows, the argument in [Rea 2] also yields the non-amenability of $\mathcal{L}(c_0)$. It is not clear at all, however, if Read's proof can be modified in such a way that it establishes the non-amenability of $\mathcal{L}(\ell^p)$ for $p \in (1, \infty)$.

The question of whether a (commutative) radical, amenable Banach algebra exists was raised for the first time by P. C. Curtis, Jr., in 1989. A first example of a radical, amenable Banach algebra was given in [Run 1]. That example, however, is not commutative and depends on deep results from abstract harmonic analysis due to J. Boidol, H. Leptin, and D. Poguntke (the radical algebra arises as the quotient of the group algebra of a peculiar amenable group). The example we present (due to C. J. Read) is from [Rea 1].

It is easy to obtain a commutative, amenable Banach algebra that is not semisimple: Let G be a non-compact, locally compact, abelian group with dual group Γ. By Malliavin's theorem ([Rud 1, 7.6.1, Theorem]), Γ contains a closed set F which is not of synthesis, i.e.

$$J(F) := \overline{\{f \in A(\Gamma) : \operatorname{supp}(f) \text{ is compact and does not intersect } F\}}$$
$$\subsetneq \{f \in A(\Gamma) : f|_F \equiv 0\}$$
$$=: I(F).$$

Then $\mathfrak{A} := A(\Gamma)/J(F)$ is amenable with radical $I(F)/J(F) \neq \{0\}$.

4 Amenability-like properties

Let $\mathfrak{A}$ be a Banach algebra, and let $\mathcal{E}$ be a class of Banach $\mathfrak{A}$-bimodules. If $\mathcal{E}$ is the class of all dual Banach $\mathfrak{A}$-bimodules, then $\mathfrak{A}$ is amenable if $\mathcal{H}^1(\mathfrak{A}, E) = \{0\}$ for all $E \in \mathcal{E}$.

What happens, if we choose another class $\mathcal{E}$ of modules? What if $\mathcal{E}$ is the class of all Banach $\mathfrak{A}$-bimodules? What if $\mathcal{E}$ consists of all finite-dimensional Banach $\mathfrak{A}$-bimodules, or of all finitely generated Banach $\mathfrak{A}$-bimodules, or just of $\mathfrak{A}$ or $\mathfrak{A}^*$ alone? For each such class $\mathcal{E}$, we can define a corresponding notion of amenability. The question, is, of course, whether the class of Banach algebras we single out through such a definition is a good one: Do we get both strong theorems and a sufficient number of interesting examples?

We discuss two such notions of amenability in this chapter: super-amenability ($\mathcal{E}$ is the class of all Banach $\mathfrak{A}$-bimodules) and weak amenability ($\mathcal{E} = \{\mathfrak{A}^*\}$).

Two more classes of Banach algebras we deal with in this chapter are the biprojective and the biflat Banach algebras. Although these Banach algebras can also be characterized through an appropriate class $\mathcal{E}$ of bimodules, we define them via the existence of certain module homomorphisms.

Finally, we study Connes-amenability — a notion of amenability that only makes sense for Banach algebras which are dual spaces. As we shall see in Chapter 6, Connes-amenability is the right notion of amenability for von Neumann algebras.

This chapter is less self-contained than the previous ones: We need some basic facts from the theory of von Neumann algebras, all of which can be found in [K–R] and [Sak], as well as Haagerup's non-commutative Grothendieck inequality ([Haa 3]); we also require some background on the representation theory of Banach algebras and on Banach algebras with non-zero socle, for which we refer to [B–D] and to [Pal].

4.1 Super-amenability

In Definition 2.1.9, we used the first Hochschild cohomology group of a Banach algebra with coefficients in a dual module to define amenable Banach algebras. Let's consider the following variant:

Definition 4.1.1 A Banach algebra $\mathfrak{A}$ is called *super-amenable* if $\mathcal{H}^1(\mathfrak{A}, E) = \{0\}$ for every Banach $\mathfrak{A}$-bimodule E.

Certainly, every super-amenable Banach algebra is amenable. But what about the converse?

Many of the basic results for amenable Banach algebras have analogues for super-amenable Banach algebras. The proofs are often similar and, in fact, are even easier in the super-amenable situation. We thus leave them to the reader as a series of exercises.

Exercise 4.1.1 Show that every super-amenable Banach algebra is unital.

Exercise 4.1.2 Let $\mathfrak{A}$ be a super-amenable Banach algebra. Show that $\mathcal{H}^n(\mathfrak{A}, E) = \{0\}$ for all $n \in \mathbb{N}$.

Remember the definition of a diagonal? If not, go back to Exercise 2.2.1.

Exercise 4.1.3 Show that a Banach algebra is super-amenable if and only if it has a diagonal. Conclude that, for $n_1, \ldots, n_k \in \mathbb{N}$, the algebra $\mathbb{M}_{n_1} \oplus \cdots \oplus \mathbb{M}_{n_k}$ is super-amenable.

Exercise 4.1.4 Prove the following hereditary properties for super-amenability:

(i) If $\mathfrak{A}$ is a super-amenable Banach algebra, $\mathfrak{B}$ is a Banach algebra, and $\theta : \mathfrak{A} \to \mathfrak{B}$ is a homomorphism with dense range, then $\mathfrak{B}$ is super-amenable.

(ii) If $\mathfrak{A}$ is a super-amenable Banach algebra, and I is a closed ideal of $\mathfrak{A}$, then I is super-amenable if and only if it has an identity and if and only if it is complemented in $\mathfrak{A}$.

(iii) If $\mathfrak{A}$ is a Banach algebra and I is a closed ideal of $\mathfrak{A}$ such that both I and $\mathfrak{A}/I$ are super-amenable, then $\mathfrak{A}$ is super-amenable.

(iv) If $\mathfrak{A}$ and $\mathfrak{B}$ are super-amenable Banach algebra, then $\mathfrak{A} \hat{\otimes} \mathfrak{B}$ is super-amenable.

Is there an analogue of Proposition 2.3.17 for super-amenable Banach algebras?

Exercise 4.1.5 Let $\mathfrak{A}$ be a super-amenable Banach algebra. Show that every admissible, short, exact sequence

$$\{0\} \to F \to E \to E/F \to \{0\}$$

of Banach $\mathfrak{A}$-modules (left-, right- or bi-) splits.

Exercise 4.1.6 Let $\mathfrak{A}$ be a unital Banach algebra.

(i) Show that the short, exact sequence

$$\{0\} \to \ker \Delta \to \mathfrak{A} \hat{\otimes} \mathfrak{A} \overset{\Delta}{\to} \mathfrak{A} \to \{0\} \tag{4.1}$$

of Banach $\mathfrak{A}$-bimodules is admissible.

(ii) Show that the following are equivalent:

(a) $\mathfrak{A}$ is super-amenable.

(b) (4.1) splits.

Hence, super-amenable Banach algebras seem to have very nice properties — nicer than amenable Banach algebras. So, why don't we study super-amenable Banach algebras in the first place? The answer to that question is that Definition 4.1.1 is so strong that, although it allows for nice theorems, all known examples are dull; *if* there are any non-dull examples, they have to be extremely pathological in terms of Banach space geometry.

We prepare the ground with the following proposition, for which we require some background on the representation theory of Banach algebras ([B–D, Chapter III]):

Proposition 4.1.2 *Let $\mathfrak{A}$ be a super-amenable Banach algebra such that all maximal ideals of $\mathfrak{A}$ have finite codimension. Then there are $n_1, \ldots, n_k \in \mathbb{N}$ such that*

$$\mathfrak{A} \cong \mathbb{M}_{n_1} \oplus \cdots \oplus \mathbb{M}_{n_k}$$

Proof Let $\Pi_\mathfrak{A}$ be the structure space of $\mathfrak{A}$, i.e. the collection of the primitive ideals of $\mathfrak{A}$ equipped with the Jacobson topology ([B–D, Definition 26.3]). We start by showing that every primitive ideal of $\mathfrak{A}$ must be maximal. Let $P \in \Pi_\mathfrak{A}$ be any primitive ideal, and consider $\overline{\{P\}}$. By the definition of the Jacobson topology, we have

$$\overline{\{P\}} = \mathrm{hull}(P) = \{Q \in \Pi_\mathfrak{A} : P \subset Q\}.$$

By Zorn's Lemma, $\overline{\{P\}}$ contains a maximal ideal M. Consider the short exact sequence

$$\{0\} \to M \to \mathfrak{A} \to \mathfrak{A}/M \to \{0\}. \tag{4.2}$$

of Banach $\mathfrak{A}$-bimodules. Since M has finite codimension, (4.2) is admissible and thus splits by Exercise 4.1.5, i.e. there is a closed $\mathfrak{A}$-submodule I of $\mathfrak{A}$ (in other words: an ideal) such that $\mathfrak{A} = M \oplus I$. Let $Q \in \Pi_\mathfrak{A} \setminus \{M\}$. Since $MI = \{0\}$, it follows from [B–D, Proposition 24.12(iii)] that $I \subset Q$, i.e. $Q \in \mathrm{hull}(I)$. Consequently, $\{M\} = \Pi_\mathfrak{A} \setminus \mathrm{hull}(I)$; by the definition of the Jacobson topology, this means that $\{M\}$ is open. Therefore, $\overline{\{P\}} \setminus \{M\}$ is closed; by the definition of $\overline{\{P\}}$ as the smallest closed set containing P, this is possible only if $M = P$.

We have thus seen that $\Pi_\mathfrak{A}$ consists entirely of maximal ideals and that the singleton subset of $\Pi_\mathfrak{A}$ corresponding to each such ideal is open, i.e. $\Pi_\mathfrak{A}$ is discrete. Since $\mathfrak{A}$ is unital, however, $\Pi_\mathfrak{A}$ is also compact ([B–D, Corollary 26.5]). Hence, $\Pi_\mathfrak{A}$ consists of only finitely many ideals, say $M_1, \ldots, M_k$, all of which have finite codimension. Hence, $\mathrm{rad}(\mathfrak{A}) = M_1 \cap \cdots \cap M_k$ has finite codimension and thus is complemented. By Exercise 4.1.4(ii), this means that $\mathrm{rad}(\mathfrak{A})$ itself is superamenable and thus has an identity by Exercise 4.1.1. This is possible only if $\mathrm{rad}(\mathfrak{A}) = \{0\}$, so that $\mathfrak{A}$ is semisimple.

It follows that

$$\mathfrak{A} \cong \mathfrak{A}/M_1 \oplus \cdots \oplus \mathfrak{A}/M_k.$$

Hence, it suffices to show that $\mathfrak{A}/M_j \cong \mathbb{M}_{n_j}$ with $n_j \in \mathbb{N}$ for $j = 1, \ldots, k$. By the definition of a primitive ideal, M_j is the kernel of an irreducible representation π of $\mathfrak{A}$ on some vector space E. Since $\mathfrak{A}/M_j$ is finite-dimensional, it follows directly from the definition of an irreducible representation ([B–D, Definition 24.3]), that E has to be finite-dimensional. With Jacobson's density theorem ([B–D, Corollary 25.5]), it is easily seen (How?) that $\mathfrak{A}/M_j \cong \mathcal{L}(E) \cong \mathbb{M}_{n_j}$, where $n_j = \dim E$. $\square$

Corollary 4.1.3 *Let $\mathfrak{A}$ be a commutative, super-amenable Banach algebra. Then there is $n \in \mathbb{N}$ such that $\mathfrak{A} \cong \mathbb{C}^n$.*

We now move on to show that the hypothesis of Proposition 4.1.2 is already satisfied under rather weak conditions. The crucial step is the following lemma:

Lemma 4.1.4 *Let $\mathfrak{A}$ be a super-amenable Banach algebra, and let L be a closed left ideal such that $\mathfrak{A}/L$ has the approximation property. Then $\mathfrak{A}/L$ is finite-dimensional.*

Proof For $a \in \mathfrak{A}$, we use $\bar{a}$ to denote its coset in $\mathfrak{A}/L$.

By Exercise 4.1.3, $\mathfrak{A}$ has a diagonal

$$\mathbf{m} = \sum_{n=1}^{\infty} a_n \otimes b_n$$

with $\sum_{n=1}^{\infty} \|a_n\| \|b_n\| < \infty$. Let $(S_\alpha)_\alpha$ be a net in $\mathcal{F}(\mathfrak{A}/L)$ that converges to $\mathrm{id}_{\mathfrak{A}/L}$ uniformly on the compact subsets of $\mathfrak{A}/L$. Define

$$T_\alpha : \mathfrak{A}/L \to \mathfrak{A}/L, \quad \bar{a} \mapsto \sum_{n=1}^{\infty} a_n \cdot S_\alpha(b_n \cdot \bar{a}).$$

Then each T_α is compact (Why?). Moreover, we have

$$T_\alpha \bar{a} = \sum_{n=1}^{\infty} a_n \cdot S_\alpha(b_n \cdot \bar{a}) = \sum_{n=1}^{\infty} a a_n \cdot S_\alpha(\bar{b}_n) = a \cdot T_\alpha \bar{e}_{\mathfrak{A}} \qquad (a \in \mathfrak{A}).$$

However, $\lim_\alpha T_\alpha \bar{e}_{\mathfrak{A}} = \bar{e}_{\mathfrak{A}}$ (Once again: Why?), so that $(T_\alpha)_\alpha$ converges to $\mathrm{id}_{\mathfrak{A}/L}$ in the norm topology. It follows that $\mathrm{id}_{\mathfrak{A}/L}$ is compact. □

Theorem 4.1.5 *Let $\mathfrak{A}$ be a super-amenable Banach algegbra such that one of the following is satisfied:*

(i) *$\mathfrak{A}$ has the approximation property.*

(ii) *For each maximal ideal M of $\mathfrak{A}$, the Banach algebra $\mathfrak{A}/M$ has the approximation property.*

(iii) *For each maximal left ideal L of $\mathfrak{A}$, the left Banach $\mathfrak{A}$-module $\mathfrak{A}/L$ has the approximation property.*

Then there are $n_1, \ldots, n_k \in \mathbb{N}$ such that

$$\mathfrak{A} \cong \mathbb{M}_{n_1} \oplus \cdots \oplus \mathbb{M}_{n_k}.$$

Proof It is obvious from Lemma 4.1.4 that $\mathfrak{A}$ satisfies the hypothesis of Proposition 4.1.2 if (i) or (ii) are true. If (iii) holds, then, for every maximal ideal M, there is a maximal left ideal L such that $M = \{a \in \mathfrak{A} : a\mathfrak{A} \subset L\}$ (by the definition of a primitive ideal). Since $\mathfrak{A}/L$ is finite-dimensional by Lemma 4.1.4, it follows that also $\mathfrak{A}/M \cong \mathcal{L}(\mathfrak{A}/L)$ is finite-dimensional. Hence, again, the hypothesis of Proposition 4.1.2 is satisfied. □

If there is an infinite-dimensional, super-amenable Banach algebra, then, by Theorem 4.1.5, the algebra as well as at least one of its quotients by a maximal ideal and by a maximal left ideal must lack the approximation property. No such algebra is known. The first Banach space without the approximation property discovered "in nature" was $\mathcal{L}(\mathfrak{H})$ for an infinite-dimensional Hilbert space $\mathfrak{H}$. This algebra, however, can never be super-amenable:

Corollary 4.1.6 *Let $\mathfrak{A}$ be a super-amenable C^*-algebra. Then there are $n_1, \ldots, n_k \in \mathbb{N}$ such that*

$$\mathfrak{A} \cong \mathbb{M}_{n_1} \oplus \cdots \oplus \mathbb{M}_{n_k},$$

where $\cong$ denotes $*$-isomorphism.

Proof If L is a closed maximal left ideal of $\mathfrak{A}$, then, by [Mur, Theorem 5.3.5], we have

$$L = \{a \in \mathfrak{A} : \langle a^* a, \phi \rangle = 0\}$$

for some pure state ϕ on $\mathfrak{A}$. By [Mur, Theorem 5.2.4], this means that $\mathfrak{A}/L$ is a Hilbert space and thus has the approximation property. By Theorem 4.1.5, we have $\mathfrak{A} \cong \mathbb{M}_{n_1} \oplus \cdots \oplus \mathbb{M}_{n_k}$ as a Banach algebra for certain $n_1, \ldots, n_k \in \mathbb{N}$; in particular, $\mathfrak{A}$ is finite-dimensional. By [Mur, Thereom 6.3.8], this means that $\mathfrak{A} \cong \mathbb{M}_{n_1} \oplus \cdots \oplus \mathbb{M}_{n_k}$ as C^*-algebras. $\quad\square$

Exercise 4.1.7 Let G be a locally compact group. Show that $L^1(G)$ is super-amenable if and only if G is finite.

We conclude this section with an analogue of Theorem 2.3.9:

Exercise 4.1.8 Let G be a locally compact group. Show that $L_0^1(G)$ has a right identity if and only if G is finite. (*Hint*: First, show that $L^1(G)$ has an identity, so that G has to be discrete. Then proceed as in the proof of Theorem 2.3.9 to construct a left invariant mean on $\ell^\infty(G)$ that belongs to $\ell^1(G)$. Finally, apply Exercise 1.1.7.)

4.2 Weak amenability

As the name suggests, weak amenability is weaker than amenability:

Definition 4.2.1 A Banach algebra $\mathfrak{A}$ is called *weakly amenable* if $\mathcal{H}^1(\mathfrak{A}, \mathfrak{A}^*) = \{0\}$.

Exercise 4.2.1 Let $\mathfrak{A}$ be a weakly amenable Banach algebra.

(i) Show that $\mathfrak{A}^2$, the linear span of products of elements in $\mathfrak{A}$, is dense in $\mathfrak{A}$. (*Hint*: Let $\phi \in \mathfrak{A}^*$ be such that $\phi|_{\mathfrak{A}^2} \equiv 0$. Then consider $\mathfrak{A} \ni a \mapsto \langle a, \phi \rangle \phi$.)

(ii) Show that, if $\mathfrak{A}$ is commutative, then $\mathcal{H}^1(\mathfrak{A}, E) = \{0\}$ for every *symmetric* Banach $\mathfrak{A}$-bimodule E, i.e. for every Banach $\mathfrak{A}$-bimodule E such that

$$a \cdot x = x \cdot a \qquad (a \in \mathfrak{A},\, x \in E).$$

Exercise 4.2.2 Let $\mathfrak{A}$ be any of the following Banach algebras:

$$A(\mathbb{D}) := \{f \in \mathcal{C}(\overline{\mathbb{D}}) : f|_{\mathbb{D}} \text{ is analytic}\};$$
$$H^\infty(\mathbb{D}) := \{f \in \ell^\infty(\mathbb{D}) : f \text{ is analytic}\};$$
$$\mathcal{C}^{(n)}([0,1]) := \{f \in \mathcal{C}([0,1]) : f \text{ is } n\text{-times continuously differentiable}\} \quad \text{with } n \geq 1.$$

Show that $\mathfrak{A}$ is not weakly amenable.

How much weaker than amenability is weak amenability? In particular, which are the locally compact groups G for which $L^1(G)$ is weakly amenable?

The following lemma is standard, but, nevertheless, we give a proof:

Lemma 4.2.2 *Let X be a locally compact Hausdorff space, and let μ be a positive, regular Borel measure on X. Then $L_{\mathbb{R}}^\infty(X, \mu)$, i.e. the real Banach space of those functions in $L^\infty(X, \mu)$ with real-valued representatives, is order complete, i.e. each non-empty, bounded subset Φ of $L_{\mathbb{R}}^\infty(X, \mu)$ has a supremum.*

Proof Let $L^1_+(X,\mu)$ denote the cone of those functions in $L^1(X,\mu)$ that have a positive representative.

Let $\varnothing \neq \Phi \subset L^\infty_{\mathbb{R}}(X,\mu)$ be bounded, i.e. there is $\phi_0 \in L^\infty_{\mathbb{R}}(X,\mu)$ such that $\phi \leq \phi_0$ for all $\phi \in \Phi$. For any $f, f_1, \ldots, f_n \in L^1_+(X,\mu)$ with $f_1 + \cdots + f_n = f$ and $\phi_1, \ldots, \phi_n \in \Phi$, we have

$$\sum_{j=1}^n \langle f_j, \phi_j \rangle \leq \sum_{j=1}^n \langle f_j, \phi_0 \rangle = \langle f, \phi_0 \rangle.$$

Consequently, the supremum

$$\sup \left\{ \sum_{j=1}^n \langle f_j, \phi_j \rangle : n \in \mathbb{N}, \ f_1, \ldots, f_n \in L^1_+(X,\mu), \ f_1 + \cdots + f_n = f, \ \phi_1, \ldots, \phi_n \in \Phi \right\}$$

$$(4.3)$$

is finite for each $f \in L^1_+(X,\mu)$. Define $\psi \colon L^1_+(X,\mu) \to \mathbb{R}$ by letting $\langle f, \psi \rangle$ be the supremum (4.3). It is immediate from this definition that ψ is positively homogeneous, i.e. $\langle tf, \psi \rangle = t\langle f, \psi \rangle$ for all $f \in L^1_+(X,\mu)$ and $t \in [0,\infty)$. Equally obvious is the superadditivity of ψ, i.e. $\langle f, \psi \rangle + \langle g, \psi \rangle \leq \langle f + g, \psi \rangle$ for $f, g \in L^1_+(X,\mu)$. To prove the converse inequality, let $f, g \in L^1_+(X,\mu)$, and let $h_1, \ldots, h_n \in L^1_+(X,\mu)$ be such that $h_1 + \cdots + h_n = f + g$. For $j = 1, \ldots, n$, define $f_j \in L^1_+(X,\mu)$ through

$$f_j(x) = \begin{cases} \dfrac{f(x)h_j(x)}{f(x) + g(x)}, & \text{if } f(x) + g(x) \neq 0, \\ 0, & \text{otherwise.} \end{cases} \qquad (x \in X)$$

Likewise, define $g_1, \ldots, g_n \in L^1_+(X,\mu)$. Then $f_1 + \cdots + f_n = f$ and $g_1 + \cdots + g_n = g$, so that, for any $\phi_1, \ldots, \phi_n \in \Phi$, we have

$$\sum_{j=1}^n \langle h_j, \phi_j \rangle = \sum_{j=1}^n \langle f_j, \phi_j \rangle + \sum_{j=1}^n \langle g_j, \phi_j \rangle \leq \langle f, \psi \rangle + \langle g, \psi \rangle.$$

It follows that $\langle f + g, \psi \rangle \leq \langle f, \psi \rangle + \langle g, \psi \rangle$.

All in all, ψ is positively homogeneous and additive on $L^1_+(X,\mu)$ and thus extends to $L^1_{\mathbb{R}}(X,\mu)$ as a real-linear functional through

$$\langle f - g, \psi \rangle := \langle f, \psi \rangle - \langle g, \psi \rangle \qquad (f, g \in L^1_+(X,\mu)).$$

Identifying ψ with an element of $L^\infty_{\mathbb{R}}(X,\mu)$, we see that it is the required supremum of Φ. $\square$

Exercise 4.2.3 Let G be a locally compact group. Let Φ be a non-empty, bounded subset of $L^\infty_{\mathbb{R}}(G)$, so that, by Lemma 4.2.2, $\phi := \sup \Phi$ exists. Show that

$$\left. \begin{cases} \sup(\delta_g \cdot \Phi \cdot \delta_{g^{-1}}) = \delta_g \cdot \phi \cdot \delta_{g^{-1}} \\ \sup(\psi + \Phi) = \psi + \phi \end{cases} \right\} \qquad (g \in G, \ \psi \in L^\infty_{\mathbb{R}}(G)).$$

Theorem 4.2.3 *Let G be a locally compact group. Then $L^1(G)$ is weakly amenable.*

Proof We identify $L^1(G)^*$ with $L^\infty(G)$.

Let $D \in \mathcal{Z}^1(L^1(G), L^\infty(G))$. Since $L^1(G)$ is pseudo-unital, D has an extension $\tilde{D} \in \mathcal{Z}^1(M(G), L^\infty(G))$ according to Proposition 2.1.6. Let $g, h \in G$. Then we have:

$$
\begin{aligned}
(\tilde{D}\delta_g) \cdot \delta_{g^{-1}} &= (\tilde{D}(\delta_h * \delta_{h^{-1}g})) \cdot \delta_{g^{-1}} \\
&= (\delta_h \cdot \tilde{D}\delta_{h^{-1}g} + (\tilde{D}\delta_h) \cdot \delta_{h^{-1}g}) \cdot \delta_{g^{-1}} \\
&= \delta_h \cdot ((\tilde{D}\delta_{h^{-1}g}) \cdot \delta_{(h^{-1}g)^{-1}}) \cdot \delta_{h^{-1}} + (\tilde{D}\delta_h) \cdot \delta_{h^{-1}}. \qquad (4.4)
\end{aligned}
$$

The set $\{\mathrm{Re}\,(\tilde{D}\delta_g) \cdot \delta_{g^{-1}} : g \in G\}$ is bounded in $L^\infty_{\mathbb{R}}(G)$ (By what?), and thus has a supremum $\phi_1 \in L^\infty_{\mathbb{R}}(G)$ by Lemma 4.2.2. Together, Exercise 4.2.3 and (4.4), imply that

$$
\phi_1 = \delta_g \cdot \phi_1 \cdot \delta_{g^{-1}} + \mathrm{Re}\,(\tilde{D}\delta_g) \cdot \delta_{g^{-1}} \qquad (g \in G)
$$

or, equivalently,

$$
\mathrm{Re}\,(\tilde{D}\delta_g) = \phi_1 \cdot \delta_g - \delta_g \cdot \phi_1 \qquad (g \in G). \qquad (4.5)
$$

Similarly, we obtain $\phi_2 \in L^\infty_{\mathbb{R}}(G)$ such that

$$
\mathrm{Im}\,(\tilde{D}\delta_g) = \phi_2 \cdot \delta_g - \delta_g \cdot \phi_2 \qquad (g \in G). \qquad (4.6)
$$

Letting $\phi := -(\phi_1 + i\phi_2)$, (4.5) and (4.6) imply that $D = \mathrm{ad}_\phi$. $\square$

So, for every non-amenable, locally compact group G, its group algebra $L^1(G)$ is weakly amenable, but not amenable.

Until now we have dealt with C^*-algebras — undoubtedly the most important and best understood class of Banach algebras — only in a rather superficial way: We haven't used any of the deep and beautiful theorems available for these algebras. This will change with our next theorem:

Theorem 4.2.4 *Let $\mathfrak{A}$ be a C^*-algebra. Then $\mathfrak{A}$ is weakly amenable.*

In view of how elementary the proof of Theorem 4.2.3 is and the fact that C^*-algebras are generally much more tractable than group algebras, one might think that proving Theorem 4.2.4 is even easier, but as we shall soon see this is not the case. For the first time, we deviate substantially from our policy of keeping our exposition self-contained. First, we require some basic facts about von Neumann algebras; for most of them, we will give references to [K–R]. We also need the fact that a von Neumann algebra is a dual space (we'll discuss this in Examples 4.4.2(c)). Most importantly, however, we rely at one point on U. Haagerup's non-commutative Grothendieck inequality ([Haa 3, Theorem 1.1 and Proposition 2.3]):

Theorem 4.2.5 (Grothendieck inequality) *Let $\mathfrak{A}$ and $\mathfrak{B}$ be C^*-algebras, and let $V \in \mathcal{L}^2(\mathfrak{A}, \mathfrak{B}; \mathbb{C})$. Then there are states ϕ_1, ϕ_2 of $\mathfrak{A}$ and ψ_1, ψ_2 of $\mathfrak{B}$ such that*

$$
|V(a, b)| \leq \|V\| \sqrt{\langle a^*a, \phi_1 \rangle + \langle aa^*, \phi_2 \rangle} \sqrt{\langle b^*b, \psi_1 \rangle + \langle bb^*, \psi_2 \rangle} \qquad (a \in \mathfrak{A},\, b \in \mathfrak{B}).
$$

If $\mathfrak{A}$ and $\mathfrak{B}$ are von Neumann algebras and V is normal in each variable, then $\phi_1, \phi_2, \psi_1,$ and ψ_2 can be chosen to be normal.

Before we begin with the actual proof of Theorem 4.2.4, let us note the following consequence of Theorem 4.2.5:

Corollary 4.2.6 *Let $\mathfrak{A}$ and $\mathfrak{B}$ be C^*-algebras, and let $T\colon \mathfrak{A} \to \mathfrak{B}^*$ be a bounded, linear map. Then T factors through a Hilbert space, i.e. there are a Hilbert space $\mathfrak{H}$ and linear operators $S \in \mathcal{L}(\mathfrak{A}, \mathfrak{H})$ and $R \in \mathcal{L}(\mathfrak{H}, \mathfrak{B}^*)$ such that $T = RS$. In particular, T is weakly compact.*

Proof The Grothendieck inequality yields states ϕ_1, ϕ_2 on $\mathfrak{A}$ and ψ_1, ψ_2 on $\mathfrak{B}$ such that

$$|\langle b, Ta\rangle| \le \|T\|\sqrt{\langle a^*a, \phi_1\rangle + \langle aa^*, \phi_2\rangle}\sqrt{\langle b^*b, \psi_1\rangle + \langle bb^*, \psi_2\rangle} \qquad (a \in \mathfrak{A}, \, b \in \mathfrak{B}),$$

so that

$$\|Ta\| \le \sqrt{2}\,\|T\|\sqrt{\langle a^*a, \phi_1\rangle + \langle aa^*, \phi_2\rangle} \qquad (a \in \mathfrak{A}). \tag{4.7}$$

Define

$$[a, b] := \langle b^*a, \phi_1\rangle + \overline{\langle ba^*, \phi_2\rangle} \qquad (a, b \in \mathfrak{A}), \tag{4.8}$$

and let $N := \{a \in \mathfrak{A} : [a, a] = 0\}$. Then (4.8) induces an inner product on $\mathfrak{A}/N$ turning it into a pre-Hilbert space; let $\mathfrak{H}$ be the completion of that pre-Hilbert space. By (4.7), $N \subset \ker T$, so that T induces an operator $R\colon \mathfrak{A}/N \to \mathfrak{B}^*$; also from (4.7), it follows that R is continuous with respect to the norm on $\mathfrak{A}/N$ induced by the inner product. Consequently, R extends to $\mathfrak{H}$. With $S\colon \mathfrak{A} \to \mathfrak{A}/N \subset \mathfrak{H}$ as the quotient map, we obtain $T = RS$. $\square$

We prepare the ground for the proof of Theorem 4.2.4 by recalling some basic notions about von Neumann algebras:

Definition 4.2.7 *Let $\mathfrak{M}$ be a von Neumann algebra, and let $p, q \in \mathfrak{M}$ be projections, i.e. self-adjoint idempotents.*

(i) *p and q are called (Murray–von Neumann) equivalent — in symbols: $p \sim q$ —, if there is $u \in \mathfrak{M}$ such that $uu^* = p$ and $u^*u = q$.*

(ii) *p is called finite if $q \sim p$ and $q \le p$ imply $p = q$.*

(iii) *$\mathfrak{M}$ is called finite if $e_{\mathfrak{M}}$ is finite.*

(iv) *$\mathfrak{M}$ is called properly infinite if 0 is the only finite projection in $Z(\mathfrak{M})$.*

If $\mathfrak{A}$ is a unital C^*-algebra, an *isometry* is an element $v \in \mathfrak{A}$ such that $v^*v = e_{\mathfrak{A}}$. The semigroup of all isometries of $\mathfrak{A}$ is denoted by $\mathcal{I}(\mathfrak{A})$.

Lemma 4.2.8 *Let $\mathfrak{M}$ be a properly infinite von Neumann algebra. Then, for each $x \in \mathfrak{M}$, zero is contained in the closed, convex hull of $\{vxv^* : v \in \mathcal{I}(\mathfrak{M})\}$.*

Proof Let $x \in \mathfrak{M}$. Since $\mathfrak{M}$ is properly infinite, there is a sequence $(p_n)_{n=1}^{\infty}$ of pairwise orthogonal projections in $\mathfrak{M}$ such that $p_n \sim e_{\mathfrak{M}}$ for all $n \in \mathbb{N}$ ([K–R, 6.3.3. Lemma]).

By the definition of $\sim$, there is a sequence $(v_n)_{n=1}^{\infty}$ in $\mathfrak{M}$ with $v_n v_n^* = p_n$ and $v_n^* v_n = e_{\mathfrak{M}}$ for all $n \in \mathbb{N}$; in particular, each v_n is an isometry. Let

$$x_n := \frac{1}{n}\sum_{j=1}^{n} v_j x v_j^* \in \mathrm{co}\{vxv^* : v \in \mathcal{I}(\mathfrak{M})\} \qquad (n \in \mathbb{N}).$$

Since $x^*x \le \|x\|^2 e_{\mathfrak{M}}$, it follows that

$$v_n x^* x v_n^* \le \|x\|^2 v_n^* v_n = \|x\|^2 p_n \qquad (n \in \mathbb{N}).$$

It follows that

$$\left\| \sum_{j=1}^{n} v_j x^* x v_j^* \right\| \leq \|x\|^2 \left\| \sum_{j=1}^{n} p_j \right\| \leq \|x\|^2 \qquad (n \in \mathbb{N}) \tag{4.9}$$

(Why?). Note also that

$$n^2 x_n^* x_n = \sum_{j=1}^{n} v_j x^* x v_j^* \qquad (n \in \mathbb{N}) \tag{4.10}$$

(Once again: Why?). Together, (4.9) and (4.10) imply that $n^2 \|x_n^* x_n\| \leq \|x\|^2$, so that $\lim_{n\to\infty} x_n = 0$. $\quad\square$

Lemma 4.2.9 *Let $\mathfrak{M}$ be a von Neumann algebra. Then there is a bounded, normal, linear map $T \colon \mathfrak{M} \to Z(\mathfrak{M})$ such that:*

(i) *$T(xy) = T(yx)$ $(x, y \in \mathfrak{M})$.*

(ii) *For each $x \in \mathfrak{M}$, the element Tx is contained in the closed, convex hull of $\{vxv^* : v \in \mathcal{I}(\mathfrak{M})\}$.*

Proof Case 1: $\mathfrak{M}$ is finite. Then [K–R, 8.3.6. Theorem] implies the existence of T as required.

Case 2: $\mathfrak{M}$ is properly infinite. Then, by Lemma 4.2.8, $T \equiv 0$ is the required map.

Case 3: $\mathfrak{M}$ is arbitrary,. By [K–R, 6.3.7. Proposition], there is a projection $p \in Z(\mathfrak{M})$ such that $\mathfrak{M}_1 := p\mathfrak{M}$ is finite and $\mathfrak{M}_2 := (e_{\mathfrak{M}} - p)\mathfrak{M}$ is properly infinite. Let $T_j \colon \mathfrak{M}_j \to Z(\mathfrak{M}_j)$ be the maps which exist according to the cases 1 and 2, and define

$$T \colon \mathfrak{M} \to \mathfrak{M}, \quad x \mapsto T_1(px) + T_2((e_{\mathfrak{M}} - p)x).$$

It is routinely checked (Do it!) that T has the required properties. $\square$

Recall that, for a C^*-algebra $\mathfrak{A}$ a positive element of $\mathcal{Z}^0(\mathfrak{A}, \mathfrak{A}^*)$ is called a *trace*.

Corollary 4.2.10 *Let $\mathfrak{M}$ be a von Neumann algebra. Then, for each $\phi \in \mathfrak{M}_*$, the w^*-closed, convex hull of $\{v^* \cdot \phi \cdot v : v \in \mathcal{I}(\mathfrak{M})\}$ contains an element $\tau \in \mathcal{Z}^0(\mathfrak{M}, \mathfrak{M}^*)$. Moreover, $\|\tau\| \leq \|\phi\|$, and τ is a trace if ϕ is positive.*

Proof Let $\phi \in \mathfrak{M}_*$, and let $T \colon \mathfrak{M} \to \mathfrak{M}$ be as in Lemma 4.2.9, and let $\tau := T^* \phi$. By Lemma 4.2.9(i), it is immediate that $\tau \in \mathcal{Z}^0(\mathfrak{M}, \mathfrak{M}^*)$. Assume now that τ is *not* in the w^*-closed convex hull of $\{v^* \cdot \phi \cdot v : v \in \mathcal{I}(\mathfrak{M})\}$. By the Hahn–Banach theorem, there is $x \in \mathfrak{M}$ such that

$$\mathrm{Re}\langle x, \tau \rangle > \sup \mathrm{co}\{\mathrm{Re}\langle x, v^* \cdot \phi \cdot v \rangle : v \in \mathcal{I}(\mathfrak{M})\}$$
$$= \sup \mathrm{co}\{\mathrm{Re}\langle vxv^*, \phi \rangle : v \in \mathcal{I}(\mathfrak{M})\}$$

Since $\langle x, \tau \rangle = \langle Tx, \phi \rangle$, this means that Tx cannot be contained in the closed, convex hull of $\{vxv^* : v \in \mathcal{I}(\mathfrak{M})\}$, which contradicts Lemma 4.2.9(ii).

The remaining claims are easily checked (by you!). $\square$

We now come to the place in the proof of Theorem 4.2.4, where Theorem 4.2.5 is invoked.

Exercise 4.2.4 Let $\mathfrak{A}$ be a unital C^*-algebra, let E be a unital Banach $\mathfrak{A}$-bimodule, and let $D \in \mathcal{Z}^1(\mathfrak{A}, E)$. Show that

$$v^* \cdot D(vav^*) - Da = a \cdot (Dv^*) \cdot v - (Dv^*) \cdot va \qquad (a \in \mathfrak{A},\ v \in \mathcal{I}(\mathfrak{A})).$$

Proposition 4.2.11 *Let $\mathfrak{M}$ be a von Neumann algebra, and let $D \in \mathcal{Z}^1(\mathfrak{M}, \mathfrak{M}^*)$. Then there is a trace* tr *on $\mathfrak{M}$ with $\|\mathrm{tr}\| \leq 1$, and $D_0 \in \mathcal{Z}^1(\mathfrak{M}, \mathfrak{M}^*)$ such that:*

- (i) $D - D_0 \in \mathcal{B}^1(\mathfrak{M}, \mathfrak{M}^*)$.
- (ii) $|\langle y, D_0 x\rangle| \leq 2\sqrt{2}\,\|D\|\|x\|\sqrt{\langle y^* y, \mathrm{tr}\rangle}$ $\qquad (x, y \in \mathfrak{M})$.

Proof Apply the Grothendieck inequality (Theorem 4.2.5) to

$$\mathfrak{M} \times \mathfrak{M} \to \mathbb{C}, \quad (x, y) \mapsto \langle y, Dx\rangle,$$

and obtain states $\phi_1, \phi_2, \psi_1, \psi_2$ on $\mathfrak{M}$ such that

$$|\langle y, Dx\rangle| \leq \|D\|\sqrt{\langle x^* x, \phi_1\rangle + \langle xx^*, \phi_2\rangle}\sqrt{\langle y^* y, \psi_1\rangle + \langle yy^*, \psi_2\rangle} \qquad (x, y \in \mathfrak{M}).$$

Letting $\psi := \frac{1}{2}(\psi_1 + \psi_2)$, we obtain

$$|\langle y, Dx\rangle| \leq 2\|D\|\|x\|\sqrt{\langle y^* y + yy^*, \psi\rangle} \qquad (x, y \in \mathfrak{M}). \tag{4.11}$$

By Corollary 4.2.10, there is a trace tr in the w^*-closed, convex hull of $\{v^* \cdot \psi \cdot v : v \in \mathcal{I}(\mathfrak{M})\}$, so that, in particular, $\|\mathrm{tr}\| \leq 1$. For each function $f : \mathcal{I}(\mathfrak{M}) \to [0, \infty)$ with finite support such that $\sum_{v \in \mathcal{I}(\mathfrak{M})} f(v) = 1$, define $T_f : \mathfrak{M}^* \to \mathfrak{M}^*$ through

$$T_f \phi := \sum_{v \in \mathcal{I}(\mathfrak{M})} f(v) v^* \cdot \phi \cdot v \qquad (\phi \in \mathfrak{M}^*).$$

Let $\mathcal{F}$ denote the collection of all finite subsets of $\mathfrak{M}$. For each finite subset $F \in \mathcal{F}$ and for each $n \in \mathbb{N}$, there is a function $f_{F,n} : \mathcal{I}(\mathfrak{M}) \to [0, \infty)$ with finite support and $\sum_{v \in \mathcal{I}(\mathfrak{M})} f(v) = 1$ such that

$$|\langle x, T_{f_{F,f}} \psi - \mathrm{tr}\rangle| < \frac{1}{n} \qquad (x \in F)$$

(this is a consequence of Corollary 4.2.10). It follows that $\mathrm{tr} := w^*\text{-}\lim_{F,n} T_{f_{F,n}}$. For $(F, n) \in \mathcal{F} \times \mathbb{N}$, let

$$\phi_{F,n} := \sum_{v \in \mathcal{I}(\mathfrak{M})} f(v) Dv^* \cdot v.$$

Let $\mathcal{U}$ be an ultrafilter on $\mathcal{F} \times \mathbb{N}$ that dominates the canonical order filter. Define $\phi := w^*\text{-}\lim_{\mathcal{U}} \phi_{F,n}$, and let $D_0 := D + \mathrm{ad}_\phi$, so that (i) is satisfied. For $x, y \in \mathfrak{M}$, we obtain:

$$\langle y, \mathrm{ad}_\phi x\rangle = \lim_{\mathcal{U}}\langle yx - xy, \phi_{F,n}\rangle$$

$$= \lim_{\mathcal{U}} \sum_{v \in \mathcal{I}(\mathfrak{M})} f_{F,n}(v)\langle yx - xy, Dv^* \cdot v\rangle$$

$$= \lim_{\mathcal{U}} \sum_{v \in \mathcal{I}(\mathfrak{M})} f_{F,n}(v)\langle y, x \cdot Dv^* \cdot v - Dv^* \cdot vx\rangle$$

$$= \lim_{\mathcal{U}} \sum_{v \in \mathcal{I}(\mathfrak{M})} f_{F,n}(v)\langle y, v^* \cdot D(vxv^*) \cdot v - Dx\rangle, \qquad \text{by Exercise 4.2.4,}$$

$$= \lim_{\mathcal{U}} \sum_{v \in \mathcal{I}(\mathfrak{M})} f_{F,n}(v)\langle y, v^* \cdot D(vxv^*) \cdot v\rangle - \langle y, Dx\rangle. \tag{4.12}$$

This yields for $x, y \in \mathfrak{M}$:

$$|\langle y, D_0 x\rangle| = \left|\lim_{\mathcal{U}} \sum_{v \in \mathcal{I}(\mathfrak{M})} f_{F,n}(v)\langle y, v^* \cdot D(vxv^*) \cdot v\rangle\right| \qquad \text{by (4.12)},$$

$$\leq \limsup_{F,n} \sum_{v \in \mathcal{I}(\mathfrak{M})} f_{F,n}(v)|\langle y, v^* \cdot D(vxv^*) \cdot v\rangle|$$

$$\leq \limsup_{F,n} \sum_{v \in \mathcal{I}(\mathfrak{M})} 2\|D\|\|x\| f_{F,n}(v)\sqrt{\langle vy^*yv^* + vyy^*v^*, \psi\rangle}, \qquad \text{by (4.11)},$$

$$\leq 2\|D\|\|x\| \limsup_{F,n} \sqrt{\sum_{v \in \mathcal{I}(\mathfrak{M})} f_{F,n}(v)\langle vy^*yv^* + vyy^*v^*, \psi\rangle} \qquad \text{(Why?)}$$

$$= 2\|D\|\|x\| \lim_{F,n} \sqrt{\langle y^*y + yy^*, T_{f_{F,n}}\psi\rangle}$$

$$= 2\|D\|\|x\| \sqrt{\langle y^*y + yy^*, \text{tr}\rangle}$$

$$\leq 2\sqrt{2}\,\|D\|\|x\| \sqrt{\langle y^*y, \text{tr}\rangle},$$

which establishes (ii). $\square$

If $\mathfrak{A}$ is a unital C^*-algebra, we write $\mathcal{U}(\mathfrak{A})$ to denote its unitary group. The following exercise is solved in the same way as Exercise 2.1.5 (and can, in fact, be derived from a solution to that exercise):

Exercise 4.2.5 Let $\mathfrak{A}$ be a unital C^*-algebra, let E be a Banach $\mathfrak{A}$-bimodule whose underlying Banach space is reflexive, and let $D \in \mathcal{Z}^1(\mathfrak{A}, E)$. Show that $D = \text{ad}_{-x}$ for some x in the closed, convex hull of $\{D(u) \cdot u^* : u \in \mathcal{U}(\mathfrak{A})\}$.

We need one more lemma:

Lemma 4.2.12 *Let $\mathfrak{A}$ be a unital C^*-algebra, let $D \in \mathcal{Z}^1(\mathfrak{A}, \mathfrak{A}^*)$, and let $\text{tr} \in \mathfrak{A}^*$ be a trace such that there is $C \geq 0$ with*

$$|\langle b, Da\rangle| \leq C\|a\|\sqrt{\langle b^*b, \text{tr}\rangle} \qquad (a, b \in \mathfrak{A}). \tag{4.13}$$

Then D is inner.

Proof Let

$$I := \{a \in \mathfrak{A} : \langle a^*a, \text{tr}\rangle = 0\}.$$

Then I is a closed ideal in $\mathfrak{A}$, and $\mathfrak{A}/I$ is a pre-Hilbert space whose inner product is given by

$$[a + I, b + I] := \langle b^*a, \text{tr}\rangle \qquad (a, b \in \mathfrak{A}).$$

Let $\mathfrak{H}$ be the completion of $\mathfrak{A}/I$ with respect to the norm induced by this inner product, and define $T \colon \mathfrak{H} \to \mathfrak{A}^*$ through

$$\langle a, T\xi\rangle := [a + I, \xi] \qquad (a \in \mathfrak{A}, \xi \in \mathfrak{H}).$$

It is clear from this definition (Why?) that T is injective. We may thus equip $T\mathfrak{H}$ with the inner product

$$\langle T\xi, T\eta \rangle := [\xi, \eta] \qquad (\xi, \eta \in \mathfrak{H}).$$

Certainly, $T\mathfrak{H}$ is a Hilbert space. For $a \in \mathfrak{A}$, let $R_a \in \mathcal{L}(\mathfrak{H})$ be the operator induced by multiplication with a from the right. For $\xi \in \mathfrak{H}$, we have

$$\langle b, a \cdot T\xi \rangle = [ba + I, \xi] = [R_a(b + I), \xi] = [b + I, R_a^*\xi] = \langle b, T(R_a^*\xi) \rangle \qquad (b \in \mathfrak{A})$$

so that $a \cdot T\xi \in T\mathfrak{H}$ and, moreover

$$\| a \cdot T\xi \| = \| R_a^*\xi \| \le \|a\| \|\xi\| = \|a\| \|T\xi\|.$$

Hence, $T\mathfrak{H}$ is a left Banach $\mathfrak{A}$-module. Analoguously, one sees that $T\mathfrak{H}$ is a right Banach $\mathfrak{A}$-module and, in fact, a Banach $\mathfrak{A}$-bimodule.

It is immediate from (4.13) that $b \in \ker Da$ for all $a \in \mathfrak{A}$ and $b \in I$. For each $a \in \mathfrak{A}$, we may thus define

$$\phi_a : \mathfrak{A}/I \to \mathbb{C}, \quad b + I \mapsto \langle b, Da \rangle.$$

It is easily seen (How?) that ϕ_a is continuous with respect to the norm induced by the inner product on $\mathfrak{A}/I$, and thus extends to $\mathfrak{H}$. Consequently, there is $\xi_a \in \mathfrak{H}$ such that

$$\langle b, Da \rangle = \langle b + I, \phi_a \rangle = [b + I, \xi_a] = \langle b, T\xi_a \rangle \qquad (b \in \mathfrak{A}).$$

It follows that $Da = T\xi_a \in T\mathfrak{H}$. Furthermore, (4.13) implies that $\|T\xi_a\| = \|\xi_a\| = \|\phi_a\| \le C$ for each $a \in \mathfrak{A}$. Hence, $D : \mathfrak{A} \to T\mathfrak{H}$ is bounded. It now follows from Exercise 4.2.5 that $D \in \mathcal{B}^1(\mathfrak{A}, \mathfrak{A}^*)$. $\square$

We are now in a position to prove Theorem 4.2.4, at least in the von Neumann algebra case:

Corollary 4.2.13 *Let $\mathfrak{M}$ be a von Neumann algebra. Then $\mathfrak{M}$ is weakly amenable.*

Proof Let $D \in \mathcal{Z}^1(\mathfrak{M}, \mathfrak{M}^*)$. By Proposition 4.2.11, there is a trace $\mathrm{tr} \in \mathfrak{M}^*$ and $D_0 \in \mathcal{Z}^1(\mathfrak{M}, \mathfrak{M}^*)$ with $D - D_0 \in \mathcal{B}^1(\mathfrak{M}, \mathfrak{M}^*)$ and

$$|\langle y, D_0 x \rangle| \le 2\sqrt{2}\, \|D\| \|x\| \sqrt{\langle y^*y, \mathrm{tr}\rangle} \qquad (x, y \in \mathfrak{M}).$$

This, however, implies that $D_0 \in \mathcal{B}^1(\mathfrak{M}, \mathfrak{M}^*)$ by Lemma 4.2.12, so that $D \in \mathcal{B}^1(\mathfrak{M}, \mathfrak{M}^*)$. $\square$

Before we treat the general case, we ask the reader to do some preparations:

Exercise 4.2.6 If E is a Banach space, then there is a canonical projection from E^{***} onto E^* (by restriction), the *Dixmier projection*.

(i) Let $\mathfrak{A}$ be a Banach algebra, and let E be a Banach $\mathfrak{A}$-module (left, right, or bi-). Show that the Dixmier projection from E^{***} onto E^* is a module homomorphism.

(ii) Let $\mathfrak{A}$ be a Banach algebra such that $\mathcal{H}^1(\mathfrak{A}, E^{**}) = \{0\}$ for each Banach $\mathfrak{A}$-bimodule E. Show that $\mathfrak{A}$ is amenable.

Exercise 4.2.7 Let $\mathfrak{A}$ be a normed algebra. For $A, B \in \mathfrak{A}^{**}$, $\phi \in \mathfrak{A}^*$, and $a \in \mathfrak{A}$, define

$$\langle a, B \cdot \phi \rangle := \langle \phi \cdot a, B \rangle \quad \text{and} \quad \langle \phi, AB \rangle := \langle B \cdot \phi, A \rangle.$$

Show that this defines a product on $\mathfrak{A}^{**}$ (the *first Arens product*) that extends the product on $\mathfrak{A}$, turns $\mathfrak{A}$ into a Banach algebra, and is w^*-continuous in the first variable. How would you define a product on $\mathfrak{A}^{**}$ turning $\mathfrak{A}^{**}$ into a Banach algebra and extending the product on $\mathfrak{A}$ which is w^*-continuous in the *second* variable (the *second Arens product*)?

If you want to know more about Arens products, [Pal] is a good source to learn from. If you want to dig deeper, have a look at [Gro] (if you aren't deterred by the fact that it's in German).

To prove Theorem 4.2.4, we need the following facts about the bidual of a C^*-algebra $\mathfrak{A}$:

- The first and the second Arens product on $\mathfrak{A}^{**}$ coincide, i.e. $\mathfrak{A}$ is *Arens regular*; in particular, the product on $\mathfrak{A}^{**}$ is separately w^*-continuous.
- More generally, whenever $\mathfrak{B}$ is another C^*-algebra, E is a Banach space, and V belongs to $\mathcal{L}^2(\mathfrak{A}, \mathfrak{B}; E)$, then V has a — necessarily unique — separately w^*-continuous extension $\tilde{V} \in \mathcal{L}^2(\mathfrak{A}^{**}, \mathfrak{B}^{**}; E^{**})$ with $\|\tilde{V}\| = \|V\|$. This follows from Corollary 4.2.6 and [Gro, Satz 1.29].
- If $\pi \colon \mathfrak{A} \to \mathcal{L}(\mathfrak{H})$ is the universal representation of $\mathfrak{A}$, then π extends to an isometric linear map from $\mathfrak{A}^{**}$ onto $\pi(\mathfrak{A})''$ which is continuous with respect to the w^*-topology on $\mathfrak{A}^{**}$ and the ultraweak topology on $\mathcal{L}(\mathfrak{H})$. It follows that $\mathfrak{A}^{**}$ with its Arens product is a von Neumann algebra ([Dal, Theorem 3.2.36]).

Proof of Theorem 4.2.4 Let $D \in \mathcal{Z}^1(\mathfrak{A}, \mathfrak{A}^*)$, and define

$$V \colon \mathfrak{A} \times \mathfrak{A} \to \mathbb{C}, \quad (a, b) \mapsto \langle b, Da \rangle.$$

Since D is a derivation, V satisfies

$$V(ab, c) = V(a, bc) + V(b, ca) \qquad (a, b, c \in \mathfrak{A}).$$

Let $\tilde{V} \in \mathcal{L}^2_{w^*}(\mathfrak{A}^{**}, \mathbb{C})$ be the separately w^*-continuous extension of V. It follows (How?) that $\tilde{V}$ satisfies

$$\tilde{V}(ab, c) = \tilde{V}(a, bc) + \tilde{V}(b, ca) \qquad (a, b, c \in \mathfrak{A}^{**}).$$

Define $\tilde{D} \colon \mathfrak{A}^{**} \to \mathfrak{A}^{***}$ through

$$\langle b, \tilde{D}a \rangle := \tilde{V}(a, b) \qquad (a, b \in \mathfrak{A}^{**}).$$

It is easily seen that $\tilde{D} \in \mathcal{Z}^1(\mathfrak{A}^{**}, \mathfrak{A}^{***})$ (Please, check this and argue that $\tilde{D} = D^{**}$). From the von Neumann algebra case, we obtain $\tilde{\phi} \in \mathfrak{A}^{***}$ such that $\tilde{D} = \mathrm{ad}_{\tilde{\phi}}$. Letting $\phi := \tilde{\phi}|_{\mathfrak{A}}$, Exercise 4.2.6 implies that $D = \mathrm{ad}_\phi$. $\quad\Box$

4.3 Biprojectivity and biflatness

Definition 4.3.1 A Banach algebra $\mathfrak{A}$ is *biprojective* if $\Delta \colon \mathfrak{A} \hat{\otimes} \mathfrak{A} \to \mathfrak{A}$ has a bounded right inverse which is an $\mathfrak{A}$-bimodule homomorphism.

Why do Banach algebras satisfying this definition carry the strange name "biprojective"? The reason will be revealed in the next chapter ...

In view of Exercise 4.1.6, it is clear that every super-amenable Banach algebra is biprojective. The following two exercises ask you to clarify the relation between amenability, super-amenability, and biprojectivity:

Exercise 4.3.1 Show that a biprojective Banach algebra $\mathfrak{A}$ is super-amenable if and only if it has an identity and if and only if $\mathfrak{A}^{\#}$ is biprojective.

Exercise 4.3.2 Let $\mathfrak{A}$ be a biprojective Banach algebra with a bounded approximate identity. Show that $\mathfrak{A}$ is amenable.

How strong is biprojectivity relative to super-amenability? Is it also so strong that we only get dull examples? Fortunately, the class of biprojective Banach algebras is much richer.

Examples 4.3.2   (a) Let $(E, F, \langle \cdot, \cdot \rangle)$ be a *dual pair of Banach spaces*, i.e. a pair of Banach spaces (E, F) with a non-degenerate, bounded, bilinear map $\langle \cdot, \cdot \rangle \colon E \times F \to \mathbb{C}$. Examples of dual pairs of Banach spaces are, for example, $(E, E^*, \langle \cdot, \cdot \rangle)$ and $(E^*, E, \langle \cdot, \cdot \rangle)$, where E is an arbitrary Banach space, and $\langle \cdot, \cdot \rangle$ is the usual duality. Define

$$(x \otimes y)(u \otimes v) := \langle u, y \rangle x \otimes v \qquad (x, u \in E, \, y, v \in F). \tag{4.14}$$

It is routinely checked that (4.14) defines a product on $E \hat{\otimes} F$ turning it into a Banach algebra (Do you believe that without actually checking it?). Fix $(x_0, y_0) \in E \times F$ such that $\langle x_0, y_0 \rangle = 1$. Define $\rho \colon E \hat{\otimes} F \to (E \hat{\otimes} F) \hat{\otimes} (E \hat{\otimes} F)$ through

$$\rho(x \otimes y) = (x \otimes y_0) \otimes (x_0 \otimes y) \qquad (x \in E, \, y \in F).$$

It is routinely checked that ρ is the desired right inverse of Δ, so that $E \hat{\otimes} F$ is biprojective. In particular, if E has the approximation property, $F = E^*$, and $\langle \cdot, \cdot \rangle$ is the canonical duality, $\mathcal{N}(E) \cong E \hat{\otimes} E^*$ is biprojective.

(b) Let G be a compact group, and identify $L^1(G) \hat{\otimes} L^1(G)$ with $L^1(G \times G)$ as in Exercise B.2.17 (How does an element of $L^1(G)$ act on $L^1(G \times G)$?). Define $\rho \colon L^1(G) \to L^1(G \times G)$ through

$$\rho(f)(g, h) := f(gh) \qquad (f \in L^1(G), \, g, h \in G).$$

If $f_1, f_2 \in L^1(G)$, then

$$\Delta(f_1 \otimes f_2)(g) := \int_G f_1(gh^{-1}) f_2(h) \, dm_G(h)$$

$$= \int_G (f_1 \otimes f_2)(gh^{-1}, h) \, dm_G(h) \qquad (g \in G),$$

so that

$$\Delta(f)(g) = \int_G f(gh^{-1}, h) \, dm_G(h) \qquad (f \in L^1(G \times G), \, g \in G).$$

It follows that $\Delta \circ \rho = \mathrm{id}_{L^1(G)}$. It is equally easy to see that ρ is an $L^1(G)$-bimodule homomorphism (but, please, check it nevertheless).

Exercise 4.3.3 Show that ℓ^1 (with pointwise multiplication) is biprojective.

Exercise 4.3.4 Let $\mathfrak{A}$ be a Banach algebra with compact multiplication, i.e. for each $a \in \mathfrak{A}$ multiplication with a both from the left and from the right is a compact operator. Show that, if $\mathfrak{A}$ is amenable, it is biprojective. (*Hint:* Let $(\mathbf{m}_\alpha)_\alpha$ be an approximate diagonal for $\mathfrak{A}$, and let $\mathcal{U}$ be an ultrafilter on the index set of $(\mathbf{m}_\alpha)_\alpha$ that dominates the order filter. Define $\rho: \mathfrak{A} \to \mathfrak{A} \hat{\otimes} \mathfrak{A}$ through $\rho(a) := \lim_\mathcal{U} a \cdot \mathbf{m}_\alpha$ for $a \in A$. Why does this limit exist?).

Exercise 4.3.5 Let $\mathfrak{A}$ be a commutative Banach algebra with character space $\Omega_\mathfrak{A}$. Show that $\Omega_\mathfrak{A}$ is discrete if $\mathfrak{A}$ is biprojective. (*Hint:* Use the fact ([B–D, Proposition 42.19]) that $\Omega_{\mathfrak{A} \hat{\otimes} \mathfrak{A}} \cong \Omega_\mathfrak{A} \times \Omega_\mathfrak{A}$. Then show that $\{(\omega, \omega) : \omega \in \Omega_\mathfrak{A}\}$ is open in $\Omega_{\mathfrak{A} \hat{\otimes} \mathfrak{A}}$.) Conversely show that, if $\mathfrak{A}$ is a commutative C^*-algebra with $\Omega_\mathfrak{A}$ discrete, then $\mathfrak{A}$ is biprojective.

The algebras in Examples 4.3.2(b) are amenable, but what about Examples 4.3.2(a)? To answer this question, we first require a definition and then a lemma:

Definition 4.3.3 Let $(E, F, \langle \cdot, \cdot \rangle)$ be a dual pair of Banach spaces. A linear operator $T: E \to E$ is called F-*nuclear* if there are sequences $(x_n)_{n=1}^\infty$ and $(y_n)_{n=1}^\infty$ in E and F, respectively, with $\sum_{n=1}^\infty \|x_n\|\|y_n\| < \infty$ such that

$$Tx = \sum_{n=1}^\infty \langle x, y_n \rangle x_n \qquad (x \in E). \qquad (4.15)$$

Its F-*nuclear norm* is the infimum over all $\sum_{n=1}^\infty \|x_n\|\|y_n\|$ with $\sum_{n=1}^\infty \|x_n\|\|y_n\| < \infty$ for which (4.15) holds. The collection of all F-nuclear operators on E is denoted by $\mathcal{N}_F(E)$.

Of course, if $F = E^*$, the F-nuclear operators are just the usual nuclear operators.

Exercise 4.3.6 Let $(E, F, \langle \cdot, \cdot \rangle)$ be a dual pair of Banach spaces. Prove the following:

(i) Show that $\mathcal{N}_F(E)$, equipped with the F-nuclear norm, is a Banach algebra which is a quotient of the Banach algebra $E \hat{\otimes} F$ in a canonical fashion.

(ii) If E or F has the approximation property, then the canonical quotient map from $E \hat{\otimes} F$ onto $\mathcal{N}_F(E)$ is an isomorphism.

Lemma 4.3.4 *Let* $(E, F, \langle \cdot, \cdot \rangle)$ *be a dual pair of Banach spaces, and let* $(T_\alpha)_\alpha$ *be a net in* $\mathcal{N}_F(E)$ *which converges to* id_E *in the strong operator topology and which is bounded in the* F-*nuclear norm. Then* E *and* F *have the same finite dimension.*

Proof The canonical embedding of $\mathcal{N}_F(E)$ into $\mathcal{A}(E)$ is continuous and thus turns $\mathcal{A}(E)$ into a left Banach $\mathcal{N}_F(E)$-module. As in the proof of Theorem 3.1.2, we get $\lim_\alpha T_\alpha S = S$ for all $S \in \mathcal{K}(E)$. From Cohen's factorization theorem in its module version ([B–D, Theorem 11.10]), each operator in $\mathcal{A}(E)$ is the product of an F-nuclear operator and another operator in $\mathcal{A}(E)$. Since $\mathcal{N}_F(E) \subset \mathcal{N}(E)$, and since $\mathcal{N}(E)$ is an ideal in $\mathcal{A}(E)$, this means that $\mathcal{A}(E) = \mathcal{N}(E)$. If $(T_\alpha)_\alpha$ is bounded in the F-nuclear norm, it is also bounded in the operator norm, so that E has the approximation property. It follows that

$$E \hat{\otimes} E^* \cong \mathcal{N}(E) = \mathcal{A}(E) \cong E \check{\otimes} E^* \qquad (4.16)$$

with the canonical identifications.

Define $\mathrm{Tr}: \mathcal{F}(E) \to \mathbb{C}$ through

$$\mathrm{Tr}(x \odot \phi) = \langle x, \phi \rangle \qquad (x \in E, \, \phi \in E^*).$$

Clearly, Tr is continuous on $\mathcal{F}(E) \cong E \otimes E^*$ with respect to the projective norm and thus, by (4.16), extends to $\mathcal{A}(E)$ as a continuous linear functional.

It is easy to see (How?) that either both E and F are infinite-dimensional or both E and F have the same finite dimension. Assume towards a contradiction that E is infinite-dimensional. Then, for each $n \in \mathbb{N}$, there is a projection $P_n \in \mathcal{F}(E)$ onto some n-dimensional subspace of E such that $\|P_n\| \leq \sqrt{n}$ (the existence of such projections a guaranteed by Theorem C.3.1). It is clear that

$$\frac{1}{\sqrt{n}}|\mathrm{Tr}(P_n)| \leq \|\mathrm{Tr}\| \qquad (n \in \mathbb{N}). \tag{4.17}$$

For each $n \in \mathbb{N}$, there is a finite, biorthogonal system $\left\{ \left(x_j^{(n)}, \phi_j^{(n)} \right) : j = 1, \ldots n \right\}$ such that $P_n = \sum_{j=1}^{n} x_j^{(n)} \odot \phi_j^{(n)}$. It follows that

$$\frac{1}{\sqrt{n}}\mathrm{Tr}(P_n) = \frac{1}{\sqrt{n}} \sum_{j=1}^{n} \left\langle x_j^{(n)}, \phi_j^{(n)} \right\rangle = \sqrt{n} \qquad (n \in \mathbb{N}),$$

which contradicts (4.17). $\quad\square$

Theorem 4.3.5 *For a dual pair $(E, F, \langle \cdot, \cdot \rangle)$ of Banach spaces, the following are equivalent:*

(i) $E \hat{\otimes} F$ *is super-amenable.*

(ii) $E \hat{\otimes} F$ *is amenable.*

(iii) $E \hat{\otimes} F$ *has a bounded approximate identity.*

(iv) $E \hat{\otimes} F$ *has a bounded left approximate identity.*

(v) $\mathcal{N}_F(E)$ *has a bounded left approximate identity.*

(vi) $\dim E = \dim F < \infty$.

Proof The implications (i) $\Longrightarrow$ (ii) $\Longrightarrow$ (iii) $\Longrightarrow$ (iv) are clear.

Exercise 4.3.6(i) yields (iv) $\Longrightarrow$ (v).

Suppose that $\mathcal{N}_F(E)$ has a bounded left approximate identity, $(T_\alpha)_\alpha$ say. As in the proof of Theorem 3.1.2, it follows that $T_\alpha \to \mathrm{id}_E$ in the strong operator topology. By Lemma 4.3.4, this implies (vi).

For (iv) $\Longrightarrow$ (i), just observe that $E \hat{\otimes} F \cong \mathbb{M}_n$, where $n = \dim E$. $\quad\square$

As a consequence of Theorem 4.3.5 and Exercise 4.3.6(ii), we get:

Corollary 4.3.6 *Let $(E, F, \langle \cdot, \cdot \rangle)$ be a dual pair of Banach spaces. Then:*

(i) $\mathcal{N}_F(E)$ *is amenable if and only if $\dim E = \dim F < \infty$.*

(ii) $\mathcal{N}_F(E)$ *is biprojective whenever E or F has the approximation property.*

Hence, for a wide range of dual pairs $(E, F, \langle \cdot, \cdot \rangle)$ of Banach spaces, the algebras $\mathcal{N}_F(E)$ are biprojective. As we shall see towards the end of this section, these algebras are the smallest "building blocks" out of which more general biprojective Banach algebras (with a few, relatively mild restrictions) are constructed. To prepare the ground for this structure theorem, we first have to consider other algebras of operators associated with dual pairs of Banach spaces.

Definition 4.3.7 Let $(E, F, \langle \cdot, \cdot \rangle)$ be a dual pair of Banach spaces. An *F-bounded operator* on E is an operator $T \in \mathcal{L}(E)$ such that $T^*F \subset F$, where F is canonically identified with a subspace of E^*. The collection of all F-bounded operators on E is denoted by $\mathcal{L}_F(E)$.

Exercise 4.3.7 Let $(E, F, \langle \cdot, \cdot \rangle)$ be a dual pair of Banach spaces.

(i) Let $T \in \mathcal{L}_F(E)$. Show that $T^*|_F$ is continuous with respect to the norm on F.

(ii) Show that $\mathcal{L}_F(E)$ equipped with the norm

$$\|T\|_{\mathcal{L}_F(E)} := \max\{\|T\|_{\mathcal{L}(E)}, \|T^*|_F\|_{\mathcal{L}(F)}\} \qquad (T \in \mathcal{L}_F(E))$$

is a Banach algebra.

(iii) Show that $\mathcal{N}_F(E)$ is an ideal of $\mathcal{L}_F(E)$ and that the canonical embedding of $\mathcal{N}_F(E)$ into $\mathcal{L}_F(E)$ is a contraction when $\mathcal{N}_F(E)$ is equipped with the F-nuclear norm.

Definition 4.3.8 Let $(E, F, \langle \cdot, \cdot \rangle)$ be a dual pair of Banach spaces.

(i) Let $\mathcal{F}_F(E) := \mathcal{F}(E) \cap \mathcal{L}_F(E)$.

(ii) A *Banach F-operator algebra* on E is a subalgebra $\mathfrak{A}$ of $\mathcal{L}_F(E)$ containing $\mathcal{F}_F(E)$ which is a Banach algebra under some norm such that the embedding of $\mathfrak{A}$ into $\mathcal{L}_F(E)$ is continuous. In case $F = E^*$, we say that $\mathfrak{A}$ is a *Banach operator algebra* on E.

Exercise 4.3.8 Let $(E, F, \langle \cdot, \cdot \rangle)$ be a dual pair of Banach spaces.

(i) Let $T \in \mathcal{F}_F(E)$. Show that there are $x_1, \ldots, x_n \in E$ and $y_1, \ldots, y_n \in F$ such that $T = \sum_{j=1}^n x_j \odot y_j$, where the notation is chosen in (obvious) analogy with Definition B.2.6.

(ii) Show that $\mathcal{F}_F(E)$ is simple. If you have done Exercise 3.1.6(i), this should pose no problems

$\ldots$

(iii) Show that $\mathcal{N}_F(E)$ is topologically simple.

For the following proposition, we require the notion of the socle of a (semiprime) algebra. If you don't know what that is, have a look a [Pal, 8.2].

Proposition 4.3.9 *For a Banach algebra $\mathfrak{A}$, the following are equivalent:*

(i) *$\mathfrak{A}$ is primitive with $\mathrm{soc}(\mathfrak{A}) \neq \{0\}$.*

(ii) *There is a dual pair $(E, F, \langle \cdot, \cdot \rangle)$ of Banach spaces such that $\mathfrak{A}$ is isomorphic to a Banach F-operator algebra on E, where the embedding of $\mathfrak{A}$ into $\mathcal{L}_F(E)$ is a contraction.*

Proof Suppose that (ii) holds. Since $\mathcal{F}_F(E)$ is easily seen to act irreducibly on E, it follows that $\mathfrak{A}$ acts irreducibly on E and thus is primitive. Also, each rank one projection in $P \in \mathcal{F}_F(E)$ is a minimal projection in $\mathfrak{A}$, so that $\mathrm{soc}(\mathfrak{A}) \neq \{0\}$.

For the converse, note first that, due to the fact that $\mathrm{soc}(\mathfrak{A}) \neq \{0\}$, there is a minimal idempotent $e \in \mathfrak{A}$. Let $E := \mathfrak{A}e$ and $F := e\mathfrak{A}$. Let E and F be equipped with their respective quotient norms, i.e., for example in the case of E,

$$\|x\| := \inf\{\|a\| : a \in \mathfrak{A}, \, ae = x\} \qquad (x \in E).$$

Define a bilinear form

$$\langle \cdot, \cdot \rangle \colon E \times F \to \mathbb{C}, \quad (x, y) \mapsto yx$$

(Why is this well defined?). In the proof of [Pal, Theorem 8.3.6], it is shown that $\langle \cdot, \cdot \rangle$ is non-degenerate. Let $\mathfrak{A}$ act on E via left multiplication. It is immediate that $\mathfrak{A} \subset \mathcal{L}_F(E)$ such that the canonical embedding is a contraction. By [Pal, Theorem 8.3.6], soc($\mathfrak{A}$) when viewed as a subalgebra of $\mathcal{L}_F(E)$ is precisely $\mathcal{F}_F(E)$. □

Exercise 4.3.9 Understand the proof of [Pal, Theorem 8.3.6].

In view of Proposition 4.3.9, our first goal is to establish that (certain) biprojective Banach algebras have non-zero socle. Towards that goal, we proceed through a series of lemmas and propositions.

Let $\mathfrak{A}$ be a Banach algebra. In Exercise 2.1.4, we introduced the notion of an essential Banach $\mathfrak{A}$-bimodule. Analoguously, we call a left Banach $\mathfrak{A}$-module E essential if the linear span of $\{a \cdot x : a \in \mathfrak{A}, x \in E\}$ is dense in E. For any left Banach $\mathfrak{A}$-module, we write $\Delta_{\mathfrak{A},E}$ for the map from $\mathfrak{A} \hat{\otimes} E$ to E defined through $\Delta_{\mathfrak{A},E}(a \otimes x) := a \cdot x$. If $\mathfrak{A} \hat{\otimes} E$ is equipped with the module operation

$$a \cdot (b \otimes x) := ab \otimes x \qquad (a, b \in \mathfrak{A}, x \in E),$$

$\Delta_{\mathfrak{A},E}$ becomes a left $\mathfrak{A}$-module homomorphism.

Lemma 4.3.10 *Let $\mathfrak{A}$ be a biprojective Banach algebra, and let L be a closed left ideal of $\mathfrak{A}$ which is essential as a left Banach $\mathfrak{A}$-module. Then the module map $\Delta_{\mathfrak{A},\mathfrak{A}/L}$ has a bounded right inverse which is also a left $\mathfrak{A}$-module homomorphism.*

Proof Let $\rho \colon \mathfrak{A} \to \mathfrak{A} \hat{\otimes} \mathfrak{A}$ be a bounded right inverse to $\Delta_{\mathfrak{A}}$ which is also an $\mathfrak{A}$-bimodule homomorphism. Let $\pi \colon \mathfrak{A} \to \mathfrak{A}/L$ denote the quotient map. Then

$$(\mathrm{id}_{\mathfrak{A}} \otimes \pi)(\rho(ab)) = (\mathrm{id}_{\mathfrak{A}} \otimes \pi)(\rho(a)b) = 0 \qquad (a \in \mathfrak{A}, b \in L).$$

Since L is essential, this means that $(\mathrm{id}_{\mathfrak{A}} \otimes \pi) \circ \rho$ vanishes on L and thus induces a bounded linear map $\tilde{\rho} \colon \mathfrak{A}/L \to \mathfrak{A} \hat{\otimes} (\mathfrak{A}/L)$. It is routinely checked (by you!) that $\tilde{\rho}$ is both a right inverse of $\Delta_{\mathfrak{A},\mathfrak{A}/L}$ and a left $\mathfrak{A}$-module homomorphism. □

Exercise 4.3.10 Let $\mathfrak{A}$ be a biprojective Banach algebra, and let I be a closed ideal which is essential as a left Banach $\mathfrak{A}$-module. Show that $\mathfrak{A}/I$ is biprojective.

Lemma 4.3.10 enables us to establish a converse to Examples 4.3.2(a):

Exercise 4.3.11 Let G be a locally compact group. Show that $L^1(G)$ is biprojective if and only if G is compact. (*Hint:* Apply Lemma 4.3.10 with $L = L_0^1(G)$.)

Proposition 4.3.11 *Let $\mathfrak{A}$ be a biprojective Banach algebra with the approximation property, and let L be a closed, essential left ideal of $\mathfrak{A}$. Then, for each $x \in \mathfrak{A} \setminus L$, there is a left $\mathfrak{A}$-module homomorphism $\theta \colon \mathfrak{A}/L \to \mathfrak{A}$ such that $\theta(x + L) \neq 0$.*

Proof Let $x \in \mathfrak{A} \setminus L$, and let $\rho \colon \mathfrak{A}/L \to \mathfrak{A} \hat{\otimes} (\mathfrak{A}/L)$ be as specified in Lemma 4.3.10. Since $x \notin L$, it follows that $\rho(x + L) \neq 0$. Since $\mathfrak{A}$ has the approximation property, the canonical map of $\mathfrak{A} \hat{\otimes} (\mathfrak{A}/L)$ into $\mathfrak{A} \check{\otimes} (\mathfrak{A}/L)$ is injective by Theorem C.1.5. From the definition of the injective norm, it follows that there are $\psi \in \mathfrak{A}^*$ and $\phi \in (\mathfrak{A}/L)^*$ such that $\langle \rho(x+L), \psi \otimes \phi \rangle \neq 0$ and thus $(\mathrm{id}_{\mathfrak{A}} \otimes \phi)(\rho(x + L)) \neq 0$. Let $\theta := (\mathrm{id}_{\mathfrak{A}} \otimes \phi) \circ \rho$. Then θ is easily seen to be a left $\mathfrak{A}$-module homomorphism; by definition, $\theta(x + L) \neq 0$. □

Exercise 4.3.12 Let $\mathfrak{A}$ be an algebra. For any subset S of $\mathfrak{A}$, the *left annhilator* of S in $\mathfrak{A}$ is defined as $\mathrm{lann}(S) := \{a \in \mathfrak{A} : aS = \{0\}\}$ (the *right annihilator* $\mathrm{rann}(S)$ is defined analoguously). Let I be an ideal of $\mathfrak{A}$.

(i) Show that both $\mathrm{lann}(I)$ and $\mathrm{rann}(I)$ are ideals of $\mathfrak{A}$.

(ii) Suppose that $\mathfrak{A}$ is semiprime. Show that $\mathrm{lann}(I) = \mathrm{rann}(I)$.

Lemma 4.3.12 *Let $\mathfrak{A}$ be a semiprime, biprojective Banach algebra with the approximation property, and let L be a closed left ideal. Then L is essential.*

Proof Let $\tilde{L}$ be the closed linear span of $\{ab : a \in \mathfrak{A}, b \in L\}$; clearly, $\tilde{L}$ is a closed, essential left ideal of $\mathfrak{A}$. Assume that there is $x \in L \setminus \tilde{L}$. By Proposition 4.3.11, there is a left $\mathfrak{A}$-module homomorphism $\theta : \mathfrak{A}/\tilde{L} \to \mathfrak{A}$ with $\theta(x + \tilde{L}) \neq 0$. For any $a \in \mathfrak{A}$, the element ax belongs to $\tilde{L}$, so that

$$0 = \theta(ax + \tilde{L}) = a\theta(x + \tilde{L}) \qquad (a \in \mathfrak{A}).$$

This means that $\theta(x + \tilde{L})$ lies in the right annihilator $\mathrm{rann}(\mathfrak{A})$. By definition, $\mathrm{rann}(\mathfrak{A})^2 = \{0\}$. Since $\mathfrak{A}$ is semiprime, this means $\mathrm{rann}(\mathfrak{A}) = \{0\}$, which contradicts $\theta(x + \tilde{L}) \neq 0$. $\square$

Proposition 4.3.13 *Let $\mathfrak{A}$ be a semiprime, biprojective Banach algebra with the approximation property, and let E be an irreducible left $\mathfrak{A}$-module. Then there is a minimal idempotent $e \in \mathfrak{A}$ such that $E \cong \mathfrak{A}e$.*

Proof By [B–D, Lemma 25.2], we may suppose that $E \cong \mathfrak{A}/L$, where L is a maximal modular left ideal of $\mathfrak{A}$. By Lemma 4.3.12, L is essential, and thus, by Proposition 4.3.11, $\Delta_{\mathfrak{A},\mathfrak{A}/L}$ has a bounded right inverse which is a left $\mathfrak{A}$-module homomorphism such that $\theta(x + L) \neq 0$ for some $x \notin L$. Since $\ker \theta \subsetneq \mathfrak{A}/L$ is a submodule of $\mathfrak{A}/L$, and since $\mathfrak{A}/L$ is irreducible, it follows that θ is injective, so that $\theta(\mathfrak{A}/L)$ is an irreducible submodule of $\mathfrak{A}$, meaning that it is a minimal left ideal of $\mathfrak{A}$. By [Pal, Proposition 8.2.2], there is a minimal idempotent $e \in \mathfrak{A}$ such that $\theta(\mathfrak{A}/L) = \mathfrak{A}e$. $\square$

Corollary 4.3.14 *Let $\mathfrak{A}$ be a semiprime, biprojective Banach algebra with the approximation property, and let E be an irreducible left Banach $\mathfrak{A}$-module. Then E has the approximation property.*

Proof By Proposition 4.3.13, E is (topologically) isomorphic to a complemented subspace of $\mathfrak{A}$ and thus has the approximation property (Exercise C.1.1). $\square$

Proposition 4.3.15 *Let $\mathfrak{A}$ be a semisimple, biprojective Banach algebra with the approximation property. Then $\mathrm{soc}(\mathfrak{A})$ is dense in $\mathfrak{A}$.*

Proof Let I be the closure of $\mathrm{soc}(\mathfrak{A})$ in $\mathfrak{A}$. Then I is a closed ideal in $\mathfrak{A}$ and thus, by Lemma 4.3.12, is essential as a left Banach $\mathfrak{A}$-module. Assume that there is $a \in \mathfrak{A} \setminus I$. By Proposition 4.3.11, there is a left $\mathfrak{A}$-module homomorphism $\theta : \mathfrak{A}/I \to \mathfrak{A}$ such that $\theta(a + I) \neq 0$. As in the proof of Lemma 4.3.12, we see that $\theta(a + I) \in \mathrm{rann}(I)$. By Exercise 4.3.12(ii), $\mathrm{rann}(I) = \mathrm{lann}(I)$, so that, in particular, $\mathrm{lann}(I) \neq \{0\}$.

Let $a \in \mathrm{lann}(\mathrm{soc}(\mathfrak{A})) = \mathrm{lann}(I)$. Let E be an irreducible left $\mathfrak{A}$-module. By Proposition 4.3.13, we may suppose that $E = \mathfrak{A}e$ for some minimal idempotent $e \in \mathfrak{A}$. Since $\mathfrak{A}e \subset$

soc($\mathfrak{A}$), it follows that $aE = \{0\}$. From the definition of the Jacobson radical of $\mathfrak{A}$, it follows that $a \in \text{rad}(\mathfrak{A})$. Since $a \in \text{lann}(I)$ was arbitrary, and since $\mathfrak{A}$ is semisimple, this means $\text{lann}(I) \subset \text{rad}(\mathfrak{A}) = \{0\}$, which is a contradiction. $\square$

Let $\mathfrak{A}$ be a primitive (and hence semisimple), biprojective Banach algebra with the approximation property. Then, by Proposition 4.3.15, we know that soc($\mathfrak{A}$) is dense in $\mathfrak{A}$ and thus, except in the obvious trivial situation, is non-zero. Hence, by Proposition 4.3.9, there is a dual pair $(E, F, \langle \cdot, \cdot \rangle)$ such that $\mathfrak{A}$ is isomorphic to a Banach F-operator algebra on E. Our next goal is thus to identify the biprojective F-operator algebras on E; to achieve it, we require another lemma:

Lemma 4.3.16 *Let $(E, F, \langle \cdot, \cdot \rangle)$ be a dual pair of Banach spaces, let $\mathfrak{A}$ be a Banach F-operator algebra, and let I be a non-zero ideal of $\mathfrak{A}$ which is a Banach algebra under some norm such that the embedding of I into $\mathfrak{A}$ is continuous. Then $\mathcal{N}_F(E)$ is contained in I, and the embedding of $\mathcal{N}_F(E)$ into I is continuous.*

Proof Let $T \in I \setminus \{0\}$. Then there are $x_0 \in E$ and $y_0 \in F$ with $\langle Tx_0, y_0 \rangle \neq 0$. Define $S := x_0 \odot y_0$. Then $S \in \mathcal{F}_F(E) \subset \mathfrak{A}$ and $STx_0 \neq 0$. Since both I and $\mathcal{F}_F(E)$ are ideals of $\mathfrak{A}$, it follows that $ST \in \mathcal{F}_F(E) \cap I$, so that, in particular, $\mathcal{F}_F(E) \cap I \neq \{0\}$. Since $\mathcal{F}_F(E)$ is simple by Exercise 4.3.8(ii), it follows that $\mathcal{F}_F(E) \subset I$. Hence, I is also a Banach F-operator algebra, and we may suppose without loss of generality that $\mathfrak{A} = I$.

Consider the bilinear map

$$V : E \times F \to \mathfrak{A}, \quad (x, y) \mapsto x \odot y.$$

Fix $x \in E$, and let $(y_n)_{n=1}^{\infty}$ be a sequence in F such that $y_n \to 0$ and $x \odot y_n \to T \in \mathfrak{A}$ with respect to the norm topology on $\mathfrak{A}$. For any $x' \in E$, we then have

$$Tx' = \lim_{n \to \infty} (x \odot y_n)(x') = \lim_{n \to \infty} \langle x', y_n \rangle x = 0$$

(Why does the first equality hold?). From the closed graph theorem, we conclude that V is continuous in its second variable. A similar argument shows that V is also continuous in its first variable, and the uniform boundedness principle yields that V is jointly continuous. From the universal property of the projective tensor product, it follows at once, that V induces a continuous linear map from $E \hat{\otimes} F$ into $\mathfrak{A}$; obviously, the range of this linear map is $\mathcal{N}_F(E)$. Consequently, $\mathcal{N}_F(E)$ is contained in $\mathfrak{A}$, and the embedding is continuous. $\square$

Theorem 4.3.17 *Let $(E, F, \langle \cdot, \cdot \rangle)$ be a dual pair of Banach spaces such that E or F has the approximation property. Then, for a Banach F-operator algebra $\mathfrak{A}$ on E, the following are equivalent:*

(i) *$\mathfrak{A}$ is biprojective.*
(ii) *$\mathfrak{A} = \mathcal{N}_F(E)$ (with an equivalent, but not necessarily the same norm).*

Proof By Corollary 4.3.6, (ii) $\Longrightarrow$ (i) is clear.

For the converse define, for $x \in E$ and $y \in F$,

$$\epsilon_x : \mathfrak{A} \to E, \quad T \mapsto Tx \quad \text{and} \quad \epsilon^y : \mathfrak{A} \to F, \quad T \mapsto T^*y.$$

Let $\rho \colon \mathfrak{A} \to \mathfrak{A} \hat{\otimes} \mathfrak{A}$ be a bounded right inverse of $\Delta_{\mathfrak{A}}$ as required in the definition of biprojectivity, and consider the following diagram:

$$\mathcal{N}_F(E) \longrightarrow \mathfrak{A} \stackrel{\rho}{\longrightarrow} \mathfrak{A} \hat{\otimes} \mathfrak{A} \stackrel{\epsilon_x \otimes \epsilon^y}{\longrightarrow} E \hat{\otimes} F$$

$$\begin{array}{ccc} & \downarrow & \downarrow \\ & \mathfrak{A} & \longrightarrow & \mathbb{C}; \end{array} \qquad (4.18)$$

here, the map from $\mathcal{N}_F(E)$ into $\mathfrak{A}$ is the canonical inclusion (justified by Lemma 4.3.16), the first vertical arrow is

$$\nabla_{\mathfrak{A}} \colon \mathfrak{A} \hat{\otimes} \mathfrak{A} \mapsto \mathfrak{A}, \quad S \otimes T \mapsto TS,$$

the second vertical arrow is the given duality between E and F, and the horizontal arrow in the second row is given by

$$\mathfrak{A} \to \mathbb{C}, \quad T \mapsto \langle Tx, y \rangle.$$

A straightforward diagram chase (which you should nevertheless carry out), shows that (4.18) is commutative for all $x \in E$ and $y \in F$.

We claim that $\nabla_{\mathfrak{A}} \circ \rho$ is non-zero on $\mathcal{N}_F(E)$. Let $x_0 \in E$ and $y_0 \in F$ with $\langle x_0, y_0 \rangle = 1$, and let $P := x_0 \odot y_0$. It follows that P is a projection in $\mathcal{F}_F(E)$. Let $(S_n)_{n=1}^{\infty}$ and $(T_n)_{n=1}^{\infty}$ be sequences in $\mathfrak{A}$, respectively, such that $\sum_{n=1}^{\infty} \|S_n\| \|T_n\| < \infty$ and $\rho(P) = \sum_{n=1}^{\infty} S_n \otimes T_n$. Since $P^3 = P$, it follows that

$$\rho(P) = \rho(P^3) = P \cdot \rho(P) \cdot P = \sum_{n=1}^{\infty} P S_n \otimes T_n P.$$

With $x_n = T_n x_0$ and $y_n = S_n^* y_0$ for $n \in \mathbb{N}$, we obtain $P S_n = x_0 \odot y_n$ and $T_n P = x_n \odot y_0$, as well as

$$\sum_{n=1}^{\infty} \langle x_n, y_n \rangle = \sum_{n=1}^{\infty} \langle T_n x_0, S_n^* y_0 \rangle = \sum_{n=1}^{\infty} \langle S_n T_n x_0, y_0 \rangle = \langle P x_0, y_0 \rangle = 1. \qquad (4.19)$$

It now follows that

$$\rho(P) = \sum_{n=1}^{\infty} (x_0 \odot y_n) \otimes (x_n \odot y_0) \in \mathcal{N}_F(E) \hat{\otimes} \mathcal{N}_F(E)$$

(Why does this series converge in $\mathcal{N}_F(E) \hat{\otimes} \mathcal{N}_F(E)$?), so that $\nabla_{\mathfrak{A}}(\rho(P)) = \sum_{n=1}^{\infty} x_n \odot y_n$. Since E or F has the approximation property, we may identify the F-nuclear operator $\sum_{n=1}^{\infty} x_n \odot y_n$ with the element $\sum_{n=1}^{\infty} x_n \otimes y_n \in E \hat{\otimes} F$ (Exercise 4.3.6(ii)). Since $\sum_{n=1}^{\infty} \langle x_n, y_n \rangle = 1$ by (4.19), it follows that $\nabla_{\mathfrak{A}}(\rho(P)) \neq 0$.

For $x \in E$ and $y \in F$ define

$$\theta_{x,y} \colon \mathcal{N}_F(E) \to E \hat{\otimes} F, \quad T \mapsto (\epsilon_x \otimes \epsilon^y)(\rho(T)),$$

i.e. $\theta_{x,y}$ is the composition of all horizontal arrows in the first row of (4.18). It is routinely checked (Do it!) that $\theta_{x,y}$ is an $E \hat{\otimes} F$- bimodule homomorphism of the Banach algebra $E \hat{\otimes} F$ (which we may identify with $\mathcal{N}_F(E)$). Let P be as in the previous paragraph and recall that $P \mathfrak{A} P \cong \mathbb{C} P$, so that

$$\theta_{x,y}(P) = \theta_{x,y}(P^3) = P\theta_{x,y}(P)P = \lambda_{x,y}P$$

for some $\lambda_{x,y} \in \mathbb{C}$. Let $u \in E$ and $v \in F$, and observe that

$$\begin{aligned}
\theta_{x,y}(u \odot v) &= \theta_{x,y}((u \odot y_0)P(x_0 \odot v)) \\
&= (u \odot y_0)\theta_{x,y}(P)(x_0 \odot v) \\
&= \lambda_{x,y}(u \odot y_0)P(x_0 \odot v) \\
&= \lambda_{x,y}(u \odot v).
\end{aligned}$$

It follows that

$$\theta_{x,y}(T) = \lambda_{x,y}T \qquad (T \in \mathcal{N}_F(E)). \tag{4.20}$$

Choose $S \in \mathcal{N}_F(E)$ such that $\nabla_{\mathfrak{A}}(\rho(S)) \neq 0$. Consequently, there are $x' \in E$ and $y' \in F$ such that $\langle \nabla_{\mathfrak{A}}(\rho(S))x', y' \rangle \neq 0$. From the commutativity of (4.18), it follows that $\theta_{x',y'}(S) \neq 0$, and consequently, $\lambda_{x',y'} \in \mathbb{C}$ as in (4.20) must be non-zero.

Define

$$\theta: \mathfrak{A} \to E \hat{\otimes} F, \quad T \mapsto \frac{1}{\lambda_{x',y'}}(\epsilon_{x'} \otimes \epsilon^{y'})(\rho(T))$$

It is routinely checked (please, don't skip it), that $\theta|_{\mathcal{N}_F(E)}$ is the identity, so that θ is a projection onto $\mathcal{N}_F(E)$. We claim that θ is injective. Being a composition of $\mathfrak{A}$-bimodule homomorphisms, θ is also a bimodule homomorphism, so that $I := \ker \theta$ is a closed ideal of $\mathfrak{A}$. By Lemma 4.3.16, $I \neq \{0\}$ implies that $\mathcal{N}_F(E) \subset I$, which contradicts $\theta|_{\mathcal{N}_F(E)} = \mathrm{id}_{\mathcal{N}_F(E)}$. It follows that $I = \{0\}$, so that θ is the desired isomorphism. $\square$

Corollary 4.3.18 *For a primitive Banach algebra $\mathfrak{A}$ with the approximation property the following are equivalent:*

(i) *$\mathfrak{A}$ is biprojective.*

(ii) *There is a dual pair $(E, F, \langle \cdot, \cdot \rangle)$ of Banach spaces such that $\mathfrak{A} \cong \mathcal{N}_F(E)$.*

Proof If $\mathfrak{A} \cong \mathcal{N}_F(E)$ has the approximation property, then both E and F have it as well (Why? *Hint*: Show that both spaces can be embedded into $\mathcal{N}_F(E)$ as complemented subspaces.). It follows that $\mathfrak{A} \cong E \hat{\otimes} F$ is biprojective.

Conversely, if $\mathfrak{A}$ is biprojective, it has non-zero socle by Proposition 4.3.15. By Proposition 4.3.9, there is a dual pair $(E, F, \langle \cdot, \cdot \rangle)$ of Banach spaces such that $\mathfrak{A}$ is isomorphic to a Banach F-operator algebra on E. Finally, Theorem 4.3.17 implies that $\mathfrak{A} \cong \mathcal{N}_F(E)$. $\square$

Now having a surprisingly concrete discription of all primitive, biprojective Banach algebras with the approximation property, we move on to general semisimple, biprojective Banach algebras (with the approximation property, of course). We're now just one more lemma away from a general structure theorem ...

If $\mathfrak{A}$ is a Banach algebra, by a minimal closed (two-sided) ideal of $\mathfrak{A}$, we mean a non-zero, closed ideal I of $\mathfrak{A}$ such that any other closed ideal of $\mathfrak{A}$ contained in I is either $\{0\}$ or I itself.

Lemma 4.3.19 *Let $\mathfrak{A}$ be a semiprime Banach algebra with dense socle. Then:*

(i) *The algebraic direct sum of the minimal closed (two-sided) ideals is dense in $\mathfrak{A}$.*

(ii) *Each minimal closed ideal of $\mathfrak{A}$ is generated by a minimal left ideal.*

Proof Let L be a minimal left ideal of $\mathfrak{A}$, and let I be the closed, two-sided ideal of $\mathfrak{A}$ generated by L. Let J be a closed ideal of $\mathfrak{A}$ which is contained in I. If $L \cap J \neq \{0\}$ the minimality of L yields $L \cap J = L$, so that $J = I$. Suppose that $L \cap J = \{0\}$. It follows that $L \subset \mathrm{rann}(J)$. Since $\mathrm{rann}(J)$ is an ideal of $\mathfrak{A}$ (by Exercise 4.3.12(i)) and necessarily closed, it follows that $I \subset \mathrm{rann}(J)$ and thus $J^2 = \{0\}$. Since $\mathfrak{A}$ is semiprime, this means $J = \{0\}$. Consequently, I is a minimal closed ideal of $\mathfrak{A}$. From the definition of $\mathrm{soc}(\mathfrak{A})$ it follows that $\mathrm{soc}(\mathfrak{A})$ is contained in the algebraic direct sum of the minimal closed ideals of $\mathfrak{A}$ (By the way: Why is the algebraic sum of the minimal closed ideals of $\mathfrak{A}$ direct?). Since $\mathrm{soc}(\mathfrak{A})$ is dense in $\mathfrak{A}$, this establishes (i).

For (ii), let I be a minimal closed ideal of $\mathfrak{A}$, and assume that I contains no minimal left ideal. Hence, for any minimal left ideal L of $\mathfrak{A}$, we have $IL \subset I \cap L = \{0\}$, i.e. $I \subset \mathrm{lann}(L)$. The definition of $\mathrm{soc}(\mathfrak{A})$ implies $I \subset \mathrm{lann}(\mathrm{soc}(\mathfrak{A}))$. Since $\mathrm{soc}(\mathfrak{A})$ is dense in $\mathfrak{A}$, we have $\mathrm{lann}(\mathrm{soc}(\mathfrak{A})) = \mathrm{lann}(\mathfrak{A})$ and thus, in particular, $I^2 = \{0\}$. Since $\mathfrak{A}$ is semiprime, this implies $I = \{0\}$, which is a contradiction. $\square$

Theorem 4.3.20 *Let $\mathfrak{A}$ be a semisimple, biprojective Banach algebra with the approximation property. Then the following hold:*

(i) *The algebraic direct sum of the minimal closed (two-sided) ideals is dense in $\mathfrak{A}$.*

(ii) *For each minimal closed ideal I of $\mathfrak{A}$, $\mathrm{lann}(I)$ is a primitive ideal of $\mathfrak{A}$ such that $\mathfrak{A} = I \oplus \mathrm{lann}(I)$. Moreover, all primitive ideals of $\mathfrak{A}$ arise as left annihilators of minimal closed ideals of $\mathfrak{A}$.*

(iii) *For each minimal closed ideal I of $\mathfrak{A}$, there is a dual pair $(E, F, \langle \cdot, \cdot \rangle)$ such that $I \cong \mathcal{N}_F(E)$.*

Proof Proposition 4.3.15 and Lemma 4.3.19(i) yield (i).

For (ii), let I be a minimal closed ideal of $\mathfrak{A}$. By Lemma 4.3.19(ii), I is generated by a minimal left ideal L of $\mathfrak{A}$. Since L is an irreducible left Banach $\mathfrak{A}$-module, $\mathrm{lann}(L)$ is a primitive ideal of $\mathfrak{A}$. It is easy to see (How?) that that $\mathrm{lann}(L) = \mathrm{lann}(I)$. Let $\pi : \mathfrak{A} \to \mathfrak{A}/\mathrm{lann}(I)$ be the quotient map. Since $\mathfrak{A}$ is semisimple, and thus semiprime, $\pi|_I$ is injective (Why?). Let $e \in \mathfrak{A}$ be a minimal idempotent such that $L = \mathfrak{A}e$ ([Pal, Corollary 8.2.3]). Then $\pi(e) \neq 0$ is a minimal idempotent in $\mathfrak{A}/\mathrm{lann}(I)$, so that $(\mathfrak{A}/\mathrm{lann}(I))\pi(e)$ is a minimal left ideal in $\mathfrak{A}/\mathrm{lann}(I)$ ([Pal, Proposition 8.2.2]); in particular, $\mathrm{soc}(\mathfrak{A}/\mathrm{lann}(I)) \neq \{0\}$. By Proposition 4.3.9, this means that there is a dual pair $(E, F, \langle \cdot, \cdot \rangle)$ of Banach spaces such that $\mathfrak{A}/\mathrm{lann}(I)$ is isomorphic to a Banach F-operator algebra on E. By the definition of a Banach F-operator algebra, $\mathfrak{A}$ acts irreducibly on E. Since any irreducible left Banach $\mathfrak{A}/\mathrm{lann}(I)$-module is automatically an irreducible left Banach $\mathfrak{A}$-module, Corollary 4.3.14 implies that E has the approximation property. By Lemma 4.3.12 and Exercise 4.3.10, $\mathfrak{A}/\mathrm{lann}(I)$ is again biprojective. It follows from Theorem 4.3.17 that $\mathfrak{A}/\mathrm{lann}(I) \cong \mathcal{N}_F(E)$. Since $\pi|_I$ maps I onto a non-zero ideal of $\mathfrak{A}/\mathrm{lann}(I)$, Lemma 4.3.16 implies (with the proper identifications made) that $\mathcal{N}_F(E)$ is contained in $\pi(I)$ meaning that $\pi|_I$ induces a (necessarily) topological isomorphism of I and $\mathfrak{A}/\mathrm{lann}(I)$. This establishes the first part of (ii), and (iii) as well.

To prove the second part of (ii), let P be a primitive ideal of $\mathfrak{A}$. By (i), there is at least one minimal closed ideal I of $\mathfrak{A}$ such that $I \not\subset P$. Since $\mathrm{lann}(I)I = \{0\} \subset P$, it follows from [B–D, Proposition 24.12] that $\mathrm{lann}(I) \subset P$. Similarly (How?), one obtains the reverse inclusion. $\square$

Exercise 4.3.13 Let $\mathfrak{A}$ be a semisimple, amenable Banach algebra with the approximation property. Show that $\mathfrak{A}$ has compact multiplication if and only if there is a family $(n_\alpha)_\alpha$ of positive integers such that $\mathfrak{A}$ contains a dense subalgebra isomorphic to the algebraic direct sum $\bigoplus_\alpha \mathbb{M}_{n_\alpha}$ with continuous projections onto the coordinates. (*Hint:* Exercise 4.3.4.)

At the end of this section, we now turn to biprojectivity's little brother:

Definition 4.3.21 A Banach algebra $\mathfrak{A}$ is *biflat* if $\Delta^* : \mathfrak{A}^* \to (\mathfrak{A} \hat{\otimes} \mathfrak{A})^*$ has a bounded left inverse which is an $\mathfrak{A}$-bimodule homomorphism.

As in the case of biprojectivity, the reason for this choice of terminology will become clear in the next chapter.

Exercise 4.3.14 Let $\mathfrak{A}$ be a biflat Banach algebra. Show that $\mathfrak{A}^2$ is dense in $\mathfrak{A}$.

Taking adjoints, one sees that every biprojective Banach algebra is biflat, but so is every amenable Banach algebra by Exercise 2.3.8. In analogy with Exercise 4.3.1 and improving Exercise 4.3.2, we have:

Exercise 4.3.15 Show that a biflat Banach algebra $\mathfrak{A}$ is amenable if and only if it has a bounded approximate identity and if and only if $\mathfrak{A}^{\#}$ is biflat.

Are there any biflat Banach algebras which are neither biprojective nor amenable? The next lemma and the following proposition will help us to uncover such algebras:

Lemma 4.3.22 *For a Banach algebra $\mathfrak{A}$ the following are equivalent:*

(i) $\mathfrak{A}$ *is biflat.*

(ii) *There is an $\mathfrak{A}$-bimodule homomorphism $\rho : \mathfrak{A} \to (\mathfrak{A} \hat{\otimes} \mathfrak{A})^{**}$ such that $\Delta^{**} \circ \rho$ is the canonical embedding of $\mathfrak{A}$ into $\mathfrak{A}^{**}$.*

Proof (i) $\Longrightarrow$ (ii) follows immediately from Definition 4.3.21 through taking adjoints.

(ii) $\Longrightarrow$ (i): Let ρ be as specified in (ii), and define $\tilde{\rho} : (\mathfrak{A} \hat{\otimes} \mathfrak{A})^* \to \mathfrak{A}^*$ to be the restriction of ρ^* to $(\mathfrak{A} \hat{\otimes} \mathfrak{A})^*$. Clearly, $\tilde{\rho}$ is an $\mathfrak{A}$-bimodule homomorphism. To see that $\tilde{\rho}$ is a left inverse of Δ^*, just observe that

$$\langle a, \tilde{\rho}(\Delta^* \phi) \rangle = \langle \Delta^* \phi, \rho(a) \rangle = \langle \phi, \Delta^{**}(\rho(a)) \rangle = \langle a, \phi \rangle \qquad (a \in \mathfrak{A}, \ \phi \in \mathfrak{A}^*).$$

This completes the proof. $\square$

Proposition 4.3.23 *Let $\mathfrak{A}$ be a Banach algebra, and let $\mathfrak{B}$ be a closed subalgebra of $\mathfrak{A}$ with the following properties:*

(i) $\mathfrak{B}$ *is amenable;*

(ii) $\mathfrak{B}$ *is a left ideal of $\mathfrak{A}$;*

(iii) $\mathfrak{B}$ *has a bounded approximate identity which is also a bounded left approximate identity for $\mathfrak{A}$.*

Then $\mathfrak{A}$ is biflat.

Proof Since $\mathfrak{B}$ is amenable, Exercise 2.3.8(ii) yields the existence of a bounded right inverse ρ of $\Delta_{\mathfrak{B}}^{**} : (\mathfrak{B}\hat{\otimes}\mathfrak{B})^{**} \to \mathfrak{B}^{**}$ which is also a $\mathfrak{B}$-bimodule homomorphism. Let $\iota: \mathfrak{B} \to \mathfrak{A}$ denote the canonical embedding, and let $\tilde{\rho} := (\iota \otimes \iota)^{**} \circ \rho$.

Let $(e_\alpha)_\alpha$ be a bounded approximate identity for $\mathfrak{B}$ which is also a bounded left approximate identity for $\mathfrak{A}$, and let $\mathcal{U}$ be an ultrafilter on the index set of $(e_\alpha)_\alpha$ that dominates the order filter. Define

$$\bar{\rho}: \mathfrak{A} \to (\mathfrak{A}\hat{\otimes}\mathfrak{A})^{**}, \quad a \mapsto w^*\text{-}\lim_{\mathcal{U}} \tilde{\rho}(e_\alpha) \cdot a.$$

It is immediate that $\bar{\rho}$ is a right $\mathfrak{A}$-module homomorphism. Since $(e_\alpha)_\alpha$ is a bounded left approximate identity for $\mathfrak{A}$, the fact that ρ is a right inverse of $\Delta_{\mathfrak{B}}^{**}$ implies

$$\Delta_{\mathfrak{A}}^{**}(\bar{\rho}(a)) = w^*\text{-}\lim_{\mathcal{U}} \Delta_{\mathfrak{A}}^{**}(\tilde{\rho}(e_\alpha)) \cdot a = w^*\text{-}\lim_{\mathcal{U}} \Delta_{\mathfrak{B}}^{**}(\rho(e_\alpha)) \cdot a = \lim_{\mathcal{U}} e_\alpha a = a \qquad (a \in \mathfrak{A}),$$

so that $\Delta_{\mathfrak{A}}^{**} \circ \bar{\rho}$ is the canonical embedding of $\mathfrak{A}$ into $\mathfrak{A}^{**}$.

To check that $\bar{\rho}$ is a left module homomorphism, note first that

$$\begin{aligned}
\bar{\rho}(axb) &= w^*\text{-}\lim_{\mathcal{U}} \tilde{\rho}(e_\alpha) \cdot axb \\
&= w^*\text{-}\lim_{\mathcal{U}} axb \cdot \tilde{\rho}(e_\alpha), \qquad \text{since } axb \in \mathfrak{B}, \\
&= w^*\text{-}\lim_{\mathcal{U}} a \cdot \tilde{\rho}(e_\alpha) \cdot xb, \qquad \text{since } xb \in \mathfrak{B}, \\
&= a \cdot \bar{\rho}(x) \cdot b \qquad (a, x \in \mathfrak{A}, \, b \in \mathfrak{B}).
\end{aligned} \qquad (4.21)$$

Let $a, x, b \in \mathfrak{A}$. Since $\mathfrak{B}$ contains a bounded approximate identity for $\mathfrak{A}$, Cohen's factorization theorem yields $c \in \mathfrak{B}$ and $d \in \mathfrak{A}$ such that $b = cd$. From (4.21) and the fact that $\bar{\rho}$ is a right $\mathfrak{A}$-module homomorphism, it follows that

$$\bar{\rho}(axb) = \bar{\rho}(axcd) = \bar{\rho}(axc) \cdot d = a \cdot \bar{\rho}(x) \cdot cd = a \cdot \bar{\rho}(x) \cdot b. \qquad (4.22)$$

Finally, let $a, x \in \mathfrak{A}$. Again, by Cohen's factorization theorem, we obtain $y, z \in \mathfrak{A}$ such that $x = yz$. From (4.22) we obtain

$$\bar{\rho}(ax) = \bar{\rho}(ayz) = a \cdot \bar{\rho}(y) \cdot z = a \cdot \bar{\rho}(x).$$

Thus $\bar{\rho}$ is an $\mathfrak{A}$-bimodule homomorphism, and Lemma 4.3.22 implies that $\mathfrak{A}$ is biflat. $\qquad\square$

In order to obtain a biflat Banach algebra which is not amenable, we should therefore search for Banach algebras which are not amenable themselves, but contain amenable Banach algebras as left ideals.

Exercise 4.3.16 Let E be a Banach space. Show that $\{T^* : T \in \mathcal{A}(E)\}$ is a closed left ideal of $\mathcal{L}(E^*)$.

Theorem 4.3.24 *Let E be a Banach space with property $(\mathbb{A})$ such that E^{**} does not have the bounded approximation property. Then $\mathcal{A}(E^*)$ is a biflat Banach algebra which is neither amenable nor biprojective.*

Proof Since E has property $(\mathbb{A})$, Theorem 3.1.9 ascertains that $\mathcal{A}(E)$ is amenable; the same is true for

$$\mathfrak{B} := \{T^* : T \in \mathcal{A}(E)\} \cong \mathcal{A}(E)^{\mathrm{op}}.$$

By Exercise 4.3.16, $\mathfrak{B}$ is a left ideal in $\mathcal{A}(E^*)$. Since $\mathcal{A}(E)$ is amenable, it has a bounded approximate identity, $(S_\alpha)_\alpha$ say. Consequently, $(S_\alpha^*)_\alpha$ is a bounded approximate identity for $\mathfrak{B}$; this implies that $S_\alpha^* \to \mathrm{id}_{E^*}$ in the strong operator topology and thus uniformly on compact subsets of E^*. An inspection of the proof of Theorem 3.1.2 shows that this implies that $(S_\alpha^*)_\alpha$ is a bounded left approximate identity for $\mathcal{A}(E^*)$. Hence, by Proposition 4.3.23, $\mathcal{A}(E^*)$ is biflat.

Assume that $\mathcal{A}(E^*)$ is amenable. Then $\mathcal{A}(E^*)$ has a bounded approximate identity. By Theorem 3.1.4, this means that E^{**} has the bounded approximation property contrary to our hypothesis.

Assume that $\mathcal{A}(E^*)$ is biprojective. Clearly, $\mathcal{A}(E^*)$ is a Banach operator algebra on E^*. Since E^* has the approximation property, Theorem 4.3.17 implies that $\mathcal{A}(E^*) = \mathcal{N}(E^*)$. In particular, $\mathcal{N}(E^*)$ has a bounded left approximate identity. This, however, is impossible by Theorem 4.3.5 $\quad\square$

Are there Banach spaces satisfying the hypotheses of Theorem 4.3.24? In fact, we have already encountered such spaces:

Example 4.3.25 Let $\mathfrak{H}$ be an infinite-dimensional Hilbert space. As observed in Remark 3.1.13, $\mathfrak{H}\hat{\otimes}\mathfrak{H}$ has property (A), whereas $(\mathfrak{H}\check{\otimes}\mathfrak{H})^{**} \cong \mathcal{L}(\mathfrak{H})$ lacks the approximation property. By Theorem 4.3.24, $\mathcal{A}(\mathfrak{H}\hat{\otimes}\mathfrak{H}) = \mathcal{A}((\mathfrak{H}\check{\otimes}\mathfrak{H})^*)$ is biflat, but neither amenable nor biprojective.

4.4 Connes-amenability

We now look at a variant of amenability that only makes sense for certain Banach algebras:

Definition 4.4.1 A Banach algebra $\mathfrak{A}$ is said to be *dual* if there is a closed submodule $\mathfrak{A}_*$ of $\mathfrak{A}^*$ such that $\mathfrak{A} = (\mathfrak{A}_*)^*$.

Exercise 4.4.1 Let $\mathfrak{A}$ be a Banach algebra which is a dual space. Show that $\mathfrak{A}$ is a dual Banach algebra if and only if multiplication in $\mathfrak{A}$ is separately w^*-continuous.

Examples 4.4.2   (a) Let G be a locally compact group. By Exercises A.1.10 and A.2.2, $\mathfrak{A} := M(G)$ is a dual Banach algebra (with $\mathfrak{A}_* = \mathcal{C}_0(G)$).

(b) Let E be a reflexive Banach space. Then $E\hat{\otimes}E^*$ becomes a Banach $\mathcal{L}(E)$-bimodule through

$$T \cdot (x \otimes \phi) := Tx \otimes \phi \quad \text{and} \quad (x \otimes \phi) \cdot T := x \otimes T^*\phi \qquad (x \in E, \phi \in E^*, T \in \mathcal{L}(E)).$$

Identifying $\mathcal{L}(E)$ with $(E\hat{\otimes}E^*)^*$ as in Exercise B.2.10, we see that $\mathcal{L}(E)$ is a dual Banach algebra.

(c) Let $\mathfrak{H}$ be a Hilbert space. By the previous example, $\mathcal{L}(\mathfrak{H})$ is a dual Banach algebra (the w^*-topology on $\mathcal{L}(\mathfrak{H})$ is often called *ultraweak topology* by people working in von Neumann algebras). Let $\mathfrak{M}$ be a von Neumann algebra acting on $\mathfrak{H}$. It is routinely checked that the w^*-topology and the weak operator topology coincide on bounded subsets of $\mathcal{L}(\mathfrak{H})$. Since the closed unit ball of $\mathfrak{M}$ is closed in $\mathcal{L}(\mathfrak{H})$ with respect to the weak operator topology, it is thus w^*-closed in $\mathcal{L}(\mathfrak{H})$. From the Kreĭn–Šmulian theorem, it follows that $\mathfrak{M}$ is w^*-closed in $\mathcal{L}(\mathfrak{H})$ and thus a dual space. It is straightforward that the predual of $\mathfrak{M}$ — the quotient of $\mathfrak{H} \hat{\otimes} \mathfrak{H}$ by $^{\perp}\mathfrak{M}$ — is a submodule of $\mathfrak{M}^*$.

(d) Let G be a locally compact group, and let $p \in (1, \infty)$. Then $PM_p(G) \cong A_q(G)^*$, where $p \in (0,1)$ is such that $\frac{1}{p} + \frac{1}{q} = 1$, is a dual Banach algebra (this follows from Theorem A.3.6).

(e) Let $\mathfrak{A}$ be any Banach algebra. In Exercise 4.2.7, we introduced two products on $\mathfrak{A}^{**}$ that extend the product on $\mathfrak{A}$ and turn $\mathfrak{A}^{**}$ into a Banach algebra. If these two products coincide, $\mathfrak{A}$ is called *Arens regular*. It is immediate from the definition that $\mathfrak{A}^{**}$ is a dual Banach algebra whenever $\mathfrak{A}$ is Arens regular.

Remark 4.4.3 Let $\mathfrak{M}$ be a W^*-*algebra*, i.e. a C^*-algebra which is a dual Banach space (it follows from [Sak, Theorem 1.13.2] that the predual of $\mathfrak{M}$ is unique). Then there is a faithful, w^*-continuous $*$-representation π of $\mathfrak{M}$ on some Hilbert space $\mathfrak{H}$ ([Sak, Theorem 1.16.7]). Since π is an isometry, it maps the closed unit ball of $\mathfrak{M}$ onto the closed unit ball of $\pi(\mathfrak{M})$. Since π is w^*-continuous, the closed unit ball of $\pi(\mathfrak{M})$ is compact and thus closed in the ultraweak topology on $\mathcal{L}(\mathfrak{H})$. Since the ultraweak topology and the weak operator topology coincide on bounded subsets of $\mathcal{L}(\mathfrak{H})$, the unit ball of $\pi(\mathfrak{M})$ is also closed in the weak operator topology and thus in the strong operator topology, which is finer. Let $\mathfrak{N}$ be the closure of $\pi(\mathfrak{M})$ in the strong operator topology. By Kaplansky's density theorem ([Mur, 4.3.3. Theorem]), the closed unit ball of $\pi(\mathfrak{M})$ is strongly operator dense in the unit ball of $\mathfrak{N}$. Since the unit ball of $\pi(\mathfrak{M})$, however, is closed in the strong operator topology, it follows that the closed unit balls of $\pi(\mathfrak{M})$ and $\mathfrak{N}$ coincide. Consequently, $\mathfrak{N} = \pi(\mathfrak{M})$, so that $\pi(\mathfrak{M})$ is a von Neumann algebra. We shall thus, from now on, use the terms "von Neumann algebra" and "W^*-algebra" interchangeably.

Which dual Banach algebras are amenable?

Examples 4.4.4   (a) Let G be a discrete, amenable group. Then $M(G) = \ell^1(G) = L^1(G)$ is amenable.

(b) Let $\mathfrak{M}$ be a subhomogeneous von Neumann algebra, i.e. the dimensions of the irreducible $*$-representations of $\mathfrak{M}$ are all less than a certain positive integer. Then $\mathfrak{M}$ is of type I ([K–R, Definition 6.5.1]). By [K–R, Theorem 6.5.2], there are von Neumann algebras $\mathfrak{M}_1, \ldots, \mathfrak{M}_k$ such that $\mathfrak{M} \cong \mathfrak{M}_1 \oplus \cdots \oplus \mathfrak{M}_k$ and $\mathfrak{M}_j$ is of type I_{n_j} with $n_j \in \mathbb{N}$ for $j = 1, \ldots, k$. By [K–R, Theorem 6.6.5], $\mathfrak{M}_j \cong \mathbb{M}_{n_j} \otimes Z(\mathfrak{M}_j)$ (as we shall see later on, there is only one norm turning $\mathbb{M}_{n_j} \otimes Z(\mathfrak{M}_j)$ into a C^*-algebra, so that we can suppose that $\mathbb{M}_{n_j} \otimes Z(\mathfrak{M}_j)$ is equipped with the injective norm). Since $Z(\mathfrak{M}_1), \ldots, Z(\mathfrak{M}_k)$ are amenable as commutative C^*-algebras, the hereditary properties of amenability (Which ones?) for Banach algebras yield at once that $\mathfrak{M}$ is amenable.

Surprisingly, a von Neumann algebra is amenable if and only if it is subhomogeneous (we'll prove the hard direction of this result in Chapter 6). It thus seems that a notion of amenability that takes the w^*-topology into account is more appropriate for dual Banach algebras.

Remark 4.4.5 Unless we are in the W^*-algebra situation, we do not know if, for a dual Banach algebra $\mathfrak{A}$, the predual $\mathfrak{A}_*$ from Definition 4.4.1 is necessarily unique. There is thus an element of ambiguity to speaking of the w^*-topology on $\mathfrak{A}$. In what follows, we shall therefore suppose that $\mathfrak{A}$ always comes with a fixed $\mathfrak{A}_*$.

Definition 4.4.6 Let $\mathfrak{A}$ be a dual Banach algebra. A dual Banach $\mathfrak{A}$-bimodule E is called *normal* if, for each $x \in E$, the maps

$$\mathfrak{A} \to E, \quad a \mapsto \begin{cases} a \cdot x \\ x \cdot a \end{cases}$$

are w^*-continuous.

Definition 4.4.7 A dual Banach algebra $\mathfrak{A}$ is *Connes-amenable* if, for every normal, dual Banach $\mathfrak{A}$-bimodule E, every w^*-continuous derivation $D \in \mathcal{Z}^1(\mathfrak{A}, E)$ is inner.

We leave it to the reader to establish a few elementary properties of Connes-amenable Banach algebras:

Exercise 4.4.2 Let $\mathfrak{A}$ be a Connes-amenable dual Banach algebra. Show that $\mathfrak{A}$ is unital.

Exercise 4.4.3 Let $\mathfrak{A}$ be a Banach algebra, let $\mathfrak{B}$ be a dual Banach algebra, and let $\theta \colon \mathfrak{A} \to \mathfrak{B}$ be a continuous homomorphism with w^*-dense range.

(i) Suppose that $\mathfrak{A}$ is amenable. Show that $\mathfrak{B}$ is Connes-amenable.
(ii) Suppose that $\mathfrak{A}$ is dual and Connes-amenable, and that θ is w^*-continuous. Show that $\mathfrak{B}$ is Connes-amenable.

Let $\mathfrak{A}$ be an amenable, Arens regular Banach algebra. Then, as an immediate consequence of Exercise 4.4.3(i), $\mathfrak{A}^{**}$ is Connes-amenable. What about the converse? Does the Connes-amenability of $\mathfrak{A}^{**}$ force $\mathfrak{A}$ to be amenable? As we shall prove in Chapter 6, this is indeed true if $\mathfrak{A}$ is a C^*-algebra.

For general Banach algebras, we have the following:

Theorem 4.4.8 *Let $\mathfrak{A}$ be an Arens regular Banach algebra which is an ideal in $\mathfrak{A}^{**}$. Then the following are equivalent:*

(i) *$\mathfrak{A}$ is amenable.*
(ii) *$\mathfrak{A}^{**}$ is Connes-amenable.*

Proof Only (ii) $\Longrightarrow$ (i) needs proof.

Since $\mathfrak{A}^{**}$ is Connes-amenable, it has an identity by Exercise 4.4.2. By [Pal, Proposition 5.1.8], this means that $\mathfrak{A}$ has a bounded approximate identity, $(e_\alpha)_\alpha$ say. By Proposition 2.1.5, it is therefore sufficient for $\mathfrak{A}$ to be amenable that $\mathcal{H}^1(\mathfrak{A}, E^*) = \{0\}$ for each pseudo-unital Banach $\mathfrak{A}$-bimodule E.

Let E be a pseudo-unital Banach $\mathfrak{A}$-bimodule, and let $D \in \mathcal{Z}^1(\mathfrak{A}, E^*)$. By Proposition 2.1.6, the bimodule action of $\mathfrak{A}$ on E^* extends canonically to $\mathfrak{A}^{**}$, and D has a unique extension $\tilde{D} \in \mathcal{Z}^1(\mathfrak{A}^{**}, E^*)$.

We claim that E^* is a normal, dual Banach $\mathfrak{A}^{**}$-bimodule. Let $(a_\alpha)_\alpha$ be a net in $\mathfrak{A}^{**}$ such that $a_\alpha \overset{w^*}{\to} 0$, let $\phi \in E^*$, and let $x \in E$. Since E is pseudo-unital, there are $b \in \mathfrak{A}$ and $y \in E$ such that $x = y \cdot b$. Since the w^*-topology of $\mathfrak{A}^{**}$ restricted to $\mathfrak{A}$ is the weak topology, we have $ba_\alpha \overset{w}{\to} 0$ (Why?), so that

$$x \cdot a_\alpha = y \cdot b a_\alpha \overset{w}{\to} 0$$

(Again: Why?) and consequently

$$\langle x, a_\alpha \cdot \phi \rangle = \langle x \cdot a_\alpha, \phi \rangle \to 0.$$

Since $x \in E$ was arbitrary, this means that $a_\alpha \cdot \phi \overset{w^*}{\to} 0$. Analoguously, one shows that $\phi \cdot a_\alpha \overset{w^*}{\to} 0$.

To see that $\tilde{D}$ is w^*-continuous, again let $(a_\alpha)_\alpha$ be a net in $\mathfrak{A}^{**}$ such that $a_\alpha \overset{w^*}{\to} 0$, let $x \in E$, and let $b \in \mathfrak{A}$ and $y \in E$ such that $x = b \cdot y$. Then we have:

$$\begin{aligned}
\langle x, \tilde{D} a_\alpha \rangle &= \langle b \cdot y, \tilde{D} a_\alpha \rangle \\
&= \langle y, (\tilde{D} a_\alpha) \cdot b \rangle \\
&= \langle y, D(a_\alpha b) - a_\alpha \cdot Db \rangle \\
&\to 0
\end{aligned}$$

because D is weakly continuous and E^* is a normal, dual Banach $\mathfrak{A}^{**}$-bimodule. From the Connes-amenability of $\mathfrak{A}^{**}$ we conclude that $\tilde{D}$, and hence D, is inner. $\square$

Example 4.4.9 If $p \in (2, \infty)$ and $q \in (1, \infty)$ is such that $\frac{1}{p} + \frac{1}{q} = 1$, then $\mathcal{A}(\ell^p \oplus \ell^q)$ is not amenable (Example 3.1.17). Since $\ell^p \oplus \ell^q$ is reflexive and has the approximation property, $\mathcal{A}(\ell^p \oplus \ell^q)$ is Arens regular with $\mathcal{A}(\ell^p \oplus \ell^q)^{**} \cong \mathcal{L}(\ell^p \oplus \ell^q)$ ([Pal, 1.7.13, Corollary]). By Theorem 4.4.8, $\mathcal{L}(\ell^p \oplus \ell^q)$ is not Connes-amenable.

What about Connes-amenability for $M(G)$, where G is a locally compact group? From Exercise 4.4.3(i) and Theorem 2.1.8, it is clear that $M(G)$ is Connes-amenable if G is a-menable. Does the converse hold? Before we turn to this question, we prove another general theorem which will help us with this problem (and which will have further applications in Chapter 6):

Definition 4.4.10 Let $\mathfrak{A}$ be a Banach algebra, and let $\mathfrak{B}$ be a closed subalgebra of $\mathfrak{A}$. A *quasi-expectation* $Q \colon \mathfrak{A} \to \mathfrak{B}$ is a projection from $\mathfrak{A}$ onto $\mathfrak{B}$ satisfying

$$Q(axb) = a(Qx)b \qquad (a, b \in \mathfrak{B}, \ x \in \mathfrak{A}).$$

If S is any subset of an algebra $\mathfrak{A}$, we use $Z_{\mathfrak{A}}(S)$ to denote the centralizer of S in $\mathfrak{A}$, i.e.

$$Z_{\mathfrak{A}}(S) := \{a \in \mathfrak{A} : as = sa \text{ for all } s \in S\}.$$

In the case where $\mathfrak{A} = \mathcal{L}(E)$ for some Banach space E, we also write S' instead of $Z_{\mathcal{L}(E)}(S)$.

Theorem 4.4.11 *Let $\mathfrak{A}$ be a Banach algebra, let $\mathfrak{B}$ be a dual Banach algebra, let $\theta \colon \mathfrak{A} \to \mathfrak{B}$ be a homomorphism, and suppose that one of the following holds:*

(i) *$\mathfrak{A}$ is amenable;*

(ii) *$\mathfrak{A}$ is Connes-amenable dual Banach algebra, and θ is w^*-continuous.*

Then there is a quasi-expectation $Q \colon \mathfrak{B} \to Z_{\mathfrak{B}}(\theta(\mathfrak{A}))$.

Proof Let $E := \mathfrak{B} \hat{\otimes} \mathfrak{B}_*$ be equipped with the $\mathfrak{A}$-bimodule operation given through

$$a \cdot (b \otimes \phi) := b \otimes \theta(a) \cdot \phi \quad \text{and} \quad (b \otimes \phi) \cdot a := b \otimes \phi \cdot \theta(a) \qquad (a \in \mathfrak{A}, \, \phi \in \mathfrak{B}_*, b \in \mathfrak{B}).$$

Identifying E^* with $\mathcal{L}(\mathfrak{B})$ as in Exercise B.2.10, we obtain as the corresponding dual $\mathfrak{B}$-bimodule operation on $\mathcal{L}(\mathfrak{B})$:

$$(a \cdot T)(b) = \theta(a)(Tb) \quad \text{and} \quad (T \cdot a)(b) = (Tb)\theta(a) \qquad (a \in \mathfrak{A}, \, T \in \mathcal{L}(\mathfrak{B}), \, b \in \mathfrak{B}).$$

It is easy to see (hopefully) that, if (ii) holds, E^* is a normal, dual Banach $\mathfrak{A}$-bimodule.

Let F be the subspace of E^* consisting of those $T \in E^*$ such that:

$$\left.\begin{cases} \langle zb \otimes \phi - b \otimes \phi \cdot z, T \rangle = 0 \\ \langle bz \otimes \phi - b \otimes z \cdot \phi, T \rangle = 0 \\ \langle z \otimes \phi, T \rangle = 0 \end{cases}\right\} \qquad (b \in \mathfrak{B}, \, \phi \in \mathfrak{B}_*, \, z \in Z_{\mathfrak{B}}(\theta(\mathfrak{A}))).$$

It is routine (but nevertheless, it should be done) to verify that F is a w^*-closed $\mathfrak{A}$-submodule of E^* and thus a dual Banach $\mathfrak{A}$-module in its own right. Obviously, if (ii) holds, F is also a normal, dual Banach $\mathfrak{A}$-bimodule.

Define $D := \mathrm{ad}_{\mathrm{id}_{\mathfrak{B}}}$. Then, if (ii) holds, D is w^*-continuous. We claim that D attains its values in F. To see this, let $a \in \mathfrak{A}$, $b \in \mathfrak{B}$, $\phi \in \mathfrak{B}_*$, and $z \in Z_{\mathfrak{B}}(\theta(\mathfrak{A}))$. We then have:

$$\langle zb \otimes \phi - b \otimes \phi \cdot z, \mathrm{ad}_{\mathrm{id}_{\mathfrak{B}}} a \rangle$$
$$= \langle zb \cdot \theta(a) \otimes \phi - b \otimes \phi \cdot z\theta(a), \mathrm{id}_{\mathfrak{B}} \rangle - \langle zb \otimes \theta(a) \cdot \phi - b \otimes \theta(a) \cdot \phi \cdot z, \mathrm{id}_{\mathfrak{B}} \rangle$$
$$= \langle \phi, \theta(a)zb \rangle - \langle \phi, z\theta(a)b \rangle - \langle \phi, zb\theta(a) \rangle + \langle \phi, zb\theta(a) \rangle$$
$$= 0 \qquad \text{(Why?)}.$$

One proves analoguously that

$$\langle bz \otimes \phi - b \otimes z \cdot \phi, \mathrm{ad}_{\mathrm{id}_{\mathfrak{B}}} a \rangle = 0.$$

Finally, we have

$$\langle z \otimes \phi, \mathrm{ad}_{\mathrm{id}_{\mathfrak{B}}} a \rangle = \langle \phi, \theta(a)z - z\theta(a) \rangle = 0.$$

If (i) or (ii) holds, the definitions of amenability and Connes-amenability, respectively, yield $\mathcal{P} \in F$ such that $D = \mathrm{ad}_{\mathcal{P}}$. We claim that $\mathcal{Q} := \mathrm{id}_{\mathfrak{B}} - \mathcal{P}$ is the desired quasi-expectation. It is immediate that $\mathcal{Q}(\mathfrak{B}) \subset Z_{\mathfrak{B}}(\theta(\mathfrak{A}))$. Since

$$\langle z \otimes \phi, \mathcal{P} \rangle = 0 \qquad (z \in Z_{\mathfrak{B}}(\theta(\mathfrak{A})), \, \phi \in \mathfrak{B}_*),$$

it follows that $\mathcal{Q}$ is the identity on $Z_{\mathfrak{B}}(\theta(\mathfrak{A}))$ and thus a projection onto $Z_{\mathfrak{B}}(\theta(\mathfrak{A}))$. For $b \in \mathfrak{B}$, $\phi \in \mathfrak{B}_*$, and $z \in Z_{\mathfrak{B}}(\theta(\mathfrak{A}))$, we have

$$0 = \langle zb \otimes \phi - b \otimes \phi \cdot z, \mathcal{P} \rangle = \langle \phi, \mathcal{P}(zb) - z(\mathcal{P}b) \rangle,$$

so that $\mathcal{P}(zb) = z(\mathcal{P}b)$. Analoguously, we see that $\mathcal{P}(bz) = (\mathcal{P}b)z$. Hence, $\mathcal{Q}$ is indeed a quasi-expectation onto $Z_{\mathfrak{B}}(\theta(\mathfrak{A}))$. $\square$

We shall now use Theorem 4.4.11 to characterize, for a locally compact group G, the Connes-amenability of $M(G)$, $VN(G)$, and, more generally, $PM_p(G)$ for $p \in (1, \infty)$.

Definition 4.4.12 A locally compact group G is *inner amenable* if there is a mean m on $L^\infty(G)$ such that

$$\langle \delta_g * \phi * \delta_{g^{-1}}, m \rangle = \langle \phi, m \rangle \qquad (g \in G, \ \phi \in L^\infty(G)).$$

By Theorem 1.1.11, every amenable, locally compact group is inner amenable, but so is every discrete group (chose $m = \delta_e$). By the way, why doesn't it make sense to define inner amenability in terms of $C_b(G)$, $LUC(G)$, etc.?

Exercise 4.4.4 Let G be a locally compact group. Show that G is inner amenable if and only if there is a net $(f_\alpha)_\alpha$ in $P(G)$ such that $\|\delta_g * f_\alpha * \delta_{g^{-1}} - f_\alpha\|_1 \to 0$ for all $g \in G$.

Exercise 4.4.5 Let G be a locally compact group, and let $p \in [1, \infty)$. Show that

$$\left(\left\| f_1^{\frac{1}{p}} - f_2^{\frac{1}{p}} \right\|_p \right)^p \leq \|f_1 - f_2\|_1 \leq p 2^{p-1} \left\| f_1^{\frac{1}{p}} - f_2^{\frac{1}{p}} \right\|_p \qquad (f_1, f_2 \in P(G)). \tag{4.23}$$

(*Hint:* Show first that

$$|t - s|^p \leq |t^p - s^p| \leq p|t - s|(s + t)^{p-1} \qquad (s, t \in \mathbb{R}, \ s, t \geq 0). \tag{4.24}$$

Then use (4.24) and Hölder's inequality to deduce (4.23).)

Theorem 4.4.13 *For a locally compact group G consider the following:*

(i) G *is amenable.*
(ii) $M(G)$ *is Connes-amenable.*
(iii) $PM_p(G)$ *is Connes-amenable for every $p \in (1, \infty)$.*
(iv) $VN(G)$ *is Connes-amenable.*
(v) $PM_p(G)$ *is Connes-amenable for one $p \in (1, \infty)$.*

Then we have:

$$\text{(i)} \implies \text{(ii)} \implies \text{(iii)} \implies \text{(iv)} \implies \text{(v)}.$$

If G is inner amenable, (v) $\implies$ (i) holds, too.

Proof We have already observed that (i) $\implies$ (ii) holds.

Let $p \in (1, \infty)$. By Exercise A.3.4(iii) $\tilde{\lambda}_p : M(G) \to PM_p(G)$ is w^*-continuous with w^*-dense range. Hence, by Exercise 4.4.3, $PM_p(G)$ is Connes amenable if $M(G)$ is. This establishes (ii) $\implies$ (iii).

Since $VN(G) = PM_2(G)$, (iii) $\implies$ (iv) and (iv) $\implies$ (v) are straightforward.

For the proof of (v) $\implies$ (i), suppose that G is inner amenable. By Exercise 4.4.4, there is a net $(f_\alpha)_\alpha$ in $P(G)$ such that

$$\|\delta_g * f_\alpha * \delta_{g^{-1}} - f_\alpha\|_1 \to 0 \qquad (g \in G),$$

or equivalently (see Remarks A.3.2(c))

$$\|\lambda_1(g^{-1})f_\alpha - \rho_1(g)_1 f_\alpha\|_1 \to 0 \qquad (g \in G). \tag{4.25}$$

Let $q \in (1, \infty)$ be such that $\frac{1}{p} + \frac{1}{q} = 1$. Let $\xi_\alpha := f_\alpha^{1/p}$, and let $\eta_\alpha := f_\alpha^{1/q}$, so that $\xi_\alpha \in L^p(G)$ and $\eta_\alpha \in L^q(G)$. It follows from (4.25) and Exercise 4.4.5 that

$$\left.\begin{cases} \|\lambda_p(g^{-1})\xi_\alpha - \rho_p(g)\xi_\alpha\|_p \to 0 \\ \|\lambda_q(g^{-1})\eta_\alpha - \rho_q(g)\eta_\alpha\|_q \to 0 \end{cases}\right\} \quad (g \in G). \tag{4.26}$$

For $\phi \in L^\infty(G)$, let $M_\phi \in \mathcal{L}(L^p(G))$ be the corresponding multiplication operator, i.e. $M_\phi \xi = \phi \xi$ for all $\xi \in L^p(G)$. By Theorem 4.4.11 — with $\mathfrak{A} = PM_p(G)$, $\mathfrak{B} = \mathcal{L}(L^p(G))$, and θ as the canonical embedding — , there is a quasi-expectation $\mathcal{Q} \colon \mathcal{L}(L^p(G)) \to PM_p(G)'$. Define $m_\alpha \in L^\infty(G)^*$ by letting

$$\langle \phi, m_\alpha \rangle := \langle \mathcal{Q}(M_\phi)\xi_\alpha, \eta_\alpha \rangle \qquad (\phi \in L^\infty(G)).$$

Let $\mathcal{U}$ be an ultrafilter on the index set of $(f_\alpha)_\alpha$ that dominates the order filter, and define

$$\langle \phi, m \rangle := \lim_{\mathcal{U}} \langle \phi, m_\alpha \rangle \qquad (\phi \in L^\infty(G)).$$

Note that $\rho_p(G) \subset PM_p(G)'$, and observe (Check it!) that

$$\rho_p(g^{-1}) M_\phi \rho_p(g) = M_{\phi*\delta_g} \qquad (g \in G, \ \phi \in L^\infty(G)).$$

We then obtain for $g \in G$ and $\phi \in L^\infty(G)$:

$$\begin{aligned}
\langle \phi * \delta_g, m \rangle &= \lim_{\mathcal{U}} \langle \phi * \delta_g, m_\alpha \rangle \\
&= \lim_{\mathcal{U}} \langle \mathcal{Q}(M_{\phi*\delta_g})\xi_\alpha, \eta_\alpha \rangle \\
&= \lim_{\mathcal{U}} \langle \mathcal{Q}(\rho_p(g^{-1}) M_\phi \rho_p(g))\xi_\alpha, \eta_\alpha \rangle \\
&= \lim_{\mathcal{U}} \langle \rho_p(g^{-1})(\mathcal{Q}M_\phi)\rho_p(g)\xi_\alpha, \eta_\alpha \rangle \\
&= \lim_{\mathcal{U}} \langle (\mathcal{Q}M_\phi)\rho_p(g)\xi_\alpha, \rho_q(g)\eta_\alpha \rangle \\
&= \lim_{\mathcal{U}} \langle (\mathcal{Q}M_\phi)\lambda_p(g^{-1})\xi_\alpha, \lambda_q(g^{-1})\eta_\alpha \rangle, \qquad \text{by (4.26)}, \\
&= \lim_{\mathcal{U}} \langle \lambda_p(g)(\mathcal{Q}M_\phi)\lambda_p(g^{-1})\xi_\alpha, \eta_\alpha \rangle \\
&= \lim_{\mathcal{U}} \langle (\mathcal{Q}M_\phi)\xi_\alpha, \eta_\alpha \rangle \\
&= \langle \phi, m \rangle.
\end{aligned}$$

Although m is not a mean, we can now easily obtain a right invariant mean by normalizing $|m|$ (Remember the proof of Theorem 2.1.8?). By Theorem 1.1.11, G is amenable. $\quad\square$

We conclude this section with an analogue of (one direction of) Theorem 2.2.4.

For any dual Banach algebra $\mathfrak{A}$, let $\mathcal{L}^2_{w^*}(\mathfrak{A}, \mathbb{C})$ denote the separately w^*-continuous elements of $\mathcal{L}^2(\mathfrak{A}, \mathbb{C}) \cong (\mathfrak{A}\hat{\otimes}\mathfrak{A})^*$. Clearly, $\mathcal{L}^2_{w^*}(\mathfrak{A}, \mathbb{C})$ is a closed submodule of $\mathcal{L}^2(\mathfrak{A}, \mathbb{C})$; in particular, $\mathcal{L}^2_{w^*}(\mathfrak{A}, \mathbb{C})^*$ carries a canonical Banach $\mathfrak{A}$-bimodule structure which makes it a quotient module of $(\mathfrak{A}\hat{\otimes}\mathfrak{A})^{**}$. Since multiplication in a dual Banach algebra is separately w^*-continuous, it follows that $\Delta_{\mathfrak{A}}^* \mathfrak{A}_* \subset \mathcal{L}^2_{w^*}(\mathfrak{A}, \mathbb{C})$, so that $\Delta_{\mathfrak{A}}^{**}$ drops to a Banach $\mathfrak{A}$-bimodule homomorphism $\Delta_{w^*} \colon \mathcal{L}^2_{w^*}(\mathfrak{A}, \mathbb{C})^* \to \mathfrak{A}$.

Definition 4.4.14 Let $\mathfrak{A}$ be a dual Banach algebra. Then $\mathbf{M} \in \mathcal{L}^2_{w^*}(\mathfrak{A}, \mathbb{C})^*$ is called a *normal, virtual diagonal* for $\mathfrak{A}$ if

$$a \cdot \mathbf{M} = \mathbf{M} \cdot a \quad \text{and} \quad a\Delta_{w^*}\mathbf{M} = a \qquad (a \in \mathfrak{A}).$$

Theorem 4.4.15 *Let $\mathfrak{A}$ be a dual Banach which has a normal, virtual diagonal. Then $\mathfrak{A}$ is Connes-amenable.*

Proof Let $\mathbf{M} \in \mathcal{L}^2_{w^*}(\mathfrak{A}, \mathbb{C})^*$ be a normal, virtual diagonal for $\mathfrak{A}$. We use the formal notation

$$\int_{\mathfrak{A} \times \mathfrak{A}} V(a,b) \, d\mathbf{M}(a,b) := \langle V, \mathbf{M} \rangle \qquad (V \in \mathcal{L}^2_{w^*}(\mathfrak{A}, \mathbb{C})).$$

Note that

$$\int_{\mathfrak{A} \times \mathfrak{A}} V(ca,b) \, d\mathbf{M}(a,b) = \int_{\mathfrak{A} \times \mathfrak{A}} V(a,bc) \, d\mathbf{M}(a,b) \qquad (c \in \mathfrak{A}). \tag{4.27}$$

Let E be a normal, dual Banach $\mathfrak{A}$-bimodule with predual E_*. It follows immediately from Definition 4.4.14 that $\mathfrak{A}$ is unital. Without loss of generality suppose that E is unital. (Why? Think of the proof of Proposition 2.1.5 ...). Let $D \in \mathcal{Z}^1(\mathfrak{A}, E)$ be w^*-continuous. Fix $x \in E_*$, and define $\phi \in E$ through

$$\langle x, \phi \rangle := \int_{\mathfrak{A} \times \mathfrak{A}} \langle x, a \cdot Db \rangle \, d\mathbf{M}(a,b).$$

Then we have for $c \in \mathfrak{A}$:

$$\langle x, c \cdot \phi \rangle = \int_{\mathfrak{A} \times \mathfrak{A}} \langle x, ca \cdot Db \rangle \, d\mathbf{M}(a,b)$$

$$= \int_{\mathfrak{A} \times \mathfrak{A}} \langle x, a \cdot D(bc) \rangle \, d\mathbf{M}(a,b), \qquad \text{by (4.27)},$$

$$= \int_{\mathfrak{A} \times \mathfrak{A}} \langle x, ab \cdot D(c) \rangle \, d\mathbf{M}(a,b) + \int_{\mathfrak{A} \times \mathfrak{A}} \langle x, a \cdot D(b) \cdot c \rangle \, d\mathbf{M}(a,b)$$

$$= \int_{\mathfrak{A} \times \mathfrak{A}} \langle x, ab \cdot D(c) \rangle \, d\mathbf{M}(a,b) + \langle x, \phi \cdot c \rangle$$

$$= \langle x, Dc \rangle + \langle x, \phi \cdot c \rangle. \qquad \text{(Why?)}$$

This means that $D = \mathrm{ad}_\phi$. $\square$

Remark 4.4.16 For an arbitrary dual Banach algebra $\mathfrak{A}$, the dual Banach $\mathfrak{A}$-bimodule $\mathcal{L}^2_{w^*}(\mathfrak{A}, \mathbb{C})^*$ need not be normal. For this reason, one cannot simply mimic the argument in the proof of Theorem 2.2.4 to prove that every Connes-amenable, dual Banach algebra has a normal, virtual diagonal.

4.5 Notes and comments

Super-amenable Banach algebras often go by the name contractible Banach algebras in the literature ([Hel 5]). The reason why we prefer to call them super-amenable is that the adjective contractible is also used in the K-theory of C^*-algebras: A C^*-algebra $\mathfrak{A}$ is called contractible if $\mathrm{id}_{\mathfrak{A}}$ is homotopic to the zero map ([W-O, Definition 6.4.1]). By Corollary 4.1.6, this is quite different from super-amenability. Theorem 4.1.5 along with Corollary 4.1.6 and Exercise 4.1.7 is due to Yu. V. Selivanov ([Sel 1]), whereas the proof we present essentially goes back to J. A. Taylor ([Tay, Proposition 5.11]).

Weakly amenable Banach algebras were introduced by W. G. Bade, P. C. Curtis, Jr., and H. G. Dales in [B–C–D]. They only consider commutative Banach algebras and use the condition stated in Exercise 4.2.1(ii) to define weak amenability. This definition, however, does not make much sense for non-commutative Banach algebras (Why?). In [B–C–D], Exercise 4.2.1 was proved (so that you can look it up there if you can't do it). Later B. E. Johnson suggested the use of Definition 4.2.1 for arbitrary Banach algebras ([Joh 4]). Theorem 4.2.3 was first proved by Johnson ([Joh 5]), but the ingeniously simple proof we present is due to M. Despić and F. Ghahramani ([D–Gh]). Theorem 4.2.4 was proved by U. Haagerup in [Haa 2], i.e. before the notion of weak amenability was even, formally introduced. The proof we give is taken from [H–L] and simpler than the original one. Does anybody see a Grothendieck-free proof? In [Haa 2], Theorem 4.2.4 is derived from the following refinement of Corollary 4.2.13: If $\mathfrak{M}$ is a von Neumann algebra with predual $\mathfrak{M}_*$, then $\mathcal{H}^1(\mathfrak{M}, \mathfrak{M}_*) = \{0\}$.

The hereditary properties of weak amenability are investigated in [Grø 2]; they are not nearly as nice as those for amenability. For a Banach algebra $\mathfrak{A}$, let $\mathfrak{A}^{n*}$ be the n-th dual of $\mathfrak{A}$; in [D–Gh–G], H. G. Dales, F. Ghahramani, and N. Grœnbæk define $\mathfrak{A}$ to be n-weakly amenable if $\mathcal{H}^1(\mathfrak{A}, \mathfrak{A}^{n*}) = \{0\}$. For example, C^*-algebras are n-weakly amenable for each $n \in \mathbb{N}$, and so is $\ell^1(G)$ for every free group G ([Joh 9]).

Biprojectivity and biflatness are notions that arise naturally in A. Ya. Helemskiĭ's Banach homology. Definitions 4.3.1 and 4.3.21 are not the original definitions, but equivalent characterizations. We shall put biprojectivity and biflatness in their proper context when we discuss Banach homology in the next chapter. Implication (iii) $\Longrightarrow$ (vi) of Theorem 4.3.5 is proved in [Grø 1] (compare also [Hel 3, p. 194]). The structure theory for biprojective Banach algebras that culminates in Theorem 4.3.20 is due to Yu. V. Selivanov ([Sel 2]), as is Example 4.3.25 along with the results on biflat Banach algebras leading to it ([Sel 5]). Examples 4.3.2(a) and Exercise 4.3.11 are [Hel 1, Theorem 51]. Like amenable, super-amenable, and weakly amenable Banach algebras, biprojective and biflat Banach algebras can be characterized through the vanishing of certain cohomology groups ([Sel 4]).

Connes-amenability was first considered for von Neumann algebras in [J–K–R] (and thus should perhaps be called Johnson–Kadison–Ringrose amenability). The reason why this notion of amenability is usually associated with A. Connes are his papers [Conn 1] (where it occurs in Remark 5.33) and [Conn 2]. The name "Connes-amenability" seems to be due to A. Ya. Helemskiĭ ([Hel 4]). Our discussion of Connes-amenability is mostly based on [Run 4]. In [C–G 1], a notion of amenability for a Banach algebra $\mathfrak{A}$ relative to a certain submodule E of $\mathfrak{A}^*$ is discussed: if $E = \mathfrak{A}^*$ this yields amenability in the usual sense; if $\mathfrak{A}$ is dual, $E = \mathfrak{A}_*$ yields Connes-amenability. Theorem 4.4.11 (with hypothesis (ii)) is [Bu–P, Theorem 2] (in [Bu–P] only von Neumann algebras are considered, but the idea of the proof carries over to the more general setting without modifications); with hypothesis (i), Theorem 4.4.11 is essentially [C–G 2, Proposition 2.2] (the proof in [C–G 2] is different from ours, and the hypotheses are somewhat more restrictive). Of course, there are examples of inner amenable, locally compact groups which are neither amenable nor discrete: for instance, all [IN]-groups ([Pal, 5.1.9, Definition]) are inner amenable. For references on to the original literature on inner amenability, see [Pat 1]. Implication (iv) $\Longrightarrow$ (i) of Theorem 4.4.13 for inner amenable groups was proved in [L–P]. It is observed in [Pat 1] that there are non-amenable, locally compact groups for which $VN(G)$ is nevertheless amenable (such as $SL(2, \mathbb{R})$); of course, such groups cannot be inner amenable. For measure algebras, Theorem 4.4.13 allows for further improvement: A locally compact group G is amenable if and only if $M(G)$ is Connes-amenable ([Run 5]). As we have observed, $M(G)$ is amenable for amenable, discrete G. Interestingly, this is the only case in which $M(G)$ can be amenable; for non-discrete G, the measure algebra $M(G)$ is not even weakly amenable: both results are proven in [D–Gh–H]. Theorem 4.4.15 is from [Eff] (in the von Neumann algebra case); the observation that the proof extends to general dual Banach algebras is made in [C–G 1]. For von Neumann algebras ([Eff]) as well as for the measure algebras of locally compact groups ([Run 5]), the converse of Theorem 4.4.15 is also true; I do not know if this is still the case for general dual Banach algebras (probably not).

5 Banach homology

At the end of Chapter 2, we mentioned that there is an alternative approach to Hochschild cohomology (due to A. Ya. Helemskiĭ's Moscow school), which is more powerful and sophisticated than the direct one presented in Section 2.4. It adapts the tools of homological algebra to the Banach algebra setting by adding certain functional analytic overtones.

This chapter is an introduction to the gospel according to Alexander (Helemskiĭ) for the not yet converted. This means that no background in homological algebra is required, and that I have avoided the jargon of category theory. You thus won't find terms like "category", "functor", "natural transformation", or "derived functor". Of course, if you have any previous exposure to category theory, you will immediately recognize how the definitions, theorems, etc., in this chapter can be reworded in terms of category theory.

Although you don't have to be familiar with homological algebra in order to be able to read this chapter, it certainly helps: Most of the concepts and results in this chapter are straightforward adaptations of results from (algebraic) homological algebra to the Banach context.

5.1 Projectivity

We begin our introduction to Banach homology with what is perhaps the most important concept in Banach homology: projectivity.

We'll often have to deal with module homomorphisms in this chapter. We therefore introduce new notation:

Definition 5.1.1 Let E and F be left Banach $\mathfrak{A}$-modules. Then ${}_{\mathfrak{A}}\mathcal{L}(E, F)$ denotes the left $\mathfrak{A}$-module homomorphisms in $\mathcal{L}(E, F)$.

Let $\mathfrak{A}$ be a Banach algebra with unitization $\mathfrak{A}^{\#}$. I would like to stress here that we always consider *unconditional* unitizations, i.e. if $\mathfrak{A}$ already has an identity, we adjoin another one. If E is a Banach space, the projective tensor product $\mathfrak{A}^{\#} \hat{\otimes} E$ becomes a left Banach $\mathfrak{A}$-module through

$$a \cdot (b \otimes x) := ab \otimes x \qquad (a \in \mathfrak{A}, \, b \in \mathfrak{A}^{\#}, \, x \in E).$$

Definition 5.1.2 Let $\mathfrak{A}$ be a Banach algebra, and let E be a Banach space. Then the left Banach $\mathfrak{A}$-module $\mathfrak{A}^{\#} \hat{\otimes} E$ is called *free*.

Exercise 5.1.1 Let E be a Banach space, let $\mathfrak{A}$ be a Banach algebra, let F be a left Banach $\mathfrak{A}$-module, and let $T \in \mathcal{L}(E, F)$. Show that there is a unique $\theta \in {}_{\mathfrak{A}}\mathcal{L}(\mathfrak{A}^{\#} \hat{\otimes} E, F)$ such that

$$\theta(e_{\mathfrak{A}^\#} \otimes x) = Tx \qquad (x \in E).$$

Already in the proof of Theorem 2.3.13, we noted that, for any Banach algebra $\mathfrak{A}$ and for any left Banach $\mathfrak{A}$-module E, the module action extends canonically to $\mathfrak{A}^\#$ (the same is true for right modules or bimodules). Therefore, the multiplication map

$$\Delta_{\mathfrak{A}^\#,E} : \mathfrak{A}^\# \hat{\otimes} E \to E, \quad a \otimes x \mapsto a \cdot x$$

is well-defined. It is obvious that $\Delta_{\mathfrak{A}^\#,E} \in {}_{\mathfrak{A}}\mathcal{L}(E,F)$ has a right inverse in $\mathcal{L}(E, \mathfrak{A}^\# \hat{\otimes} E)$ (Which one?).

Definition 5.1.3 Let $\mathfrak{A}$ be a Banach algebra. A left Banach $\mathfrak{A}$-module P is called *projective* if the multiplication map $\Delta_{\mathfrak{A}^\#,P} : \mathfrak{A}^\# \hat{\otimes} P \to P$ has a right inverse in ${}_{\mathfrak{A}}\mathcal{L}(P, \mathfrak{A}^\# \hat{\otimes} P)$.

Why do we use the multiplication map $\Delta_{\mathfrak{A}^\#,P} : \mathfrak{A}^\# \hat{\otimes} P \to P$ instead of $\Delta_{\mathfrak{A},P} : \mathfrak{A} \hat{\otimes} P \to P$? The answer is simplicity itself: In general, $\Delta_{\mathfrak{A},P}$ is not surjective, and without surjectivity, it makes no sense to require right inverses to exist. Nevertheless, we have:

Exercise 5.1.2 Let $\mathfrak{A}$ be a Banach algebra, and let P be an essential left Banach $\mathfrak{A}$-module. Show that P is projective if and only if $\Delta_{\mathfrak{A},P} : \mathfrak{A} \hat{\otimes} P \to P$ has a right inverse in ${}_{\mathfrak{A}}\mathcal{L}(P, \mathfrak{A} \hat{\otimes} P)$.

Exercise 5.1.3 Let $\mathfrak{A}$ be a Banach algebra, and let P be a left Banach $\mathfrak{A}$-module. Show that P is projective as a left Banach $\mathfrak{A}$-module if and only if P is projective as a left Banach $\mathfrak{A}^\#$-module.

Examples 5.1.4   (a) Every free left Banach module is projective. (Is that obvious to you?)

(b) Let E be a Banach space, and let $\mathfrak{A}$ be a Banach operator algebra on E. Then E is an essential left Banach $\mathfrak{A}$-module through

$$T \cdot x := Tx \qquad (T \in \mathfrak{A}, \, x \in E).$$

Fix $y \in E$ and $\phi \in E^*$ such that $\langle y, \phi \rangle = 1$. Define $\theta : E \to \mathfrak{A} \hat{\otimes} E$ through

$$\theta(x) := (x \odot \phi) \otimes y \qquad (x \in E).$$

It is clear from this definition that $\theta \in {}_{\mathfrak{A}}\mathcal{L}(E, \mathfrak{A} \hat{\otimes} E)$ is such that $\Delta_{\mathfrak{A},E} \circ \theta = \mathrm{id}_E$. By Exercise 5.1.2, E is projective.

Next, we give a characterization of projective left Banach modules, for which we require yet another definition:

Definition 5.1.5 Let $\mathfrak{A}$ be a Banach algebra, and let E and F be left Banach $\mathfrak{A}$-modules. Then $\theta \in {}_{\mathfrak{A}}\mathcal{L}(E,F)$ is called *admissible* if $\theta(E)$ and $\ker \theta$ are closed, complemented subspaces of F and E, respectively.

We have previously defined what an admissible, short, exact sequence of Banach modules is supposed to mean. How do these two uses of the adjective "admissible" fit together? Perfectly: A short exact sequence

$$\{0\} \to F \to E \to E/F \to \{0\}$$

of left Banach $\mathfrak{A}$-modules is admissible in the sense of Definition 2.3.12 if and only if the projection map from E onto E/F is admissible (or — equivalently — if and only if the inclusion of F into E is admissible).

Proposition 5.1.6 *Let $\mathfrak{A}$ be a Banach algebra, and let P be a left Banach $\mathfrak{A}$-module. Then the following are equivalent:*

(i) *P is projective.*

(ii) *There is a left Banach $\mathfrak{A}$-module Q such that $Q \oplus P$ is free.*

(iii) *If E and F are left Banach $\mathfrak{A}$-modules, if $\theta \in {}_{\mathfrak{A}}\mathcal{L}(E, F)$ is surjective and admissible, and if $\sigma \in {}_{\mathfrak{A}}\mathcal{L}(P, F)$, then there is $\rho \in {}_{\mathfrak{A}}\mathcal{L}(P, E)$ such that $\sigma = \theta \circ \rho$, i.e. the diagram*

$$
\begin{array}{ccc}
 & & P \\
 & \swarrow & \downarrow \\
E & \longrightarrow & F.
\end{array}
$$

commutes.

(iv) *Every admissible, short, exact sequence*

$$\{0\} \to F \to E \to P \to \{0\}$$

of left Banach $\mathfrak{A}$-modules splits.

Proof (i) $\Longrightarrow$ (ii): By definition, $\Delta_{\mathfrak{A}^{\#}, P} : \mathfrak{A}^{\#} \hat{\otimes} P \to P$ has a right inverse in ${}_{\mathfrak{A}}\mathcal{L}(P, \mathfrak{A}^{\#} \hat{\otimes} P)$, say ρ. Then $Q = \ker \rho$ satisfies $Q \oplus P \cong \mathfrak{A}^{\#} \hat{\otimes} P$.

(ii) $\Longrightarrow$ (iii): We may suppose without loss of generality (Why?) that P is free. Let $T \in \mathcal{L}(F, E)$ be a right inverse of θ, and define $R := T \circ \sigma$. By the definition of a free left Banach module, there is a Banach space B such that $P \cong \mathfrak{A}^{\#} \hat{\otimes} B$, and by Exercise 5.1.1, there is a unique $\rho \in {}_{\mathfrak{A}}\mathcal{L}(P, E)$ such that

$$\rho(e_{\mathfrak{A}^{\#}} \otimes x) = Rx \qquad (x \in B).$$

We then have:

$$
\begin{aligned}
\theta(\rho(a \otimes x)) &= \theta(a \cdot \rho(e_{\mathfrak{A}^{\#}} \otimes x)) \\
&= a \cdot \theta(Rx) \\
&= a \cdot \sigma(e_{\mathfrak{A}^{\#}} \otimes x) \\
&= \sigma(a \otimes x) \qquad (a \in \mathfrak{A}^{\#}, \, x \in B).
\end{aligned}
$$

This means that $\sigma = \theta \circ \rho$.

(iv) is a particular case of (iii).

(iv) $\Longrightarrow$ (i): Apply (iv) to the short, exact sequence

$$\{0\} \to \ker \Delta_{\mathfrak{A}^{\#}, P} \to \mathfrak{A}^{\#} \hat{\otimes} P \overset{\Delta_{\mathfrak{A}^{\#}, P}}{\to} P \to \{0\}.$$

This yields the claim. $\square$

Combining Proposition 5.1.6 and Exercise 4.1.5, we obtain:

Corollary 5.1.7 *Let $\mathfrak{A}$ be a super-amenable Banach algebra. Then every left Banach $\mathfrak{A}$-module is projective.*

Of course, one can also define projective right Banach modules and projective Banach bimodules. We will now look at the bimodule situation.

Let $\mathfrak{A}$ be a Banach algebra, and let E and F be Banach $\mathfrak{A}$-bimodules. Then $\mathfrak{A}\mathcal{L}_\mathfrak{A}(E, F)$ denotes the $\mathfrak{A}$-bimodule homomorphisms in $\mathcal{L}(E, F)$.

Definition 5.1.8 Let $\mathfrak{A}$ be a Banach algebra. A Banach $\mathfrak{A}$-bimodule is called *projective* if the multiplication map

$$\Delta_{\mathfrak{A}^\#,P,\mathfrak{A}^\#} : \mathfrak{A}^\#\hat{\otimes}P\hat{\otimes}\mathfrak{A}^\# \to P, \quad a \otimes x \otimes b \mapsto a \cdot x \cdot b.$$

has a right inverse in $\mathfrak{A}\mathcal{L}_\mathfrak{A}(P, \mathfrak{A}^\#\hat{\otimes}P\hat{\otimes}\mathfrak{A}^\#)$.

In analogy with the left module situation, we call a Banach bimodule over a Banach algebra $\mathfrak{A}$ *free* if it is of the form $\mathfrak{A}^\#\hat{\otimes}E\hat{\otimes}\mathfrak{A}^\#$ for some Banach space E; of course, any free Banach $\mathfrak{A}$-bimodule is projective.

Exercise 5.1.4 Formulate (and, if you want, prove) an analogue of Proposition 5.1.6 for projective Banach bimodules.

We announced earlier in these notes that the reason why biprojective Banach algebras are called "biprojective" would become clear in the framework of Banach homology. The time has come . . .

Lemma 5.1.9 *The following are equivalent for a Banach algebra $\mathfrak{A}$:*

(i) $\mathfrak{A}$ *is a projective Banach $\mathfrak{A}$-bimodule.*
(ii) *Both multiplication maps $\Delta_{\mathfrak{A}^\#,\mathfrak{A}} : \mathfrak{A}^\#\hat{\otimes}\mathfrak{A} \to \mathfrak{A}$ and $\Delta_{\mathfrak{A},\mathfrak{A}^\#} : \mathfrak{A}\hat{\otimes}\mathfrak{A}^\# \to \mathfrak{A}$ have right inverses in $\mathfrak{A}\mathcal{L}_\mathfrak{A}(\mathfrak{A}, \mathfrak{A}^\#\hat{\otimes}\mathfrak{A})$ and $\mathfrak{A}\mathcal{L}_\mathfrak{A}(\mathfrak{A}, \mathfrak{A}\hat{\otimes}\mathfrak{A}^\#)$, respectively.*

Proof (i) $\implies$ (ii): The short exact sequence

$$\{0\} \to \ker \Delta_{\mathfrak{A}^\#,\mathfrak{A}} \to \mathfrak{A}^\#\hat{\otimes}\mathfrak{A} \overset{\Delta_{\mathfrak{A}^\#,\mathfrak{A}}}{\to} \mathfrak{A} \to \{0\}$$

is clearly admissible and thus splits by (the bimodule analogue of) Proposition 5.1.6. The same is true for

$$\{0\} \to \ker \Delta_{\mathfrak{A},\mathfrak{A}^\#} \to \mathfrak{A}\hat{\otimes}\mathfrak{A}^\# \overset{\Delta_{\mathfrak{A},\mathfrak{A}^\#}}{\to} \mathfrak{A} \to \{0\}.$$

(ii) $\implies$ (i): Let $\rho \in \mathfrak{A}\mathcal{L}_\mathfrak{A}(\mathfrak{A}, \mathfrak{A}^\#\hat{\otimes}\mathfrak{A})$ be a right inverse of $\Delta_{\mathfrak{A}^\#,\mathfrak{A}}$, and let $\sigma \in \mathfrak{A}\mathcal{L}_\mathfrak{A}(\mathfrak{A}, \mathfrak{A}\hat{\otimes}\mathfrak{A}^\#)$ be a right inverse of $\Delta_{\mathfrak{A},\mathfrak{A}^\#}$. It is routinely checked that $(\mathrm{id}_{\mathfrak{A}^\#} \otimes \sigma) \circ \rho \in \mathfrak{A}\mathcal{L}_\mathfrak{A}(\mathfrak{A}, \mathfrak{A}^\#\hat{\otimes}\mathfrak{A}\hat{\otimes}\mathfrak{A}^\#)$ is a right inverse of $\Delta_{\mathfrak{A}^\#,\mathfrak{A},\mathfrak{A}^\#}$. $\square$

Theorem 5.1.10 *For a Banach algebra $\mathfrak{A}$ the following are equivalent:*

(i) $\mathfrak{A}$ *is biprojective.*
(ii) $\mathfrak{A}$ *is a projective $\mathfrak{A}$-bimodule.*

Proof (i) $\implies$ (ii): Let $\rho \in \mathfrak{A}\mathcal{L}_\mathfrak{A}(\mathfrak{A}, \mathfrak{A}\hat{\otimes}\mathfrak{A})$ be a right inverse of $\Delta_\mathfrak{A} : \mathfrak{A}\hat{\otimes}\mathfrak{A} \to \mathfrak{A}$. It is obvious that ρ is also a right inverse of both $\Delta_{\mathfrak{A}^\#,\mathfrak{A}} : \mathfrak{A}^\#\hat{\otimes}\mathfrak{A} \to \mathfrak{A}$ and $\Delta_{\mathfrak{A},\mathfrak{A}^\#} : \mathfrak{A}\hat{\otimes}\mathfrak{A}^\# \to \mathfrak{A}$. By Lemma 5.1.9, $\mathfrak{A}$ is a projective Banach $\mathfrak{A}$-bimodule.

(ii) $\implies$ (i): By Lemma 5.1.9, there is a right inverse $\rho \in \mathfrak{A}\mathcal{L}_\mathfrak{A}(\mathfrak{A}, \mathfrak{A}^\#\hat{\otimes}\mathfrak{A})$ of $\Delta_{\mathfrak{A}^\#,\mathfrak{A}}$. We have

$$\rho(ab) = a \cdot \rho(b) \in \mathfrak{A} \hat{\otimes} \mathfrak{A} \qquad (a, b \in \mathfrak{A}),$$

so that $\rho(\mathfrak{A}^2) \subset \mathfrak{A} \hat{\otimes} \mathfrak{A}$. In order to establish that $\mathfrak{A}$ is biprojective, it is thus sufficient to show that $\mathfrak{A}^2$ is dense in $\mathfrak{A}$.

Fix $a \in \mathfrak{A}$. Then there are $b \in \mathfrak{A}$ and $\mathbf{a} \in \mathfrak{A} \hat{\otimes} \mathfrak{A}$ such that $\rho(a) = e_{\mathfrak{A}^\#} \otimes b + \mathbf{a}$. Since $\Delta_{\mathfrak{A}^\#, \mathfrak{A}} \circ \rho = \mathrm{id}_{\mathfrak{A}}$, it follows that $a = b + \Delta_{\mathfrak{A}^\#, \mathfrak{A}} \mathbf{a}$, so that

$$\rho(a) = e_{\mathfrak{A}^\#} \otimes a - e_{\mathfrak{A}^\#} \otimes \Delta_{\mathfrak{A}^\#, \mathfrak{A}} \mathbf{a} + \mathbf{a}.$$

Consequently, we have

$$\rho(a^2) = a \cdot \rho(a) = a \otimes a - a \otimes \Delta_{\mathfrak{A}^\#, \mathfrak{A}} \mathbf{a} + a \cdot \mathbf{a} \tag{5.1}$$

and

$$\rho(a^2) = \rho(a) \cdot a = e_{\mathfrak{A}^\#} \otimes a^2 - e_{\mathfrak{A}^\#} \otimes (\Delta_{\mathfrak{A}^\#, \mathfrak{A}} \mathbf{a}) a + \mathbf{a} \cdot a. \tag{5.2}$$

Choose $\phi \in \mathfrak{A}^*$ with $\phi|_{\mathfrak{A}^2} \equiv 0$, and define an extension $\phi^\#$ of ϕ on $\mathfrak{A}^\#$ by letting $\langle e_{\mathfrak{A}^\#}, \phi^\# \rangle = 1$. Then (5.1) yields

$$\langle \rho(a^2), \phi^\# \otimes \phi \rangle = \langle a, \phi^\# \rangle \langle a, \phi \rangle = \langle a, \phi \rangle^2$$

whereas, from (5.2), we obtain

$$\langle \rho(a^2), \phi^\# \otimes \phi \rangle = \langle a^2, \phi \rangle = 0,$$

so that $\langle a, \phi \rangle = 0$. Since $a \in \mathfrak{A}$ was arbitrary, we have $\phi \equiv 0$. The Hahn–Banach theorem implies that $\mathfrak{A}^2$ is dense in $\mathfrak{A}$. This completes the proof. $\square$

At the beginning of this section, we claimed that projectivity is probably the most important notion in Banach homology. In this section, however, we did nothing to corroborate this claim: We defined projective modules and proved a few facts about them (none of which was deep). This will change in the next section ...

5.2 Resolutions and Ext-groups

The reason why projective modules are so important is that they can be used quite effectively to compute the Hochschild cohomology groups of a Banach algebra.

We start with the basic definitions:

Definition 5.2.1 A *cochain complex* $\mathcal{E} = (E_n, \delta^n)_{n \in \mathbb{Z}}$ of Banach spaces is a sequence $(E_n)_{n \in \mathbb{Z}}$ of Banach spaces along with bounded, linear maps

$$\cdots \to E_n \xrightarrow{\delta^n} E_{n+1} \to \cdots$$

such that $\delta^n \circ \delta^{n-1} = 0$ for $n \in \mathbb{Z}$.

Example 5.2.2 Let $\mathfrak{A}$ be a Banach algebra, and let E be a Banach $\mathfrak{A}$-bimodule. For $n \in \mathbb{N}_0$, let $E_n := \mathcal{L}^n(\mathfrak{A}, E)$, and let δ^n be the n-coboundary operator. For $n \in \mathbb{Z}$ with $n < 0$, let $E_n := \{0\}$ and $\delta^n := 0$. Then $(E_n, \delta^n)_{n \in \mathbb{Z}}$ is a cochain complex by Lemma 2.4.2. This puts the name Hochschild cochain complex into perspective.

In this example, the spaces E_n are $\{0\}$ if $n < 0$. In such a situation, we shall simply forget about the uninteresting spaces and denote the complex by $(E_n, \delta^n)_{n=0}^{\infty}$.

Definition 5.2.3 Let $\mathcal{E} = (E_n, \delta^n)_{n \in \mathbb{Z}}$ be a complex of Banach spaces, and let $n \in \mathbb{Z}$. For $n \in \mathbb{Z}$, let $\mathcal{Z}^n(\mathcal{E}) := \ker \delta^n$, let $\mathcal{B}^n(\mathcal{E}) := \operatorname{ran} \delta^{n-1}$, and define

$$\mathcal{H}^n(\mathcal{E}) := \mathcal{Z}^n(\mathcal{E})/\mathcal{B}^n(\mathcal{E}).$$

Then $\mathcal{H}^n(\mathcal{E})$ is the *n-th cohomology group* of the complex $\mathcal{E}$.

Remarks 5.2.4   (a) The cohomology "groups" of a cochain complex of Banach spaces are in fact linear spaces. Equipped with the quotient topology, each such cohomology group becomes even a topological vector space (although not necessarily Hausdorff).

(b) If we choose $\mathcal{E}$ as in Example 5.2.2, we have $\mathcal{H}^n(\mathcal{E}) = \mathcal{H}^n(\mathfrak{A}, E)$ for $n \in \mathbb{N}_0$ and $\mathcal{H}^n(\mathcal{E}) = \{0\}$ for $n < 0$. If we forget about the uninteresting groups $\mathcal{H}^n(\mathcal{E}) = \{0\}$ for $n < 0$, we obtain Hochschild cohomology as a particular case of Definition 5.2.3.

The first results we prove in this section are general results on cohomology in the sense of Definition 5.2.3, which — in particular — apply to Hochschild cohomology. We first deal with the question of when two cochain complexes have the same cohomology.

Definition 5.2.5 Let $\mathcal{E} = (E_n, \delta_{\mathcal{E}}^n)_{n \in \mathbb{Z}}$ and $\mathcal{F} = (F_n, \delta_{\mathcal{F}}^n)_{n \in \mathbb{Z}}$ be cochain complexes of Banach spaces. A *morphism* $\phi \colon \mathcal{E} \to \mathcal{F}$ is a family $(\phi_n)_{n \in \mathbb{Z}}$ of bounded linear maps $\phi_n \colon E_n \to F_n$ such that

$$\delta_{\mathcal{F}}^n \circ \phi_n = \phi_{n+1} \circ \delta_{\mathcal{E}}^n \qquad (n \in \mathbb{Z}).$$

Exercise 5.2.1 Let $\mathcal{E}$ and $\mathcal{F}$ be cochain complexes of Banach spaces, and let $\phi \colon \mathcal{E} \to \mathcal{F}$ be a morphism. Show that ϕ induces a sequence $\bar{\phi} = (\bar{\phi}_n)_{n \in \mathbb{Z}}$ of group homomorphisms — in fact, linear maps — $\bar{\phi}_n \colon \mathcal{H}^n(\mathcal{E}) \to \mathcal{H}^n(\mathcal{F})$.

Definition 5.2.6 Let $\mathcal{E} = (E_n, \delta_{\mathcal{E}}^n)_{n \in \mathbb{Z}}$ and $\mathcal{F} = (F_n, \delta_{\mathcal{F}}^n)_{n \in \mathbb{Z}}$ be cochain complexes of Banach spaces. Two morphism $\phi, \psi \colon \mathcal{E} \to \mathcal{F}$ are called *homotopic* if there is a family $\tau = (\tau_n)_{n \in \mathbb{Z}}$ of bounded linear maps $\tau_n \colon E_{n+1} \to F_n$ such that

$$\phi_n - \psi_n = \delta_{\mathcal{F}}^{n-1} \circ \tau_{n-1} + \tau_n \circ \delta_{\mathcal{E}}^n \qquad (n \in \mathbb{Z}). \tag{5.3}$$

The family τ is called a *homotopy* of ϕ and ψ.

The reason why homotopy is an important concept when dealing with cohomology is the following theorem:

Theorem 5.2.7 *Let $\mathcal{E}$ and $\mathcal{F}$ be cochain complexes of Banach spaces, and let $\phi, \psi \colon \mathcal{E} \to \mathcal{F}$ be homotopic morphisms. Then $\bar{\phi} = \bar{\psi}$.*

Proof Fix $n \in \mathbb{Z}$, and let $x \in \mathcal{Z}^n(\mathcal{E})$. By (5.3), it is immediate that $\phi_n(x) - \psi_n(x) \in \mathcal{B}^n(\mathcal{F})$ and thus $\bar{\phi}_n(x + \mathcal{B}^n(\mathcal{E})) = \bar{\psi}_n(x + \mathcal{B}^n(\mathcal{E}))$. $\square$

That was painless . . .

Definition 5.2.8 Two cochain complexes $\mathcal{E}$ and $\mathcal{F}$ of Banach spaces are called *homotopically equivalent* if there are morphisms $\phi \colon \mathcal{E} \to \mathcal{F}$ and $\psi \colon \mathcal{F} \to \mathcal{E}$ such that $\phi \circ \psi$ and $\psi \circ \phi$ are homotopic to the identity morphism on $\mathcal{F}$ and $\mathcal{E}$, respectively.

We haven't formally defined what we mean by the composition of two morphisms of cochain complexes of Banach spaces, but it should be obvious.

As an immediate consequence of Theorem 5.2.7, we obtain:

Corollary 5.2.9 *Let $\mathcal{E}$ and $\mathcal{F}$ be homotopically equivalent cochain complexes of Banach spaces. Then*

$$\mathcal{H}^n(\mathcal{E}) \cong \mathcal{H}^n(\mathcal{F}) \qquad (n \in \mathbb{Z}).$$

Exercise 5.2.2 Use Corollary 5.2.9 to derive Theorem 2.4.6.

Definition 5.2.10 Let $\mathcal{E} = (E_n, \delta_{\mathcal{E}}^n)_{n \in \mathbb{Z}}$ be a cochain complex of Banach spaces.

(i) A *subcomplex* of $\mathcal{E}$ is a cochain complex $\mathcal{F} = (F_n, \delta_{\mathcal{F}}^n)_{n \in \mathbb{Z}}$ of Banach spaces such that, for each $n \in \mathbb{Z}$, the space F_n is a closed subspace of E_n and $\delta_{\mathcal{F}}^n = \delta_{\mathcal{E}}^n|_{F_n}$.

(ii) Given a subcomplex $\mathcal{F}$ of $\mathcal{E}$, the corresponding *quotient complex* $\mathcal{E}/\mathcal{F}$ is the complex $(E_n/F_n, \delta_{\mathcal{E}/\mathcal{F}}^n)_{n \in \mathbb{Z}}$, where the maps $\delta_{\mathcal{E}/\mathcal{F}}^n$ are those induced by the maps $\delta_{\mathcal{E}}^n$.

(iii) The subcomplex $\mathcal{F}$ is called *complemented* if, for each $n \in \mathbb{Z}$, the space F_n is complemented in E_n.

Theorem 5.2.11 (long, exact sequence) *Let $\mathcal{E}$ be a cochain complex of Banach spaces, and let $\mathcal{F}$ be a complemented subcomplex. Then we have a long, exact sequence*

$$\cdots \to \mathcal{H}^{n-1}(\mathcal{E}/\mathcal{F}) \to \mathcal{H}^n(\mathcal{F}) \to \mathcal{H}^n(\mathcal{E}) \to \mathcal{H}^n(\mathcal{E}/\mathcal{F}) \to \mathcal{H}^{n+1}(\mathcal{F}) \to \cdots \qquad (5.4)$$

of cohomology groups.

Proof For each $n \in \mathbb{Z}$, let $\rho_n : E_n/F_n \to E_n$ be a bounded, linear right inverse of the quotient map $\pi_n : E_n \to E_n/F_n$.

For each $n \in \mathbb{Z}$, the natural maps

$$\{0\} \to F_n \overset{\iota_n}{\to} E_n \overset{\pi_n}{\to} E_n/F_n \to \{0\}$$

induce group homomorphisms — in fact: linear maps —

$$\mathcal{H}^n(\mathcal{F}) \overset{\bar{\iota}_n}{\to} \mathcal{H}^n(\mathcal{E}) \overset{\bar{\pi}_n}{\to} \mathcal{H}^n(\mathcal{E}/\mathcal{F}).$$

It is routinely checked that $\operatorname{ran} \bar{\iota}_n \subset \ker \bar{\pi}_n$. For the converse inclusion, let $x \in \mathcal{Z}^n(\mathcal{E})$ be such that $\bar{\pi}_n(x + \mathcal{B}^n(\mathcal{E})) = 0$, i.e. there is $y \in E_{n-1}/F_{n-1}$ such that $\pi_n(x) = \delta_{\mathcal{E}/\mathcal{F}}^n y$. Let $z := x - \delta_{\mathcal{E}}^{n-1}(\rho_{n-1}(y))$. It is immediate that $z \in \mathcal{Z}^n(\mathcal{E})$. Moreover, we have

$$\pi_n(z) = \pi_n(x) - \pi_n(\delta_{\mathcal{E}}^{n-1}(\rho_{n-1}(y))) = \pi_n(x) - \delta_{\mathcal{E}/\mathcal{F}}^{n-1} y = \pi_n(x) - \pi_n(x) = 0,$$

so that, in fact, $z \in \mathcal{Z}^n(\mathcal{F})$. From the definition of z, it is clear that x and z belong to the same equivalence class in $\mathcal{H}^n(\mathcal{E})$, so that $x + \mathcal{B}^n(\mathcal{E}) \in \operatorname{ran} \bar{\iota}_n$.

For $n \in \mathbb{Z}$, define

$$\sigma_n : E_n/F_n \to E_{n+1}, \qquad x \mapsto \delta_{\mathcal{E}}^n(\rho_n(x)) - \rho_{n+1}(\delta_{\mathcal{E}/\mathcal{F}}^n x).$$

First, note that

$$\pi_{n+1}(\sigma_n(x)) = \pi_{n+1}(\delta^n_\mathcal{E}(\rho_n(x))) - \pi_{n+1}(\rho_{n+1}(\delta^n_{\mathcal{E}/\mathcal{F}}x))$$
$$= \delta^n_{\mathcal{E}/\mathcal{F}}(\pi_n(\rho_n(x))) - \delta^n_{\mathcal{E}/\mathcal{F}}x$$
$$= \delta^n_{\mathcal{E}/\mathcal{F}}x - \delta^n_{\mathcal{E}/\mathcal{F}}x$$
$$= 0,$$

so that σ_n is, in fact, a map into F_{n+1}. Let $x \in \mathcal{Z}^n(\mathcal{E}/\mathcal{F})$. Then we obtain:

$$\delta^{n+1}_\mathcal{F}(\sigma_n(x)) = \delta^{n+1}_\mathcal{E}(\delta^n_\mathcal{E}(\rho_n(x))) = 0,$$

i.e. σ_n maps $\mathcal{Z}^n(\mathcal{E}/\mathcal{F})$ into $\mathcal{Z}^{n+1}(\mathcal{F})$. If $x \in \mathcal{B}^n(\mathcal{E}/\mathcal{F})$, i.e. if there is $y \in E_{n-1}/F_{n-1}$ such that $\delta^{n-1}_{\mathcal{E}/\mathcal{F}}y = x$, we obtain

$$\sigma_n(x) = \delta^n_\mathcal{E}(\rho_n(\delta^{n-1}_{\mathcal{E}/\mathcal{F}}y)) = \delta^n_\mathcal{E}(\rho_n(\delta^{n-1}_{\mathcal{E}/\mathcal{F}}y) - \delta^{n-1}_\mathcal{E}(\rho_{n-1}(y))) = \delta^n_\mathcal{F}(-\sigma_{n-1}(y)) \in \mathcal{B}^{n+1}(\mathcal{F}).$$

Hence, σ_n induces a group homomorphism $\bar\sigma_n \colon \mathcal{H}^n(\mathcal{E}/\mathcal{F}) \to \mathcal{H}^{n+1}(\mathcal{F})$.

We claim that $\operatorname{ran}\bar\sigma_n = \ker\bar\iota_{n+1}$ and that $\ker\bar\sigma_n = \operatorname{ran}\bar\pi_n$.

Let $x \in F_n$ be such that $x + \mathcal{B}^n(\mathcal{F}) \in \operatorname{ran}\bar\sigma_n$, i.e. there is $y \in \mathcal{Z}^n(E_n/F_n)$ such that $x = \sigma_n(y) = \delta^n_\mathcal{E}(\rho_n(y)) \in \mathcal{B}^{n+1}(\mathcal{E})$. It follows that $x + \mathcal{B}^n(\mathcal{F}) \in \ker\bar\iota_{n+1}$. Hence, $\operatorname{ran}\bar\sigma_n \subset \ker\bar\iota_{n+1}$ holds. Conversely, let $x \in \mathcal{Z}^{n+1}(\mathcal{F})$ be such that $x + \mathcal{B}^{n+1}(\mathcal{F}) \in \ker\bar\iota_{n+1}$. This means, there is $y \in E_n$ such that $x = \delta^n_\mathcal{E}y$. Let $z := y - \rho_n(\pi_n(y))$; it is immediate that $\pi_n(z) = 0$, so that $z \in F_n$. Furthermore, we have

$$x = \delta^n_\mathcal{E}(\rho_n(\pi_n(y))) - \rho_{n+1}(\pi_n(x)) + \delta^n_\mathcal{E}y - \delta^n_\mathcal{E}(\rho_n(\pi_n(y)))$$
$$= \delta^n_\mathcal{E}(\rho_n(\pi_n(y))) - \rho_{n+1}(\delta^n_{\mathcal{E}/\mathcal{F}}(\pi_n(y))) + \delta^n_\mathcal{E}z$$
$$= \sigma_n(\pi_n(y)) + \delta^n_\mathcal{F}z.$$

This yields $\ker\bar\iota_{n+1} \subset \operatorname{ran}\bar\sigma_n$.

Let $x \in \mathcal{Z}^n(\mathcal{E})$. We have:

$$\sigma_n(\pi_n(x)) = \delta^n_\mathcal{E}(\rho_n(\pi_n(x))) - \rho_{n+1}(\delta^n_{\mathcal{E}/\mathcal{F}}(\pi_n(x)))$$
$$= \delta^n_\mathcal{E}(\rho_n(\pi_n(x))) - \rho_{n+1}(\pi_{n+1}(\delta^n_\mathcal{E}x))$$
$$= \delta^n_\mathcal{E}(\rho_n(\pi_n(x)))$$
$$= \delta^n_\mathcal{E}(\rho_n(\pi_n(x))) - \delta^n_\mathcal{E}x$$
$$= \delta^n_\mathcal{F}(\rho_n(\pi_n(x)) - x), \qquad \text{since } \rho_n(\pi_n(x)) - x \in F_n,$$
$$\in \mathcal{B}^{n+1}(\mathcal{F}).$$

It follows that $\operatorname{ran}\bar\pi_n \subset \ker\bar\sigma_n$. Conversely, let $x \in \mathcal{Z}^n(\mathcal{E}/\mathcal{F})$ be such that $\sigma_n(x) \in \mathcal{B}^{n+1}(\mathcal{F})$. Then there is $y \in F$ such that $\sigma_n(x) = \delta^n_\mathcal{F}y$. Let $z := \rho_n(x) - y$. Since

$$\delta^n_\mathcal{E}x = \delta^n_\mathcal{E}(\rho_n(x)) - \delta^n_\mathcal{E}y = \delta^n_\mathcal{E}(\rho_n(x)) - \sigma_n(x) = \rho_{n+1}(\delta^n_{\mathcal{E}/\mathcal{F}}x) = 0,$$

we have $z \in \mathcal{Z}^n(\mathcal{E})$. It is immediate that $\pi_n(z) = x$ (since $y \in F_n$). Hence, we obtain $\ker\bar\sigma_n \subset \operatorname{ran}\bar\pi_n$.

This completes the proof. $\square$

Remark 5.2.12 The connecting maps in (5.4) are not only group homomorphisms, but linear. If each cohomology group is equipped with its (possibly non-Hausdorff) quotient topology, these maps are even continuous.

We now move from cochain complexes of Banach spaces to chain complexes of left Banach modules.

Definition 5.2.13 Let $\mathfrak{A}$ be a Banach algebra. A *chain complex* $\mathcal{E} = (E_n, d_n)_{n \in \mathbb{Z}}$ of left Banach $\mathfrak{A}$-modules is a sequence $(E_n)_{n \in \mathbb{Z}}$ of left Banach $\mathfrak{A}$-modules along with bounded, $\mathfrak{A}$-module homomorphisms

$$\cdots \leftarrow E_n \xleftarrow{d_n} E_{n+1} \leftarrow \cdots$$

such that $d_n \circ d_{n+1} = 0$ for $n \in \mathbb{Z}$.

We make the distinction between chain and cochain complexes mainly for our convenience: It doesn't really matter in which direction the arrows point.

We have analogues of Definitions 5.2.5, 5.2.6, and 5.2.8:

Definition 5.2.14 Let $\mathfrak{A}$ be a Banach algebra, and let $\mathcal{E} = (E_n, d_n^{\mathcal{E}})_{n \in \mathbb{Z}}$ and $\mathcal{F} = (F_n, d_n^{\mathcal{F}})_{n \in \mathbb{Z}}$ be chain complexes of left Banach $\mathfrak{A}$-modules. A *morphism* $\phi : \mathcal{E} \to \mathcal{F}$ is a family $(\phi_n)_{n \in \mathbb{Z}}$ of bounded $\mathfrak{A}$-module homomorphisms $\phi_n : E_n \to F_n$ such that

$$d_n^{\mathcal{F}} \circ \phi_{n+1} = \phi_n \circ d_n^{\mathcal{E}} \qquad (n \in \mathbb{Z}).$$

Definition 5.2.15 Let $\mathfrak{A}$ be a Banach algebra, and let $\mathcal{E} = (E_n, d_n^{\mathcal{E}})_{n \in \mathbb{Z}}$ and $\mathcal{F} = (F_n, d_n^{\mathcal{F}})_{n \in \mathbb{Z}}$ be chain complexes of left Banach $\mathfrak{A}$-modules. Two morphism $\phi, \psi : \mathcal{E} \to \mathcal{F}$ are called *homotopic* if there is a family $\tau = (\tau_n)_{n \in \mathbb{Z}}$ of bounded $\mathfrak{A}$-module homomorphisms $\tau_n : E_n \to F_{n+1}$ such that

$$\phi_n - \psi_n = d_n^{\mathcal{F}} \circ \tau_n + \tau_{n-1} \circ d_{n-1}^{\mathcal{E}} \qquad (n \in \mathbb{Z}).$$

The family τ is called a *homotopy* of ϕ and ψ.

Definition 5.2.16 Let $\mathfrak{A}$ be a Banach algebra. Two chain complexes $\mathcal{E}$ and $\mathcal{F}$ of left Banach $\mathfrak{A}$-modules are called *homotopically equivalent* if there are morphisms $\phi : \mathcal{E} \to \mathcal{F}$ and $\psi : \mathcal{F} \to \mathcal{E}$ such that $\phi \circ \psi$ and $\psi \circ \phi$ are homotopic to the identity morphism on $\mathcal{F}$ and $\mathcal{E}$, respectively.

We're not finished with definitions yet ...

Definition 5.2.17 Let $\mathfrak{A}$ be a Banach algebra. A chain complex $\mathcal{E} = (E_n, d_n)_{n \in \mathbb{Z}}$ of left Banach $\mathfrak{A}$-modules is called *admissible* if:

(i) $\ker d_n = \operatorname{ran} d_{n+1}$ for $n \in \mathbb{Z}$;
(ii) each module homomorphism d_n is admissible.

Definition 5.2.18 Let $\mathfrak{A}$ be a Banach algebra, and let E be a left Banach $\mathfrak{A}$-module. A *resolution* for E is an admissible chain complex

$$\{0\} \leftarrow E \xleftarrow{\varepsilon} E_0 \xleftarrow{d_0} E_1 \xleftarrow{d_1} E_2 \leftarrow \cdots$$

of left Banach $\mathfrak{A}$-modules. If the modules E_0, E_1, etc., are projective, the resolution is called *projective*.

After so many definitions, we deserve an example:

Example 5.2.19 Let $\mathfrak{A}$ be a Banach algebra, and let E be a left Banach $\mathfrak{A}$-module. For $n \in \mathbb{N}_0$, let

$$\mathcal{B}_n(E) := \mathfrak{A}^\# \hat{\otimes} \underbrace{\mathfrak{A} \hat{\otimes} \cdots \hat{\otimes} \mathfrak{A}}_{n\text{-times}} \hat{\otimes} E.$$

Define

$$\epsilon \colon \mathcal{B}_0(E) \to E, \quad a \otimes x \mapsto a \cdot x,$$

and, for $n \in \mathbb{N}_0$, let $d_n \colon \mathcal{B}_{n+1}(E) \to \mathcal{B}_n(E)$ be given by

$$d_n(a \otimes a_1 \otimes \cdots \otimes a_{n+1} \otimes x) = aa_1 \otimes \cdots \otimes a_{n+1} \otimes x$$
$$+ \sum_{k=1}^{n} (-1)^k a \otimes a_1 \otimes \cdots \otimes a_k a_{k+1} \otimes \cdots \otimes a_{n+1} \otimes x$$
$$+ (-1)^{n+1} a \otimes a_1 \otimes \cdots \otimes a_{n+1} \cdot x$$
$$(a \in \mathfrak{A}^\#, \, a_1, \ldots, a_{n+1} \in \mathfrak{A}, \, x \in E).$$

It is immediate that ϵ and d_0, d_1, etc., are left $\mathfrak{A}$-module homomorphisms, and a dull, albeit tedious calculation yields $\epsilon \circ d_0 = 0$ and $d_n \circ d_{n+1} = 0$ for $n \in \mathbb{N}_0$. The modules $\mathcal{B}_0(E)$, $\mathcal{B}_1(E)$, etc., are free and therefore projective. Obviously, ϵ is admissible. Let $\phi \colon \mathfrak{A}^\# \to \mathbb{C}$ be the character with kernel $\mathfrak{A}$. For $n \in \mathbb{N}_0$, define

$$\tau_n \colon \mathcal{B}_n(E) \to \mathcal{B}_{n+1}(E),$$
$$a \otimes a_1 \otimes \cdots \otimes a_n \otimes x \mapsto e_{\mathfrak{A}^\#} \otimes (a - \phi(a)e_{\mathfrak{A}^\#}) \otimes a_1 \otimes \cdots \otimes a_n \otimes x.$$

It is then routinely checked (Do it!) that

$$d_n \circ \tau_n + \tau_{n-1} \circ d_{n-1} = \mathrm{id}_{\mathcal{B}_n(E)} \quad (n \in \mathbb{N}_0),$$

where formally $d_{-1} = \epsilon$. For $n \in \mathbb{N}_0$, the map

$$\mathcal{B}_n(E) \to \ker d_{n-2} \oplus \ker d_{n-1}, \quad x \mapsto (d_{n-1}x, d_n(\tau_n(x))) \tag{5.5}$$

is a topological isomorphism (formally let $d_{-2} := 0$): An inverse of (5.5) is given by

$$\ker d_{n-2} \oplus \ker d_{n-1} \to \mathcal{B}_n(E), \quad (x, y) \mapsto \tau_{n-1}x + y. \tag{5.6}$$

It follows (How precisely?) that $\ker d_{n-1} = \mathrm{ran}\, d_n$ for $n \in \mathbb{N}_0$ and that d_n is admissible for $n \in \mathbb{N}_0$. Hence, the complex

$$\{0\} \leftarrow E \overset{\epsilon}{\leftarrow} \mathcal{B}_0(E) \overset{d_0}{\leftarrow} \mathcal{B}_1(E) \overset{d_1}{\leftarrow} \mathcal{B}_2(E) \leftarrow \cdots$$

is a projective resolution of E, the *Bar resolution* of E. We use $\mathcal{B}(E)$ to denote the chain complex

$$\{0\} \leftarrow \mathcal{B}_0(E) \overset{d_0}{\leftarrow} \mathcal{B}_1(E) \overset{d_1}{\leftarrow} \mathcal{B}_2(E) \leftarrow \cdots$$

Exercise 5.2.3 Show that (5.6) is indeed an inverse of (5.5).

The following is an immediate consequence of Example 5.2.19:

Corollary 5.2.20 *Let $\mathfrak{A}$ be a Banach algebra, and let E be a left Banach $\mathfrak{A}$-module. Then E has a projective resolution.*

Of course, one module can have several projective resolutions: If $\mathfrak{A}$ is a Banach algebra and P is a projective left Banach $\mathfrak{A}$-module, then

$$\{0\} \leftarrow P \stackrel{\Delta_{\mathfrak{A}^{\#},P}}{\leftarrow} \mathfrak{A}^{\#} \hat{\otimes} P \leftarrow \ker \Delta_{\mathfrak{A}^{\#},P} \leftarrow \{0\}$$

is a projective resolution different from $\mathcal{B}(P)$.

Nevertheless, two resolutions of the same module are not completely unrelated provided one is projective:

Theorem 5.2.21 (comparison theorem) *Let $\mathfrak{A}$ be a Banach algebra, let E and F be left Banach $\mathfrak{A}$-modules, and let $\theta \in {}_{\mathfrak{A}}\mathcal{L}(E,F)$. Furthermore, let*

$$\{0\} \leftarrow E \stackrel{\epsilon}{\leftarrow} P_0 \stackrel{d_0}{\leftarrow} P_1 \stackrel{d_1}{\leftarrow} P_2 \leftarrow \cdots \tag{5.7}$$

be a projective resolution of E, and let

$$\{0\} \leftarrow F \stackrel{\epsilon'}{\leftarrow} E_0 \stackrel{d'_0}{\leftarrow} E_1 \stackrel{d'_1}{\leftarrow} E_2 \leftarrow \cdots \tag{5.8}$$

be an arbitrary resolution of F. Then there is a morphism $\phi = (\phi_n)_{n=0}^{\infty}$ from the chain complex (5.7) to the chain complex (5.8) such that the diagram

$$\begin{array}{ccccccccc}
\{0\} & \leftarrow & E & \stackrel{\epsilon}{\leftarrow} & P_0 & \leftarrow \cdots \leftarrow & P_n & \stackrel{d_n}{\leftarrow} & P_{n+1} & \leftarrow \cdots \\
& & \downarrow & & \downarrow & & \downarrow & & \downarrow & \\
\{0\} & \leftarrow & F & \stackrel{\epsilon'}{\leftarrow} & E_0 & \leftarrow \cdots \leftarrow & E_n & \stackrel{d'_n}{\leftarrow} & E_{n+1} & \leftarrow \cdots
\end{array}$$

commutes, where the vertical arrows represent θ and ϕ_0, ϕ_1, etc., respectively.

Proof We proceed inductively. Suppose that we have already found $\phi_0, \phi_1, \dots \phi_n$ with $\phi_k \in {}_{\mathfrak{A}}\mathcal{L}(P_k, E_k)$ such that the diagram

$$\begin{array}{ccccccccc}
\{0\} & \leftarrow & E & \stackrel{\epsilon}{\leftarrow} & P_0 & \leftarrow \cdots \leftarrow & P_n & \stackrel{d_n}{\leftarrow} & P_{n+1} & \leftarrow \cdots \\
& & \downarrow & & \downarrow & & \downarrow & & & \\
\{0\} & \leftarrow & F & \stackrel{\epsilon'}{\leftarrow} & E_0 & \leftarrow \cdots \leftarrow & E_n & \stackrel{d'_n}{\leftarrow} & E_{n+1} & \leftarrow \cdots
\end{array} \tag{5.9}$$

commutes. Let $F_n := \ker d'_{n-1} = \operatorname{ran} d'_n$, and let $\tilde{\theta} := d'_n$. Then $\tilde{\theta} : E_{n+1} \to F_n$ is an admissible epimorphism of left Banach $\mathfrak{A}$-modules. Define $\sigma : P_{n+1} \to E_n$ as $\phi_n \circ d_n$. Since

$$d'_{n-1} \circ \phi_n \circ d_n = \phi_{n-1} \circ d_{n-1} \circ d_n = 0,$$

the $\mathfrak{A}$-module homomorphism σ attains its values in F_n. By Proposition 5.1.6(iii), there is $\phi_{n+1} : P_{n+1} \to E_{n+1}$ such that $\sigma := \tilde{\theta} \circ \phi_{n+1}$. Clearly, if we extend the diagram (5.9) by ϕ_{n+1}, this extended diagram still commutes. $\square$

Exercise 5.2.4 Show that any two morphisms as in Theorem 5.2.21 are homotopic.

Corollary 5.2.22 *Let $\mathfrak{A}$ be a Banach algebra, and let E be a left Banach $\mathfrak{A}$-module. Then any two projective resolutions for E are homotopically equivalent.*

Proof Apply Theorem 5.2.21 with $\theta = \mathrm{id}_E$. $\square$

We shall now see that projective resolutions can be used efficiently to calculate cohomology groups.

Lemma 5.2.23 *Let $\mathfrak{A}$ be a Banach algebra, let F be a left Banach $\mathfrak{A}$-module and let $\mathcal{E} = (E_n, d_n)_{n \in \mathbb{Z}}$ be a chain complex of left Banach $\mathfrak{A}$-modules. Define*

$$\delta^n : \mathfrak{A}\mathcal{L}(E_n, F) \to \mathfrak{A}\mathcal{L}(E_{n+1}, F), \quad T \mapsto T \circ d_n.$$

Then $(\mathfrak{A}\mathcal{L}(E_n, F), \delta^n)_{n \in \mathbb{Z}}$ is a cochain complex of Banach spaces.

Proof Obvious. $\square$

We suggestively denote the cochain complex described in Lemma 5.2.23 by $\mathfrak{A}\mathcal{L}(\mathcal{E}, F)$.

The following lemma is equally easy to check:

Lemma 5.2.24 *Let $\mathfrak{A}$ be a Banach algebra, let F be a left Banach $\mathfrak{A}$-module, and let $\mathcal{E}_1$ and $\mathcal{E}_2$ be homotopically equivalent chain complexes of left Banach $\mathfrak{A}$-modules. Then the cochain complexes of Banach spaces $\mathfrak{A}\mathcal{L}(\mathcal{E}_1, F)$ and $\mathfrak{A}\mathcal{L}(\mathcal{E}_2, F)$ are homotopically equivalent.*

We can now define Ext-groups:

Definition 5.2.25 Let $\mathfrak{A}$ be a Banach algebra, let E and F be left Banach $\mathfrak{A}$-modules, and let a projective resolution for E be given:

$$\{0\} \leftarrow E \xleftarrow{\epsilon} P_0 \xleftarrow{d_0} P_1 \xleftarrow{d_1} P_2 \leftarrow \cdots .$$

Let $\mathcal{P}$ denote the chain complex of left Banach $\mathfrak{A}$-modules

$$\{0\} \leftarrow P_0 \xleftarrow{d_0} P_1 \xleftarrow{d_1} P_2 \leftarrow \cdots . \tag{5.10}$$

Then, for $n \in \mathbb{N}_0$, the *n-th* Ext-*group* of E by F is defined as

$$\mathrm{Ext}_{\mathfrak{A}}^n(E, F) := \mathcal{H}^n(\mathfrak{A}\mathcal{L}(\mathcal{P}, F)).$$

Remark 5.2.26 Let two projective resolutions for E given. By Corollary 5.2.22, these two resolutions are homotopically equivalent, and so are the corresponding chain complexes in (5.10), which we denote by $\mathcal{P}_1$ and $\mathcal{P}_2$. By Lemma 5.2.24, the cochain complexes of Banach spaces $\mathfrak{A}\mathcal{L}(\mathcal{P}_1, F)$ and $\mathfrak{A}\mathcal{L}(\mathcal{P}_2, F)$ are also homotopically equivalent. By Corollary 5.2.9, this means that

$$\mathcal{H}^n(\mathfrak{A}\mathcal{L}(\mathcal{P}_1, F)) \cong \mathcal{H}^n(\mathfrak{A}\mathcal{L}(\mathcal{P}_2, F)) \qquad (n \in \mathbb{N}_0).$$

Hence, $\mathrm{Ext}_{\mathfrak{A}}^n(E, F)$ is independent of a particular projective resolution for E.

Exercise 5.2.5 Let $\mathfrak{A}$ be a Banach algebra, and let E and F be left Banach $\mathfrak{A}$-modules. Show that

$$\mathrm{Ext}_{\mathfrak{A}}^0(E, F) \cong \mathfrak{A}\mathcal{L}(E, F).$$

canonically.

Exercise 5.2.6 Let $\mathfrak{A}$ be a Banach algebra, and let E be a left Banach $\mathfrak{A}$-module. Show that E is projective if and only if $\text{Ext}^1_{\mathfrak{A}}(E, F) = \{0\}$ for all left Banach $\mathfrak{A}$-modules F if and only if $\text{Ext}^n_{\mathfrak{A}}(E, F) = \{0\}$ for all $n \in \mathbb{N}$ and for all left Banach $\mathfrak{A}$-modules F.

The following is extremely useful when it comes to calculating Ext-groups:

Theorem 5.2.27 *Let $\mathfrak{A}$ be a Banach algebra, let E and F be left Banach $\mathfrak{A}$-modules, and let*

$$\{0\} \to E_2 \to E_1 \to E_1/E_2 \to \{0\}$$

and

$$\{0\} \to F_2 \to F_1 \to F_1/F_2 \to \{0\}$$

be admissible, short, exact sequences of left Banach $\mathfrak{A}$-modules. Then we have long exact sequences

$$\{0\} \to {}_{\mathfrak{A}}\mathcal{L}(E, F_2) \to {}_{\mathfrak{A}}\mathcal{L}(E, F_1) \to {}_{\mathfrak{A}}\mathcal{L}(E, F_1/F_2) \to \text{Ext}^1_{\mathfrak{A}}(E, F_2) \to$$
$$\cdots \to \text{Ext}^n_{\mathfrak{A}}(E, F_2) \to \text{Ext}^n_{\mathfrak{A}}(E, F_1) \to \text{Ext}^n_{\mathfrak{A}}(E, F_1/F_2) \to \text{Ext}^{n+1}_{\mathfrak{A}}(E, F_2) \to \cdots$$

and

$$\{0\} \to {}_{\mathfrak{A}}\mathcal{L}(E_1/E_2, F) \to {}_{\mathfrak{A}}\mathcal{L}(E_1, F) \to {}_{\mathfrak{A}}\mathcal{L}(E_2, F) \to \text{Ext}^1_{\mathfrak{A}}(E_1/E_2, F) \to$$
$$\cdots \to \text{Ext}^n_{\mathfrak{A}}(E_1/E_2, F) \to \text{Ext}^n_{\mathfrak{A}}(E_1, F) \to \text{Ext}^n_{\mathfrak{A}}(E_2, F) \to \text{Ext}^{n+1}_{\mathfrak{A}}(E_1/E_2, F) \to \cdots.$$

Proof Both exact sequences follow from Theorem 5.2.11.

Since, for $n \in \mathbb{N}_0$, each module $\mathcal{B}_n(E)$ is projective, we may view ${}_{\mathfrak{A}}\mathcal{L}(\mathcal{B}(E), F_2)$ as a complemented subcomplex of ${}_{\mathfrak{A}}\mathcal{L}(\mathcal{B}(E), F_1)$ such that the resulting quotient complex is canonically isomorphic to ${}_{\mathfrak{A}}\mathcal{L}(\mathcal{B}(E), F_1/F_2)$ (Please, check this!). The first exact sequence is thus a consequence of Theorem 5.2.11, Definition 5.2.25, and Exercise 5.2.5.

We may canonically view $\mathcal{B}(E_2)$ as a complemented subcomplex of $\mathcal{B}(E_1)$ such that the resulting quotient complex is isomorphic to $\mathcal{B}(E_1/E_2)$. Applying Definition 5.2.25 and Theorem 5.2.11, we obtain the second exact sequence. $\square$

We shall now express Hochschild cohomology groups in terms of Ext-groups.

For a Banach algebra $\mathfrak{A}$, let $\mathfrak{A}^{\text{env}} := \mathfrak{A}^{\#} \hat{\otimes} (\mathfrak{A}^{\text{op}})^{\#}$ denote the *enveloping algebra* of $\mathfrak{A}$. Any Banach $\mathfrak{A}$-bimodule is a unital, left Banach $\mathfrak{A}^{\text{env}}$-module in a canonical fashion (How?).

Exercise 5.2.7 Let $\mathfrak{A}$ be a Banach algebra. Show that a Banach $\mathfrak{A}$-bimodule is projective as an $\mathfrak{A}$-bimodule if and only if it is projective as a left Banach $\mathfrak{A}^{\text{env}}$-module.

Theorem 5.2.28 *Let $\mathfrak{A}$ be a Banach algebra, and let E be a Banach $\mathfrak{A}$-bimodule. Then we have:*

$$\mathcal{H}^n(\mathfrak{A}, E) \cong \text{Ext}^n_{\mathfrak{A}^{\text{env}}}(\mathfrak{A}^{\#}, E) \qquad (n \in \mathbb{N}_0).$$

Proof The Bar resolution of the left Banach $\mathfrak{A}^{\#}$-module $\mathfrak{A}^{\#}$ is in fact a projective resolution of the left Banach $\mathfrak{A}^{\text{env}}$-module $\mathfrak{A}^{\#}$, so that we can use $\mathcal{B}(\mathfrak{A}^{\#})$ to calculate the groups $\text{Ext}^n_{\mathfrak{A}^{\text{env}}}(\mathfrak{A}^{\#}, E)$ for $n \in \mathbb{N}$. Since

$$\mathcal{B}_n(\mathfrak{A}^\#) = \mathfrak{A}^\# \hat{\otimes} \underbrace{\mathfrak{A}\hat{\otimes}\cdots\hat{\otimes}\mathfrak{A}}_{n\text{-times}} \hat{\otimes}\mathfrak{A}^\# \qquad (n \in \mathbb{N}_0),$$

the universal property of the projective tensor product produces canonical isomorphisms

$$_{\mathfrak{A}^{\mathrm{env}}}\mathcal{L}(\mathcal{B}_n(\mathfrak{A}^\#), E) = {}_{\mathfrak{A}^\#}\mathcal{L}_{\mathfrak{A}^\#}(\mathcal{B}_n(\mathfrak{A}^\#), E) \cong \mathcal{L}^n(\mathfrak{A}, E) \qquad (n \in \mathbb{N}_0).$$

A tedious, but not difficult calculation shows that these isomorphisms induce an isomorphism of the cochain complex $_{\mathfrak{A}^{\mathrm{env}}}\mathcal{L}(\mathcal{B}(\mathfrak{A}^\#), E)$ and the Hochschild cochain complex.   □

Exercise 5.2.8 Let $\mathfrak{A}$ be a unital Banach algebra, and let E be a unital Banach $\mathfrak{A}$-bimodule. Show that

$$\mathcal{H}^n(\mathfrak{A}, E) \cong \mathrm{Ext}^n_{\mathfrak{A}\hat{\otimes}\mathfrak{A}^{\mathrm{op}}}(\mathfrak{A}, E) \qquad (n \in \mathbb{N}).$$

Together, Theorems 5.2.27 and 5.2.28 yield:

Corollary 5.2.29 *Let $\mathfrak{A}$ be a Banach algebra, and let*

$$\{0\} \to F \to E \to E/F \to \{0\}$$

be a short, exact sequence of Banach $\mathfrak{A}$-bimodules. Then we have a long exact sequence:

$$\{0\} \to \mathcal{H}^0(\mathfrak{A}, F) \to \mathcal{H}^0(\mathfrak{A}, E) \to \mathcal{H}^0(\mathfrak{A}, E/F) \to \mathcal{H}^1(\mathfrak{A}, F) \to$$
$$\cdots \to \mathcal{H}^n(\mathfrak{A}, F) \to \mathcal{H}^n(\mathfrak{A}, E) \to \mathcal{H}^n(\mathfrak{A}, E/F) \to \mathcal{H}^{n+1}(\mathfrak{A}, F) \to \cdots.$$

You might ask yourself: What's the purpose of all this? We can now compute Hochschild cohomology groups using Ext-groups. But who would want to do this? Definition 2.4.3 was straightforward whereas, just to be able to define Ext-groups, we had to go through quite extensive preparations. Why trade a simple definition for a complicated characterization? The answer lies in Definition 5.2.25: Ext-groups can be computed based on *any* projective resolution. In certain situations, we can perhaps choose a projective resolution for which the Ext-groups — and hence the Hochschild cohomology groups — are particularly easy to compute.

We conclude this section with an application of Theorems 5.2.27 and 5.2.28, which shows that certain cohomology groups of biprojective Banach algebras are trivial:

Theorem 5.2.30 *Let $\mathfrak{A}$ be a biprojective Banach algebra. Then $\mathcal{H}^n(\mathfrak{A}, E) = \{0\}$ for all $n \geq 3$ and for all Banach $\mathfrak{A}$-bimodules E.*

Proof Consider the following admissible, short, exact sequences of Banach $\mathfrak{A}$-bimodule, i.e. left Banach $\mathfrak{A}^{\mathrm{env}}$-modules:

$$\{0\} \to \mathfrak{A} \to \mathfrak{A}^\# \to \mathfrak{A}^\#/\mathfrak{A} \to \{0\},$$
$$\{0\} \to J \to \mathfrak{A}^{\mathrm{env}} \to \mathfrak{A}^{\mathrm{env}}/J \to \{0\},$$

where J is the closed linear span of $(\mathfrak{A}^\#\hat{\otimes}\mathfrak{A}) \cup (\mathfrak{A}\hat{\otimes}\mathfrak{A}^\#)$, and

$$\{0\} \to \mathfrak{A}\hat{\otimes}\mathfrak{A} \to P \to J \to \{0\}$$

with $P := (\mathfrak{A}\hat{\otimes}\mathfrak{A}^\#) \oplus (\mathfrak{A}^\#\hat{\otimes}\mathfrak{A})$, where

$$\mathfrak{A} \hat{\otimes} \mathfrak{A} \to P, \quad \mathbf{a} \mapsto (\mathbf{a}, \mathbf{a})$$

and

$$P \to J, \quad (\mathbf{a}, \mathbf{b}) \mapsto \mathbf{a} - \mathbf{b}.$$

We apply the second of the long exact sequences from Theorem 5.2.27 to these short, exact sequences and consider the following segments of the resulting long exact sequences:

$$\operatorname{Ext}^2_{\mathfrak{A}^{env}}(\mathfrak{A}, E) \to \operatorname{Ext}^3_{\mathfrak{A}^{env}}(\mathfrak{A}^\#/\mathfrak{A}, E) \to \operatorname{Ext}^3_{\mathfrak{A}^{env}}(\mathfrak{A}^\#, E) \to \operatorname{Ext}^3_{\mathfrak{A}^{env}}(\mathfrak{A}, E),$$

$$\operatorname{Ext}^2_{\mathfrak{A}^{env}}(\mathfrak{A}^{env}, E) \to \operatorname{Ext}^2_{\mathfrak{A}^{env}}(J, E) \to \operatorname{Ext}^3_{\mathfrak{A}^{env}}(\mathfrak{A}^{env}/J, E) \to \operatorname{Ext}^3_{\mathfrak{A}^{env}}(\mathfrak{A}^{env}, E),$$

and

$$\operatorname{Ext}^1_{\mathfrak{A}^{env}}(\mathfrak{A} \hat{\otimes} \mathfrak{A}, E) \to \operatorname{Ext}^2_{\mathfrak{A}^{env}}(J, E) \to \operatorname{Ext}^2_{\mathfrak{A}^{env}}(P, E) \to \operatorname{Ext}^2(\mathfrak{A} \hat{\otimes} \mathfrak{A}, E).$$

Since the left Banach $\mathfrak{A}^{env}$-modules $\mathfrak{A}$, $\mathfrak{A}^{env}$, and $\mathfrak{A} \hat{\otimes} \mathfrak{A}$ are projective (Why?), it follows from Exercise 5.2.6 that the endpoints of these segments are all $\{0\}$. We thus obtain isomorphisms:

$$\operatorname{Ext}^3_{\mathfrak{A}^{env}}(\mathfrak{A}^\#, E) \cong \operatorname{Ext}^3_{\mathfrak{A}^{env}}(\mathfrak{A}^\#/\mathfrak{A}, E)$$
$$\cong \operatorname{Ext}^3_{\mathfrak{A}^{env}}(\mathfrak{A}^{env}/J, E), \qquad \text{(Why does this hold?)}$$
$$\cong \operatorname{Ext}^2_{\mathfrak{A}^{env}}(J, E)$$
$$\cong \operatorname{Ext}^2_{\mathfrak{A}^{env}}(P, E). \tag{5.11}$$

Since the left $\mathfrak{A}^{env}$-modules $\mathfrak{A}^\# \hat{\otimes} \mathfrak{A}$ and $\mathfrak{A} \hat{\otimes} \mathfrak{A}^\#$ are projective (again: Why?), so is P. It follows, again from Exercise 5.2.6, that $\operatorname{Ext}^2_{\mathfrak{A}^{env}}(P, E) = \{0\}$. By (5.11) and Theorem 5.2.28, this means $\mathcal{H}^3(\mathfrak{A}, E) = \{0\}$. The claim for arbitrary $n \geq 3$, now follows from Theorem 2.4.6.
□

I don't know of any proof of Theorem 5.2.30 that only uses Definition 2.4.3.

5.3 Flatness and injectivity

We conclude this chapter with a discussion of two further important properties of Banach modules: flatness and injectivity.

The following definition is a prerequisite for the definition of flatness.

Definition 5.3.1 Let $\mathfrak{A}$ be a Banach algebra, let E be a right Banach $\mathfrak{A}$-module, and let F be a left Banach $\mathfrak{A}$-module. The *projective module tensor product* $E \hat{\otimes}_{\mathfrak{A}} F$ of E and F is defined as the quotient of $E \hat{\otimes} F$ by the closed linear span of $\{x \cdot a \otimes y - x \otimes a \cdot y : x \in E, y \in F, a \in \mathfrak{A}\}$.

There is no need, in general, for $E \hat{\otimes}_{\mathfrak{A}} F$ to again be some sort of Banach module.

Exercise 5.3.1 Let $\mathfrak{A}$ be a Banach algebra, let E be a right Banach $\mathfrak{A}$-module, and let F be a left Banach $\mathfrak{A}$-module. Show that the isomorphism between $(E \hat{\otimes} F)^*$ and $\mathcal{L}(E, F^*)$ (Exercise B.2.10) induces an isometric isomorphism of $(E \hat{\otimes}_{\mathfrak{A}} F)^*$ and $_{\mathfrak{A}}\mathcal{L}(E, F^*)$.

Proposition 5.3.2 *Let $\mathfrak{A}$ be a Banach algebra, let R be a closed right ideal of $\mathfrak{A}^{\#}$ with a bounded left approximate identity, and let E be a left Banach $\mathfrak{A}$-module. Then the multiplication map*

$$\Delta_{R,E} \colon R\hat{\otimes}E \to E, \quad r \otimes x \mapsto r \cdot x$$

induces a topological isomorphism of the Banach spaces $R\hat{\otimes}_{\mathfrak{A}}E$ and $R \cdot E$.

Proof By Cohen's factorization theorem, $R \cdot E$ is indeed a Banach space, so that $\Delta_{R,E}$ is an open map onto $R \cdot E$. Let $\mathbf{x} = \sum_{n=1}^{\infty} r_n \otimes x_n \in \ker \Delta_{R,E}$, where $\sum_{n=1}^{\infty} \|r_n\|\|x_n\| < \infty$. Let

$$\mathcal{R} := \left\{ (s_n)_{n=1}^{\infty} \in R^{\mathbb{N}} : \sum_{n=1}^{\infty} \|s_n\|\|x_n\| < \infty \text{ and } \sum_{n=1}^{\infty} s_n \cdot x_n = 0 \right\}$$

For $(s_n)_{n=1}^{\infty} \in \mathcal{R}$, let

$$\||(s_n)_{n=1}^{\infty}\|| := \sum_{n=1}^{\infty} \|s_n\|\|x_n\|.$$

Then $(\mathcal{R}, \||\cdot\||)$ is a left Banach R-mpdule, and any bounded left approximate identity for R is a bounded left approximate identity for $\mathcal{R}$. By definition, $(r_n)_{n=1}^{\infty} \in \mathcal{R}$. By Cohen's factorization theorem, there is $r \in R$ and $(t_n)_{n=1}^{\infty} \in \mathcal{R}$ such that $r_n = rt_n$ for $n \in \mathbb{N}$. Consequently,

$$\mathbf{x} = \sum_{n=1}^{\infty} rt_n \otimes x_n = \sum_{n=1}^{\infty} (rt_n \otimes x_n - r \otimes t_n \cdot x_n)$$

lies in the closed linear span of $\{r' \cdot a \otimes x - r' \otimes a \cdot x : r' \in R, \, x \in E, \, a \in \mathfrak{A}\}$. $\quad\square$

Definition 5.3.3 Let $\mathfrak{A}$ be a Banach algebra. A left Banach $\mathfrak{A}$-module F is called *flat* if for every admissible, short, exact sequence

$$\{0\} \to E_2 \to E_1 \to E_1/E_2 \to \{0\}$$

of right Banach $\mathfrak{A}$-modules, the sequence

$$\{0\} \to E_2\hat{\otimes}_{\mathfrak{A}}F \to E_1\hat{\otimes}_{\mathfrak{A}}F \to (E_1/E_2)\hat{\otimes}_{\mathfrak{A}}F \to \{0\}$$

is exact.

Of course, one can define flatness for right Banach modules analoguously.

Example 5.3.4 Let $\mathfrak{A}$ be a Banach algebra, and let L be a closed left ideal of $\mathfrak{A}^{\#}$ with a bounded right approximate identity. Let

$$\{0\} \to E_2 \to E_1 \to E_1/E_2 \to \{0\}$$

be an admissible, short, exact sequence of right Banach $\mathfrak{A}$-modules. Consider the commutative diagram

$$\{0\} \to E_2 \hat{\otimes}_{\mathfrak{A}} L \to E_1 \hat{\otimes}_{\mathfrak{A}} L \to (E_1/E_2) \hat{\otimes}_{\mathfrak{A}} L \to \{0\}$$
$$\downarrow \qquad \downarrow \qquad \downarrow \qquad \qquad (5.12)$$
$$\{0\} \to E_2 \cdot L \to E_1 \cdot L \to (E_1/E_2) \cdot L \to \{0\},$$

where the vertical arrows are the maps induced by the respective multiplication operators. By (the right module version of) Proposition 5.3.2, the columns are all isomorphisms. Since the second row of (5.12) is trivially exact, the same is true for the first row. Hence, L is a flat left Banach $\mathfrak{A}$-module.

We shall soon encounter further examples of flat Banach modules.

Let $\mathfrak{A}$ a Banach algebra, and let E be an arbitrary Banach space. Then $\mathcal{L}(\mathfrak{A}^{\#}, E)$ becomes a left Banach $\mathfrak{A}$-module through

$$(a \cdot T)(b) := T(ba) \qquad (a \in \mathfrak{A}, \, b \in \mathfrak{A}^{\#}, \, T \in \mathcal{L}(\mathfrak{A}^{\#}, E)).$$

Modules of this type are called *cofree*. If E itself is a left Banach $\mathfrak{A}$-module, we have a canonical homomorphism $\Delta^{\mathfrak{A}^{\#}, E} : E \to \mathcal{L}(\mathfrak{A}^{\#}, E)$ of left Banach $\mathfrak{A}$-modules defined by

$$(\Delta^{\mathfrak{A}^{\#}, E} x)(a) := a \cdot x \qquad (a \in \mathfrak{A}, \, x \in E).$$

The following is an analogue of Proposition 5.1.6:

Proposition 5.3.5 *Let $\mathfrak{A}$ be a Banach algebra, and let I be a left Banach $\mathfrak{A}$-module. Then the following are equivalent:*

(i) *The canonical homomorphism $\Delta^{\mathfrak{A}^{\#}, I}$ has a left inverse in $_{\mathfrak{A}}\mathcal{L}(\mathcal{L}(\mathfrak{A}^{\#}, I), I)$.*

(ii) *There is a left Banach $\mathfrak{A}$-module J such that $I \oplus J$ is cofree.*

(iii) *If E and F are left Banach $\mathfrak{A}$-modules, if $\theta \in {}_{\mathfrak{A}}\mathcal{L}(F, E)$ is injective and admissible, and if $\sigma \in {}_{\mathfrak{A}}\mathcal{L}(F, I)$, then there is $\rho \in {}_{\mathfrak{A}}\mathcal{L}(E, I)$ such that $\rho \circ \theta = \sigma$, i.e. the diagram*

$$
\begin{array}{ccc}
 & & I \\
 & \nearrow & \uparrow \\
F & \longrightarrow & E.
\end{array}
$$

commutes.

(iv) *Every admissible, short, exact sequence*

$$\{0\} \to I \to E \to E/I \to \{0\}$$

of left Banach $\mathfrak{A}$-modules splits.

Exercise 5.3.2 Prove Proposition 5.3.5.

Definition 5.3.6 Let $\mathfrak{A}$ be a Banach algebra. A left Banach $\mathfrak{A}$-module I satisfying the equivalent conditions of Proposition 5.3.5 is called *injective*.

Of course, there is an analoguous notion of injectivity for right Banach modules.

Examples 5.3.7   (a) Let $\mathfrak{A}$ be an amenable Banach algebra, and let

$$\{0\} \to F \to E \to E/F \to \{0\} \qquad (5.13)$$

be an admissible, short, exact sequence of left or right Banach $\mathfrak{A}$-modules, where F is a dual module. By Theorem 2.3.13 (or rather its right module version), (5.13) splits, so that Proposition 5.3.5(iv) is satisfied. Consequently, F is injective.

(b) Let $\mathfrak{A}$ be an arbitrary Banach algebra, and let P be a projective right Banach $\mathfrak{A}$-module. With the canonical identification $\mathcal{L}(\mathfrak{A}^{\#}, P^*) \cong (\mathfrak{A}^{\#} \hat{\otimes} P)^*$, we see easily that $\Delta^{\mathfrak{A}^{\#}, P^*} = \Delta^*_{P, \mathfrak{A}^{\#}}$. From the definition of projectivity, it is immediate that Proposition 5.3.5(i) is satisfied. Hence, P^* is an injective left Banach $\mathfrak{A}$-module.

The following is an analogue of Exercise 5.2.6.

Exercise 5.3.3 Let $\mathfrak{A}$ be a Banach algebra, and let F be a left Banach $\mathfrak{A}$-module. Show that F is injective if and only if $\mathrm{Ext}^1_{\mathfrak{A}}(E, F) = \{0\}$ for all left Banach $\mathfrak{A}$-modules E if and only if $\mathrm{Ext}^n_{\mathfrak{A}}(E, F) = \{0\}$ for all $n \in \mathbb{N}$ and for all left Banach $\mathfrak{A}$-modules E.

Next, we shall see that flatness and injectivity are dual to each other:

Theorem 5.3.8 *Let $\mathfrak{A}$ be a Banach algebra, and let E be a left Banach $\mathfrak{A}$-module. Then the following are equivalent:*

(i) *E is flat.*

(ii) *E^* is an injective right Banach $\mathfrak{A}$-module.*

Proof (i) $\Longrightarrow$ (ii): Let F_1 and F_2 be right Banach $\mathfrak{A}$-modules, and let $\theta \in {}_{\mathfrak{A}}\mathcal{L}(F_2, F_1)$ be injective and admissible, so that the short, exact sequence

$$\{0\} \to F_2 \xrightarrow{\theta} F_1 \to F_1/\theta(F_2) \to \{0\}$$

of right Banach $\mathfrak{A}$-modules is admissible. Since E is flat, the sequence

$$\{0\} \to F_2 \hat{\otimes}_{\mathfrak{A}} E \xrightarrow{\theta \otimes \mathrm{id}_E} F_1 \hat{\otimes}_{\mathfrak{A}} E \to (F_1/\theta(F_2)) \hat{\otimes}_{\mathfrak{A}} E \to \{0\}$$

is exact, and so is the dual sequence

$$\{0\} \leftarrow {}_{\mathfrak{A}}\mathcal{L}(F_2, E^*) \xleftarrow{(\theta \otimes \mathrm{id}_E)^*} {}_{\mathfrak{A}}\mathcal{L}(F_1, E^*) \leftarrow {}_{\mathfrak{A}}\mathcal{L}(F_1/\theta(F_2), E^*) \leftarrow \{0\}. \qquad (5.14)$$

Let $\sigma \in {}_{\mathfrak{A}}\mathcal{L}(F_2, E^*)$. The exactness of (5.14) — more precisely: the surjectivity of $(\theta \otimes \mathrm{id}_E)^*$ — yields $\rho \in {}_{\mathfrak{A}}\mathcal{L}(F_1, E^*)$ with $\sigma = (\theta \otimes \mathrm{id}_E)^* \rho$, i.e. $\rho \circ \theta = \sigma$. Hence, E^* satisfies (the right module analogue of) Proposition 5.3.5(iii).

(ii) $\Longrightarrow$ (i): This is proved by reversing the arguments from (i) $\Longrightarrow$ (ii). $\square$

Exercise 5.3.4 Work out the proof of Theorem 5.3.8 (ii) $\Longrightarrow$ (i) in detail.

With Theorem 5.3.8 and Examples 5.3.7, we obtain more examples of flat Banach modules:

Examples 5.3.9   (a) Every left or right Banach module over an amenable Banach algebra is flat.

(b) Every projective left or right Banach module is flat.

So far, we have treated flatness and injectivity for one-sided Banach modules only. We leave it to the reader to do the same for bimodules:

Exercise 5.3.5 Define flatness and injectivity for Banach bimodules such that analogues of Proposition 5.3.5 and Theorem 5.3.8 hold.

In Theorem 5.1.10, we showed that a Banach algebra is biprojective if and only if it is a projective Banach bimodule over itself. Guess what's true for biflat Banach algebras ...

We start with an analogue of Lemma 5.1.9:

Lemma 5.3.10 *The following are equivalent for a Banach algebra* $\mathfrak{A}$*:*

(i) $\mathfrak{A}$ *is a flat Banach* $\mathfrak{A}$*-bimodule.*

(ii) *Both maps* $\Delta^*_{\mathfrak{A}^\#,\mathfrak{A}}: \mathfrak{A}^* \to (\mathfrak{A}^\# \hat{\otimes} \mathfrak{A})^*$ *and* $\Delta^*_{\mathfrak{A},\mathfrak{A}^\#}: \mathfrak{A}^* \to (\mathfrak{A} \hat{\otimes} \mathfrak{A}^\#)^*$ *have left inverses in* $_\mathfrak{A}\mathcal{L}_\mathfrak{A}((\mathfrak{A}^\# \hat{\otimes} \mathfrak{A})^*, \mathfrak{A}^*)$ *and* $_\mathfrak{A}\mathcal{L}_\mathfrak{A}((\mathfrak{A} \hat{\otimes} \mathfrak{A}^\#)^*, \mathfrak{A}^*)$*, respectively.*

Proof (i) $\Longrightarrow$ (ii): By (the bimodule analogue of) Theorem 5.3.8, $\mathfrak{A}^*$ is an injective Banach $\mathfrak{A}$-bimodule. The claim then follows immediately from (the bimodule analogue of) Proposition 5.3.5(iv).

(ii) $\Longrightarrow$ (i): Let $\theta \in {}_\mathfrak{A}\mathcal{L}_\mathfrak{A}((\mathfrak{A}^\# \hat{\otimes} \mathfrak{A})^*, \mathfrak{A}^*)$ and $\sigma \in {}_\mathfrak{A}\mathcal{L}_\mathfrak{A}((\mathfrak{A} \hat{\otimes} \mathfrak{A}^\#)^*, \mathfrak{A}^*)$ be left inverses of $\Delta^*_{\mathfrak{A}^\#,\mathfrak{A}}$ and $\Delta^*_{\mathfrak{A},\mathfrak{A}^\#}$, respectively. For $\phi \in (\mathfrak{A}^\# \hat{\otimes} \mathfrak{A} \hat{\otimes} \mathfrak{A}^\#)^*$ and $c \in \mathfrak{A}^\#$, define $\phi[c] \in (\mathfrak{A}^\# \hat{\otimes} \mathfrak{A})^*$ by letting

$$\langle a \otimes b, \phi[c] \rangle := \langle a \otimes b \otimes c, \phi \rangle \qquad (a \in \mathfrak{A}^\#, b \in \mathfrak{A}).$$

Define $\rho \in {}_\mathfrak{A}\mathcal{L}_\mathfrak{A}((\mathfrak{A}^\# \hat{\otimes} \mathfrak{A} \hat{\otimes} \mathfrak{A}^\#)^*, (\mathfrak{A} \hat{\otimes} \mathfrak{A}^\#)^*)$ through

$$\langle b \otimes c, \rho(\phi) \rangle := \langle b, \theta(\phi[c]) \rangle \qquad (b \in \mathfrak{A}, c \in \mathfrak{A}^\#).$$

Then $\sigma \circ \rho \in {}_\mathfrak{A}\mathcal{L}_\mathfrak{A}((\mathfrak{A}^\# \hat{\otimes} \mathfrak{A} \hat{\otimes} \mathfrak{A}^\#)^*, \mathfrak{A}^*)$ is a left inverse of $\Delta^*_{\mathfrak{A}^\#,\mathfrak{A},\mathfrak{A}^\#}: \mathfrak{A}^* \to (\mathfrak{A}^\# \hat{\otimes} \mathfrak{A} \hat{\otimes} \mathfrak{A}^\#)^*$, so that $\mathfrak{A}^*$ is injective. Hence, $\mathfrak{A}$ is flat. $\quad\square$

Lemma 5.3.11 *Let* $\mathfrak{A}$ *be a Banach algebra which is a flat Banach* $\mathfrak{A}$*-bimodule. Then* $\mathfrak{A}^2$ *is dense in* $\mathfrak{A}$*.*

Proof Let $\phi \in \mathfrak{A}^*$ be such that $\phi|_{\mathfrak{A}^2} \equiv 0$.

By (the bimodule analogue of) Theorem 5.3.8, the Banach $\mathfrak{A}$-bimodule $\mathfrak{A}^*$ is injective. Consequently — by (the bimodule analogue of) Proposition 5.3.5(iv) —, the injective $\mathfrak{A}$-bimodule homomorphism $\Delta^*_{\mathfrak{A}^\#,\mathfrak{A}}: \mathfrak{A}^* \to (\mathfrak{A}^\# \hat{\otimes} \mathfrak{A})^*$ has a left inverse $\theta \in {}_\mathfrak{A}\mathcal{L}_\mathfrak{A}((\mathfrak{A}^\# \hat{\otimes} \mathfrak{A})^*, \mathfrak{A}^*)$. Define

$$\tilde{\psi}: \mathfrak{A}^\# \hat{\otimes} \mathfrak{A} \to \mathbb{C}, \quad (\lambda e_{\mathfrak{A}^\#} + a) \otimes b \mapsto \langle a, \phi \rangle \langle b, \phi \rangle,$$

and let $\psi := \theta(\tilde{\psi})$. Let $a, b \in \mathfrak{A}$ and $c \in \mathfrak{A}^\#$. Since $\langle ab, \phi \rangle = 0$, it follows that $\langle c \otimes a, b \cdot \tilde{\psi} \rangle = 0$, i.e. $b \cdot \tilde{\psi} = 0$ and consequently $b \cdot \psi = 0$ for all $b \in \mathfrak{B}$. This means that $\psi|_{\mathfrak{A}^2} \equiv 0$ as well. On the other hand, we have

$$\langle (\lambda e_{\mathfrak{A}^\#} + a) \otimes c, \tilde{\psi} \cdot b \rangle = \langle \lambda b + ba, \phi \rangle \langle c, \phi \rangle$$
$$= \lambda \langle b, \phi \rangle \langle c, \phi \rangle$$
$$= \langle b, \phi \rangle \langle \lambda c + ac, \phi \rangle$$
$$= \langle b, \phi \rangle \langle (\lambda e_{\mathfrak{A}^\#} + a) \otimes c, \Delta^*_{\mathfrak{A}^\#,\mathfrak{A}} \phi \rangle \qquad (\lambda \in \mathbb{C}, a, b, c \in \mathfrak{A}).$$

It follows that

$$\tilde{\psi} \cdot b = \langle b, \phi \rangle \Delta^*_{\mathfrak{A}^\#,\mathfrak{A}} \phi \qquad (b \in \mathfrak{A})$$

and thus

$$\psi \cdot b = \langle b, \phi \rangle \phi \qquad (b \in \mathfrak{A}).$$

This, however, means that

$$\langle b, \phi \rangle^2 = \langle b, \psi \cdot b \rangle = \langle b^2, \psi \rangle = 0 \qquad (b \in \mathfrak{A}),$$

so that $\phi = 0$. The Hahn–Banach theorem yields that $\mathfrak{A}^2$ is dense in $\mathfrak{A}$. $\square$

Theorem 5.3.12 *Let $\mathfrak{A}$ be a Banach algebra. Then the following are equivalent:*

(i) *$\mathfrak{A}$ is biflat.*

(ii) *$\mathfrak{A}$ is a flat Banach $\mathfrak{A}$-bimodule.*

Proof (i) $\implies$ (ii): Suppose that $\Delta_{\mathfrak{A}}^* : \mathfrak{A}^* \to (\mathfrak{A} \hat{\otimes} \mathfrak{A})^*$ has a left inverse in $_\mathfrak{A}\mathcal{L}_\mathfrak{A}((\mathfrak{A} \hat{\otimes} \mathfrak{A})^*, \mathfrak{A}^*)$. It follows easily (How?) that then $\Delta_{\mathfrak{A}^\#, \mathfrak{A}}^* : \mathfrak{A}^* \to (\mathfrak{A}^\# \hat{\otimes} \mathfrak{A})^*$ and $\Delta_{\mathfrak{A}, \mathfrak{A}^\#}^* : \mathfrak{A}^* \to (\mathfrak{A} \hat{\otimes} \mathfrak{A}^\#)^*$ have left inverses in $_\mathfrak{A}\mathcal{L}_\mathfrak{A}((\mathfrak{A}^\# \hat{\otimes} \mathfrak{A})^*, \mathfrak{A}^*)$ and $_\mathfrak{A}\mathcal{L}_\mathfrak{A}((\mathfrak{A} \hat{\otimes} \mathfrak{A}^\#)^*, \mathfrak{A}^*)$, respectively. It follows from Lemma 5.3.10 that $\mathfrak{A}$ is a flat Banach $\mathfrak{A}$-bimodule.

(ii) $\implies$ (i): By Lemma 5.3.10, $\Delta_{\mathfrak{A}^\#, \mathfrak{A}}^*$ has a left inverse $\theta \in {}_\mathfrak{A}\mathcal{L}_\mathfrak{A}((\mathfrak{A}^\# \hat{\otimes} \mathfrak{A})^*, \mathfrak{A}^*)$. Let $\phi \in (\mathfrak{A}^\# \hat{\otimes} \mathfrak{A})^*$ be such that $\phi|_{\mathfrak{A} \hat{\otimes} \mathfrak{A}} \equiv 0$. It follows that $\phi \cdot a = 0$ for all $a \in \mathfrak{A}$. Hence (Why?), $\theta(\phi)$ vanishes on $\mathfrak{A}^2$. By Lemma 5.3.11, this means that $\theta(\phi) = 0$. Hence, θ drops to an $\mathfrak{A}$-bimodule homomorphism from $(\mathfrak{A} \hat{\otimes} \mathfrak{A})^*$ to $\mathfrak{A}^*$, which is clearly a left inverse of $\Delta_{\mathfrak{A}}^*$. $\square$

With the help of Theorem 5.3.12, we can now compute further Hochschild cohomology groups:

Theorem 5.3.13 *Let $\mathfrak{A}$ be a biflat Banach algebra. Then $\mathcal{H}^n(\mathfrak{A}, \mathfrak{A}^*) = \{0\}$ for all $n \in \mathbb{N}$. In particular, $\mathfrak{A}$ is weakly amenable.*

Proof By Theorem 5.3.12, $\mathfrak{A}$ is a flat Banach $\mathfrak{A}$-bimodule, so that $\mathfrak{A}^*$ is an injective Banach $\mathfrak{A}$-bimodule. It is immediate that then $\mathfrak{A}^*$ is also in injective left Banach $\mathfrak{A}^{\mathrm{env}}$-module. From Theorem 5.2.28 and Exercise 5.3.3, we conclude that

$$\mathcal{H}^n(\mathfrak{A}, \mathfrak{A}^*) \cong \mathrm{Ext}_{\mathfrak{A}^{\mathrm{env}}}^n(\mathfrak{A}^\#, \mathfrak{A}^*) = \{0\} \qquad (n \in \mathbb{N}).$$

This was it already! $\square$

5.4 Notes and comments

Banach homology — or more generally: topological homology — was initiated independently by A. Ya. Helemskiĭ and J. L. Taylor ([Tay]). For many applications, e.g. to several variable spectral theory, it is necessary to not only consider Banach algebras and modules, but also more general topological algebras and modules (see [E–P]).

This chapter presents the very basics of Helemkiĭ's approach to topological homology. It is intended as a "prequel" to his textbook [Hel 5] and his monograph [Hel 3]. Essentially all the material in this chapter is from these two sources. Helemskiĭ has also written several survey articles on topological homology — such as [Hel 1], [Hel 2], and [Hel 7] — which offer an excellent overview of the field.

Comparing topological homology with homological algebra in the purely algebraic setting, as expounded in [C–E], [Lang 2], [MacL], or [Wei], one cannot help getting the impression that topological homology is about nothing more than adding a few functional analytic overtones to the concepts and results from homological algebra. Indeed, topological homology is a particular case of relative homological algebra in the sense of [MacL]. Doesn't this mean that topological homology is nothing more than a remotely interesting footnote to general homological algebra?

Surprisingly, however, one encounters novel and interesting phenomena in Banach homology that have no analogue in algebraic homological algebra (and not even in general topological homology).

Forbidden values for homological dimensions

Let $\mathfrak{A}$ be a Banach algebra. Then the *global homological dimension* of $\mathfrak{A}$ is defined as

$$\operatorname{dg} \mathfrak{A} := \inf\{n \in \mathbb{N}_0 : \operatorname{Ext}_{\mathfrak{A}}^{n+1}(E, F) = \{0\} \text{ for all left Banach } \mathfrak{A}\text{-modules } E \text{ and } F\}$$

and the *homological bidimension* of $\mathfrak{A}$ is defined as

$$\operatorname{db} \mathfrak{A} := \inf\{n \in \mathbb{N}_0 : \mathcal{H}^{n+1}(\mathfrak{A}, E) = \{0\} \text{ for all Banach } \mathfrak{A}\text{-bimodules } E\}.$$

It is not hard to see that $\operatorname{db} \mathfrak{A} \geq \operatorname{dg} \mathfrak{A}$. Helemskiĭ's global dimension theorem asserts that $\operatorname{dg} \mathfrak{A} \geq 2$ whenever $\mathfrak{A}$ is commutative with infinite character space (see [Pot] for an exposition of this theorem that should be accessible to someone who has worked through this chapter). Consequently, whenver $\mathfrak{A}$ is a commutative Banach algebra with infinite character space, there is a Banach $\mathfrak{A}$-bimodule E such that $\mathcal{H}^2(\mathfrak{A}, E) \neq \{0\}$. Similar results hold for certain radical Banach algebras ([Gh-S]).

Additivity formulae for homological dimensions

Let $\mathfrak{A}$ be a unital Banach algebra, and let $\mathfrak{B}$ be a commutative, biprojective Banach algebra with infinite character space. Then

$$\operatorname{dg} \mathfrak{A} \hat{\otimes} \mathfrak{B}^{\#} = \operatorname{dg} \mathfrak{A} + \operatorname{dg} \mathfrak{B}^{\#} = \operatorname{dg} \mathfrak{A} + 2$$

holds as proved in [Sel 3]. The *weak homological bidimension* of a Banach algebra $\mathfrak{A}$ is defined as

$$\operatorname{wdb} \mathfrak{A} := \inf\{n \in \mathbb{N}_0 : \mathcal{H}^{n+1}(\mathfrak{A}, E^*) = \{0\} \text{ for all Banach } \mathfrak{A}\text{-bimodules } E\}.$$

According to an apparently still unpublished result by Yu. V. Selivanov (see [Hel 7]), we have

$$\operatorname{wdb} \mathfrak{A} \hat{\otimes} \mathfrak{B} = \operatorname{wdb} \mathfrak{A} + \operatorname{wdb} \mathfrak{B}$$

for any two Banach algebras $\mathfrak{A}$ and $\mathfrak{B}$ with bounded approximate identities.

6 C^*- and W^*-algebras

The most important class of Banach algebras are certainly the C^*-algebras. Over the past few decades, the theory of C^*-algebras has thrived and grown, and now is an area of mathematics pretty much independent of general Banach algebra theory. Most researchers in Banach algebras are fairly ignorant of what's currently going on in C^*-algebras (and vice versa).

Amenability brings C^*-algebras back into the framework of general Banach algebras. The purely banach algebraic notion of amenability turns out to be equivalent to an important C^*-algebraic property: *nuclearity*. This chapter is devoted to establishing this equivalence.

Very often, the only — or at least the most convenient — way of proving a result for C^*-algebras is to make the detour through their enveloping von Neumann algebras. Most of this chapter will thus deal with W^*-algebras. As it turns out, amenability in the sense of Definition 2.1.9 is too strong to yield an interesting theory for W^*-algebras: For example, if $\mathfrak{H}$ is a Hilbert space, then $\mathcal{L}(\mathfrak{H})$ is amenable if and only if $\mathfrak{H}$ is finite-dimensional. The right notion of amenability for W^*-algebras is Connes-amenability. It is equivalent to a number of important W^*-algebraic properties, some of which, at first glance, have little in common. In this chapter, we shall prove the equivalence of the following properties:

- *Injectivity*: A von Neumann algebra $\mathfrak{M}$ acting on some Hilbert space $\mathfrak{H}$ is injective if its commutant is complemented in $\mathcal{L}(\mathfrak{H})$, such that the corresponding projection onto $\mathfrak{M}'$ has norm one.
- *Semidiscreteness*: This is some kind of approximation property for W^*-algebras.
- *Existence of a normal, virtual diagonal*: For W^*-algebras, the converse of Theorem 4.4.15 is true.

For a general C^*-algebra $\mathfrak{A}$, the Connes-amenability (or injectivity, etc.) of $\mathfrak{A}^{**}$ is equivalent to $\mathfrak{A}$ being nuclear/amenable.

It won't come as a surprise that to prove these equivalences requires substantial background from the theory of C^*- and, in particular, W^*-algebras. We thus definitely shift gears in this chapter as far as the self-containment of these notes is concerned: There are plenty of textbooks and well written monographs available on this subject — [Dixm 2], [Dixm 3], [Sak], [Ped], [Tak 2], [Mur], [K–R], and many more —, so that it would be a waste of time and space to duplicate material, e.g. on the type decomposition of von Neumann algebras, which might not be standard in Banach algebras, but certainly is in von Neumann algebras. Some of the background we require from von Neumann algebra theory might be considered non-standard even there, e.g. Tomita–Takesaki theory or continuous crossed products, but there are still texts and monographs — [K–R] and [Dae], for example — through which this material is accessible.

Some concepts we use in this chapter, such as complete positivity and the Haagerup tensor product, have their proper place in the framework of operator space theory (see [E–R]). Nevertheless, we refrain from making any references to operator spatiology here (this will come in the next chapter).

6.1 Amenable W^*-algebras

The W^*-algebras are the most important (and best understood) class of C^*-algebras, so that we'll discuss their amenability first. One might expect that, if the important class of W^*-algebras teams up with a property as important as amenability, then the resulting class of amenable W^*-algebras will be even more important. If you really think so, this section will be a source of considerable disappointment for you ...

There is an obvious way of obtaining amenable (but otherwise uninteresting) W^*-algebras: Choose $n_1, \ldots, n_k \in \mathbb{N}$ along with compact, hyperstonean Hausdorff spaces $\Omega_1, \ldots, \Omega_k$. If you don't know what compact, hyperstonean spaces are: these spaces are formally defined in [Tak 2, Definition 1.14]; all we need to know is that a compact Hausdorff space Ω is hyperstonean if and only if $\mathcal{C}(\Omega)$ is a W^*-algebra (this follows from [Tak 2, Theorem 1.18]). For $j = 1, \ldots, n$, the algebra

$$\mathbb{M}_{n_j} \otimes \mathcal{C}(\Omega_j) \cong \mathcal{C}(\Omega_j) \otimes \mathbb{M}_{n_j} \cong \mathcal{C}(\Omega_j) \check{\otimes} \mathbb{M}_{n_j}$$

is a C^*-algebra in a canonical fashion (Exercise 2.3.6(iii)), and is, in fact, a W^*-algebra (Why?). It follows that

$$\mathbb{M}_{n_1} \otimes \mathcal{C}(\Omega_1) \oplus \cdots \oplus \mathbb{M}_{n_k} \otimes \mathcal{C}(\Omega_k) \tag{6.1}$$

is also a W^*-algebra, and the hereditary properties of amenability for Banach algebras yield at once that it is amenable. In Examples 4.4.4(b), we saw that every subhomogeneous W^*-algebra has this form, and as we mentioned immediately after Examples 4.4.4, every amenable W^*-algebra is already subhomogeneous (and thus has the rather dull form (6.1)). In this section, we will prove it.

We start with a definition:

Definition 6.1.1 Let $\mathfrak{A}$ be a Banach *-algebra. We say that $\mathfrak{A}$ is of *type* (QE) if, for each *-representation π of $\mathfrak{A}$ on a Hilbert space $\mathfrak{H}$, there is a quasi-expectation $\mathcal{Q} \colon \mathcal{L}(\mathfrak{H}) \to \pi(\mathfrak{A})''$.

By a Banach *-algebra, we mean a Banach algebra equipped with a continuous involution.

Exercise 6.1.1 Let G be a locally compact, inner amenable group which is not amenable. Show that $VN(G)$ is not of type (QE). (*Hint:* Proceed as in the proof of Theorem 4.4.13, but with the rôles of λ_2 and ρ_2 interchanged.)

As we shall soon see, amenability forces any Banach *-algebra to be of type (QE), but before we can prove it, we need a lemma. For the sake of brevity, from now on, we shall call a w^*-continuous *-representation of a W^*-algebra on some Hilbert space a W^*-*representation*.

Lemma 6.1.2 *Let $\mathfrak{M}$ be a von Neumann algebra acting on a Hilbert space $\mathfrak{H}$. Then the following are equivalent:*

(i) *There is a quasi-expectation* $Q\colon \mathcal{L}(\mathfrak{H}) \to \mathfrak{M}$.

(ii) *For every faithful W^*-representation π of $\mathfrak{M}$ on a Hilbert space $\mathfrak{K}$, there is a quasi-expectation* $Q\colon \mathcal{L}(\mathfrak{K}) \to \pi(\mathfrak{M})$.

Proof Of course, only (i) $\Longrightarrow$ (ii) needs proof.

Let π be a faithful W^*-representation of $\mathfrak{M}$ on a Hilbert space $\mathfrak{K}$, and let $\mathfrak{N} := \pi(\mathfrak{M})$. Using the idea of the proof of [Dixm 3, Théorème 3, §I.4] (How precisely?), we can choose a third Hilbert space $\mathfrak{L}$ such that $\mathfrak{M} \otimes \mathrm{id}_\mathfrak{L}$, i.e. the algebra $\{x \otimes \mathrm{id}_\mathfrak{L} : x \in \mathfrak{M}\}$, on $\mathfrak{H} \bar\otimes \mathfrak{L}$ and $\mathfrak{N} \otimes \mathrm{id}_\mathfrak{L}$ on $\mathfrak{K} \bar\otimes \mathfrak{L}$ are spatially isomorphic. Fix $\xi_0, \eta_0 \in \mathfrak{L}$ such that $\langle \xi_0, \eta_0 \rangle = 1$, and define $\mathcal{P}_0\colon \mathcal{L}(\mathfrak{H} \bar\otimes \mathfrak{L}) \to \mathcal{L}(\mathfrak{H})$ through

$$\langle (\mathcal{P}_0 T)\xi, \eta \rangle := \langle T(\xi \otimes \xi_0), \eta \otimes \eta_0 \rangle \qquad (T \in \mathcal{L}(\mathfrak{H}),\ \xi, \eta \in \mathfrak{H}). \tag{6.2}$$

Identifying $\mathcal{L}(\mathfrak{H})$ with $\mathcal{L}(\mathfrak{H}) \otimes \mathrm{id}_\mathfrak{L}$, we see that $\mathcal{P}_0$ is a projection onto $\mathcal{L}(\mathfrak{H})$. Furthermore, for $T \in \mathcal{L}(\mathfrak{H} \bar\otimes \mathfrak{L})$ and $R, S \in \mathcal{L}(\mathfrak{H})$, we have

$$\begin{aligned}
\langle \mathcal{P}_0((R \otimes \mathrm{id}_\mathfrak{L})T(S \otimes \mathrm{id}_\mathfrak{L}))\xi, \eta \rangle &= \langle (R \otimes \mathrm{id}_\mathfrak{L})T(S \otimes \mathrm{id}_\mathfrak{L})(\xi \otimes \xi_0), \eta \otimes \eta_0 \rangle \\
&= \langle T(S\xi \otimes \xi_0), R^*\eta \otimes \eta_0 \rangle \\
&= \langle (\mathcal{P}_0 T)S\xi, R^*\eta \rangle \\
&= \langle R(\mathcal{P}_0 T)S\xi, \eta \rangle \qquad (\xi, \eta \in \mathfrak{H}),
\end{aligned}$$

so that $\mathcal{P}_0$ is a quasi-expectation. Let $\mathcal{P}\colon \mathcal{L}(\mathfrak{H}) \to \mathfrak{M}$ be a quasi-expectation, and define

$$\tilde{\mathcal{P}}\colon \mathcal{L}(\mathfrak{H} \bar\otimes \mathfrak{L}) \to \mathfrak{M} \otimes \mathrm{id}_\mathfrak{L}, \quad T \mapsto (\mathcal{P} \circ \mathcal{P}_0)T \otimes \mathrm{id}_\mathfrak{L};$$

it is clear that $\tilde{\mathcal{P}}$ is also a quasi-expectation. Since $\mathfrak{M} \otimes \mathrm{id}_\mathfrak{L}$ and $\mathfrak{N} \otimes \mathrm{id}_\mathfrak{L}$ are spatially isomorphic, there is a quasi-expecation $\tilde{\mathcal{Q}}\colon \mathcal{L}(\mathfrak{K} \bar\otimes \mathfrak{L}) \to \mathfrak{N} \otimes \mathrm{id}_\mathfrak{L}$ (Why?). Fix again $\xi_0, \eta_0 \in \mathfrak{L}$ such that $\langle \xi_0, \eta_0 \rangle = 1$, and define $\mathcal{Q}_0\colon \mathcal{L}(\mathfrak{K} \bar\otimes \mathfrak{L}) \to \mathcal{L}(\mathfrak{K})$ as in (6.2). Then

$$\mathcal{Q}\colon \mathcal{L}(\mathfrak{K}) \to \mathfrak{N}, \quad T \mapsto (\mathcal{Q}_0 \circ \tilde{\mathcal{Q}})(T \otimes \mathrm{id}_\mathfrak{L})$$

is the desired quasi-expectation. $\square$

With Lemma 6.1.2 at hand, we can now bring amenability into the picture. For the proof of the following proposition, we depend on Tomita–Takesaki theory as expounded, for example, in [K–R].

Proposition 6.1.3 *Every amenable Banach *-algebra is of type (QE).*

Proof Let π be a *-representation of $\mathfrak{A}$ on a Hilbert space $\mathfrak{H}$. Then, by [K–R, Exercise 7.6.46], there is a faithful, normal, semifinite weight on the von Neumann algebra $\pi(\mathfrak{A})''$. From this weight, we can construct a Hilbert space $\mathfrak{K}$ and a faithful W^*-representation ρ of $\pi(\mathfrak{A})''$ on $\mathfrak{K}$ ([K–R, Theorem 7.5.3]). By Theorem 4.4.11(i) — applied to $\rho \circ \pi\colon \mathfrak{A} \to \mathcal{L}(\mathfrak{K})$ —, there is a quasi-expectation $\mathcal{P}\colon \mathcal{L}(\mathfrak{K}) \to (\rho \circ \pi)(\mathfrak{A})'$. Let $J\colon \mathfrak{K} \to \mathfrak{K}$ be the conjugate linear isometry from [K–R, Theorem 9.2.37], i.e.

$$J^2 = \mathrm{id}_\mathfrak{K} \quad \text{and} \quad J(\rho \circ \pi)(\mathfrak{A})'J = (\rho \circ \pi)(\mathfrak{A})''.$$

Define

$$\tilde{Q} \colon \mathcal{L}(\mathfrak{K}) \to (\rho \circ \pi)(\mathfrak{A})'', \quad T \mapsto JP(JTJ)J.$$

It is immediate (Really?) that $\tilde{Q}$ is a quasi-expectation. Since ρ is a faithful W^*-representation, it follows from Lemma 6.1.2 that there is a quasi-expectation $Q \colon \mathcal{L}(\mathfrak{H}) \to \pi(\mathfrak{A})''$. Hence, $\mathfrak{A}$ is of type (QE). $\square$

If we want to show that every amenable W^*-algebra is subhomogeneous, it is thus sufficient to prove that every W^*-algebra which is not of the form (6.1) is not of type (QE) (as if that would make things much easier ...). We proceed indirectly and by reduction: Assuming that a W^*-algebra which is not subhomogeneous is of type (QE), we obtain another W^*-algebra which then should be of type (QE) as well, but for which we can show that it isn't. For this reason, we have to establish two hereditary properties.

The first one is easy:

Exercise 6.1.2 Let $\mathfrak{A}$ be a Banach *-algebra of type (QE), and let I be a closed *-ideal of $\mathfrak{A}$. Show that $\mathfrak{A}/I$ is also of type (QE).

The second hereditary property is harder to prove. We leave the proof of the following preparatory assertion to the reader:

Exercise 6.1.3 Let $\mathfrak{A}$ be a C^*-algebra, let $\mathfrak{B}$ be a C^*-subalgebra of $\mathfrak{A}$, and let $Q \colon \mathfrak{A} \to \mathfrak{B}$ be a quasi-expectation. Show that $Q^{**} \colon \mathfrak{A}^{**} \to \mathfrak{B}^{**}$ is also a quasi-expectation.

Lemma 6.1.4 *Let $\mathfrak{A}$ be a unital C^*-algebra of type (QE), and let $\mathfrak{B}$ be a unital C^*-subalgebra of $\mathfrak{A}$ such that there is a quasi-expectation $Q \colon \mathfrak{A} \to \mathfrak{B}$. Then $\mathfrak{B}$ is of type (QE).*

Proof Let $\mathfrak{A}^{**}$ act as a von Neumann algebra on the Hilbert space $\mathfrak{H}$, and let π be a *-representation of $\mathfrak{B}$ on some Hilbert space $\mathfrak{K}$. Then there is a projection $p \in Z(\mathfrak{B}^{**})$ such that $p\mathfrak{B}^{**} \cong \pi(\mathfrak{B})''$ (Why?). By hypothesis, there is a quasi-expectation $P \colon \mathcal{L}(\mathfrak{H}) \to \mathfrak{A}^{**}$. Define

$$\mathcal{R} \colon \mathcal{L}(p\mathfrak{H}) \to \mathfrak{B}^{**}, \quad T \mapsto (Q^{**} \circ P)(Tp).$$

Viewing $\mathfrak{B}^{**}$ as a W^*-subalgebra of $\mathfrak{A}^{**}$, we obtain from Exercise 6.1.3 that

$$\mathcal{R}(Tp) = (\mathcal{R}T)p \in \mathfrak{B}^{**}p \quad (T \in \mathcal{L}(p\mathfrak{H})),$$

so that $\mathcal{R}$ attains its values in $\mathfrak{B}^{**}p$. It is easily checked that $\mathcal{R} \colon \mathcal{L}(p\mathfrak{H}) \to \mathfrak{B}^{**}p$ is a quasi-expectation. It now follows from Lemma 6.1.2 (How exactly?) that there is a quasi-expectation from $\mathcal{L}(\mathfrak{K})$ onto $\pi(\mathfrak{B})''$. $\square$

So far, the only W^*-algebras for which we definitively know that they are not of type (QE) are the algebras $VN(G)$, where G is a locally compact, inner amenable group G which fails to be amenable (Exercise 6.1.1); for instance, $VN(\mathbb{F}_2)$ is not of type (QE). With the help of Lemma 6.1.4 and Exercise 6.1.2, we shall now exhibit another example of a W^*-algebra which is not of type (QE):

Example 6.1.5 Let $\mathfrak{M} := \ell^\infty \text{-} \bigoplus_{n=1}^\infty \mathbb{M}_n$. Clearly, $\mathfrak{M}$ is a W^*-algebra (What is $\mathfrak{M}_*$?). From the type decomposition of von Neumann algebras ([K–R, 6.2.5. Theorem]), it is also clear that $\mathfrak{M}$ is a finite type I von Neumann algebra, whose center is isomorphic to ℓ^∞. We claim that $\mathfrak{M}$ is not of type (QE) (and thus not amenable by Proposition 6.1.3). Assume towards a contradiction that $\mathfrak{M}$ is of type (QE). Since $Z(\mathfrak{M}) \cong \ell^\infty$, the character space of $Z(\mathfrak{M})$ is isomorphic to $\beta\mathbb{N}$, the Stone–Čech compactification of $\mathbb{N}$. Let $\mathcal{U}$ be a free ultrafilter on $\mathbb{N}$, so that $\mathcal{U}$ corresponds to a point of $\beta\mathbb{N} \setminus \mathbb{N}$. Define

$$M_\mathcal{U} := \left\{ (x_n)_{n=1}^\infty \in \mathfrak{M} : \lim_\mathcal{U} \mathrm{tr}_n(x_n^* x_n) = 0 \right\},$$

where tr_n is the canonical normalized trace on M_n. It is easily seen that $M_\mathcal{U}$ is a closed *-ideal of $\mathfrak{M}$. By Exercise 6.1.2, this means that $\mathfrak{M}/M_\mathcal{U}$ is of type (QE) as well. In fact, $M_\mathcal{U}$ is a maximal ideal of $\mathfrak{M}$ by [Tak 2, Corollary 4.10]. By [Tak 2, Theorem 5.2], this means that $\mathfrak{M}/M_\mathcal{U}$ is a finite factor whose faithful, normal trace is given by

$$\mathrm{tr} \colon \mathfrak{M}/M_\mathcal{U} \to \mathbb{C}, \quad (x_n)_{n=1}^\infty + M_\mathcal{U} \mapsto \lim_\mathcal{U} \mathrm{tr}_n x_n.$$

There are two possibilities:

– $\mathfrak{M}/M_\mathcal{U}$ is a type I_n factor for some $n \in \mathbb{N}$ (and thus finite-dimensional), or
– $\mathfrak{M}/M_\mathcal{U}$ is a type II_1 factor.

For each $n \in \mathbb{N}$, choose a decreasing sequence $\left(p_k^{(n)} \right)_{k=1}^\infty$ of projections in M_n such that

$$\dim p_k^{(n)} \mathrm{M} p_k^{(n)} = \left\lfloor \frac{n}{2^k} \right\rfloor \quad (n, k \in \mathbb{N}) \tag{6.3}$$

(How can we accomplish that?), where $\lfloor \cdot \rfloor$ denotes the Gauß bracket. Note that $p_k^{(n)} = 0$ whenever $2^k > n$. For $k \in \mathbb{N}$, let

$$p_k := \left(p_k^{(1)}, p_k^{(2)}, p_k^{(3)}, \dots \right) \in \mathfrak{M}.$$

Then $(p_k)_{n=1}^\infty$ is a decreasing sequence of projections in $\mathfrak{M}$. Assume that $\mathfrak{M}/M_\mathcal{U}$ is finite-dimensional. Then the sequence $(p_k + M_\mathcal{U})_{n=1}^\infty$ must become constant eventually, i.e. there is $K \in \mathbb{N}$ such that

$$\lim_\mathcal{U} \mathrm{tr}_n \left(p_k^{(n)} \right) = \lim_\mathcal{U} \mathrm{tr}_n \left(p_m^{(n)} \right) \quad (k, m \geq K). \tag{6.4}$$

However, it is clear from (6.3) that

$$\lim_{n \to \infty} \mathrm{tr}_n \left(p_k^{(n)} \right) = \frac{1}{2^k} \quad (k \in \mathbb{N}),$$

which contradicts (6.4). Hence, $\mathfrak{M}/M_\mathcal{U}$ is of type II_1.

We now claim that we can choose $\mathcal{U}$ in such a fashion that $\mathfrak{M}/M_\mathcal{U}$ contains a W^*-subalgebra $\mathfrak{N}$ which is isomorphic to $VN(\mathbb{F}_2)$.

Choose a strictly decreasing sequence $(N_j)_{j=1}^\infty$ of normal subgroups of $\mathbb{F}_2$ such that $n_j := [\mathbb{F}_2 : N_j] < \infty$ for $j \in \mathbb{N}$ and $\bigcap_{j=1}^\infty N_j = \{e_{\mathbb{F}_2}\}$; the existence of such a sequence follows from [H–R, (2.9) Theorem]. For $j \in \mathbb{N}$, identify M_{n_j} with $\mathcal{L}(\ell^2(\mathbb{F}_2/N_j))$. Then the left regular representation of $\mathbb{F}_2/N_j$ induces a group homomorphism π_j from $\mathbb{F}_2$ into the unitaries of M_{n_j} such that $\ker \pi_j = N_j$ and $\mathrm{tr}_{n_j} \pi_j(g) = 0$ for $g \in \mathbb{F}_2 \setminus N_j$ (Why does the second property hold?). Choose $\mathcal{U}$ such that $\{n_1, n_2, n_3, \dots\} \in \mathcal{U}$. Define a group homomorphism π from $\mathbb{F}_2$ into the unitaries of $\mathfrak{M}/M_\mathcal{U}$ by letting $\pi(g) := (x_n)_{n=1}^\infty + M_\mathcal{U}$ with

$$x_n := \begin{cases} \pi_j(g), \text{ if } n = n_j, \\ \quad 0, \quad \text{otherwise} \end{cases}$$

(Why does this yield unitaries in $\mathfrak{M}/M_{\mathcal{U}}$?). We claim that π is injective. Let $g \in \mathbb{F}_2 \setminus \{e_{\mathbb{F}_2}\}$. Then there is $j_0 \in \mathbb{N}$ such that $g \notin N_j$ and thus $\mathrm{tr}_{n_j} \pi_j(g) = 0$ for all $j \geq j_0$. It follows (Why?) that $e_{\mathbb{F}_2} - \pi(g) \notin M_{\mathcal{U}}$. Let $\mathfrak{N}$ be the W^*-subalgebra of $\mathfrak{M}/M_{\mathcal{U}}$ generated by the set $\{\pi(g) : g \in \mathbb{F}_2\}$.

We claim that $\mathfrak{N}$ and $VN(\mathbb{F}_2)$ are isomorphic as W^*-algebras. To see this, apply the GNS-construction construction to $\mathfrak{M}/M_{\mathcal{U}}$ and the normal state tr to obtain a faithful W^*-representation of $\mathfrak{M}/M_{\mathcal{U}}$ on some Hilbert space $\mathfrak{H}$, and simply view $\mathfrak{M}/M_{\mathcal{U}}$ as a von Neumann algebra acting on $\mathfrak{H}$. It follows (how?) that $\mathfrak{H}$ contains a cyclic and separating trace vector ξ for $\mathfrak{M}$; this means, in particular, that

$$\mathrm{tr}\, x = \langle x\xi, \xi \rangle \qquad (x \in \mathfrak{M}/M_{\mathcal{U}}). \tag{6.5}$$

Let $\mathfrak{K}$ be the closure in $\mathfrak{H}$ of $\{x\xi : x \in \mathfrak{N}\}$. From (6.5), it follows at once that $(\pi(g)\xi)_{g \in \mathbb{F}_2}$ is an orthonormal basis for $\mathfrak{K}$. We thus have a unitary operator

$$U : \ell^2(\mathbb{F}_2) \to \mathfrak{K}, \quad \sum_{g \in \mathbb{F}_2} \lambda_g \delta_g \mapsto \sum_{g \in \mathbb{F}_2} \lambda_g \pi(g)\xi$$

(slightly abusing notation ...). Let $P \in \mathcal{L}(\mathfrak{H})$ be the orthogonal projection onto $\mathfrak{K}$. Then $P \in \mathfrak{N}'$, and

$$\mathfrak{N} \to \mathfrak{N}P, \quad x \mapsto xP \tag{6.6}$$

is a homomorphism of W^*-algebras. Let $x \in \mathfrak{N}$ be such that $xP = 0$. Since $xP\xi = x\xi$, and since ξ is a separating vector for $\mathfrak{M}/M_{\mathcal{U}}$, it follows that $x = 0$. Hence, (6.6) is even an isomorphism. Finally, note that

$$U^*(\pi(g)|_{\mathfrak{K}})U\delta_h = U^*\pi(g)\pi(h)\xi = U^*\pi(gh)\xi = \delta_{gh} = \lambda_2(g)\delta_g \qquad (g, h \in \mathbb{F}_2).$$

It follows that

$$\mathcal{L}(\mathfrak{K}) \to \mathcal{L}(\ell^2(\mathbb{F}_2)), \quad T \mapsto U^*TU$$

induces a spatial isomorphism of the W^*-algebras $\mathfrak{N}P$ and $VN(\mathbb{F}_2)$. Consequently, $\mathfrak{N}$ and $VN(\mathbb{F}_2)$ are isomorphic.

As we have already observed, $\mathfrak{M}/M_{\mathcal{U}}$ is a finite W^*-algebra. Since $\mathfrak{M}/M_{\mathcal{U}}$ is a factor, its center is trivially countably decomposable. Hence, $\mathfrak{M}/M_{\mathcal{U}}$ itself is countably decomposable ([Sak, 2.4.1. Proposition]). It then follows from [Sak, 4.4.23. Proposition] that there is a (norm one) quasi-expectation $\mathcal{Q} : \mathfrak{M}/M_{\mathcal{U}} \to \mathfrak{N}$. Since $\mathfrak{M}/M_{\mathcal{U}}$ is of type (QE), Lemma 6.1.4 implies that the same is true for $\mathfrak{N}$. Since $\mathfrak{N} \cong VN(\mathbb{F}_2)$, this is a contradiction.

That was quite a lot of work for an algebra looking as innocent as $\ell^\infty\text{-}\bigoplus_{n=1}^\infty \mathbb{M}_n$, but the general result is now just one (relatively easy) lemma away. To prepare the ground for its proof, please do the following three exercises; they will turn out to be useful in some of the following sections of this chapter.

Exercise 6.1.4 Let $\mathfrak{A}$ and $\mathfrak{B}$ be algebras with involution.

(i) Show that there is a unique involution on $\mathfrak{A} \otimes \mathfrak{B}$ such that

$$(a \otimes b)^* = a^* \otimes b^* \qquad (a \in \mathfrak{A}, b \in \mathfrak{B}).$$

(ii) Suppose that $\mathfrak{A}$ and $\mathfrak{B}$ are Banach *-algebras. Show that the involution defined in (i) extends to $\mathfrak{A} \hat{\otimes} \mathfrak{B}$ turning it into a Banach *-algebra.

Exercise 6.1.5 Let $\mathfrak{A}$ be a C^*-algebra, and let $n \in \mathbb{N}$. Show that there is a unique norm on $\mathbb{M}_n \otimes \mathfrak{A}$ turning it into a C^*-algebra. (*Hint:* Identify $\mathbb{M}_n \otimes \mathfrak{A}$ with $\mathbb{M}_n(\mathfrak{A})$ as in Exercise B.1.7, and note that there is a canonical involution on $\mathbb{M}_n(\mathfrak{A})$ such that (B.5) is a *-isomorphism. Let π be a faithful *-representation of $\mathfrak{A}$ on a Hilbert space $\mathfrak{H}$. Construct a faithful *-representation of $\mathbb{M}_n(\mathfrak{A})$ on $\mathfrak{H}^n$.)

Exercise 6.1.6 Let $n \in \mathbb{N}$, and let $\mathfrak{A}$ be a unital C^*-algebra containing $\mathbb{M}_n$ as a unital C^*-subalgebra. Show that

$$\mathbb{M}_n \otimes Z_{\mathfrak{A}}(\mathbb{M}_n) \to \mathfrak{A}, \quad a \otimes b \mapsto ab$$

is a *-isomorphism.

Lemma 6.1.6 *Let $\mathfrak{A}$ be a unital C^*-algebra containing ℓ^∞-$\bigoplus_{n=1}^\infty \mathbb{M}_n$ as a C^*-subalgebra. Then $\mathfrak{A}$ is not of type (QE).*

Proof Let $n \in \mathbb{N}$, and let $p_n \in \mathfrak{A}$ be the projection corresponding to $E_n \in \mathbb{M}_n$. Then $p_n \mathfrak{A} p_n$ is a C^*-subalgebra of $\mathfrak{A}$ whose identity is p_n and which contains $\mathbb{M}_n$ as a unital C^*-subalgebra. Let ϕ_n be a state on $p_n \mathfrak{A} p_n$. Define

$$\tilde{Q}_n \colon \mathbb{M}_n \otimes Z_{p_n \mathfrak{A} p_n}(\mathbb{M}_n) \to \mathbb{M}_n, \quad a \otimes b \mapsto \langle b, \phi_n \rangle a.$$

Identifying $\mathbb{M}_n$ with $\mathbb{M}_n \otimes p_n$, we see easily (Hopefully!) that $\tilde{Q}_n$ is a quasi-expectation with $\|\tilde{Q}_n\| = 1$. By Exercise 6.1.6, $\mathbb{M}_n \otimes Z_{p_n \mathfrak{A} p_n}(\mathbb{M}_n)$ and $p_n \mathfrak{A} p_n$ are isomorphic as C^*-algebras, so that we have a quasi-expectation $Q_n \colon p_n \mathfrak{A} p_n \to \mathbb{M}_n$ with $\|Q_n\| = 1$.
Define

$$Q \colon \mathfrak{A} \to \ell^\infty\text{-}\bigoplus_{n=1}^\infty \mathbb{M}_n, \quad a \mapsto (Q_1(p_1 a p_1), Q_2(p_2 a p_2), Q_3(p_3 a p_3), \dots).$$

Then Q is a quasi-expectation. As we saw in Example 6.1.5, ℓ^∞-$\bigoplus_{n=1}^\infty \mathbb{M}_n$ is not of type (QE). Lemma 6.1.4 implies that $\mathfrak{A}$ is not of type (QE) either. $\square$

Theorem 6.1.7 *For a W^*-algebra $\mathfrak{M}$, the following are equivalent:*

(i) $\mathfrak{M}$ *is amenable.*

(ii) $\mathfrak{M}$ *is of type (QE).*

(iii) $\mathfrak{M}$ *does not contain ℓ^∞-$\bigoplus_{n=1}^\infty \mathbb{M}_n$ as a C^*-subalgebra.*

(iv) *There are compact, hyperstonean Hausdorff spaces $\Omega_1, \dots, \Omega_k$ and $n_1, \dots, n_k \in \mathbb{N}$ such that*

$$\mathfrak{M} \cong \bigoplus_{j=1}^k \mathbb{M}_{n_j} \otimes \mathcal{C}(\Omega_j).$$

Proof (i) $\Longrightarrow$ (ii) follows from Proposition 6.1.3, and (iv) $\Longrightarrow$ (i) was explained at the very beginning of this section.

(ii) $\Longrightarrow$ (iii) is Lemma 6.1.6.

For (iii) $\Longrightarrow$ (iv), assume that $\mathfrak{M}$ is not of the form (6.1). The type decomposition of von Neumann algebras ([K–R, 6.5.2. Theorem]) then leaves three alternatives:

Case 1: There is a strictly increasing sequence $(n_k)_{k=1}^\infty$ in $\mathbb{N}$ such that $\mathfrak{M}$ has, for each $k \in \mathbb{N}$, a direct summand of type I_{n_k}. In this case it is obvious that $\mathfrak{M}$ contains $\ell^\infty\text{-}\bigoplus_{n=1}^\infty \mathbb{M}_n$ as a C^*-subalgebra.

Case 2: $\mathfrak{M}$ has a properly infinite direct summand. It then follows from [Sak, 2.2.4. Proposition] that this summand contains $\mathcal{L}(\ell^2)$ as a C^*-subalgebra (How?). It is not hard to show (So, do it!) that $\mathcal{L}(\ell^2)$ contains $\ell^\infty\text{-}\bigoplus_{n=1}^\infty \mathbb{M}_n$ as a C^*-subalgebra.

Case 3: $\mathfrak{M}$ has a direct summand of type II_1. Then, by [Sak, 2.2.13. Proposition], there is a sequence $(p_n)_{n=1}^\infty$ of non-zero, mutually orthogonal projections in that summand. Consequently, the W^*-algebras $p_n\mathfrak{M}p_n$ have no direct summand of type I. Hence, it follows from [K–R, 6.5.6. Lemma] that each W^*-subalgebra $p_n\mathfrak{M}p_n$ contains $\mathbb{M}_n$ as C^*-subalgebra. Again, $\ell^\infty\text{-}\bigoplus_{n=1}^\infty \mathbb{M}_n$ is contained in $\mathfrak{M}$ as a C^*-subalgebra.

Thus, in all three cases, $\mathfrak{M}$ contains $\ell^\infty\text{-}\bigoplus_{n=1}^\infty \mathbb{M}_n$ as a C^*-subalgebra. $\square$

Corollary 6.1.8 *Let $\mathfrak{H}$ be a Hilbert space. Then $\mathcal{L}(\mathfrak{H})$ is amenable if and only if $\dim \mathfrak{H} < \infty$.*

Remark 6.1.9 As a consequence of [Ch–S, Theorem 1.1], the Banach spaces $\mathcal{L}(\ell^2)$ and $\ell^\infty\text{-}\bigoplus_{n=1}^\infty \mathbb{M}_n$ are isomorphic. Since $\mathcal{L}(\ell^2)$ lacks the approximation property by [Sza], the same is true for $\ell^\infty\text{-}\bigoplus_{n=1}^\infty \mathbb{M}_n$. Let $\mathfrak{M}$ be a W^*-algebra which does not satisfy the equivalent conditions of Theorem 6.1.7; in particular, it contains $\ell^\infty\text{-}\bigoplus_{n=1}^\infty \mathbb{M}_n$ as a C^*-subalgebra. In the proof of Lemma 6.1.6, we saw that then $\mathfrak{M}$ contains $\ell^\infty\text{-}\bigoplus_{n=1}^\infty \mathbb{M}_n$ as a complemented subspace. It follows from Exercise C.1.1 that $\mathfrak{M}$ does not have the approximation property either. On the other hand, it is not hard to see that every W^*-algebra of the form (6.1) has the approximation property. We can thus add a fifth equivalent statement to Theorem 6.1.7:

(v) $\mathfrak{M}$ *has the approximation property.*

The bottom line of this section is: Amenability as introduced in Definition 2.1.9 is not a good concept when dealing with von Neumann algebras — we only get dull examples. As we shall see in the following sections, the right notion of amenability for von Neumann algebras is Connes-amenability.

6.2 Injective W^*-algebras

As we have seen in the previous section, Definition 2.1.9 imposes too stringent a condition to yield an interesting theory for W^*-algebras. The class of Connes-amenable W^*-algebra is certainly richer: $\mathcal{L}(\mathfrak{H})$ is Connes-amenable for *every* Hilbert space (Why?). Is Connes-amenability a useful concept in the theory of W^*-algebras? Indeed it is: As we have already mentioned in the introduction to this chapter, Connes-amenability is equivalent to a number of important properties of W^*-algebras. In this section, we discuss the first of these properties: *injectivity*.

To define what an injective W^*-algebra is, we introduce the notion of an expectation.

Definition 6.2.1 Let $\mathfrak{A}$ be a C^*-algebra, and let $\mathfrak{B}$ be a C^*-subalgebra of $\mathfrak{A}$. An *expectation* $\mathcal{E}: \mathfrak{A} \to \mathfrak{B}$ is a norm one projection from $\mathfrak{A}$ onto $\mathfrak{B}$.

Definition 6.2.2 Let $\mathfrak{M}$ be a von Neumann algebra acting on a Hilbert space $\mathfrak{H}$. Then $\mathfrak{M}$ is called *injective* if there is an expectation $\mathcal{E}: \mathcal{L}(\mathfrak{H}) \to \mathfrak{M}'$.

Example 6.2.3 Since $\mathcal{L}(\mathfrak{H})' = \mathbb{C}\,\mathrm{id}_{\mathfrak{H}}$, the von Neumann algebra $\mathcal{L}(\mathfrak{H})$ is injective for every Hilbert space $\mathfrak{H}$.

If our choice of terminology is to make any sense, every expectation ought to be a quasi-expectation. This is indeed the case:

Theorem 6.2.4 *Let $\mathfrak{A}$ be C^*-algebra, let $\mathfrak{B}$ be a C^*-subalgebra of $\mathfrak{A}$ containing a bounded approximate identity for $\mathfrak{A}$, and let $\mathcal{E}: \mathfrak{A} \to \mathfrak{B}$ be an expectation. Then:*

(i) *$\mathcal{E}$ is positive, i.e. it maps the positive elements of $\mathfrak{A}$ into the positive elements of $\mathfrak{B}$.*

(ii) *$\mathcal{E}$ is a quasi-expectation.*

(iii) *$(\mathcal{E}a)^*\mathcal{E}a \leq \mathcal{E}(a^*a)$ $(a \in \mathfrak{A})$.*

Proof Passing to $\mathfrak{A}^{**}$, $\mathfrak{B}^{**}$, and $\mathcal{E}^{**}: \mathfrak{A}^{**} \to \mathfrak{B}^{**}$, we may suppose that $\mathfrak{A}$ and $\mathfrak{B}$ are W^*-algebras such that $\mathfrak{B}$ is w^*-closed and contains $e_{\mathfrak{A}}$, and that $\mathcal{E}$ is w^*-continuous.

To establish (i), it suffices to show that $\mathcal{E}^*$ maps the states of $\mathfrak{B}$ into the states of $\mathfrak{A}$ (Why?). Let ϕ be a state of $\mathfrak{B}$. Then

$$\langle e_{\mathfrak{A}}, \mathcal{E}^*\phi \rangle = \langle e_{\mathfrak{A}}, \phi \rangle = 1 = \|\phi\| \geq \|\mathcal{E}^*\phi\|.$$

Since $|\langle e_{\mathfrak{A}}, \mathcal{E}^*\phi \rangle| \leq \|\mathcal{E}^*\phi\|$ holds trivially, $\mathcal{E}^*\phi$ is a state of $\mathfrak{A}$.

For (ii), let $p \in \mathfrak{B}$ be a projection, and let $a \in \mathfrak{A}$ be positive with $\|a\| \leq 1$, so that $pap \leq p$. The positivity of $\mathcal{E}$ yields $\mathcal{E}(pap) \leq p$ and thus $\mathcal{E}(pap) \leq p(\mathcal{E}(pap))p$. Since the converse inequality holds trivially, we have $\mathcal{E}(pap) = p(\mathcal{E}(pap))p$ and thus (Why?)

$$\mathcal{E}(pxp) = p(\mathcal{E}(pxp))p \qquad (x \in \mathfrak{A}). \tag{6.7}$$

Fix $x \in \mathfrak{A}$ with $\|x\| \leq 1$, and let $y := \mathcal{E}((e_{\mathfrak{A}} - p)xp)$. Note that

$$
\begin{aligned}
\|(e_{\mathfrak{A}} - p)xp + np\|^2 &= \|((e_{\mathfrak{A}} - p)xp + np)^*((e_{\mathfrak{A}} - p)xp + np)\| \\
&= \|px^*(e_{\mathfrak{A}} - p)xp + n^2 p\| \\
&\leq 1 + n^2 \qquad (n \in \mathbb{N}).
\end{aligned}
\tag{6.8}
$$

Assume that $\frac{1}{2}(pyp + py^*p) \neq 0$, and suppose without loss of generality that the spectrum of this element contains some $t > 0$. We obtain

$$
\begin{aligned}
\|y + ne_{\mathfrak{A}}\| &\geq \|p(y + ne_{\mathfrak{A}})p\| \\
&\geq \frac{1}{2}\|pyp + np + ey^*p + np\| \\
&= \left\| \frac{1}{2}(pyp + py^*p) + np \right\| \\
&\geq t + n \qquad (n \in \mathbb{N}).
\end{aligned}
$$

Combining this with (6.8), we obtain for sufficiently large n:

$$\|\mathcal{E}((e_{\mathfrak{A}} - p)xp + np)\| = \|y + np\| \geq t + n > \sqrt{1 + n^2} \geq \|(e_{\mathfrak{A}} - p)xp + np\|.$$

This is a contradiction, so that $\frac{1}{2}(pyp + py^*p) = 0$; in a similar vein, we obtain $\frac{1}{2i}(pyp + py^*p) = 0$. All in all, $pyp = 0$. Analoguously, we see that $(e_{\mathfrak{A}} - p)y(e_{\mathfrak{A}} - p) = 0$ holds. Assume next that $py(e_{\mathfrak{A}} - p) \neq 0$. For $n \in \mathbb{N}$ sufficiently large, we have

$$\|y + npy(e_{\mathfrak{A}} + p)\| = \|(e_{\mathfrak{A}} - p)yp + (n + e_{\mathfrak{A}})py(e_{\mathfrak{A}} - p)\|$$
$$= \max\{\|(e_{\mathfrak{A}} - p)yp\|, \|(n + e_{\mathfrak{A}})py(e_{\mathfrak{A}} - p)\| \qquad \text{(Why?)}$$
$$= (1 + n)\|py(e_{\mathfrak{A}} - p)\|. \qquad (6.9)$$

On the other hand,

$$\|y + npy(e_{\mathfrak{A}} - p)\| \leq \|(e_{\mathfrak{A}} - p)xp + npy(e_{\mathfrak{A}} - p)\|$$
$$= \max\{\|(e_{\mathfrak{A}} - p)yp\|, \|npy(e_{\mathfrak{A}} - p)\|\}$$
$$= n\|py(e_{\mathfrak{A}} - p)\|$$

holds for sufficiently large n, which contradicts (6.9). It follows that $py(e_{\mathfrak{A}} - p) = 0$ and thus

$$y = \mathcal{E}((e_{\mathfrak{A}} - p)xp) = (e_{\mathfrak{A}} - p)yp.$$

Combining this equality with (6.7), we get

$$(e_{\mathfrak{A}} - p)(\mathcal{E}x)p = (e_{\mathfrak{A}} - p)yp = \mathcal{E}((e_{\mathfrak{A}} - p)xp) \qquad \text{and} \qquad p(\mathcal{E}x)p = \mathcal{E}(pxp).$$

It follows that $\mathcal{E}(xp) = (\mathcal{E}x)p$. Since every element in a von Neumann algebra is the w^*-limit of linear combinations of projections, we obtain at once that

$$\mathcal{E}(xa) = (\mathcal{E}x)a \qquad (x \in \mathfrak{A},\, a \in \mathfrak{B}).$$

By (i), $\mathcal{E}$ is positive and thus $*$-preserving. It follows that

$$\mathcal{E}(ax) = a\mathcal{E}x \qquad (x \in \mathfrak{A},\, a \in \mathfrak{B}).$$

Hence, $\mathcal{E}$ is indeed a quasi-expectation.

To see that (iii) holds, just note that

$$0 \leq \mathcal{E}((x - \mathcal{E}x)^*(x - \mathcal{E}x))$$
$$= \mathcal{E}(x^*x - x^*\mathcal{E}x - (\mathcal{E}x)^*x + (\mathcal{E}x)^*\mathcal{E}x)$$
$$= \mathcal{E}(x^*x) - (\mathcal{E}x)^*\mathcal{E}x \qquad (x \in \mathfrak{A}).$$

This completes the proof. $\square$

Exercise 6.2.1 Let $\mathfrak{A}$ be C^*-algebra, let $\mathfrak{B}$ be a C^*-subalgebra of $\mathfrak{A}$ containing a bounded approximate identity for $\mathfrak{A}$, and let $\mathcal{E}: \mathfrak{A} \to \mathfrak{B}$ be a positive projection. Show that $\mathcal{E}$ is an expectation.

Definition 6.2.2 is in terms of a von Neumann algebra acting on a particular Hilbert space. If we want to speak of injective W^*-algebras, we first have to make sure that this definition is independent of a particular W^*-representation:

Exercise 6.2.2 Show that a von Neumann algebra $\mathfrak{M}$ is injective if and only if $\pi(\mathfrak{M})$ is injective for every faithful W^*-representation of $\mathfrak{M}$. (*Hint:* Proceed as in the proof of Lemma 6.1.2, and utilize the easily verified fact that, if two von Neumann algebras are spatially isomorphic, then so are their commutants.)

Injectivity is defined in terms of the commutant of a von Neumann algebra $\mathfrak{M}$. Do we get a different class of von Neumann algebras if we instead demand the existence of an expectation onto $\mathfrak{M}$ itself? See for yourself ...

Exercise 6.2.3 Let $\mathfrak{M}$ be a von Neumann algebra acting on some Hilbert space. Show that $\mathfrak{M}$ is injective if and only $\mathfrak{M}'$ is. (*Hint:* Use Tomita–Takesaki theory as in the proof of Proposition 6.1.3.)

The following hereditary properties of injectivity are useful:

Exercise 6.2.4 Let $\mathfrak{M}$ be an injective W^*-algebra, and let $p \in \mathfrak{M}$ be a projection. Show that $p\mathfrak{M}p$ is injective.

Exercise 6.2.5 Let $(\mathfrak{M}_\alpha)_\alpha$ be a family of injective W^*-algebras. Show that $\ell^\infty\text{-}\bigoplus_\alpha \mathfrak{M}_\alpha$ is an injective W^*-algebra.

It is easy to come up with quasi-expectations which are not expectations (How?). So, if in Definition 6.2.2 we demanded the existence not of an expectation, but merely of a quasi-expectation, we would should get a larger class of von Neumann algebras ... The remainder of this section, is devoted to showing that this is not the case.

Lemma 6.2.5 *Let $\mathfrak{A}$ be a unital C^*-algebra, and let $\mathfrak{M}$ be a unital C^*-subalgebra of $\mathfrak{A}$ which is also a finite, countably decomposable W^*-algebra. Then the following are equivalent:*

(i) *There is $\phi \in \mathcal{Z}^0(\mathfrak{M}, \mathfrak{A}^*)$ such that $\phi|_{\mathfrak{M}}$ is positive and faithful.*

(ii) *There is a positive functional $\phi \in \mathcal{Z}^0(\mathfrak{M}, \mathfrak{A}^*)$ such that $\phi|_{\mathfrak{M}}$ is faithful.*

(iii) *There is an expectation $\mathcal{E} : \mathfrak{A} \to \mathfrak{M}$.*

Proof (i) $\Longrightarrow$ (ii): Without loss of generality, suppose that $\phi^* = \phi$. Let $\phi = \phi^+ - \phi^-$ be the Jordan decomposition ([Mur, 3.3.10. Theorem]) of ϕ with $\phi^+, \phi^- \geq 0$ and $\|\phi\| = \|\phi^+\| + \|\phi^-\|$. For any $u \in \mathcal{U}(\mathfrak{A})$, we have $u^* \cdot \phi \cdot u = \phi$. Uniqueness of the Jordan decomposition implies that $u^* \cdot \phi^+ \cdot u = \phi^+$ for all $u \in \mathcal{U}(\mathfrak{A})$, so that $\phi^+ \in \mathcal{Z}^0(\mathfrak{M}, \mathfrak{A}^*)$ (Why?). Since $\phi^+|_{\mathfrak{M}} \geq \phi|_{\mathfrak{M}}$, it follows that ϕ^+ is faithful.

(ii) $\Longrightarrow$ (iii): For any functional $\psi \in \mathfrak{M}^*$, let ψ_n denote its normal and ψ_s its singular part ([K–R, 10.4.2. Proposition]; see also [Tak 2]). Let $\phi \in \mathfrak{A}^*$ be as in (ii), and define

$$\theta : \mathfrak{A} \to \mathfrak{M}_*, \quad a \mapsto (\phi \cdot a|_{\mathfrak{M}})_n.$$

We claim that $\theta(a) \geq 0$ for all positive $a \in \mathfrak{A}$: For any positive $a \in \mathfrak{A}$, we have

$$\langle x^* x, \phi \cdot a \rangle = \langle ax^* x, \phi \rangle = \langle |x|a|x|, \phi \rangle \geq 0 \qquad (x \in \mathfrak{M});$$

since forming the normal or singular part, respectively, of a functional respects positivity ([K–R, 10.4.3. Proposition]), it follows that θ is positive. Let $\mathrm{tr} := \theta(e_\mathfrak{A})$. It is easy to see (How?) that tr is a normal trace on $\mathfrak{M}$. We claim that tr is faithful. Let $\psi := \phi|_{\mathfrak{M}}$, so that $\psi = \mathrm{tr} + \psi_s$. It suffices (Why?) to show that $\mathrm{tr}\, p \neq 0$ for every non-zero projection $p \in \mathfrak{M}$.

Let $p \in \mathfrak{M}$ be such a projection. By [K–R, 10.5.15 Exercise], there is a non-zero projection $q \in \mathfrak{M}$ with $q \leq p$ such that $\langle q, \psi_s \rangle = 0$. It follows that

$$\operatorname{tr} p \geq \operatorname{tr} q = \langle q, \phi \rangle > 0$$

because ϕ is faithful. Let $a \in \mathfrak{A}$ be positive. It follows that $0 \leq \theta(a) \leq \|a\| \operatorname{tr}$. The Radon–Nikodým type theorem [Sak, 1.24.3. Theorem] yields a positive $\mathcal{E}a \in \mathfrak{M}$ such that

$$\theta(a) = (\mathcal{E}a)^{\frac{1}{2}} \cdot \operatorname{tr} \cdot (\mathcal{E}a)^{\frac{1}{2}} = (\mathcal{E}a) \cdot \operatorname{tr};$$

since tr is faithful, $\mathcal{E}a$ is uniquely determined. Extend $\mathcal{E}$ to a linear map from $\mathfrak{A}$ into $\mathfrak{M}$. The uniqueness of the decomposition of a functional into its normal and its singular part (once again [K–R, 10.4.3. Proposition]), yields that

$$\theta(x) = (\phi \cdot x|_{\mathfrak{M}})_n = (\phi|_{\mathfrak{M}})_n \cdot x = \operatorname{tr} \cdot x = x \cdot \operatorname{tr} = (\mathcal{E}x) \cdot \operatorname{tr} \qquad (x \in \mathfrak{M}).$$

Hence, $\mathcal{E}$ is a projection onto $\mathfrak{M}$. Since $\mathcal{E}$ is positive and satisfies $\mathcal{E}e_{\mathfrak{A}} = e_{\mathfrak{A}}$, it follows that $\|\mathcal{E}\| = 1$ by Exercise 6.2.1.

(iii) $\Longrightarrow$ (i): Let tr be a faithful, normal trace on $\mathfrak{M}$, and let $\phi := \operatorname{tr} \circ \mathcal{E}$. $\quad\square$

For the main theorem of this section, we require the theory of *continuous crossed products* of W^*-algebras. A covariant system is a triple $(\mathfrak{M}, G, \alpha)$, where $\mathfrak{M}$ is a von Neumann algebra, G is a locally compact group, and α is a homomorphism from G into the w^*-continuous *-automorphisms of $\mathfrak{M}$ such that, for each $x \in \mathfrak{M}$ the map $G \ni g \mapsto \alpha_g(x)$ is continuous with respect to the given topology on G and the weak operator topolgy on $\mathfrak{M}$ (arguing as in the proof of Lemma 1.4.2, one can see that this already implies continuity with respect to the strong operator topology on $\mathfrak{M}$). To each covariant system $(\mathfrak{M}, G, \alpha)$, one can associate another von Neumann algebra $W^*(\mathfrak{M}, G, \alpha)$, the *continuous crossed product* of $\mathfrak{M}$ by G. Our sources on continuous crossed products are [Dae] and (to a lesser extent) [K–R]; in [Dae], the notation $\mathfrak{N} \otimes_\alpha \mathbb{R}$ is used instead of $W^*(\mathfrak{N}, \mathbb{R}, \alpha)$.

Theorem 6.2.6 *For a von Neumann algebra $\mathfrak{M}$ acting on a Hilbert space $\mathfrak{H}$ the following are equivalent:*

(i) *$\mathfrak{M}$ is injective.*

(ii) *There is a quasi-expectation $\mathcal{Q}: \mathcal{L}(\mathfrak{H}) \to \mathfrak{M}'$.*

Proof By Theorem 6.2.4, (i) $\Longrightarrow$ (ii) is clear.

Let $\mathcal{Q}: \mathcal{L}(\mathfrak{H}) \to \mathfrak{M}'$ be a quasi-expectation.

Suppose first that $\mathfrak{M}'$ is finite and countably decomposable. Let tr be a faithful, normal trace on $\mathfrak{M}'$, and define $\phi := \operatorname{tr} \circ \mathcal{Q}$. Then ϕ satisfies Lemma 6.2.5(i), so that there is an expectation $\mathcal{E}: \mathcal{L}(\mathfrak{H}) \to \mathfrak{M}'$.

Next, let $\mathfrak{M}'$ be semifinite, i.e. it has no direct summand of type III. By [K–R, 6.3.8. Theorem and 5.7.45. Exercise], there is an increasing net $(p_\alpha)_\alpha$ of projections in $\mathfrak{M}'$ such that

– $p_\alpha \nearrow \operatorname{id}_{\mathfrak{H}}$ in the strong operator topology, and

– each $p_\alpha \mathfrak{M}' p_\alpha$ is finite and countably decomposable.

Since $\mathcal{Q}(p_\alpha T p_\alpha) = p_\alpha(\mathcal{Q}T)p_\alpha$ for $T \in \mathcal{L}(\mathfrak{H})$, restricting $\mathcal{Q}$ to $p_\alpha \mathcal{L}(\mathfrak{H}) p_\alpha$ induces a quasi-expectation $\mathcal{Q}_\alpha : p_\alpha \mathcal{L}(\mathfrak{H}) p_\alpha \to p_\alpha \mathfrak{M}' p_\alpha$. By the foregoing, we then have an expectation $\mathcal{E}_\alpha : p_\alpha \mathcal{L}(\mathfrak{H}) p_\alpha \to p_\alpha \mathfrak{M}' p_\alpha$. Let $\mathcal{U}$ be an ultrafilter on the index set of $(p_\alpha)_\alpha$ that dominates the order filter. Define

$$\mathcal{E} : \mathcal{L}(\mathfrak{H}) \to \mathcal{L}(\mathfrak{H}), \quad T \mapsto w^*\text{-}\lim_{\mathcal{U}} \mathcal{E}_\alpha T.$$

It is easily seen that $\mathcal{E}$ is a norm one projection onto $\mathfrak{M}'$, i.e. an expectation.

We now turn to the general case.

Since we have already settled the semifinite case, we may suppose by Exercise 6.2.2 that $\mathfrak{M}'$ is of type III. We first suppose that $\mathfrak{M}'$ is also countably decomposable. By [Dae, II.4.8 Theorem], there is a covariant system $(\mathfrak{N}, \mathbb{R}, \alpha)$ such that $\mathfrak{M}' \cong W^*(\mathfrak{N}, \mathbb{R}, \alpha)$, where $\mathfrak{N}$ is a von Neumann algebra of type II$_\infty$. Consider the dual action $\hat{\alpha}$ of $\hat{\mathbb{R}} = \mathbb{R}$ on $\mathfrak{M}'$ ([Dae, Section I.4]). By [Dae, I.4.12 Proposition], the fixed point algebra of this dual action

$$\{x \in W^*(\mathfrak{N}, \mathbb{R}, \alpha) : \hat{\alpha}_t(x) = x \text{ for all } t \in \mathbb{R}\}$$

is (canonically) isomorphic to $\mathfrak{N}$. Let $m : \ell^\infty(\mathbb{R}) \to \mathbb{C}$ be an invariant mean. For $\xi, \eta \in \mathfrak{H}$ define

$$\phi_{\xi,\eta}(t, x) := \langle \hat{\alpha}_t(x)\xi, \eta \rangle \qquad (t \in \mathbb{R}, \; x \in \mathfrak{M}').$$

It is immediate that $\phi_{\xi,\eta}(\cdot, x) \in \ell^\infty(\mathbb{R})$ for all $x \in \mathfrak{M}'$. Define $\mathcal{P} : \mathfrak{M}' \to \mathfrak{N}$ by letting

$$\langle (\mathcal{P}x)\xi, \eta \rangle := \langle \phi_{\xi,\eta}(\cdot, x), m \rangle \qquad (x \in \mathfrak{M}', \; \xi, \eta \in \mathfrak{H}).$$

It is routinely checked (hopefully) that $\mathcal{P}$ is a norm one projection onto $\mathfrak{N}$. From Theorem 6.2.4(ii), it follows that $\mathcal{P} \circ \mathcal{Q} : \mathcal{L}(\mathfrak{H}) \to \mathfrak{N}$ is a quasi-expectation. Since $\mathfrak{N}$ is of type II$_\infty$ and thus semifinite, it follows that there is an expectation from $\mathcal{L}(\mathfrak{H})$ onto $\mathfrak{N}$. Exercise 6.2.3 implies that there is an expectation $\tilde{\mathcal{E}} : \mathcal{L}(\mathfrak{H}) \to \mathfrak{M}'$. The construction of $W^*(\mathfrak{N}, \mathbb{R}, \alpha)$ ([Dae, I.2.10 Definition]) shows that $\mathfrak{M}'$ is generated by its subalgebra $\mathfrak{N}$ and a one-parameter group $\{u_t : t \in \mathbb{R}\} \subset \mathcal{U}(\mathfrak{M}')$ such that

$$u_t x u_t^* = \alpha_t(x) \qquad (t \in \mathbb{R}, \; x \in \mathfrak{N}). \tag{6.10}$$

In particular, it follows that

$$\mathfrak{M} = \mathfrak{M}'' = \mathfrak{N}' \cap \{u_t : t \in \mathbb{R}\}'. \tag{6.11}$$

For $\xi, \eta \in \mathfrak{H}$, let

$$\psi_{\xi,\eta}(t, T) := \langle u_t(\tilde{\mathcal{E}}T)u_t^* \xi, \eta \rangle \qquad (t \in \mathbb{R}, \; T \in \mathcal{L}(\mathfrak{H})).$$

Define $\mathcal{E} : \mathcal{L}(\mathfrak{H}) \to \mathcal{L}(\mathfrak{H})$ through

$$\langle (\mathcal{E}T)\xi, \eta \rangle := \langle \psi_{\xi,\eta}(\cdot, T), m \rangle \qquad (T \in \mathcal{L}(\mathfrak{H}), \; \xi, \eta \in \mathfrak{H}).$$

It is immediate that $\mathcal{E}$ has norm one. From (6.11), it follows at once that $\mathcal{E}|_{\mathfrak{M}} = \mathrm{id}_{\mathfrak{M}}$. Since m is an invariant mean, it is easy to see that $\mathcal{E}$ attains its values in $\{u_t : t \in \mathbb{R}\}'$; with (6.10), it is also straightforward to verify that $\mathcal{E}(\mathcal{L}(\mathfrak{H})) \subset \mathfrak{N}'$. Hence, (6.11) again implies that the range of $\mathcal{E}$ is contained in $\mathfrak{M}$. Thus, $\mathcal{E} : \mathcal{L}(\mathfrak{H}) \to \mathfrak{M}$ is an expectation, and $\mathfrak{M}'$ is injective. By Exercise 6.2.3, $\mathfrak{M}$ is injective.

Finally, suppose that $\mathfrak{M}'$ is of type III and arbitrary. Once again invoking [K–R, Exercise 5.7.45], we obtain a net $(p_\alpha)_\alpha$ of projections in $\mathfrak{M}'$ such that $p_\alpha \nearrow \mathrm{id}_{\mathfrak{H}}$ in the strong operator topology, and such that each $p_\alpha \mathfrak{M}' p_\alpha$ is countably decomposable. It is routinely checked that each of the W^*-algebras $p_\alpha \mathfrak{M}' p_\alpha$ is of type III as well. As in our treatment of the semifinite case, we conclude that each $p_\alpha \mathfrak{M} p_\alpha$ and eventually $\mathfrak{M}$ itself is injective. $\Box$

Remark 6.2.7 In the proof of Theorem 6.2.6, it is not necessary, in the type III case, to treat countably decomposable W^*-algebras first and then reduce the general situation to the countably decomposable case: [Dae, II.Theorem 4.8] can be extended to hold for arbitrary W^*-algebras of type III; in the general situation, one has to construct the modular automorphism group with respect to a weight instead of a state. The reason why [Dae], as well as [K–R], confines itself to the countably decomposable case is evident: this simply makes the construction much easier. For the general situation, see [Stra].

In conjunction with Theorem 4.4.11(ii), Theorem 6.2.6 immediately yields the first connection between injectivity and Connes-amenability:

Corollary 6.2.8 *Let $\mathfrak{M}$ be a Connes-amenable W^*-algebra. Then $\mathfrak{M}$ is injective.*

Example 6.2.9 Let $\mathfrak{M}$ be a commutative W^*-algebra. By Example 2.3.4, $\mathfrak{M}$ is amenable and thus, in particular, Connes-amenable. By Corollary 6.2.8, it follows that $\mathfrak{M}$ is injective.

We have just seen that

$$\text{Connes-amenability} \quad \Longrightarrow \quad \text{injectivity}.$$

We already mentioned at the beginning of this section that Connes-amenability and injectivity are equivalent. The proof, however, will have to wait a while.

6.3 Tensor products of C^*- and W^*-algebras

Let $\mathfrak{A}$ and $\mathfrak{B}$ be C^*-algebras. By Exercise 6.1.4(i), their algebraic tensor product $\mathfrak{A} \otimes \mathfrak{B}$ carries a canonical involution. What we shall discuss in this chapter are ways of turning (a completion of) $\mathfrak{A} \otimes \mathfrak{B}$ into a C^*-algebra. Unfortunately, the projective norm won't do ...

Definition 6.3.1 Let $\mathfrak{A}$ be an algebra with involution. A C^*-*norm* on $\mathfrak{A}$ is a submultiplicative norm $\|\cdot\|$ on $\mathfrak{A}$ such that

$$\|a^*a\| = \|a\|^2 \qquad (a \in \mathfrak{A}).$$

Of course, if $\mathfrak{A}$ is an algebra with involution and $\|\cdot\|$ is a C^*-norm on $\mathfrak{A}$, then the completion of $\mathfrak{A}$ with respect to $\|\cdot\|$ is a C^*-algebra.

For two C^*-algebras $\mathfrak{A}$ and $\mathfrak{B}$, we thus look for a C^*-norm on $\mathfrak{A} \otimes \mathfrak{B}$. So far, we only know that such norms exist in rather specific cases, such as if one of the algebras is commutative (Exercise 2.3.6(iii)) or a full matrix algebra (Exercise 6.1.5). In the general situation, the existence of C^*-norms on $\mathfrak{A} \otimes \mathfrak{B}$ is an immediate consequence of the following proposition:

Proposition 6.3.2 *Let $\mathfrak{A}$ and $\mathfrak{B}$ be C^*-algebras, and let π and ρ be $*$-representations of $\mathfrak{A}$ and $\mathfrak{B}$, respectively, on Hilbert spaces $\mathfrak{H}$ and $\mathfrak{K}$, respectively. Then there is a unique $*$-representation $\pi \otimes \rho$ of $\mathfrak{A} \otimes \mathfrak{B}$ on $\mathfrak{H} \bar{\otimes} \mathfrak{K}$ such that*

$$(\pi \otimes \rho)(a \otimes b) = \pi(a) \otimes \rho(b) \qquad (a \in \mathfrak{A}, b \in \mathfrak{B}). \tag{6.12}$$

Moreover, if both π and ρ are faithful, then so is $\pi \otimes \rho$.

Proof Use (6.12) to define $\pi \otimes \rho$ (this can be done by Exercise B.3.3). It is clear that (6.12) defines $\pi \otimes \rho$ uniquely.

Suppose that both π and ρ are faithful, and let $a_1, \ldots, a_n \in \mathfrak{A}$ and $b_1, \ldots, b_n \in \mathfrak{B}$ be such that $\sum_{j=1}^{n} \pi(a_j) \otimes \rho(b_j) = 0$. Without loss of generality suppose that $a_1, \ldots, a_n$ are linearly independent, so that $\pi(a_1), \ldots, \pi(a_n)$ are linearly independent, too, by the faithfulness of π. Let $\eta \in \mathfrak{K}$ be arbitrary, and observe that

$$0 = \sum_{j=1}^{n} \langle \pi(a_j)\xi, \xi \rangle \langle \rho(b_j)\eta, \eta \rangle \qquad (\xi \in \mathfrak{H}).$$

It follows that $\sum_{j=1}^{n} \langle \rho(b_j)\eta, \eta \rangle \pi(a_j) = 0$. Since $\pi(a_1), \ldots, \pi(a_n)$ are linearly independent, this implies

$$\langle \rho(b_j)\eta, \eta \rangle = 0 \qquad (j = 1, \ldots, n).$$

Since $\eta \in \mathfrak{K}$ is arbitrary, and ρ is faithful, we obtain that $b_1 = \cdots = b_n = 0$ and thus, in particular, $\sum_{j=1}^{n} a_j \otimes b_j = 0$. $\square$

With this proposition at hand, we have an obvious way of defining C^*-norms on tensor products of C^*-algebras:

Definition 6.3.3 Let $\mathfrak{A}$ and $\mathfrak{B}$ be C^*-algebras with universal $*$-representations $\pi_{\mathfrak{A}}$ and $\pi_{\mathfrak{B}}$, respectively. Then the *minimal* or *spatial* C^*-*norm* $\| \cdot \|_{\min}$ on $\mathfrak{A} \otimes \mathfrak{B}$ is defined through

$$\|\mathbf{a}\|_{\min} := \|(\pi_{\mathfrak{A}} \otimes \pi_{\mathfrak{B}})(\mathbf{a})\| \qquad (\mathbf{a} \in \mathfrak{A} \otimes \mathfrak{B}).$$

We write $\mathfrak{A} \otimes_{\min} \mathfrak{B}$ for $\mathfrak{A} \otimes \mathfrak{B}$ equipped with $\| \cdot \|_{\min}$ and $\mathfrak{A} \bar{\otimes}_{\min} \mathfrak{B}$ for the completion of $\mathfrak{A} \otimes_{\min} \mathfrak{B}$.

We haven't introduced a particular symbol for the state space of a C^*-algebra yet, but for the sake of brevity we shall denote, from now on, the states of a C^*-algebra $\mathfrak{A}$ by $S(\mathfrak{A})$.

Exercise 6.3.1 Let $\mathfrak{A}$ and $\mathfrak{B}$ be C^*-algebras, and let $\mathbf{a} \in \mathfrak{A} \otimes \mathfrak{B}$ be self-adjoint. Show that

$$\|\mathbf{a}\|_{\min} = \sup\{|\langle \mathbf{a}, \phi \otimes \psi \rangle| : \phi \in S(\mathfrak{A}), \psi \in S(\mathfrak{B})\}.$$

There are several question that come up naturally with this definition:

- Is $\| \cdot \|_{\min}$ the only C^*-norm on $\mathfrak{A} \otimes \mathfrak{B}$?
- How does $\| \cdot \|_{\min}$ relate to $\| \cdot \|_\epsilon$ and $\| \cdot \|_\pi$?
- Why is $\| \cdot \|_{\min}$ called the minimal C^*-norm?

The first question leads us immediately to the notion of *nuclearity*:

Definition 6.3.4 A C^*-algebra $\mathfrak{A}$ is called *nuclear* if there is only one C^*-norm on $\mathfrak{A} \otimes \mathfrak{B}$ for every C^*-algebra $\mathfrak{B}$.

As we have already mentioned in the introduction to this chapter, nuclearity will turn out to be same as amenability for C^*-algebras, so that we will be a little short on examples here.

Example 6.3.5 By Exercise 6.1.5, $\mathbb{M}_n$ is nuclear for each $n \in \mathbb{N}$. Using the structure theorem [Mur, 6.3.8. Theorem], it can easily be seen (How exactly?) that all finite-dimensional C^*-algebras are nuclear.

From the (still unproven) equivalence of nuclearity and amenability, it follows immediately that every commutative C^*-algebra is nuclear. Since we require this result already in this section — in order to prove that $\|\cdot\|_{\min}$ is indeed minimal —, we give a direct proof. The following exercise will be useful:

Exercise 6.3.2 Let $\mathfrak{A}$ and $\mathfrak{B}$ be C^*-algebras such that $\mathfrak{B}$ is non-unital, and let $\|\cdot\|$ be a C^*-norm on $\mathfrak{A} \otimes \mathfrak{B}$. Show that $\|\cdot\|$ extends to $\mathfrak{A} \otimes \mathfrak{B}^\#$ as a C^*-norm. (*Hint*: Take a faithful *-representation of $\mathfrak{A} \otimes \mathfrak{B}$ on some Hilbert space, and use the fact that $\mathfrak{B}$ has a bounded approximate identity to extend it to $\mathfrak{A} \otimes \mathfrak{B}^\#$.)

Of course, we may interchange the rôles of $\mathfrak{A}$ and $\mathfrak{B}$ in Exercise 6.3.2.

Proposition 6.3.6 *Let $\mathfrak{A}$ and $\mathfrak{B}$ be commutative C^*-algebras. Then there is only one C^*-norm on $\mathfrak{A} \otimes \mathfrak{B}$.*

Proof By Exercise 6.3.2, we may suppose without loss of generality that both $\mathfrak{A}$ and $\mathfrak{B}$ are unital. Let $\|\cdot\|$ be a C^*-norm, and let $\mathfrak{A} \tilde{\otimes} \mathfrak{B}$ denote the completion of $\mathfrak{A} \otimes \mathfrak{B}$ with respect to $\|\cdot\|$. Let $\Omega_\mathfrak{A}$, $\Omega_\mathfrak{B}$, and $\Omega_{\mathfrak{A} \tilde{\otimes} \mathfrak{B}}$ be the character spaces of $\mathfrak{A}$, $\mathfrak{B}$, and $\mathfrak{A} \tilde{\otimes} \mathfrak{B}$, respectively. Let

$$K := \{(\phi, \psi) : \phi \in \Omega_\mathfrak{A}, \psi \in \Omega_\mathfrak{B} \text{ such that } \phi \otimes \psi \in \Omega_{\mathfrak{A} \tilde{\otimes} \mathfrak{B}}\}.$$

Let $\phi \in \Omega_{\mathfrak{A} \tilde{\otimes} \mathfrak{B}}$, and define $\phi_\mathfrak{A} : \mathfrak{A} \to \mathbb{C}$ and $\phi_\mathfrak{B} : \mathfrak{B} \to \mathbb{C}$ through

$$\langle a, \phi_\mathfrak{A} \rangle := \langle a \otimes e_\mathfrak{B}, \phi \rangle \quad (a \in \mathfrak{A}) \qquad \text{and} \qquad \langle b, \phi_\mathfrak{B} \rangle := \langle e_\mathfrak{A} \otimes b, \phi \rangle \quad (b \in \mathfrak{B}).$$

It is clear that $\phi_\mathfrak{A} \in \Omega_\mathfrak{A}$, $\phi_\mathfrak{B} \in \Omega_\mathfrak{B}$, and that $\phi = \phi_\mathfrak{A} \otimes \phi_\mathfrak{B}$. It follows that

$$\|\mathbf{a}\| = \sup\{|\langle \mathbf{a}, \phi \otimes \psi \rangle| : (\phi, \psi) \in K\} \qquad (\mathbf{a} \in \mathfrak{A} \otimes \mathfrak{B}). \tag{6.13}$$

Assume that $K \subsetneq \Omega_\mathfrak{A} \times \Omega_\mathfrak{B}$. It is easily seen that K is closed in $\Omega_\mathfrak{A} \times \Omega_\mathfrak{B}$, so that there are non-void, open sets $U_\mathfrak{A} \subset \Omega_\mathfrak{A}$ and $U_\mathfrak{B} \subset \Omega_\mathfrak{B}$ with $(U_\mathfrak{A} \times U_\mathfrak{B}) \cap K = \varnothing$. Choose non-zero functions $f \in \mathcal{C}(\Omega_\mathfrak{A})$ and $g \in \mathcal{C}(\Omega_\mathfrak{B})$ with $\text{supp}(f) \subset U_\mathfrak{A}$ and $\text{supp}(g) \subset U_\mathfrak{B}$. Identifying $\mathcal{C}(\Omega_\mathfrak{A})$ and $\mathcal{C}(\Omega_\mathfrak{B})$ with $\mathfrak{A}$ and $\mathfrak{B}$, respectively, via Gelfand theory, we may view $f \otimes g$ as an element of $\mathfrak{A} \tilde{\otimes} \mathfrak{B}$. According to (6.13) and the choices of f and g, we have $\|f \otimes g\| = 0$. But this is impossible because both $f \neq 0$ and $g \neq 0$.

It follows that $K = \Omega_\mathfrak{A} \times \Omega_\mathfrak{B}$ and hence

$$\|\mathbf{a}\| = \sup\{|\langle \mathbf{a}, \phi \otimes \psi \rangle| : \phi \in \Omega_\mathfrak{A}, \psi \in \Omega_\mathfrak{B}\} \qquad (\mathbf{a} \in \mathfrak{A} \otimes \mathfrak{B}).$$

Obviously, the right hand side of this equation is independent of $\|\cdot\|$. $\square$

That's not quite the nuclearity of commutative C^*-algebras yet ...

Lemma 6.3.7 *Let $\mathfrak{A}$ and $\mathfrak{B}$ be unital C^*-algebras, let $\|\cdot\|$ be a C^*-norm on $\mathfrak{A}\otimes\mathfrak{B}$, and let $\mathfrak{A}\tilde{\otimes}\mathfrak{B}$ denote the completion of $\mathfrak{A}\otimes\mathfrak{B}$ with respect to $\|\cdot\|$. Furthermore, let $\phi\in S(\mathfrak{A})$, and define*

$$S_\phi := \{\psi\in S(\mathfrak{B}) : \phi\otimes\psi \text{ has norm one on } \mathfrak{A}\otimes\mathfrak{B} \text{ with respect to } \|\cdot\|\}.$$

Then:

(i) *For each $\psi\in S_\phi$, the functional $\phi\otimes\psi$ on $\mathfrak{A}\otimes\mathfrak{B}$ has a unique extension to an element of $S(\mathfrak{A}\tilde{\otimes}\mathfrak{B})$.*

(ii) *S_ϕ is a w^*-compact, convex subset of $S(\mathfrak{B})$.*

(iii) *If ϕ is pure, and if $b\in\mathfrak{B}$ is self-adjoint such that there is only one C^*-norm on $\mathfrak{A}\otimes\mathcal{C}(\sigma(b))$, then*

$$\|b\| = \sup\{|\langle b,\psi\rangle| : \psi\in S_\phi\}.$$

(iv) *If every self-adjoint element $b\in\mathfrak{B}$ has the property that there is only one C^*-norm on $\mathfrak{A}\otimes\mathcal{C}(\sigma(b))$, then $S_\phi = S(\mathfrak{B})$.*

Proof (i) and (ii) are routine.

For the proof of (iii), let $C^*(b)$ be the unital C^*-subalgebra of $\mathfrak{B}$ generated by b; by Gelfand theory, we have $C^*(b)\cong\mathcal{C}(\sigma(b))$. There is $\chi\in\Omega_{C^*(b)}$ such that $\|b\| = |\langle b,\chi\rangle|$. The functional $\phi\otimes\chi$ on $\mathfrak{A}\otimes C^*(b)\cong\mathfrak{A}\otimes\mathcal{C}(\sigma(b))$ is continuous with respect to $\|\cdot\|_{\min}$. Since we are supposing that there is only one C^*-norm on $\mathfrak{A}\otimes C^*(b)$, we have that $\phi\otimes\chi$ is also continuous with respect to $\|\cdot\|$; it thus extends to an element $\psi\in S(\mathfrak{A}\tilde{\otimes}\mathfrak{B})$. Define $\psi_\mathfrak{B}\in S(\mathfrak{B})$ through

$$\langle c,\psi_\mathfrak{B}\rangle := \langle e_\mathfrak{A}\otimes c,\psi\rangle \qquad (c\in\mathfrak{B}).$$

From the definition, it is immediate that $\|b\| = |\langle b,\psi_\mathfrak{B}\rangle|$. We claim that $\psi_\mathfrak{B}\in S_\phi$ (which establishes (iii)). Let $c\in\mathfrak{B}$ be such that $0\le c\le e_\mathfrak{B}$, and define positive functionals $\phi_1,\phi_2\in\mathfrak{A}^*$ through

$$\langle a,\phi_1\rangle := \langle a\otimes c,\psi\rangle \quad\text{and}\quad \langle a,\phi_2\rangle := \langle a\otimes(e_\mathfrak{B}-c),\psi\rangle \qquad (a\in\mathfrak{A}).$$

We have

$$\langle a,\phi\rangle = \langle a\otimes e_\mathfrak{B},\psi\rangle = \langle a,\phi_1\rangle + \langle a,\phi_2\rangle \qquad (a\in\mathfrak{A}).$$

Since ϕ is an extreme point of $S(\mathfrak{A})$, it follows that there is $t\ge0$ such that $\phi_1 = t\phi$ (Why?). Since

$$t = t\langle e_\mathfrak{A},\phi\rangle = \langle e_\mathfrak{A},\phi_1\rangle = \langle e_\mathfrak{A}\otimes c,\psi\rangle = \langle c,\psi_\mathfrak{B}\rangle,$$

it follows that

$$\langle a\otimes c,\psi\rangle = \langle a,\phi_1\rangle = t\langle a,\phi\rangle = \langle a,\phi\rangle\langle c,\psi_\mathfrak{B}\rangle \qquad (a\in\mathfrak{A}).$$

By linearity, we obtain eventually that $\psi = \phi\otimes\psi_\mathfrak{B}$, so that $\psi_\mathfrak{B}\in S_\phi$.

To prove (iv), suppose first again that ϕ is pure. We claim that $S_\phi = S(\mathfrak{B})$ in this particular case. Assume towards a contradiction that there is $\psi_0 \in S(\mathfrak{B}) \setminus S_\phi$. By (ii), an application of the Hahn–Banach theorem yields $a \in \mathfrak{A}$ such that

$$\sup\{\operatorname{Re}\langle a, \psi\rangle : \psi \in S_\phi\} < \operatorname{Re}\langle a, \psi_0\rangle.$$

Letting $b := \|a\|e_\mathfrak{A} + \frac{1}{2}(a + a^*)$, we obtain

$$\sup\{|\langle b, \psi\rangle| : \psi \in S_\phi\} < |\langle b, \psi_0\rangle| \leq \|b\|.$$

This contradicts (iii). For $\psi \in S(\mathfrak{B})$, let

$$S^\psi := \{\phi' \in S(\mathfrak{A}) : \psi \in S_{\phi'}\}.$$

It is easily checked that S^ψ is a w^*-compact, convex subset of $S(\mathfrak{A})$. By the foregoing, every pure state of $\mathfrak{A}$ is in S^ψ. By the Kreĭn–Milman theorem, every element of $S(\mathfrak{A})$ is the w^*-limit of convex combinations of pure states; it follows that $S(\mathfrak{A}) = S^\psi$. In particular, we have $\psi \in S_\phi$. $\square$

Exercise 6.3.3 Let $\mathfrak{A}$ be a C^*-algebra, and let $\phi \in S(\mathfrak{A})$ be pure. Show that $\phi|_{Z(\mathfrak{A})}$ is multiplicative.

Theorem 6.3.8 *Let $\mathfrak{A}$ be a commutative C^*-algebra. Then $\mathfrak{A}$ is nuclear.*

Proof Let $\mathfrak{B}$ be another C^*-algebra. By Exercise 6.3.2, there is no loss of generality if we suppose that both $\mathfrak{A}$ and $\mathfrak{B}$ are unital.

Let $\|\cdot\|$ be a C^*-norm on $\mathfrak{A} \otimes \mathfrak{B}$, and let $\mathfrak{A}\tilde{\otimes}\mathfrak{B}$ be the completion of $\mathfrak{A} \otimes \mathfrak{B}$ with respect to $\|\cdot\|$. Let $\mathbf{a} \in \mathfrak{A} \otimes \mathfrak{B}$ be self-adjoint. Then there is a pure state χ of $\mathfrak{A}\tilde{\otimes}\mathfrak{B}$ such that $\|\mathbf{a}\| = |\langle\mathbf{a}, \chi\rangle|$. Define $\chi_\mathfrak{A} \in S(\mathfrak{A})$ and $\chi_\mathfrak{B} \in S(\mathfrak{B})$ through

$$\langle a, \chi_\mathfrak{A}\rangle := \langle a \otimes e_\mathfrak{B}, \chi\rangle \quad (a \in \mathfrak{A}) \qquad \text{and} \qquad \langle b, \chi_\mathfrak{B}\rangle := \langle e_\mathfrak{A} \otimes b, \chi\rangle \quad (b \in \mathfrak{B}).$$

By Exercise 6.3.3, $\chi_\mathfrak{A}$ is multiplicative and thus pure. As in the proof of Lemma 6.3.7(iii) (How exactly?), we see that $\chi = \chi_\mathfrak{A} \otimes \chi_\mathfrak{B}$, so that, in particular, $\chi \in \{\phi \otimes \psi : \phi \in S(\mathfrak{A}), \psi \in S(\mathfrak{B})\}$.

By Proposition 6.3.6, the hypothesis of Lemma 6.3.7(iv) is satisfied, so that

$$\{\phi \otimes \psi : \phi \in S(\mathfrak{A}), \psi \in S(\mathfrak{B})\} \subset S(\mathfrak{A}\tilde{\otimes}\mathfrak{B}).$$

It follows that

$$\|\mathbf{a}\| = |\langle\mathbf{a}, \chi\rangle| = \sup\{|\langle\mathbf{a}, \phi \otimes \psi\rangle| : \phi \in S(\mathfrak{A}), \psi \in S(\mathfrak{B})\} \leq \|\mathbf{a}\|,$$

and hence

$$\|\mathbf{a}\| = \sup\{|\langle\mathbf{a}, \phi \otimes \psi\rangle| : \phi \in S(\mathfrak{A}), \psi \in S(\mathfrak{B})\}.$$

Since the right hand side of the last equality is independent of $\|\cdot\|$, this completes the proof. $\square$

So, why is $\|\cdot\|_{\min}$ called the minimal norm? With Theorem 6.3.8 at hand, this is no longer difficult to answer:

Theorem 6.3.9 *Let $\mathfrak{A}$ and $\mathfrak{B}$ be C^*-algebras, and let $\|\cdot\|$ be a C^*-norm on $\mathfrak{A} \otimes \mathfrak{B}$. Then we have:*

$$\|\mathbf{a}\|_\epsilon \leq \|\mathbf{a}\|_{\min} \leq \|\mathbf{a}\| \qquad (\mathbf{a} \in \mathfrak{A} \otimes \mathfrak{B}).$$

Proof To prove the first inequality, let $\phi \in B_1[0, \mathfrak{A}^*]$ and $\psi \in B_1[0, \mathfrak{B}^*]$, and show that $\phi \otimes \psi$ has norm at most one with respect to $\|\cdot\|_{\min}$. This is clear if both ϕ and ψ are states. Since $B_1[0, \mathfrak{A}^*]$ is the absolutely convex hull of $S(\mathfrak{A})$ and $B_1[0, \mathfrak{B}^*]$ is the absolutely convex hull of $S(\mathfrak{B})$, the general claim follows at once.

For the second inequality, suppose that both $\mathfrak{A}$ and $\mathfrak{B}$ are unital. Let $\mathfrak{A}\tilde{\otimes}\mathfrak{B}$ be the completion of $\mathfrak{A} \otimes \mathfrak{B}$. By Theorem 6.3.8, the hypothesis of Lemma 6.3.7(iv) is satisfied, so that

$$\{\phi \otimes \psi : \phi \in S(\mathfrak{A}), \, \psi \in S(\mathfrak{B})\} \subset S(\mathfrak{A}\tilde{\otimes}\mathfrak{B}).$$

By Exercise 6.3.1, this establishes the claim. $\square$

Exercise 6.3.4 Let $\mathfrak{A}$ and $\mathfrak{B}$ be C^*-algebras, and let π and ρ be faithful *-representations of $\mathfrak{A}$ and $\mathfrak{B}$, respectively. Show that

$$\|\mathbf{a}\|_{\min} = \|(\pi \otimes \rho)(\mathbf{a})\| \qquad (\mathbf{a} \in \mathfrak{A} \otimes \mathfrak{B}).$$

Exercise 6.3.5 Show that $\|\cdot\|_{\min}$ respects C^*-subalgebras: If $\mathfrak{A}$, $\mathfrak{B}$, and $\mathfrak{C}$ are C^*-algebra with $\mathfrak{C}$ a C^*-subalgebra of $\mathfrak{B}$, then the canonical inclusion

$$\mathfrak{A} \otimes_{\min} \mathfrak{C} \hookrightarrow \mathfrak{A} \otimes_{\min} \mathfrak{B}$$

is an isometry.

Hence, $\|\cdot\|_{\min}$ is indeed the minimal C^*-norm on tensor products of C^*-algebras. Is there also a maximal one? There is — and its existence is quite easy to establish:

Definition 6.3.10 Let $\mathfrak{A}$ and $\mathfrak{B}$ be C^*-algebras. Then the *maximal C^*-norm* $\|\cdot\|_{\max}$ on $\mathfrak{A} \otimes \mathfrak{B}$ is defined through

$$\|\mathbf{a}\|_{\max} := \sup\{\|\mathbf{a}\| : \|\cdot\| \text{ is a } C^*\text{-norm on } \mathfrak{A} \otimes \mathfrak{B}\} \qquad (\mathbf{a} \in \mathfrak{A} \otimes \mathfrak{B}). \qquad (6.14)$$

We write $\mathfrak{A} \otimes_{\max} \mathfrak{B}$ for $\mathfrak{A} \otimes \mathfrak{B}$ equipped with $\|\cdot\|_{\max}$ and $\mathfrak{A}\tilde{\otimes}_{\max}\mathfrak{B}$ for the completion of $\mathfrak{A} \otimes_{\max} \mathfrak{B}$.

Theorem 6.3.11 *Let $\mathfrak{A}$ and $\mathfrak{B}$ be C^*-algebras. Then $\|\cdot\|_{\max}$ is a C^*-norm on $\mathfrak{A} \otimes \mathfrak{B}$. If $\|\cdot\|$ is another C^*-norm on $\mathfrak{A}$, then*

$$\|\mathbf{a}\| \leq \|\mathbf{a}\|_{\max} \leq \|\mathbf{a}\|_\pi \qquad (\mathbf{a} \in \mathfrak{A} \otimes \mathfrak{B}).$$

Proof It is sufficient to prove that $\|\cdot\|_\pi$ dominates an arbitrary C^*-norm $\|\cdot\|$ on $\mathfrak{A} \otimes \mathfrak{B}$ (this establishes at once that the supremum (6.14) is finite).

Let $\mathfrak{A}\tilde{\otimes}\mathfrak{B}$ be the completion of $\mathfrak{A}\otimes\mathfrak{B}$ with respect to $\|\cdot\|$. Let $a, b \in B_1[0, \mathfrak{B}]$ be positive, so that, by the continuous functional calculus, $a^2 \leq a$ and $b^2 \leq b$. Certainly, $a \otimes b$ is positive in $\mathfrak{A}\tilde{\otimes}\mathfrak{B}$. Since

$$a \otimes b - (a \otimes b)^2 = a \otimes b - a^2 \otimes b + a^2 \otimes b - a^2 \otimes b^2$$
$$= (a - a^2) \otimes b + a^2 \otimes (b - b^2)$$
$$\geq 0.$$

Continuous functional calculus, again, yields that $\|a \otimes b\| \leq 1$. It follows that $\|a \otimes b\| \leq \|a\| \|b\|$ for all positive $a \in \mathfrak{A}$ and $b \in \mathfrak{B}$. Finally, we have

$$\|a \otimes b\|^2 = \|a^*a \otimes b^*b\| \leq \|a^*a\| \|b^*b\| = \|a\|^2 \|b\|^2 \qquad (a \in \mathfrak{A}, \, b \in \mathfrak{B}).$$

The desired inequality now follows from Exercise B.2.8. □

Exercise 6.3.6 Let $\mathfrak{A}$ and $\mathfrak{B}$ be unital C^*-algebras, and let $\phi \colon \mathfrak{A} \otimes \mathfrak{B} \to \mathbb{C}$ be an algebraic state, i.e. $\langle e_{\mathfrak{A}} \otimes e_{\mathfrak{B}}, \phi \rangle = 1$ and $\langle \mathbf{a}^* \bullet \mathbf{a}, \phi \rangle \geq 0$ for all $\mathbf{a} \in \mathfrak{A} \otimes \mathfrak{B}$. Show that ϕ is continuous with respect to $\| \cdot \|_{\max}$. (*Hint:* GNS-construction.)

Together Theorems 6.3.9 and 6.3.11 yield:

Corollary 6.3.12 *A C^*-algebra $\mathfrak{A}$ is nuclear if and only if, for each C^*-algebra $\mathfrak{B}$, the identity map*

$$\mathfrak{A} \otimes_{\max} \mathfrak{B} \to \mathfrak{A} \otimes_{\min} \mathfrak{B}$$

is an isometry.

Corollary 6.3.13 *Let $\mathfrak{A}$ and $\mathfrak{B}$ be C^*-algebras, and let $\| \cdot \|$ be a C^*-norm on $\mathfrak{A} \otimes \mathfrak{B}$. Then $\| \cdot \|$ is a cross norm.*

We haven't said anything about W^*-algebras so far. Of course, given two W^*-algebras, we can equip their algebraic tensor product with some C^*-norm and form the completion; the resulting C^*-algebra, however, will rarely be a W^*-algebra again.

Let $\mathfrak{A}$ and $\mathfrak{B}$ be C^*-algebras. In the proof of Theorem 6.3.9, we saw that $\phi \otimes \psi$ is continuous with respect to $\| \cdot \|_{\min}$ for all $\phi \in \mathfrak{A}^*$ and $\psi \in \mathfrak{B}^*$. We may thus view $\mathfrak{A}^* \otimes \mathfrak{B}^*$ as a subspace of $(\mathfrak{A} \otimes_{\min} \mathfrak{B})^*$.

Exercise 6.3.7 Let $\mathfrak{M}$ and $\mathfrak{N}$ be W^*-algebras. Show that:

(i) $\mathfrak{M}_* \otimes \mathfrak{N}_*$ is an $\mathfrak{M} \otimes \mathfrak{N}$-submodule of $(\mathfrak{M} \otimes_{\min} \mathfrak{N})^*$.
(ii) $(\mathfrak{M}_* \otimes \mathfrak{N}_*)^\perp$ is a w^*-closed ideal of $(\mathfrak{M} \otimes_{\min} \mathfrak{N})^{**}$.

Definition 6.3.14 Let $\mathfrak{M}$ and $\mathfrak{N}$ be W^*-algebras. Then the W^*-*tensor product* $\mathfrak{M} \bar{\otimes}_{W^*} \mathfrak{N}$ of $\mathfrak{M}$ and $\mathfrak{N}$ is defined as $(\mathfrak{M} \otimes_{\min} \mathfrak{N})^{**}/(\mathfrak{M}_* \otimes \mathfrak{N}_*)^\perp$.

In practice, it is often convenient to have a more concrete description of W^*-tensor products at hand:

Proposition 6.3.15 *Let $\mathfrak{M}$ and $\mathfrak{N}$ be W^*-algebras, and let π and ρ be W^*-representations of $\mathfrak{M}$ and $\mathfrak{N}$ on Hilbert spaces $\mathfrak{H}$ and $\mathfrak{K}$, respectively. Then $\pi \otimes \rho$ extends uniquely to a W^*-representation $\pi \bar{\otimes}_{W^*} \rho$ of $\mathfrak{M} \bar{\otimes}_{W^*} \mathfrak{N}$ on $\mathfrak{H} \bar{\otimes} \mathfrak{K}$. Moreover, if both π and ρ are faithful, then so is $\pi \bar{\otimes}_{W^*} \rho$.*

Proof Let $\| \cdot \|_{\min^*}$ denote the norm on $\mathfrak{M}_* \otimes \mathfrak{N}_*$ it inherits as a subspace of $(\mathfrak{M} \otimes_{\min} \mathfrak{N})^*$, and write $\mathfrak{M}_* \tilde{\otimes}_{\min^*} \mathfrak{N}_*$ for the completion of $\mathfrak{M}_* \otimes \mathfrak{N}_*$ with respect to $\| \cdot \|_{\min^*}$.

It is not hard to see that $\pi \otimes \rho \colon \mathfrak{M} \otimes \mathfrak{N} \to \mathcal{L}(\mathfrak{H} \tilde{\otimes} \mathfrak{K})$ is continuous with respect to $\| \cdot \|_{\min}$ (you should have seen this already when doing Exercise 6.3.4). We claim that $(\pi \otimes \rho)^*$ maps $\mathcal{L}(\mathfrak{H} \tilde{\otimes} \mathfrak{K})_*$ into the completion of $\mathfrak{M}_* \otimes \mathfrak{N}_*$ with respect to $\| \cdot \|_{\min^*}$. Let $\phi \in \mathcal{L}(\mathfrak{H} \tilde{\otimes} \mathfrak{K})_*$; without loss of generality (Why?) suppose that there are $\xi_1, \xi_2 \in \mathfrak{H}$ and $\eta_1, \eta_2 \in \mathfrak{K}$ such that

$$\langle T, \phi \rangle = \langle T(\xi_1 \otimes \xi_2), \eta_1 \otimes \eta_2 \rangle \qquad (T \in \mathcal{L}(\mathfrak{H} \tilde{\otimes} \mathfrak{K})).$$

Define $\phi_{\mathfrak{M}} \in \mathfrak{M}^*$ and $\phi_{\mathfrak{N}} \in \mathfrak{N}^*$ through

$$\langle T, \phi_{\mathfrak{M}} \rangle := \langle \pi(T)\xi_1, \xi_2 \rangle \quad (T \in \mathfrak{M}) \qquad \text{and} \qquad \langle T, \phi_{\mathfrak{N}} \rangle := \langle \rho(T)\eta_1, \eta_2 \rangle \quad (T \in \mathfrak{N}).$$

It is clear that $\phi_{\mathfrak{M}} \in \mathfrak{M}_*$, $\phi_{\mathfrak{N}} \in \mathfrak{N}_*$, and $(\pi \otimes \rho)^* \phi = \phi_{\mathfrak{M}} \otimes \phi_{\mathfrak{N}}$. Letting

$$\pi \tilde{\otimes}_{W^*} \rho := ((\pi \otimes \rho)^*|_{\mathcal{L}(\mathfrak{H} \tilde{\otimes} \mathfrak{K})_*})^*,$$

we obtain a (necessarily unique) W^*-representation of $\mathfrak{M} \tilde{\otimes}_{W^*} \mathfrak{N}$ that extends $\pi \otimes \rho$.

Suppose that π and ρ are faithful. Let $\mathfrak{M}_* \tilde{\otimes}_{\min^*} \mathfrak{N}_*$ denote the completion of $\mathfrak{M}_* \otimes \mathfrak{N}_*$ in $(\mathfrak{M} \otimes_{\min} \mathfrak{N})^*$. From the foregoing, it follows that $(\pi \otimes \rho)^*$ maps $\mathcal{L}(\mathfrak{H} \tilde{\otimes} \mathfrak{K})_*$ into a norm dense subset of $\mathfrak{M}_* \tilde{\otimes}_{\min^*} \mathfrak{N}_*$, so that $\pi \tilde{\otimes}_{W^*} \rho$ is necessarily faithful (Where did we need here that π and ρ are injective?). $\square$

Corollary 6.3.16 *Let $\mathfrak{M}$ and $\mathfrak{N}$ be W^*-algebras, and let π and ρ be faithful W^*-representations of $\mathfrak{M}$ and $\mathfrak{N}$, respectively. Then we have*

$$\mathfrak{M} \tilde{\otimes}_{W^*} \mathfrak{N} \cong (\pi \otimes \rho)(\mathfrak{M} \otimes N)''$$

as W^*-algebras.

Exercise 6.3.8 Let G and H be locally compact groups. Show that

$$VN(G) \tilde{\otimes}_{W^*} VN(H) \cong VN(G \times H).$$

(*Hint*: Exercises B.3.5 and A.3.6.)

From the definition of the W^*-tensor product, we immediately get a hereditary property for Connes-amenability:

Proposition 6.3.17 *Let $\mathfrak{M}$ and $\mathfrak{N}$ be Connes-amenable W^*-algebras. Then $\mathfrak{M} \tilde{\otimes}_{W^*} \mathfrak{N}$ is Connes-amenable.*

Proof Let E^* be a normal, dual Banach $\mathfrak{M} \tilde{\otimes}_{W^*} \mathfrak{N}$-bimodule, and let $D \colon \mathfrak{M} \tilde{\otimes}_{W^*} \mathfrak{N} \to E^*$ be a w^*-continuous derivation. We may suppose that E^* is unital. The W^*-subalgebras $\mathfrak{M} \otimes e_{\mathfrak{N}}$ and $e_{\mathfrak{M}} \otimes \mathfrak{N}$ of $\mathfrak{M} \tilde{\otimes}_{W^*} \mathfrak{N}$ are canonically isomorphic to $\mathfrak{M}$ and $\mathfrak{N}$, respectively, and thus Connes-amenable. The Connes-amenability of $e_{\mathfrak{M}} \otimes \mathfrak{N}$ yields $\phi \in E^*$ such that $D|_{e_{\mathfrak{M}} \otimes \mathfrak{N}} = \mathrm{ad}_\phi|_{e_{\mathfrak{M}} \otimes \mathfrak{N}}$. Let $\tilde{D} := D - \mathrm{ad}_\phi$. It easy to see that $\tilde{D}$ attains its values in $\mathcal{Z}^0(e_{\mathfrak{M}} \otimes \mathfrak{N}, E^*)$. Since $e_{\mathfrak{M}} \otimes \mathfrak{N}$ and $\mathfrak{M} \otimes e_{\mathfrak{N}}$ commute with each other, $\mathcal{Z}^0(e_{\mathfrak{M}} \otimes \mathfrak{N}, E^*)$ is a w^*-closed $\mathfrak{M} \otimes e_{\mathfrak{N}}$-submodule of E^*. Since $\mathfrak{M} \otimes e_{\mathfrak{N}}$ is Connes-amenable, there is $\psi \in \mathcal{Z}^0(e_{\mathfrak{M}} \otimes \mathfrak{N}, E^*)$ such that $\tilde{D}|_{\mathfrak{M} \otimes e_{\mathfrak{N}}} = \mathrm{ad}_\psi|_{\mathfrak{M} \otimes e_{\mathfrak{N}}}$. It follows that $D|_{\mathfrak{M} \otimes \mathfrak{N}} = \mathrm{ad}_{\phi + \psi}|_{\mathfrak{M} \otimes \mathfrak{N}}$. By w^*-continuity, we have $D = \mathrm{ad}_{\phi + \psi}$. $\square$

Together with the following hereditary property, Proposition 6.3.17 enables us to enlarge our stock of examples of injective W^*-algebras:

Example 6.3.18 Let $\mathfrak{M}$ be a W^*-algebra of type I. We claim that $\mathfrak{M}$ is injective. First, note that, by [K–R, 6.5.2. Theorem] and Exercise 6.2.5, we may suppose that $\mathfrak{M}$ is of type I_α for some cardinal number α. By [K–R, 6.5.5. Theorem], $\mathfrak{M}$ is isomorphic to $Z(\mathfrak{M})\bar{\otimes}_{W^*}\mathcal{L}(\mathfrak{H})$, where $\mathfrak{H}$ is a Hilbert space of dimension α. Since both $Z(\mathfrak{M})$ and $\mathcal{L}(\mathfrak{H}) \cong \mathcal{K}(\mathfrak{H})^{**}$ are Connes-amenable, it follows that $\mathfrak{M}$ is Connes-amenable and thus injective by Corollary 6.2.8.

6.4 Semidiscrete W^*-algebras

Let E and F be linear spaces, and let $n \in \mathbb{N}$. There is an obvious way of amplifying a linear map $T: E \to F$ to a linear map $T^{(n)}: \mathbb{M}_n(E) \to \mathbb{M}_n(F)$: just apply T to each entry of a matrix in $\mathbb{M}_n(E)$. If we identify $\mathbb{M}_n(E)$ and $\mathbb{M}_n(F)$ with $\mathbb{M}_n \otimes E$ and $\mathbb{M}_n \otimes F$, respectively (as in Exercise B.1.6(i)), we see that $T^{(n)} = \mathrm{id}_{\mathbb{M}_n} \otimes T$.

Let $\mathfrak{A}$ be a C^*-algebra, and let $n \in \mathbb{N}$. As we have seen in Exercise 6.1.5, $\mathbb{M}_n(\mathfrak{A})$ carries a unique C^*-algebra norm. In particular, each subspace of $\mathbb{M}_n(\mathfrak{A})$ inherits a canonical order structure from $\mathbb{M}_n(\mathfrak{A})$. Let $A = [a_{j,k}]_{\substack{j=1,\dots,n \\ k=1,\dots,n}} \in \mathbb{M}_n(\mathfrak{A})$ and $\Phi = [\phi_{j,k}]_{\substack{j=1,\dots,n \\ k=1,\dots,n}} \in \mathbb{M}_n(\mathfrak{A}^*)$, and define

$$\langle A, \Phi \rangle := \sum_{j,k=1}^n \langle a_{j,k}, \phi_{j,k}\rangle. \tag{6.15}$$

Then (6.15) induces an isomorphism between $\mathbb{M}_n(\mathfrak{A}^*)$ and $\mathbb{M}_n(\mathfrak{A})^*$. Via this isomorphism, every subspace of $\mathbb{M}_n(\mathfrak{A}^*)$ also inherits a canonical order structure.

This makes the following definition meaningful:

Definition 6.4.1 Let E and F be linear spaces such that is either a subspace of a C^*-algebra or of the dual of a C^*-algebra. A linear map $T: E \to F$ is called *completely positive* if $T^{(n)}: \mathbb{M}_n(E) \to \mathbb{M}_n(F)$ is positive for each $n \in \mathbb{N}$, i.e. maps positive elements to positive elements.

Exercise 6.4.1 Let $\mathfrak{A}$ be a C^*-aglebra, and let $n \in \mathbb{N}$.

(i) Show that an element of $\mathbb{M}_n(\mathfrak{A})$ is positive if and only if it is a finite linear combination of matrices $[a_j^* a_k]_{\substack{j=1,\dots,n \\ k=1,\dots,n}}$, where $a_1,\dots,a_n \in \mathfrak{A}$.

(ii) Let E be a linear subspace of a C^*-algebra or of the dual of a C^*-algebra. Conclude that a linear map $T: \mathfrak{A} \to E$ is completely positive if and only if $[T(a_j^* a_k)]_{\substack{j=1,\dots,n \\ k=1,\dots,n}}$ is positive in $\mathbb{M}_n(E)$, for all $n \in \mathbb{N}$ and for all $a_1,\dots,a_n \in \mathfrak{A}$.

Examples 6.4.2   (a) Let $\mathfrak{A}$ and $\mathfrak{B}$ be C^*-algebras, and let $\pi: \mathfrak{A} \to \mathfrak{B}$ be a $*$-homomorphism. Then $\pi^{(n)}: \mathbb{M}_n(\mathfrak{A}) \to \mathbb{M}_n(\mathfrak{B})$ is, for each $n \in \mathbb{N}$, also a $*$-homomorphism and thus positive. It follows that π is completely positive.

(b) Let $\mathfrak{A}$ be a unital C^*-algebra, let $\mathfrak{B}$ be a unital C^*-subalgebra of $\mathfrak{A}$, and let $\mathcal{E}: \mathfrak{A} \to \mathfrak{B}$ be an expectation. Let $n \in \mathbb{N}$. Then $\mathbb{M}_n(\mathfrak{B})$ is a unital C^*-subalgebra of $\mathbb{M}_n(\mathfrak{A})$, and $\mathcal{E}^{(n)}: \mathbb{M}_n(\mathfrak{A}) \to \mathbb{M}_n(\mathfrak{B})$ is a projection onto $\mathbb{M}_n(\mathfrak{B})$. Let $\phi \in S(\mathbb{M}_n)$, and let $\psi \in S(\mathfrak{A})$. Then $(\mathrm{id}_{\mathbb{M}_n} \otimes \mathcal{E})^*(\phi \otimes \psi) = \phi \otimes \mathcal{E}^*\psi$, and $\mathcal{E}^*\psi \in S(\mathfrak{A})$. It follows from Exercise 6.3.1 that $\mathcal{E}^{(n)} = \mathrm{id}_{\mathbb{M}_n} \otimes \mathcal{E}$ has norm one (with respect to $\|\cdot\|_{\min}$ and thus with respect to every

C^*-norm on $\mathbb{M}_n(\mathfrak{A})$). Hence, $\mathcal{E}^{(n)}: \mathbb{M}_n(\mathfrak{A}) \to \mathbb{M}_n(\mathfrak{B})$ is an expectation. By Theorem 6.2.4(i), $\mathcal{E}^{(n)}$ is positive. Since $n \in \mathbb{N}$ is arbitrary, $\mathcal{E}$ is completely positive.

(c) Let $\mathfrak{A}$ be a unital C^*-algebra, and let $\phi \in S(\mathfrak{A})$. Then ϕ can be viewed as an expectation from $\mathfrak{A}$ onto $\mathbb{C}e_{\mathfrak{A}}$. From the previous example, it follows that ϕ is completely positive.

(d) The transpose map

$$T: \mathbb{M}_2 \to \mathbb{M}_2, \quad A \mapsto A^t$$

is clearly positive. Consider

$$T^{(2)} \left(\begin{bmatrix} 1 & 0 & 0 & 1 \\ 0 & 0 & 0 & 0 \\ 0 & 0 & 0 & 0 \\ 1 & 0 & 0 & 1 \end{bmatrix} \right) = \begin{bmatrix} 1 & 0 & 0 & 0 \\ 0 & 0 & 1 & 0 \\ 0 & 1 & 0 & 0 \\ 0 & 0 & 0 & 1 \end{bmatrix}.$$

The matrix on the left hand side of this equation is positive whereas the one on the right hand side is not. Hence, T is not completely positive.

Exercise 6.4.2 Let $\mathfrak{A}$ and $\mathfrak{B}$ be C^*-algebras, let $S: \mathfrak{A} \to \mathfrak{B}$ be completely positive, and let $p \in \mathfrak{A}$ and $q \in \mathfrak{B}$ be projections. Show that

$$T: \mathfrak{A} \to \mathfrak{B}, \quad a \mapsto qS(pap)q$$

is completely positive.

Definition 6.4.3 A W^*-algebra $\mathfrak{M}$ is called *semidiscrete* if there is a net $(S_\alpha)_\alpha$ of unit preserving, completely positive, w^*-continuous maps in $\mathcal{F}(\mathfrak{M})$ such that

$$\langle \phi, x \rangle = \lim_\alpha \langle \phi, S_\alpha x \rangle \qquad (x \in \mathfrak{M}, \ \phi \in \mathfrak{M}_*).$$

We'll need the following hereditary properties later on:

Exercise 6.4.3 Let $\mathfrak{M}$ be a semidiscrete W^*-algebra, and let $p \in \mathfrak{M}$ be a projection. Show that $p\mathfrak{M}p$ is semidiscrete as well.

Exercise 6.4.4 Let $(\mathfrak{M}_\alpha)_\alpha$ be a family of semidiscrete W^*-algebras. Show that $\ell^\infty\text{-}\bigoplus_\alpha \mathfrak{M}_\alpha$ is a semidiscrete W^*-algebra.

Exercise 6.4.5 Let $\mathfrak{M}$ be a semidiscrete von Neumann algebra acting on a Hilbert space $\mathfrak{H}$. Show that $\mathfrak{M}'$ is also semidiscrete. (*Hint:* How did you solve the corresponding problem for injectivity?).

The definition of semidiscreteness bears some resemblance to Definition C.1.1. For the predual of a semidiscrete W^*-algebra, we actually obtain the approximation property:

Proposition 6.4.4 *Let $\mathfrak{M}$ be a semidiscrete W^*-algebra. Then $\mathfrak{M}_*$ has the metric approximation property.*

Proof Let $(S_\alpha)_\alpha$ be a net as in Definition 6.4.3. Since each S_α is positive and unit preserving, each S_α has norm one. Since each S_α is w^*-continuous, it follows that $(S_\alpha^*|_{\mathfrak{M}_*})_\alpha$ is a net in $\mathcal{F}(\mathfrak{M}_*)$ that converges to $\mathrm{id}_{\mathfrak{M}_*}$ in the weak operator topology. Passing to convex combinations, we obtain a net of operators in $\mathcal{F}(\mathfrak{M}_*)$ bounded by one that converges to $\mathrm{id}_{\mathfrak{M}_*}$ in the strong operator topology. $\square$

Together with Exercise C.3.1, Proposition 6.4.4 yields:

Corollary 6.4.5 *Let $\mathfrak{A}$ be a C^*-algebra such that $\mathfrak{A}^{**}$ is semidiscrete. Then both $\mathfrak{A}^*$ and $\mathfrak{A}$ have the metric approximation property.*

We shall see in this section that injectivity is the same as semidiscreteness. We begin with the proof of

$$\text{semidiscreteness} \implies \text{injectivity}.$$

We leave some of the preparations to the reader:

Exercise 6.4.6 Let $\mathfrak{A}$ and $\mathfrak{B}$ be unital C^*-algebras. For $\phi \in (\mathfrak{A} \tilde{\otimes}_{\max} \mathfrak{B})^*$, define $T_\phi : \mathfrak{B} \to \mathfrak{A}^*$ through

$$\langle a, T_\phi b \rangle := \langle a \otimes b, \phi \rangle \qquad (a \in \mathfrak{A}, \, b \in \mathfrak{B}).$$

(i) Show that $\phi \in S(\mathfrak{A} \tilde{\otimes}_{\max} \mathfrak{B})$ if and only if T_ϕ is completely positive and $T_\phi e_{\mathfrak{B}} \in S(\mathfrak{A})$.

(ii) Conversely, let $T : \mathfrak{B} \to \mathfrak{A}^*$ be completely positive with $Te_{\mathfrak{B}} \in S(\mathfrak{A})$. Show that there is $\phi \in S(\mathfrak{A} \tilde{\otimes}_{\max} \mathfrak{B})$ such that $T = T_\phi$.

(iii) Let $T \in \mathcal{F}(\mathfrak{B}, \mathfrak{A}^*)$ be completely positive with $Te_{\mathfrak{B}} \in S(\mathfrak{A})$. Show that there is ϕ in the convex hull of $\{\psi_{\mathfrak{A}} \otimes \psi_{\mathfrak{B}} : \psi_{\mathfrak{A}} \in S(\mathfrak{A}), \, \psi_{\mathfrak{B}} \in S(\mathfrak{B})\}$ such that $T = T_\phi$.

(iv) Show that $\phi \in S(\mathfrak{A} \tilde{\otimes}_{\min} \mathfrak{B})$ if and only if there is a net of completely positive maps $(T_\alpha)_\alpha$ in $\mathcal{F}(\mathfrak{B}, \mathfrak{A}^*)$ with $T_\alpha e_{\mathfrak{B}} \in S(\mathfrak{A})$ such that

$$\langle a, T_\phi b \rangle = \lim_\alpha \langle a, T_\alpha b \rangle \qquad (a \in \mathfrak{A}, \, b \in \mathfrak{B}).$$

(*Hint:* The convex hull of $\{\psi_{\mathfrak{A}} \otimes \psi_{\mathfrak{B}} : \psi_{\mathfrak{A}} \in S(\mathfrak{A}), \, \psi_{\mathfrak{B}} \in S(\mathfrak{B})\}$ is w^*-dense in $S(\mathfrak{A} \tilde{\otimes}_{\min} \mathfrak{B})$.)

Definition 6.4.6 Let $\mathfrak{M}$ be a W^*-algebra, and let $\mathfrak{B}$ be a unital C^*-algebra.

(i) Let

$$S_{\text{nor}}(\mathfrak{M} \otimes \mathfrak{B}) := \{\phi \in S(\mathfrak{M} \tilde{\otimes}_{\max} \mathfrak{N}) : T_\phi \mathfrak{B} \subset \mathfrak{M}_*\}.$$

(ii) Let

$$\|\mathbf{x}\|_{\text{nor}} := \sup \left\{ \sqrt{\langle \mathbf{x}^* \bullet \mathbf{x}, \phi \rangle} : \phi \in S_{\text{nor}}(\mathfrak{M} \otimes \mathfrak{B}) \right\} \qquad (\mathbf{x} \in \mathfrak{M} \otimes \mathfrak{B}).$$

We write $\mathfrak{M} \otimes_{\text{nor}} \mathfrak{B}$ for $\mathfrak{M} \otimes \mathfrak{B}$ equipped with $\| \cdot \|_{\text{nor}}$ and $\mathfrak{M} \tilde{\otimes}_{\text{nor}} \mathfrak{B}$ for the completion of $\mathfrak{M} \otimes_{\text{nor}} \mathfrak{B}$.

Exercise 6.4.7 Let $\mathfrak{M}$ be a W^*-algebra, and let $\mathfrak{B}$ be a unital C^*-algebra. Show that $\| \cdot \|_{\text{nor}}$ is a C^*-norm on $\mathfrak{M} \otimes \mathfrak{B}$.

Proposition 6.4.7 *Let $\mathfrak{M}$ be a semidiscrete W^*-algebra. Then, for each unital C^*-algebra $\mathfrak{B}$, the identity map*

$$\mathfrak{M} \otimes_{\text{nor}} \mathfrak{B} \to \mathfrak{M} \otimes_{\min} \mathfrak{B}$$

is an isometry.

Proof Let $\phi \in S_{\mathrm{nor}}(\mathfrak{M} \otimes \mathfrak{B})$; it is sufficient to show that $\phi \in S(\mathfrak{M} \tilde{\otimes}_{\min} \mathfrak{B})$.

Let $T_\phi : \mathfrak{B} \to \mathfrak{M}^*$ be defined as in Exercise 6.4.6; note that $T_\phi \mathfrak{B} \subset \mathfrak{M}_*$ because $\phi \in S_{\mathrm{nor}}(\mathfrak{M} \otimes \mathfrak{B})$. Let $(S_\alpha)_\alpha$ be a net as in Definition 6.4.3, and let $T_\alpha := S_\alpha^* T_\phi$. Then $(T_\alpha)_\alpha$ is a net of completely positive maps in $\mathcal{F}(\mathfrak{B}, \mathfrak{M}_*)$ with $T_\alpha e_{\mathfrak{B}} \in S(\mathfrak{M})$ such that

$$\langle x, T_\phi b \rangle = \lim_\alpha \langle x, T_\alpha b \rangle \qquad (x \in \mathfrak{M}, \, b \in \mathfrak{B}). \tag{6.16}$$

By Exercise 6.4.6(iii), there is a net $(\phi_\alpha)_\alpha$ in $S(\mathfrak{A} \tilde{\otimes}_{\min} \mathfrak{B})$ such that $T_\alpha = T_{\phi_\alpha}$. It is then immediate from (6.16) that ϕ is the w^*-limit of $(\phi_\alpha)_\alpha$ in $(\mathfrak{M} \tilde{\otimes}_{\mathrm{nor}} \mathfrak{B})^*$.

Due to the minimality of $\| \cdot \|_{\min}$, the C^*-algebra $\mathfrak{M} \tilde{\otimes}_{\min} \mathfrak{B}$ is a quotient of $\mathfrak{M} \tilde{\otimes}_{\mathrm{nor}} \mathfrak{B}$; let I denote the kernel of the quotient map. Each ϕ_α vanishes on I, and so does ϕ. It follows that $\phi \in S(\mathfrak{M} \tilde{\otimes}_{\min} \mathfrak{B})$. $\square$

As an immediate consequence of Proposition 6.4.7, we obtain a first connection between semidiscreteness and nuclearity:

Corollary 6.4.8 *Let $\mathfrak{A}$ be a C^*-algebra such that $\mathfrak{A}^{**}$ is semidiscrete. Then $\mathfrak{A}$ is nuclear.*

Proof Let $\mathfrak{B}$ be a C^*-algebra. Without loss of generality suppose that both $\mathfrak{A}$ and $\mathfrak{B}$ are unital (Why may we do this?). Let $\phi \in S(\mathfrak{A} \tilde{\otimes}_{\max} \mathfrak{B})$ it suffices to show that that $\phi \in S(\mathfrak{A} \tilde{\otimes}_{\min} \mathfrak{B})$. By w^*-continuity, ϕ has a unique extension $\tilde{\phi} \in S_{\mathrm{nor}}(\mathfrak{A}^{**} \otimes \mathfrak{B})$, and by Proposition 6.4.7, $\tilde{\phi} \in S(\mathfrak{A}^{**} \tilde{\otimes}_{\min} \mathfrak{B})$. From Exercise 6.3.5, we conclude that $\phi \in S(\mathfrak{A} \tilde{\otimes}_{\min} \mathfrak{B})$. $\square$

A second consequence of Proposition 6.4.7 is even more immediate:

Corollary 6.4.9 *Let $\mathfrak{M}$ be a semidiscrete W^*-algebra. Then, for each unital C^*-algebra $\mathfrak{B}$ and for each unital C^*-subalgebra $\mathfrak{C}$ of $\mathfrak{B}$, the inclusion map*

$$\mathfrak{M} \otimes_{\mathrm{nor}} \mathfrak{C} \hookrightarrow \mathfrak{M} \otimes_{\mathrm{nor}} \mathfrak{B}$$

is an isometry.

Proof This follows at once from Exercise 6.3.5 and Proposition 6.4.7. $\square$

We shall see next that the conclusion of Corollary 6.4.9 characterizes the injective W^*-algebras (so that semidiscrete W^*-algebras are injective).

Lemma 6.4.10 *Let $\mathfrak{A}$ be a C^*-algebra acting on a Hilbert space $\mathfrak{H}$ such that there is a cyclic unit vector $\xi \in \mathfrak{H}$ for $\mathfrak{A}$. Let ϕ_ξ denote the vector state given by ξ, let E be the linear span of $\{\phi \in \mathfrak{A}^* : 0 \leq \phi \leq \phi_\xi\}$ in $\mathfrak{A}^*$, and let $\theta : \mathfrak{A}' \to E$ be given by*

$$\langle a, \theta(x) \rangle := \langle x a \xi, \xi \rangle \qquad (x \in \mathfrak{A}', \, a \in \mathfrak{A}).$$

Then θ is a completely positive bijection with a completely positive inverse.

Proof It follows immediately from [Dixm 3, 2.5.1. Proposition] or [Ped, 3.3.5. Proposition] that $\theta : \mathfrak{A}' \to E$ is a positive bijection.

Let $x_1, \ldots, x_n \in \mathfrak{A}'$ and let $a_1, \ldots, a_n \in \mathfrak{A}$. Then we have for $\Theta := [\theta(x_j^* x_k)]_{\substack{j=1,\ldots,n \\ k=1,\ldots,n}}$ and $A = [a_j^* a_k]_{\substack{j=1,\ldots,n \\ k=1,\ldots,n}}$:

$$\langle A, \Theta \rangle = \sum_{j,k=1}^{n} \langle a_j^* a_k, \theta(x_j^* x_k) \rangle$$

$$= \sum_{j,k=1}^{n} \langle x_j^* x_k a_j^* a_k \xi, \xi \rangle$$

$$= \left\| \sum_{j=1}^{n} x_j a_j \xi \right\|^2$$

$$\geq 0.$$

By Exercise 6.4.1(i), this is sufficient to conclude that θ is completely positive.

To show that θ^{-1} is completely positive, let $\Phi = [\phi_{j,k}]_{\substack{j=1,\ldots,n \\ k=1,\ldots,n}} \in \mathbb{M}_n(\mathfrak{A}^*)$ be positive. For $a_1, \ldots, a_n$, let $\Xi := (a_1 \xi, \ldots, a_n \xi) \in \mathfrak{H}^n$. Since θ^{-1} maps into $\mathfrak{A}'$, we have:

$$\langle ((\theta^{-1})^{(n)} \Phi) \Xi, \Xi \rangle = \sum_{j,k=1}^{n} \langle \theta^{-1}(\phi_{j,k}) a_j \xi, a_k \xi \rangle$$

$$= \sum_{j,k=1}^{n} \langle \theta(\phi_{j,k}) a_j^* a_k \xi, \xi \rangle$$

$$= \sum_{j,k=1}^{n} \langle a_j^* a_k, \phi_{j,k} \rangle$$

$$\geq 0.$$

Since $a_1, \ldots, a_n \in \mathfrak{A}$ were arbitrary, and since $\xi \in \mathfrak{H}$ is cyclic for $\mathfrak{A}$, this implies that $(\theta^{-1})^{(n)} \Phi$ is positive in $\mathbb{M}_n(\mathfrak{A}')$.   □

Exercise 6.4.8 Let $\mathfrak{M}$ be a von Neumann algebra acting on Hilbert space $\mathfrak{H}$ such that there is a cyclic unit vector $\xi \in \mathfrak{H}$ for $\mathfrak{A}$. Show that the map θ defined in Lemma 6.4.10 attains its values in $\mathfrak{M}_*$.

Remark 6.4.11 In general, θ^{-1} need not be continuous.

Lemma 6.4.12 *For a W^*-algebra $\mathfrak{M}$, the following are equivalent:*

(i) *For each unital C^*-algebra $\mathfrak{B}$ and for each unital C^*-subalgebra $\mathfrak{C}$ of $\mathfrak{B}$, the inclusion map*

$$\mathfrak{M} \otimes_{\mathrm{nor}} \mathfrak{C} \hookrightarrow \mathfrak{M} \otimes_{\mathrm{nor}} \mathfrak{B}$$

is an isometry.

(ii) *For each unital C^*-algebra $\mathfrak{B}$ and for each unital C^*-subalgebra $\mathfrak{C}$ of $\mathfrak{B}$, the restriction map*

$$S_{\mathrm{nor}}(\mathfrak{M} \otimes \mathfrak{B}) \to S_{\mathrm{nor}}(\mathfrak{M} \otimes \mathfrak{C})$$

is surjective.

(iii) *For each unital C^*-algebra $\mathfrak{B}$ and for each unital C^*-subalgebra $\mathfrak{C}$ of $\mathfrak{B}$, every completely positive map $T : \mathfrak{C} \to \mathfrak{M}_*$ with $T e_{\mathfrak{B}} \in S(\mathfrak{M})$ has a completely positive extension to $\mathfrak{B}$.*

Proof The equivalence of (i) and (ii) is straightforward (Do you believe this?).

(ii) $\implies$ (iii): Let $\mathfrak{B}$ be a unital C^*-algebra, let $\mathfrak{C}$ be a unital C^*-subalgebra of $\mathfrak{B}$, and let $T \colon \mathfrak{C} \to \mathfrak{M}_*$ be completely positive. By Exercise 6.4.6(ii), there is $\phi \in S(\mathfrak{M} \tilde{\otimes}_{\max} \mathfrak{C})$ such that $T = T_\phi$. Since $T\mathfrak{C} \subset \mathfrak{M}_*$, it follows that $\phi \in S_{\mathrm{nor}}(\mathfrak{M} \otimes \mathfrak{C})$. By (ii), there is $\psi \in S_{\mathrm{nor}}(\mathfrak{M} \otimes \mathfrak{B})$ such that $\psi|_{\mathfrak{M} \otimes \mathfrak{C}} = \phi$. Then $T_\psi \colon \mathfrak{B} \to \mathfrak{M}_*$ is completely positive and extends T.

(iii) $\implies$ (ii): Let $\mathfrak{B}$ be a unital C^*-algebra, let $\mathfrak{C}$ be a unital C^*-subalgebra of $\mathfrak{B}$, and let $\phi \in S_{\mathrm{nor}}(\mathfrak{M} \otimes \mathfrak{C})$. Then $T_\phi \colon \mathfrak{C} \to \mathfrak{M}_*$ is completely positive and thus has a completely positive extension $T \colon \mathfrak{B} \to \mathfrak{M}_*$. Again by Exercise 6.4.6(ii), T is of the form T_ψ for some $\psi \in S_{\mathrm{nor}}(\mathfrak{M} \otimes \mathfrak{B})$. It is clear that ψ extends ϕ. $\square$

Theorem 6.4.13 *Let $\mathfrak{M}$ be a W^*-algebra such that, for each unital C^*-algebra $\mathfrak{B}$ and for each unital C^*-subalgebra $\mathfrak{C}$ of $\mathfrak{B}$, the inclusion map*

$$\mathfrak{M} \otimes_{\mathrm{nor}} \mathfrak{C} \hookrightarrow \mathfrak{M} \otimes_{\mathrm{nor}} \mathfrak{B}$$

is an isometry. Then $\mathfrak{M}$ is injective.

Proof Certainly, $\mathfrak{M}$ satisfies Lemma 6.4.12(i). Fix $\phi \in S(\mathfrak{M}) \cap \mathfrak{M}_*$. Through the GNS-construction, we obtain a Hilbert space $\mathfrak{H}_\phi$ and a W^*-representation π_ϕ of $\mathfrak{M}$ on $\mathfrak{H}_\phi$. The von Neumann algebra $\pi_\phi(\mathfrak{M})$ then has a cyclic vector in $\mathfrak{H}_\phi$ by construction. Let $\mathfrak{B} := \mathcal{L}(\mathfrak{H}_\phi)$, let $\mathfrak{C} := \pi_\phi(\mathfrak{M})'$, and let $\theta \colon \mathfrak{C} \to \mathfrak{M}^*$ be as in Lemma 6.4.10; since $\phi \in \mathfrak{M}_*$, it follows that θ attains its values in $\mathfrak{M}_*$. Since θ is completely positive by Lemma 6.4.10, it has a completely positive extension $T \colon \mathfrak{B} \to \mathfrak{M}_*$. Let $b \in \mathfrak{B}$ be such that $0 \le b \le e_{\mathfrak{B}}$. Since T is positive, $0 \le Tb \le Te_{\mathfrak{B}} = \theta(e_{\mathfrak{B}}) = \phi$. By Lemma 6.4.10, this means that $T\mathfrak{B} = \theta(\mathfrak{C})$, so that $\mathcal{E} := \theta^{-1} \circ T$ is well defined. Clearly, $\mathcal{E}$ is a (completely) positive projection onto $\pi_\phi(\mathfrak{M})'$, so that, by Exercise 6.2.1, it is an expectation. Consequently, $\pi_\phi(\mathfrak{M})$ is injective.

Let $(p_\alpha)_\alpha$ be a maximal, orthogonal family of projections in $Z(\mathfrak{M})$ with the following property: For each p_α, there is $\phi_\alpha \in S(\mathfrak{M}) \cap \mathfrak{M}_*$ such that $\pi_{\phi_\alpha}|_{\mathfrak{M} p_\alpha}$ is an isomorphism (Why do such projections exist?). By Exercise 6.4.3, each W^*-algebra $\mathfrak{M} p_\alpha \cong \pi_{\phi_\alpha}(\mathfrak{M})$ is semidiscrete and thus injective by the foregoing. Since

$$\mathfrak{M} \cong \ell^\infty\text{-}\bigoplus_\alpha \mathfrak{M} p_\alpha,$$

$\mathfrak{M}$ is injective by Exercise 6.2.5. $\square$

Together, Corollary 6.4.9 and Theorem 6.4.13 yield:

Corollary 6.4.14 *Let $\mathfrak{M}$ be a semidiscrete W^*-algebra. Then $\mathfrak{M}$ is injective.*

As another consequence of Theorem 6.4.13, we obtain:

Corollary 6.4.15 *Let $\mathfrak{A}$ be a nuclear C^*-algebra. Then $\mathfrak{A}^{**}$ is injective.*

Proof Let $\mathfrak{B}$ be a unital C^*-algebra, and let $\phi \in S_{\mathrm{nor}}(\mathfrak{A}^{**} \otimes \mathfrak{B})$. It follows that ϕ is uniquely determined by its values on $\mathfrak{A} \otimes \mathfrak{B}$. Since $\mathfrak{A}$ is nuclear, $\phi|_{\mathfrak{A} \otimes \mathfrak{B}}$ is continuous with respect to $\|\cdot\|_{\min}$, so that $\phi \in S(\mathfrak{A}^{**} \tilde{\otimes}_{\min} \mathfrak{B})$. It follows that the identity map

$$\mathfrak{A}^{**} \otimes_{\mathrm{nor}} \mathfrak{B} \to \mathfrak{A}^{**} \otimes_{\min} \mathfrak{B}$$

is an isometry. The claim then is an immediate consequence of Exercise 6.3.5 and Theorem 6.4.13. □

So far, we know that, for a C^*-algebra $\mathfrak{A}$,

$$\mathfrak{A}^{**} \text{ semidiscrete} \implies \mathfrak{A} \text{ nuclear} \implies \mathfrak{A}^{**} \text{ injective.}$$

To close the circle, we now turn to proving

$$\text{injectivity} \implies \text{semidiscreteness.}$$

The following characterization (Theorem 6.4.17) will be useful. But first, we need a lemma:

Lemma 6.4.16 *Let $\mathfrak{M}$ be a W^*-algebra, let $\mathfrak{B}$ be a unital C^*-algebra, and let $\phi \in S(\mathfrak{M} \tilde{\otimes}_{\min} \mathfrak{B})$ such that $T_\phi e_\mathfrak{B} \in \mathfrak{M}_*$. Then there is a net $(\phi_\alpha)_\alpha$ in the convex hull of $\{\phi_\mathfrak{M} \otimes \phi_\mathfrak{B} : \phi_\mathfrak{M} \in S(\mathfrak{M}) \cap \mathfrak{M}_*, \, \phi_\mathfrak{B} \in S(\mathfrak{B})\}$ converging to ϕ in the w^*-topology such that $T_{\phi_\alpha} e_\mathfrak{B} = T_\phi$ for each ϕ_α.*

Proof Clearly, we may find a net $(\phi_\alpha)_\alpha$ in the convex hull of $\{\phi_\mathfrak{M} \otimes \phi_\mathfrak{B} : \phi_\mathfrak{M} \in S(\mathfrak{M}) \cap \mathfrak{M}_*, \, \phi_\mathfrak{B} \in S(\mathfrak{B})\}$ such that $\phi = w^*\text{-}\lim_\alpha \phi_\alpha$. We shall modify this net in such a way that the additional requirement is satisfied.

For the sake of simplicity, let $\psi := T_\phi e_\mathfrak{B}$ and $\psi_\alpha := T_{\phi_\alpha} e_\mathfrak{B}$. We immediately have that $\psi = w\text{-}\lim_\alpha \psi_\alpha$. Passing to convex combinations, we can choose $(\phi_\alpha)_\alpha$ in such a way that $\lim_\alpha \|\psi_\alpha - \psi\| = 0$. Let $\phi_\mathfrak{B} \in S(\mathfrak{B})$ be arbitrary, and let

$$\tilde{\phi}_\alpha := \phi_\alpha + |\psi_\alpha - \psi| \otimes \phi_\mathfrak{B}.$$

Then $(\tilde{\phi}_\alpha)_\alpha$ satisfies:

- $w^*\text{-}\lim_\alpha \tilde{\phi}_\alpha = \phi$,
- $T_{\tilde{\phi}_\alpha} e_\mathfrak{B} \geq \psi$, and
- $\lim_\alpha \|T_{\tilde{\phi}_\alpha} e_\mathfrak{B} - \psi\| = 0$.

Sakai's Radon–Nikodým theorem [Sak, Theorem 1.24.3], yields a net $(h_\alpha)_\alpha$ in $\mathfrak{M}$ with $0 \leq h_\alpha \leq e_\mathfrak{M}$ such that

$$\langle x, \psi \rangle = \langle h_\alpha x h_\alpha, T_{\tilde{\phi}_\alpha} e_\mathfrak{B} \rangle \qquad (x \in \mathfrak{M}).$$

Define a net $(\bar{\phi}_\alpha)_\alpha$ through

$$\langle x \otimes b, \bar{\phi}_\alpha \rangle = \langle h_\alpha x h_\alpha \otimes b, \tilde{\phi}_\alpha \rangle \qquad (x \in \mathfrak{M}, \, b \in \mathfrak{B}),$$

so that $T_{\bar{\phi}_\alpha} = \psi$; also, $(\bar{\phi}_\alpha)_\alpha$ is certainly a net of states contained in the right set. To see that $\phi = w^*\text{-}\lim_\alpha \bar{\phi}_\alpha = \phi$, note that $\bar{\phi}_\alpha \leq \tilde{\phi}_\alpha$. Let $x \in \mathfrak{M}$ be positive, and let $b \in \mathfrak{B}$ such that $0 \leq b \leq e_\mathfrak{B}$. Then

$$0 \leq \langle x \otimes b, \tilde{\phi}_\alpha - \bar{\phi}_\alpha \rangle \leq \langle x \otimes e_\mathfrak{B}, \tilde{\phi}_\alpha - \bar{\phi}_\alpha \rangle = \langle x, T_{\tilde{\phi}_\alpha} - \psi \rangle \to 0,$$

so that indeed $\phi = w^*\text{-}\lim_\alpha \bar{\phi}_\alpha$.

So, the net $(\phi_\alpha)_\alpha$ whose existence we wanted to prove is our $(\bar{\phi}_\alpha)_\alpha$ ... □

Theorem 6.4.17 *Let $\mathfrak{A}$ be a von Neumann algebra acting on a Hilbert space $\mathfrak{H}$. Then the following are equivalent:*

(i) *$\mathfrak{M}$ is semidiscrete.*

(ii) *The $*$-homomorphism*

$$\pi\colon \mathfrak{M} \otimes \mathfrak{M}' \to \mathcal{L}(\mathfrak{H}), \quad x \otimes y \mapsto xy$$

is continuous with respect to $\|\cdot\|_{\min}$.

Proof (i) $\Longrightarrow$ (ii): Let $\xi \in \mathfrak{H}$ be such that $\|\xi\| = 1$. Then

$$\phi\colon \mathfrak{M} \otimes \mathfrak{M}' \to \mathbb{C}, \quad x \otimes y \mapsto \langle xy\xi, \xi \rangle$$

belongs to $S_{\mathrm{nor}}(\mathfrak{M} \otimes \mathfrak{M}')$. It follows that π is continuous with respect to $\|\cdot\|_{\mathrm{nor}}$. Since $\|\cdot\|_{\min} = \|\cdot\|_{\mathrm{nor}}$ on $\mathfrak{M} \otimes \mathfrak{M}'$ by Corollary 6.4.9, the claim follows.

(ii) $\Longrightarrow$ (i): We first treat the case in which there is a cyclic unit vector $\xi \in \mathfrak{H}$ for $\mathfrak{M}$. Let $\phi_\xi \in \mathcal{L}(\mathfrak{H})_*$ be the vector state associated with ξ, and let $\phi := \pi^* \phi_\xi$; note that $T_\phi e_{\mathfrak{M}'} = \phi_\xi|_{\mathfrak{M}}$. Since π is continuous with respect to $\|\cdot\|_{\min}$, we have $\phi \in S(\mathfrak{M} \tilde{\otimes}_{\min} \mathfrak{M}')$. By Lemma 6.4.16, there is a net $(\phi_\alpha)_\alpha$ in the convex hull of $\{\phi_{\mathfrak{M}} \otimes \phi_{\mathfrak{M}'} : \phi_{\mathfrak{M}} \in S(\mathfrak{M}) \cap \mathfrak{M}_*, \, \phi_{\mathfrak{M}'} \in S(\mathfrak{M}')\}$ that converges to ϕ in the w^*-topology and satisfies $T_{\phi_\alpha} e_{\mathfrak{M}'} = \phi_\xi|_{\mathfrak{M}}$. For the corresponding w^*-continuous, completely positive maps $T_{\phi_\alpha}\colon \mathfrak{M}' \to \mathfrak{M}_*$ the w^*-convergence of $(\phi_\alpha)_\alpha$ to ϕ means

$$\langle T_\phi y, x \rangle = \lim_\alpha \langle T_{\phi_\alpha} y, x \rangle \qquad (x \in \mathfrak{M}' \, y \in \mathfrak{M}').$$

Since $T_{\phi_\alpha} e_{\mathfrak{M}'} = \phi_\xi|_{\mathfrak{M}}$, it follows that all T_{ϕ_α} along with T_ϕ attain their values in the linear span of $\{\psi \in \mathfrak{M}_* : 0 \le \psi \le \phi_\xi|_{\mathfrak{M}}\}$. Let $\theta\colon \mathfrak{M}' \to \mathfrak{M}_*$ be as defined in Lemma 6.4.10. It then makes sense to define $S_\alpha := \theta^{-1} \circ T_{\phi_\alpha}$. Clearly, each S_α is w^*-continuous, completely positive and unit preserving. Passing to a subnet, we may suppose that, for all $x \in \mathfrak{M}'$, the w^*-limit $Sx \in \mathfrak{M}'$ of $(S_\alpha x)_\alpha$ exists. Since ξ is cyclic for $\mathfrak{M}$, it is separating for $\mathfrak{M}'$. Since

$$\begin{aligned}
\langle (Sx)\xi, \xi \rangle &= \lim_\alpha \langle (S_\alpha x)\xi, \xi \rangle \\
&= \lim_\alpha \langle \theta^{-1}(T_{\phi_\alpha} x)\xi, \xi \rangle \\
&= \lim_\alpha \langle T_{\phi_\alpha} x, e_{\mathfrak{M}} \rangle \\
&= \langle T_\phi x, e_{\mathfrak{M}} \rangle \\
&= \langle x\xi, \xi \rangle \qquad (x \in \mathfrak{M}'),
\end{aligned}$$

we have $S = \mathrm{id}_{\mathfrak{M}'}$, and therefore $\mathfrak{M}'$ is semidiscrete. By Exercise 6.4.5, $\mathfrak{M}'' = \mathfrak{M}$ is also semidiscrete.

Now, let $\mathfrak{M}$ be arbitrary. Fix $\xi \in \mathfrak{H}$ with $\|\xi\| = 1$. Let $P \in \mathcal{L}(\mathfrak{H})$ be the orthogonal projection onto the closed linear subspace $\mathfrak{K} = \overline{\mathfrak{M}\xi}$ of $\mathfrak{H}$, so that $P \in \mathfrak{M}'$. Then $\xi \in \mathfrak{K}$ is cyclic for the von Neumann algebra $P\mathfrak{M}$ acting on $\mathfrak{K}$. The commutant of $P\mathfrak{M}$ in $\mathcal{L}(\mathfrak{K})$ is easily seen to be (isomorphic to) $P\mathfrak{M}'$. It is not hard (Really?) to see that the $*$-homomorphism

$$P\mathfrak{M} \otimes P\mathfrak{M}' \to \mathcal{L}(\mathfrak{K}), \quad x \otimes y \to xy$$

is continuous with respect to $\|\cdot\|_{\min}$. Consequently, $P\mathfrak{M}$ is semidiscrete by the foregoing.

For each $P \in \mathfrak{M}'$, the map

$$\pi_P \colon \mathfrak{M} \to P\mathfrak{M}, \qquad x \mapsto Px$$

is a w^*-homomorphism. By the foregoing, the family consisting of those π_P such that $P\mathfrak{M}$ is semidiscrete separates the points of $\mathfrak{M}$. Through an argument similar to the one given at the end of the proof of Theorem 6.4.13 (Which one exactly?), we see that $\mathfrak{M}$ is semidiscrete. $\square$

The first steps of the implication from injectivity to semidiscreteness, will be the following three relatively easy lemmas:

Lemma 6.4.18 *Let $\mathfrak{M}$ be a finite, injective W^*-algebra acting on a Hilbert space $\mathfrak{H}$, and let tr be a faithful, normal trace on $\mathfrak{M}$. Then for any $u_1, \ldots, u_n \in \mathcal{U}(\mathfrak{M})$ and $\epsilon > 0$, there is $\phi \in S(\mathcal{L}(\mathfrak{H})) \cap \mathcal{L}(\mathfrak{H})_*$ such that*

$$\|u_j \cdot \phi - \phi \cdot u_j\| \le \epsilon \qquad (j = 1, \ldots, n)$$

and

$$\|\phi|_{\mathfrak{M}} - \mathrm{tr}\| \le \epsilon.$$

Proof Let $\epsilon > 0$, and let $u_1, \ldots, u_n \in \mathcal{U}(\mathfrak{M})$. Let $\mathcal{E} \colon \mathcal{L}(\mathfrak{H}) \to \mathfrak{M}$ be an expectation, and let $\psi := \mathrm{tr} \circ \mathcal{E}$. Then $\psi \in S(\mathcal{L}(\mathfrak{H}))$ satisfies $u^* \cdot \psi \cdot u = \psi$ for all $u \in \mathcal{U}(\mathfrak{M})$ (Why?). Since $S(\mathcal{L}(\mathfrak{H})) \cap \mathcal{L}(\mathfrak{H})_*$ is w^*-dense in $S(\mathcal{L}(\mathfrak{H}))$, there is a net $(\psi_\alpha)_\alpha$ in $S(\mathcal{L}(\mathfrak{H})) \cap \mathcal{L}(\mathfrak{H})_*$ such that $\psi_\alpha \to \psi$ in the w^*-topology. This means, in particular, that

$$w\text{-}\lim_\alpha (\psi_\alpha - u^* \cdot \psi_\alpha \cdot u) = 0 \qquad (j = 1, \ldots, n)$$

and

$$w\text{-}\lim_\alpha \psi_\alpha|_{\mathfrak{M}} = \mathrm{tr}.$$

Passing to convex combinations, we obtain $\phi \in S(\mathcal{L}(\mathfrak{H})) \cap \mathcal{L}(\mathfrak{H})_*$ with the desired properties. $\square$

For any Hilbert space $\mathfrak{H}$, we write $\mathcal{HS}(\mathfrak{H})$ to denote the Hilbert–Schmidt operators on $\mathfrak{H}$.

Exercise 6.4.9 Let $\mathfrak{H}$ be a Hilbert space $\mathfrak{H}$. Show that the linear map

$$\mathfrak{H} \otimes \mathfrak{H} \to \mathcal{L}(\mathfrak{H}), \quad \xi \otimes \eta \mapsto \xi \odot \eta,$$

where

$$(\xi \odot \eta)(x) := \langle x, \eta \rangle \xi \qquad (\xi, \eta, x \in \mathfrak{H}),$$

extends to an isometric isomorphism of $\mathfrak{H} \bar\otimes \mathfrak{H}$ and $\mathcal{HS}(\mathfrak{H})$.

Lemma 6.4.19 *Let $\mathfrak{H}$ be a Hilbert space, and let $J \colon \mathfrak{H} \to \mathfrak{H}$ be a conjugate-linear, isometric isomorphism. Then, for any $S_1, T_1, \ldots, S_n, T_n \in \mathcal{L}(\mathfrak{H})$, we have*

$$\left\| \sum_{j=1}^n S_j \otimes JT_j J^{-1} \right\| = \sup \left\{ \left\| \sum_{j=1}^n S_j H T_j^* \right\|_{\mathcal{HS}(\mathfrak{H})} : H \in \mathcal{HS}(\mathfrak{H}), \|H\|_{\mathcal{HS}(\mathfrak{H})} \le 1 \right\}.$$

Proof Identifying $\mathfrak{H} \bar{\otimes} \mathfrak{H}$ and $\mathcal{HS}(\mathfrak{H})$ as in Exercise 6.4.19, the operator $S \otimes T \in \mathcal{L}(\mathfrak{H} \bar{\otimes} \mathfrak{H})$ with $S, T \in \mathcal{L}(\mathfrak{H})$ becomes

$$\mathcal{HS}(\mathfrak{H}) \to \mathcal{HS}(\mathfrak{H}), \quad H \mapsto SHT^*.$$

Since J is an isometry, we obtain for $S_1, T_1, \ldots, S_n, T_n \in \mathcal{L}(\mathfrak{H})$:

$$\left\| \sum_{j=1}^n S_j \otimes JT_j J^{-1} \right\| = \left\| \sum_{j=1}^n S_j \otimes T_j \right\| \quad \text{(Why?)}$$

$$= \sup \left\{ \left\| \sum_{j=1}^n S_j HT_j^* \right\|_{\mathcal{HS}(\mathfrak{H})} : H \in \mathcal{HS}(\mathfrak{H}), \|H\|_{\mathcal{HS}(\mathfrak{H})} \leq 1 \right\}.$$

This is the claim. $\square$

The next lemma is known as the *Powers–Størmer inequality*:

Lemma 6.4.20 *Let $\mathfrak{H}$ be a Hilbert space, and let $S, T \in \mathcal{N}(\mathfrak{H})$ be positive. Then $S^{\frac{1}{2}}, T^{\frac{1}{2}} \in \mathcal{HS}(\mathfrak{H})$ and*

$$\left\| S^{\frac{1}{2}} - T^{\frac{1}{2}} \right\|_{\mathcal{HS}(\mathfrak{H})}^2 \leq \|S - T\|_{\mathcal{N}(\mathfrak{H})}.$$

Proof It is clear from the definition of $\mathcal{HS}(\mathfrak{H})$ that $S^{\frac{1}{2}}, T^{\frac{1}{2}} \in \mathcal{HS}(\mathfrak{H})$.

Let

$$A := S^{\frac{1}{2}} - T^{\frac{1}{2}} \quad \text{and} \quad B := S^{\frac{1}{2}} + T^{\frac{1}{2}}.$$

Note that

$$B \geq \pm A \quad \text{and} \quad \frac{1}{2}(AB + BA) = S - T.$$

Since $\mathcal{HS}(\mathfrak{H}) \subset \mathcal{K}(\mathfrak{H})$, the self-adjoint operator A is compact. In particular, there is an orthonormal basis $(e_\alpha)_\alpha$ of $\mathfrak{K}$ such that each e_α is an eigenvector of A corresponding to some eigenvalue λ_α. Let Tr denote the canonical trace on $\mathcal{N}(\mathfrak{H})$; then we have:

$$\|S - T\|_{\mathcal{N}(\mathfrak{H})} = \text{Tr}(|S - T|)$$

$$= \sum_\alpha \frac{1}{2} \langle |AB + BA| e_\alpha, e_\alpha \rangle$$

$$\geq \sum_\alpha \frac{1}{2} |\langle (AB + BA) e_\alpha, e_\alpha \rangle|$$

$$= \sum_\alpha |\lambda_\alpha \langle Be_\alpha, e_\alpha \rangle| \quad \text{(Why?)}$$

$$\geq \sum_\alpha \lambda_\alpha^2$$

$$= \sum_\alpha \langle A^2 e_\alpha, e_\alpha \rangle$$

$$= \text{Tr} \left(\left(S^{\frac{1}{2}} - T^{\frac{1}{2}} \right)^2 \right)$$

$$= \left\| S^{\frac{1}{2}} - T^{\frac{1}{2}} \right\|_{\mathcal{HS}(\mathfrak{H})}^2.$$

This establishes the claim. $\square$

Corollary 6.4.21 *Let $\mathfrak{H}$ be a Hilbert space, and let $\epsilon > 0$. Let $N \in \mathcal{N}(\mathfrak{H})$, and let $U_1, \ldots, U_n \in \mathcal{L}(\mathfrak{H})$ be unitary operators such that*

$$\|U_j N - N U_j\|_{\mathcal{N}(\mathfrak{H})} < \epsilon^2 \qquad (j = 1, \ldots, n).$$

Then $H := N^{\frac{1}{2}} \in \mathcal{HS}(\mathfrak{H})$ satisfies

$$\|U_j H - H U_j\|_{\mathcal{HS}(\mathfrak{H})} < \epsilon \qquad (j = 1, \ldots, n).$$

Proof Note that $U_j H U_j^* = (U_j N U_j^*)^{\frac{1}{2}}$. We then obtain:

$$\begin{aligned}
\|U_j H - H U_j\|^2_{\mathcal{HS}(\mathfrak{H})} &\leq \|U_j H U_j^* - H\|^2_{\mathcal{HS}(\mathfrak{H})} \\
&\leq \|U_j N U_j^* - N\|_{\mathcal{N}(\mathfrak{H})}, \qquad \text{by Lemma 6.4.20,} \\
&= \|U_j N - N U_j\|_{\mathcal{N}(\mathfrak{H})} \\
&< \epsilon^2 \qquad (j = 1, \ldots, n).
\end{aligned}$$

Taking square roots on both sides yields the claim.   □

We start proving

$$\text{injectivity} \quad \Longrightarrow \quad \text{semidiscreteness}$$

with the finite case:

Proposition 6.4.22 *Let $\mathfrak{M}$ be a finite, injective W^*-algebra. Then $\mathfrak{M}$ is semidiscrete.*

Proof The W^*-algebra $\mathfrak{M}$ is of the form $\ell^\infty\text{-}\bigoplus_\alpha \mathfrak{M}_\alpha$, where each $\mathfrak{M}_\alpha$ is a W^*-algebra with a faithful, normal trace (Why?). By Exercises 6.2.4 and 6.4.4, we may thus suppose that $\mathfrak{M}$ has a faithful, normal trace tr.

Applying the GNS-construction with tr, we obtain a Hilbert space $\mathfrak{H}$ on which $\mathfrak{M}$ acts as a von Neumann algebra, and a cyclic, separating trace vector ξ for $\mathfrak{M}$. The conjugate linear map

$$J : \mathfrak{M}\xi \to \mathfrak{M}\xi, \quad x\xi \mapsto x^*\xi$$

is then well-defined and extends to a conjugate-linear isometric isomorphism of $\mathfrak{H}$ into itself.

We want to apply Theorem 6.4.17. Let $x_1, \ldots, x_n \in \mathfrak{M}$, and let $y_1, \ldots, y_n \in \mathfrak{M}'$. In order to establish that Theorem 6.4.17(ii) holds, it is sufficient (Why?) to show that

$$\left| \sum_{j=1}^n \langle x_j y_j \xi, \xi \rangle \right| \leq \left\| \sum_{j=1}^n x_j \otimes y_j \right\|_{\min}.$$

Let $z_1, \ldots, z_n \in \mathfrak{M}$ be such that $y_j = J z_j^* J$ for $j = 1, \ldots, n$ (Why do such $z_1, \ldots, z_n$ exist?), and suppose without loss of generality that $x_1, z_1, \ldots, x_n, z_n \in B_1(0, \mathfrak{M})$.

Let $\epsilon \in (0,1)$. By the Russo–Dye theorem ([Aup, Corollary 6.2.14]), there are elements $u_1, \ldots, u_m \in \mathcal{U}(\mathfrak{M})$ such that $z_1, \ldots, z_n$ lie in the convex hull of $\{u_1, \ldots, u_m\}$. By Lemma 6.4.18, there is $\phi \in S(\mathcal{L}(\mathfrak{H})) \cap \mathcal{L}(\mathfrak{H})_*$ such that

$$\|u_j \cdot \phi - \phi \cdot u_j\| \leq \epsilon^2 \qquad (j = 1, \ldots, m) \tag{6.17}$$

and

$$\|\phi|_{\mathfrak{M}} - \operatorname{tr}\| \leq \epsilon^2 < \epsilon. \tag{6.18}$$

The trace duality between $\mathcal{N}(\mathfrak{H})$ and $\mathcal{L}(\mathfrak{H})$ yields a positive operator $N \in \mathcal{N}(\mathfrak{H})$ with $\operatorname{Tr} N = 1$ and

$$\langle \phi, T \rangle = \operatorname{Tr}(TN) \qquad (T \in \mathcal{L}(\mathfrak{H})).$$

The inequalities (6.17) then translate into

$$\|u_j N - N u_j\| \leq \epsilon^2 \qquad (j = 1, \dots, m)$$

Let $H := N^{\frac{1}{2}}$. Then $H \in \mathcal{HS}(\mathfrak{H})$ with $\|H\|_{\mathcal{HS}(\mathfrak{H})} = 1$, and Corollary 6.4.21 implies that

$$\|u_j H - H u_j\|_{\mathcal{HS}(\mathfrak{H})} \leq \epsilon \qquad (j = 1, \dots, m).$$

Since each z_j is a convex combination of some of the u_k's, this yields:

$$\|z_j H - H z_j\|_{\mathcal{HS}(\mathfrak{H})} \leq \epsilon \qquad (j = 1, \dots, n).$$

Consequently, we obtain:

$$\left\| \sum_{j=1}^n x_j H z_j \right\|_{\mathcal{HS}(\mathfrak{H})} \geq \left| \left\langle \sum_{j=1}^n x_j H z_j, H \right\rangle_{\mathcal{HS}(\mathfrak{H})} \right|$$

$$\geq \left| \left\langle \sum_{j=1}^n x_j z_j H, H \right\rangle_{\mathcal{HS}(\mathfrak{H})} \right| - n\epsilon. \tag{6.19}$$

On the other hand, we have:

$$\left| \left\langle \sum_{j=1}^n x_j z_j H, H \right\rangle_{\mathcal{HS}(\mathfrak{H})} - \operatorname{tr}\left(\sum_{j=1}^n x_j z_j \right) \right| = \left| \operatorname{Tr}\left(\left(\sum_{j=1}^n x_j z_j \right) N \right) - \operatorname{tr}\left(\sum_{j=1}^n x_j z_j \right) \right|$$

$$= \left| \phi\left(\sum_{j=1}^n x_j z_j \right) - \operatorname{tr}\left(\sum_{j=1}^n x_j z_j \right) \right|$$

$$\leq \epsilon \left\| \sum_{j=1}^n x_j z_j \right\|, \qquad \text{by (6.18),}$$

$$\leq n\epsilon. \tag{6.20}$$

Ultimately, we obtain:

$$\left\|\sum_{j=1}^{n} x_j \otimes y_j\right\|_{\min} = \left\|\sum_{j=1}^{n} x_j \otimes J z_j^* J\right\|_{\min}$$

$$\geq \left\|\sum_{j=1}^{n} x_j H z_j\right\|_{\mathcal{HS}(\mathfrak{H})} , \qquad \text{by Lemma 6.4.19,}$$

$$\geq \left| \operatorname{tr}\left(\sum_{j=1}^{n} x_j z_j\right)\right| - 2n\epsilon, \qquad \text{by (6.19) and (6.20),}$$

$$= \left|\sum_{j=1}^{n} \langle x_j z_j \xi, \xi\rangle\right| - 2n\epsilon$$

$$= \left|\sum_{j=1}^{n} \langle x_j J z_j^* J \xi, \xi\rangle\right| - 2n\epsilon, \qquad \text{(Why?),}$$

$$= \left|\sum_{j=1}^{n} \langle x_j y_j \xi, \xi\rangle\right| - 2n\epsilon.$$

Since $\epsilon \in (0,1)$ was arbitrary, we have

$$\left|\sum_{j=1}^{n} \langle x_j y_j \xi, \xi\rangle\right| \leq \left\|\sum_{j=1}^{n} x_j \otimes y_j\right\|_{\min} ,$$

so that $\mathfrak{M}$ is semidiscrete by Theorem 6.4.17. □

With Proposition 6.4.22 proved, the route we have to take lies clearly ahead of us. Next, we tackle the semifinite case:

Proposition 6.4.23 *Let $\mathfrak{M}$ be a semifinite, injective W^*-algebra. Then $\mathfrak{M}$ is semidiscrete.*

Proof Suppose that $\mathfrak{M}$ acts as a von Neumann algebra on a Hilbert space $\mathfrak{H}$.

Invoking [K–R, 6.3.8. Theorem] as in the proof of Theorem 6.2.6, we obtain a net $(p_\alpha)_\alpha$ of projections in $\mathfrak{M}$ such that $p_\alpha \nearrow e_{\mathfrak{M}}$ in the strong operator topology, and each W^*-algebra $p_\alpha \mathfrak{M} p_\alpha$ is finite. By Exercise 6.2.4, each $p_\alpha \mathfrak{M} p_\alpha$ is also injective and thus semidiscrete by Proposition 6.4.22. Let p_α, $\epsilon > 0$, $x_1, \ldots, x_n \in \mathfrak{M}$, and $\phi_1, \ldots, \phi_m \in \mathfrak{M}_*$ be arbitrary. Then there is $p_\beta \geq p_\alpha$ such that

$$|\langle \phi_j, x_k - p_\beta x_k p_\beta\rangle| < \frac{\epsilon}{2} \qquad (j = 1, \ldots, m,\ k = 1, \ldots, n). \tag{6.21}$$

Using the semidiscreteness of $p_\beta \mathfrak{M} p_\beta$, we obtain a unit-preserving, completely positive, w^*-continuous map $S \in \mathcal{F}(p_\beta \mathfrak{M} p_\beta)$ such that

$$|\langle \phi_j, p_\beta x_k p_\beta - S(p_\beta x_k p_\beta)\rangle| < \frac{\epsilon}{2} \qquad (j = 1, \ldots, m,\ k = 1, \ldots, n). \tag{6.22}$$

Define $T: \mathfrak{M} \to \mathfrak{M}$ by letting $Tx := S(p_\beta x p_\beta)$ for $x \in \mathfrak{M}$. Then (6.21) and (6.22) together yield:

$$|\langle \phi_j, x_k - Tx_k\rangle| < \epsilon \qquad (j = 1, \ldots, m,\ k = 1, \ldots, n).$$

We thus obtain a net $(T_\gamma)_\gamma$ of completely positive, w^*-continuous maps in $\mathcal{F}(\mathfrak{M})$ such that

$$x = w^*\text{-}\lim_\gamma T_\gamma x \qquad (x \in \mathfrak{M}),$$

and $(T_\gamma e_{\mathfrak{M}})_\gamma$ is an increasing net of projections converging to $e_{\mathfrak{M}}$ in the w^*-topology. Let $\phi \in S(\mathfrak{M}) \cap \mathfrak{M}_*$, and define

$$S_\gamma := T_\gamma x + \langle x, \phi \rangle (e_{\mathfrak{M}} - T_\gamma e_{\mathfrak{M}}) \qquad (x \in \mathfrak{M}).$$

Clearly, each S_γ is unit preserving, w^*-continuous, and has finite rank, and, by Examples 6.4.2(c), it is also completely positive. From the properties of $(T_\gamma)_\gamma$, it is also clear that

$$x = w^*\text{-}\lim_\gamma S_\gamma x \qquad (x \in \mathfrak{M}).$$

By definition, $\mathfrak{M}$ is semidiscrete. $\square$

Having established

$$\text{injectivity} \quad \Longrightarrow \quad \text{semidiscreteness}$$

in the semifinite case, the structure theory of W^*-algebras along with the hereditary properties of injectivity and semidiscreteness imply that, for the general case, we may confine ourselves to W^*-algebras of type III. As in the proof of Theorem 6.2.6, we will rely on the fact that a countably decomposable W^*-algebra of type III can be described as the W^*-algebra generated by an appropriate covariant system. We'll need the following lemma:

Lemma 6.4.24 *Let $(\mathfrak{M}, G, \alpha)$ be a covariant system, such that $W^*(\mathfrak{M}, G, \alpha)$ is semidiscrete. Then $\mathfrak{M}$ is semidiscrete.*

Proof Suppose that $\mathfrak{M}$ acts as a von Neumann algebra on a Hilbert space $\mathfrak{H}$. By the definition of $W^*(\mathfrak{M}, G, \alpha)$ ([Dae, I.2.10 Definition]), this W^*-algebra acts as a von Neumann algebra on the Hilbert space $L^2(G, \mathfrak{H}) \cong L^2(G) \bar{\otimes} \mathfrak{H}$. Let $\pi \colon \mathfrak{M} \to \mathcal{L}(L^2(G, \mathfrak{H}))$ be the faithful W^*-representation defined in [Dae, I.2.4 Definition], and let

$$\rho \colon \mathfrak{M}' \to \mathcal{L}(L^2(G, \mathfrak{H})) \cong \mathcal{L}(L^2(G) \bar{\otimes} \mathfrak{H}), \quad y \mapsto \mathrm{id}_{L^2(G)} \otimes y;$$

clearly, ρ is a faithful W^*-representation. Let $x_1, \ldots, x_n \in \mathfrak{M}$ and $y_1, \ldots, y_n \in \mathfrak{M}'$. Then $\pi(x_1), \ldots, \pi(x_n) \in W^*(\mathfrak{M}, G, \alpha)$ and $\rho(y_1), \ldots, \rho(y_n) \in W^*(\mathfrak{M}, G, \alpha)'$ (Do you believe that without checking it for yourself?). Since $W^*(\mathfrak{M}, G, \alpha)$ is semidiscrete, we have:

$$\left\| \sum_{j=1}^n \pi(x_j) \rho(y_j) \right\| \leq \left\| \sum_{j=1}^n \pi(x_j) \otimes \rho(y_j) \right\| = \left\| \sum_{j=1}^n x_j \otimes y_j \right\|_{\min}. \qquad (6.23)$$

By the definition of π, the left hand side of (6.23) is equal to

$$\sup_{g \in G} \left\| \sum_{j=1}^n \alpha_g^{-1}(x_j) y_j \right\| \geq \left\| \sum_{j=1}^n x_j y_j \right\|.$$

Hence, by Theorem 6.4.17, $\mathfrak{M}$ is semidiscrete. $\square$

Ready for the general case?

Theorem 6.4.25 *Let $\mathfrak{M}$ be an injective W^*-algebra. Then $\mathfrak{M}$ is semidiscrete.*

Proof As we already stated, we may suppose that $\mathfrak{M}$ is of type III.

As in the proof of Theorem 6.2.6, we suppose first that $\mathfrak{M}$ is also countably decomposable. Let $\{\sigma_t : t \in \mathbb{R}\}$ be the modular automorphism group of $\mathfrak{M}$ corresponding to some faithful $\phi \in S(\mathfrak{M}) \cap \mathfrak{M}_*$. Then $(\mathfrak{M}, \mathbb{R}, \sigma)$ is a covariant system, and $W^*(\mathfrak{M}, \mathbb{R}, \sigma)$ is of type II_∞ ([Dae, II.4.7 Theorem]). Let $\hat{\sigma}$ denote the dual action of $\hat{\mathbb{R}} = \mathbb{R}$ on $W^*(\mathfrak{M}, \mathbb{R}, \sigma)$. By [K–R, 13.2.10 Corollary], we have

$$W^*(W^*(\mathfrak{M}, \mathbb{R}, \sigma), \mathbb{R}, \hat{\sigma}) \cong \mathfrak{M},$$

so that we may view $W^*(\mathfrak{M}, \mathbb{R}, \sigma)$ as a W^*-subalgebra of $\mathfrak{M}$. Using the amenability of $\mathbb{R}_d$ as in the proof of Theorem 6.2.6, we obtain an expectation $\mathcal{E} : \mathfrak{M} \to W^*(\mathfrak{M}, \mathbb{R}, \sigma)$. It follows that $W^*(\mathfrak{M}, \mathbb{R}, \sigma)$ is injective and thus semidiscrete by Proposition 6.4.23. From Lemma 6.4.24, we finally conclude that $\mathfrak{M}$ is semidiscrete.

For the general type III case, suppose that $\mathfrak{M}$ acts on a Hilbert space as a von Neumann algebra, and choose again, as in the proof of Theorem 6.2.6, a net of projections $(p_\alpha)_\alpha$ such that $p_\alpha \nearrow e_{\mathfrak{M}}$ in the strong operator topology, and each $p_\alpha \mathfrak{M} p_\alpha$ is countably decomposable. Since each $p_\alpha \mathfrak{M} p_\alpha$ is of type III and injective, the foregoing implies that each $p_\alpha \mathfrak{M} p_\alpha$ is also semidiscrete. In the same way as in the proof of Proposition 6.4.23, we conclude that $\mathfrak{M}$ itself is semidiscrete. $\square$

Remarks 6.4.26   (a) For the same reason as pointed out in Remark 6.2.7, one doesn't have to deal with the countably decomposable case first in the proof of Theorem 6.4.25.

(b) Type I von Neumann algebras are also called *discrete* in the literature. The prefix "semi-" suggests that semidiscreteness is something weaker than discreteness. In view of Example 6.3.18, Theorem 6.4.25 shows that this is indeed the case.

(c) Let $\mathfrak{M}$ be a von Neumann algebra acting on a Hilbert space $\mathfrak{H}$, and let $\mathcal{E} : \mathcal{L}(\mathfrak{H}) \to \mathfrak{M}$ be an expectation. Since $\mathcal{L}(\mathfrak{H})$ is easily seen to be semidiscrete (How?), the semidiscreteness of $\mathfrak{M}$ follows much more easily in the case where $\mathcal{E}$ is w^*-continuous. This, however, is possible only if $\mathfrak{M}$ is atomic ([Tak 2, Exercise V.2.8]).

It follows that nuclear C^*-algebras are characterized by the injectivity/semidiscreteness of their biduals:

Corollary 6.4.27 *For a C^*-algebra $\mathfrak{A}$, the following are equivalent:*

(i) $\mathfrak{A}^{**}$ *is semidiscrete.*

(ii) $\mathfrak{A}$ *is nuclear.*

(iii) $\mathfrak{A}^{**}$ *is injective.*

Together with Corollary 6.4.5, this yields:

Corollary 6.4.28 *Let $\mathfrak{A}$ be a nuclear C^*-algebra. Then both $\mathfrak{A}$ and $\mathfrak{A}^*$ have the metric approximation property.*

And from Corollary 6.2.8, we conclude:

Corollary 6.4.29 *Let $\mathfrak{A}$ be an amenable C^*-algebra. Then $\mathfrak{A}$ is nuclear.*

We give more examples of nuclear C^*-algebras:

Example 6.4.30 A positive element a of a C^*-algebra $\mathfrak{A}$ is called *abelian* if the norm closure of $a\mathfrak{A}a$ is a commutative C^*-algebra. The algebra $\mathfrak{A}$ is said to be of *type* I_0 if it is generated by its abelian elements ([Ped]). For each abelian element $a \in \mathfrak{A}$, let $p \in \mathfrak{A}^{**}$ be its range projection; then $p\mathfrak{A}^{**}p$ is a commutative von Neumann algebra, i.e. p is an abelian projection, containing $a\mathfrak{A}a$. It follows that the von Neumann algebra $\mathfrak{A}^{**}$ is generated by its abelian projections and thus is of type I. As we have seen in Example 6.3.18, this means that $\mathfrak{A}^{**}$ is injective. By Corollary 6.4.27, $\mathfrak{A}$ is nuclear.

6.5 Normal, virtual diagonals

We're almost finished.

What is left to be shown is that we can get back from injectivity/semidiscreteness to Connes-amenability. In this section, we shall see that every injective W^*-algebra has a normal, virtual diagonal (and is thus Connes-amenable by Theorem 4.4.15). This will also yield that a C^*-algebra $\mathfrak{A}$ is amenable if and only if its bidual $\mathfrak{A}^{**}$ is Connes-amenable.

We start with defining yet another norm on the tensor product of two unital C^*-algebras:

Definition 6.5.1 Let $\mathfrak{A}$ and $\mathfrak{B}$ be unital C^*-algebras. Then the *Haagerup norm* on $\mathfrak{A} \otimes \mathfrak{B}$ is defined as

$$\|\mathbf{x}\|_h := \inf \left\{ \sqrt{\left\| \sum_{j=1}^{n} a_j a_j^* \right\|} \sqrt{\left\| \sum_{j=1}^{n} b_j^* b_j \right\|} : \mathbf{x} = \sum_{j=1}^{n} a_j \otimes b_j \right\} \qquad (\mathbf{x} \in \mathfrak{A} \otimes \mathfrak{B}).$$

Just because we call something a norm, it doesn't have to be one. See for yourself that $\| \cdot \|_h$ is at least a seminorm:

Exercise 6.5.1 Let $\mathfrak{A}$ and $\mathfrak{B}$ be unital C^*-algebras.

(i) Show that

$$\sqrt{xy} = \frac{1}{2} \inf \left\{ tx + \frac{y}{t} : t > 0 \right\} \qquad (x, y \geq 0).$$

(ii) Conclude that $\| \cdot \|_h$ is a seminorm on $\mathfrak{A} \otimes \mathfrak{B}$.
(iii) Show that $\| \cdot \|_h \leq \| \cdot \|_\pi$ on $\mathfrak{A} \otimes \mathfrak{B}$.

Remark 6.5.2 The Haagerup norm is *not* a C^*-norm.

Definition 6.5.3 Let $\mathfrak{A}$ and $\mathfrak{B}$ be unital C^*-algebras.

(i) We write $\mathfrak{A} \otimes_h \mathfrak{B}$ for $\mathfrak{A} \otimes \mathfrak{B}$ equipped with $\| \cdot \|_h$ and $\mathfrak{A} \tilde{\otimes}_h \mathfrak{B}$ for the completion of $\mathfrak{A} \otimes_h \mathfrak{B}$ with respect to $\| \cdot \|_h$.
(ii) The Banach space $\mathfrak{A} \tilde{\otimes}_h \mathfrak{B}$ is called the *Haagerup tensor product* of $\mathfrak{A}$ and $\mathfrak{B}$.
(iii) We write $\| \cdot \|_{h^*}$ for the norm on $(\mathfrak{A} \tilde{\otimes}_h \mathfrak{B})^*$.

Remark 6.5.4 The Haagerup tensor product can be defined more generally for operator spaces (Definition D.3.10). For unital C^*-algebras the two definitions coincide.

Exercise 6.5.2 Let $\mathfrak{A}$ be a unital C^*-algebra. Show that $\mathfrak{A} \tilde{\otimes}_h \mathfrak{A}$ is a Banach $\mathfrak{A}$-bimodule in the canonical fashion.

For what lies ahead we have to characterize the separately w^*-continuous, bilinear forms on a von Neumann algebra that are bounded with respect to $\| \cdot \|_h$:

Theorem 6.5.5 *Let $\mathfrak{M}$ and $\mathfrak{N}$ be W^*-algebras, and let $V \in \mathcal{L}^2_{w^*}(\mathfrak{M}, \mathfrak{N}; \mathbb{C})$. Then the following are equivalent:*

(i) $\|V\|_{h^*} \leq 1$.

(ii) *There are $\phi \in S(\mathfrak{M}) \cap \mathfrak{M}_*$ and $\psi \in S(\mathfrak{N}) \cap \mathfrak{N}_*$ such that*

$$|V(x,y)| \leq \sqrt{\langle xx^*, \phi \rangle} \sqrt{\langle y^*y, \psi \rangle} \qquad (x \in \mathfrak{M}, \, y \in \mathfrak{N}). \tag{6.24}$$

If $\mathfrak{M}$ and $\mathfrak{N}$ are von Neumann algebras acting on Hilbert spaces $\mathfrak{H}$ and $\mathfrak{K}$, respectively, such that each normal state of $\mathfrak{M}$ and $\mathfrak{N}$, respectively, is a vector state, then (i) and (ii) are equivalent to:

(iii) *There are unit vectors $\xi \in \mathfrak{H}$ and $\eta \in \mathfrak{K}$, and $T \in \mathcal{L}(\mathfrak{K}, \mathfrak{H})$ with $\|T\| \leq 1$ such that*

$$V(x,y) = \langle xTy\eta, \xi \rangle \qquad (x \in \mathfrak{M}, \, y \in \mathfrak{N}).$$

In this situation, if $\phi \in S(\mathfrak{M}) \cap \mathfrak{M}_*$ and $\psi \in S(\mathfrak{N}) \cap \mathfrak{N}_*$ are as in (ii), the vectors $\xi \in \mathfrak{H}$ and $\eta \in \mathfrak{K}$ in (iii) can be chosen such that

$$\langle x, \phi \rangle = \langle x\xi, \xi \rangle \quad (x \in \mathfrak{M}) \qquad and \qquad \langle y, \psi \rangle = \langle y\eta, \eta \rangle \quad (y \in \mathfrak{N}).$$

Proof (i) $\Longrightarrow$ (ii): We first claim that there are $\phi \in S(\mathfrak{M})$ and $\psi \in S(\mathfrak{M})$ (not necessarily normal) such that (6.24) holds. Using Exercise 6.5.1(i), we see (How?) that it is enough to find $\phi \in S(\mathfrak{M})$ and $\psi \in S(\mathfrak{M})$ such that

$$|V(x,y)| \leq \frac{1}{2}(\langle xx^*, \phi \rangle + \langle y^*y, \psi \rangle) \qquad (x \in \mathfrak{M}, \, y \in \mathfrak{N}).$$

In fact, it is sufficient (Why?) to find $\phi \in S(\mathfrak{M})$ and $\psi \in S(\mathfrak{M})$ such that

$$\mathrm{Re}\, V(x,y) \leq \frac{1}{2}(\langle xx^*, \phi \rangle + \langle y^*y, \psi \rangle) \qquad (x \in \mathfrak{M}, \, y \in \mathfrak{N}).$$

Let C be the cone in $\ell^\infty_{\mathbb{R}}(S(\mathfrak{M}) \times S(\mathfrak{N}))$ consisting of all functions of the form

$$f_{x_1, y_1, \ldots, x_n, y_n}(\phi, \psi) = \frac{1}{2}\sum_{j=1}^n (\langle x_j x_j^*, \phi \rangle + \langle y_j^* y_j, \psi \rangle) - \mathrm{Re} \sum_{j=1}^n V(x_j, y_j)$$

$$(x_1, \ldots, x_n \in \mathfrak{M}, \, y_1, \ldots, y_n \in \mathfrak{N}, \, \phi \in S(\mathfrak{M}), \, \psi \in S(\mathfrak{N})).$$

For $x_1, \ldots, x_n \in \mathfrak{M}$ and $y_1, \ldots, y_n \in \mathfrak{N}$ fix $\phi_0 \in S(\mathfrak{M})$ and $\psi_0 \in S(\mathfrak{N})$ such that

$$\left\langle \sum_{j=1}^n x_j x_j^*, \phi_0 \right\rangle = \left\| \sum_{j=1}^n x_j x_j^* \right\| \qquad and \qquad \left\langle \sum_{j=1}^n y_j^* y_j, \phi_0 \right\rangle = \left\| \sum_{j=1}^n y_j^* y_j \right\|.$$

It follows that

$$f_{x_1,y_1,\ldots,x_n,y_n}(\phi,\psi) = \frac{1}{2}\sum_{j=1}^{n}(\langle x_j x_j^*, \phi_0\rangle + \langle y_j^* y_j, \psi_0\rangle) - \mathrm{Re}\sum_{j=1}^{n} V(x_j,y_j)$$

$$= \frac{1}{2}\left\|\sum_{j=1}^{n} x_j x_j^*\right\| + \frac{1}{2}\left\|\sum_{j=1}^{n} y_j^* y_j\right\| - \mathrm{Re}\sum_{j=1}^{n} V(x_j,y_j)$$

$$\geq \sqrt{\left\|\sum_{j=1}^{n} x_j x_j^*\right\|}\sqrt{\left\|\sum_{j=1}^{n} y_j^* y_j\right\|} - \left|\sum_{j=1}^{n} V(x_j,y_j)\right|$$

$$\geq 0.$$

Hence, C is disjoint from the open cone U of all strictly negative functions in $\ell_{\mathbb{R}}^{\infty}(S(\mathfrak{M}) \times S(\mathfrak{N}))$. Clearly, U is open in the w^*-topology on $\ell_{\mathbb{R}}^{\infty}(S(\mathfrak{M}) \times S(\mathfrak{N}))$. The Hahn–Banach separation theorem ([Rud 2, 3.4 Theorem(a)]) yields $\mu \in \ell_{\mathbb{R}}^{1}(S(\mathfrak{M}) \times S(\mathfrak{N}))$ — we may view μ as a discrete measure on $S(\mathfrak{M}) \times S(\mathfrak{N})$ — such that

$$\int_{S(\mathfrak{M}) \times S(\mathfrak{N})} f(\phi,\psi)\, d\mu(\phi,\psi) \geq 0 \qquad (f \in C) \tag{6.25}$$

and

$$\int_{S(\mathfrak{M}) \times S(\mathfrak{N})} g(\phi,\psi)\, d\mu(\phi,\psi) < 0 \qquad (g \in U) \tag{6.26}$$

From (6.26), it is immediate that $\mu \geq 0$, so that we may suppose that μ is a probability measure. Define $\tilde{\phi} \in S(\mathfrak{M})$ and $\tilde{\psi} \in S(\mathfrak{N})$ through

$$\langle x, \tilde{\phi}\rangle := \int_{S(\mathfrak{M}) \times S(\mathfrak{N})} \langle x, \phi\rangle\, d\mu(\phi,\psi)$$

$$\text{and } \langle y, \tilde{\psi}\rangle := \int_{S(\mathfrak{M}) \times S(\mathfrak{N})} \langle y, \psi\rangle\, d\mu(\phi,\psi) \quad (x \in \mathfrak{M}, y \in \mathfrak{N}).$$

From (6.25), we conclude that

$$\mathrm{Re}\, V(x,y) \leq \frac{1}{2}(\langle xx^*, \tilde{\phi}\rangle + \langle y^* y, \tilde{\psi}\rangle) \qquad (x \in \mathfrak{M}, y \in \mathfrak{N}). \tag{6.27}$$

To complete the proof of (i) $\Longrightarrow$ (ii), we claim that in (6.27) the states $\tilde{\phi}$ and $\tilde{\psi}$ can be replaced by their normal parts $\tilde{\phi}_n$ and $\tilde{\psi}_n$. Let $\tilde{\phi}_s$ be the singular part of $\tilde{\phi}$. By [K–R, 10.5.15 Exercise], there is a net $(p_\alpha)_\alpha$ of projections in $\mathfrak{M}$ such that $p_\alpha \nearrow e_{\mathfrak{M}}$ in the strong operator topology and $\langle p_\alpha, \tilde{\phi}_s\rangle = 0$ for all p_α. Since

$$0 \leq \langle p_\alpha x p_\alpha, \tilde{\phi}_s\rangle \leq \|x\| \langle p_\alpha, \tilde{\psi}_s\rangle = 0$$

for all p_α and for all positive $x \in \mathfrak{M}$, it follows that $\langle p_\alpha x p_\alpha, \tilde{\phi}_s\rangle = 0$ for all p_α and for all $x \in \mathfrak{M}$. Fix $x,y \in \mathfrak{M}$. Since $x = w^*\text{-}\lim_\alpha p_\alpha x p_\alpha$ and $xx^* = w^*\text{-}\lim_\alpha p_\alpha x p_\alpha x^* p_\alpha$, we have

$$|V(x,y)| = \lim_\alpha |V(p_\alpha x p_\alpha, y)|$$

$$\leq \limsup_\alpha \sqrt{\langle p_\alpha x p_\alpha x^* p_\alpha, \tilde{\phi} \rangle} \sqrt{\langle y^* y, \tilde{\psi} \rangle}$$

$$\leq \limsup_\alpha \sqrt{\langle p_\alpha x p_\alpha x^* p_\alpha, \tilde{\phi}_n + \tilde{\phi}_s \rangle} \sqrt{\langle y^* y, \tilde{\psi} \rangle}$$

$$= \limsup_\alpha \sqrt{\langle p_\alpha x p_\alpha x^* p_\alpha, \tilde{\phi}_n \rangle} \sqrt{\langle y^* y, \tilde{\psi} \rangle}$$

$$= \sqrt{\langle xx^*, \tilde{\phi}_n \rangle} \sqrt{\langle y^* y, \tilde{\psi} \rangle}.$$

The analoguous treatment of $\tilde{\psi}$ then yields the claim.

Finally, note that both $\|\tilde{\phi}_n\| \leq 1$ and $\|\tilde{\psi}\|_n \leq 1$. We can thus find (How?) $\phi' \in S(\mathfrak{M}) \cap \mathfrak{M}_*$ and $\psi' \in S(\mathfrak{N}) \cap \mathfrak{N}_*$ with $\phi' \geq \tilde{\phi}_n$ and $\psi' \geq \tilde{\psi}_n$. This proves (ii).

The implication (ii) $\Longrightarrow$ (i) is obvious.

Suppose now that (ii) holds, and that $\mathfrak{M}$ and $\mathfrak{N}$ are von Neumann algebras acting on Hilbert spaces $\mathfrak{H}$ and $\mathfrak{K}$, respectively, such that the respective normal states of $\mathfrak{M}$ and $\mathfrak{N}$ are vector states. In particular, there are unit vectors $\xi \in \mathfrak{H}$ and $\eta \in \mathfrak{K}$ be such that

$$\langle x, \phi \rangle = \langle x\xi, \xi \rangle \quad (x \in \mathfrak{M}) \qquad \text{and} \qquad \langle y, \psi \rangle = \langle y\eta, \eta \rangle \quad (y \in \mathfrak{N}).$$

Define a sesquilinear form $[\cdot, \cdot]$ on $\mathfrak{M}\xi \times \mathfrak{N}\eta$ through

$$[x\xi, y\eta] := V(x^*, y) \qquad (x \in \mathfrak{M}, \ y \in \mathfrak{N}).$$

Then

$$|[x\xi, y\eta]| = |V(x^*, y)| \leq \sqrt{\langle x^* x, \phi \rangle} \sqrt{\langle y^* y, \eta \rangle} = \|x\xi\|\|y\eta\| \qquad (x \in \mathfrak{M}, \ y \in \mathfrak{N}). \qquad (6.28)$$

Let $\mathfrak{H}_0 := \overline{\mathfrak{M}\xi}$ and $\mathfrak{K}_0 := \overline{\mathfrak{N}\eta}$. By (6.28), $[\cdot, \cdot]$ extends to a sesquilinear form on $\mathfrak{H}_0 \times \mathfrak{K}_0$ of norm at most one. By elementary Hilbert space theory, there is thus a contraction $S \in \mathcal{L}(\mathfrak{K}_0, \mathfrak{H}_0)$ such that

$$V(x^*, y) = \langle S\eta, x\xi \rangle \qquad (x \in \mathfrak{M}, \ y \in \mathfrak{N}).$$

Let $P \in \mathcal{L}(\mathfrak{K})$ denote the orthogonal projection onto $\mathfrak{K}_0$. Then $T := SP$ is the required operator.

The implication (iii) $\Longrightarrow$ (ii) is again obvious (for this implication, no hypothesis on the normal states of $\mathfrak{M}$ and $\mathfrak{N}$ is needed). $\quad\square$

There is a variant of Theorem 6.5.5 for general unital C^*-algebras:

Exercise 6.5.3 Let $\mathfrak{A}$ and $\mathfrak{B}$ be unital C^*-algebras.

(i) Show that the following are equivalent for $V \in \mathcal{L}^2(\mathfrak{A}, \mathfrak{B}; \mathbb{C})$:

(a) $\|V\|_{h^*} \leq 1$;

(b) There are $\phi \in S(\mathfrak{A})$ and $\psi \in S(\mathfrak{B})$ such that

$$|V(a, b)| \leq \sqrt{\langle aa^*, \phi \rangle} \sqrt{\langle b^* b, \psi \rangle} \qquad (a \in \mathfrak{A}, \ b \in \mathfrak{B}).$$

(c) There are $*$-representations π of $\mathfrak{A}$ and ρ of $\mathfrak{B}$ on Hilbert spaces $\mathfrak{H}$ and $\mathfrak{K}$, respectively, as well as unit vectors $\xi \in \mathfrak{H}$ and $\eta \in \mathfrak{K}$, and $T \in \mathcal{L}(\mathfrak{K}, \mathfrak{H})$ with $\|T\| \leq 1$ such that

$$V(a, b) = \langle \pi(a) T \rho(b) \eta, \xi \rangle \qquad (a \in \mathfrak{A}, \ b \in \mathfrak{B}).$$

(ii) Conclude that $\| \cdot \|_h$ is a cross norm on $\mathfrak{A} \otimes \mathfrak{B}$.

Let $\mathfrak{M}$ be a von Neumann algebra acting on a Hilbert space $\mathfrak{H}$. Then $_{\mathfrak{M}'}\mathcal{L}_{\mathfrak{M}'}(\mathcal{L}(\mathfrak{H}))$ is a Banach $\mathfrak{M}$-bimodule through

$$(x \cdot \mathcal{T})(T) := x\mathcal{T}(T)$$
$$\text{and } (\mathcal{T} \cdot x)(T) := \mathcal{T}(T)x \qquad (x \in \mathfrak{M}, \, \mathcal{T} \in {}_{\mathfrak{M}'}\mathcal{L}_{\mathfrak{M}'}(\mathcal{L}(\mathfrak{H})), \, T \in \mathcal{L}(\mathfrak{H})).$$

Lemma 6.5.6 *Let $\mathfrak{M}$ be a von Neumann algebra acting on a Hilbert space $\mathfrak{H}$. Then the map $\Theta_0 : \mathfrak{M} \otimes \mathfrak{M} \to \mathcal{L}(\mathcal{L}(\mathfrak{H}))$ defined through*

$$\Theta_0(x \otimes y)(T) := xTy \qquad (x, y \in \mathfrak{M}, \, T \in \mathcal{L}(\mathfrak{H}))$$

attains its values in $_{\mathfrak{M}'}\mathcal{L}_{\mathfrak{M}'}(\mathcal{L}(\mathfrak{H}))$ and is an $\mathfrak{M}$-bimodule homomorphism.

Proof Obvious. $\square$

Next, we wish to extend the map Θ_0 from Lemma 6.5.6 to a certain "completion" of $\mathfrak{M} \otimes \mathfrak{M}$. To this end, we need a description of $_{\mathfrak{M}'}\mathcal{L}_{\mathfrak{M}'}(\mathcal{L}(\mathfrak{H}))$ as a dual Banach $\mathfrak{M}$-bimodule:

Exercise 6.5.4 Let $\mathfrak{M}$ be a von Neumann algebra acting on a Hilbert space $\mathfrak{H}$, and let the projective tensor product $\mathcal{L}(\mathfrak{H}) \hat{\otimes} \mathcal{N}(\mathfrak{H})$ be equipped with the following Banach $\mathfrak{M}$-bimodule action:

$$x \cdot (T \otimes N) := T \otimes xN \quad \text{and} \quad (T \otimes N) \cdot x := T \otimes Nx \qquad (x \in \mathfrak{M}, \, T \in \mathcal{L}(\mathfrak{H}), \, N \in \mathcal{N}(\mathfrak{H})).$$

(i) Let F be the closed linear subspace of $\mathcal{L}(\mathfrak{H}) \hat{\otimes} \mathcal{N}(\mathfrak{H})$ spanned by all elements of the form

$$\left\{ \begin{array}{l} yT \otimes N - T \otimes Ny \\ Ty \otimes N - T \otimes yN \end{array} \right\} \qquad (y \in \mathfrak{M}', \, T \in \mathcal{L}(\mathfrak{H}), \, N \in \mathcal{N}(\mathfrak{H})).$$

Check that F is a closed $\mathfrak{M}$-submodule of $\mathcal{L}(\mathfrak{H}) \hat{\otimes} \mathcal{N}(\mathfrak{H})$.

(ii) Show that the canonical isomorphism between $(\mathcal{L}(\mathfrak{H}) \hat{\otimes} \mathcal{N}(\mathfrak{H}))^*$ and $\mathcal{L}(\mathcal{L}(\mathfrak{H}))$ induces an isometric isomorphism of the Banach $\mathfrak{M}$-bimodules $(\mathcal{L}(\mathfrak{H}) \hat{\otimes} \mathcal{N}(\mathfrak{H}))/F$ and $_{\mathfrak{M}'}\mathcal{L}_{\mathfrak{M}'}(\mathcal{L}(\mathfrak{H}))$.

For any W^*-algebra, let

$$\mathcal{L}^2_{h,w^*}(\mathfrak{M}, \mathbb{C}) := \{V \in \mathcal{L}^2_{w^*}(\mathfrak{M}, \mathbb{C}) : V \text{ is continuous with respect to } \| \cdot \|_h\}.$$

It is clear that $\mathcal{L}^2_{h,w^*}(\mathfrak{M}, \mathbb{C})$ is a closed submodule of $(\mathfrak{M} \tilde{\otimes}_h \mathfrak{M})^*$. Let $\mathcal{L}^2_{h,w^*}(\mathfrak{M}, \mathbb{C})^*$ be equipped with the corresponding dual $\mathfrak{M}$-bimodule action. The canonical embedding of $\mathfrak{M} \otimes \mathfrak{M}$ into $\mathcal{L}^2_{h,w^*}(\mathfrak{M}, \mathbb{C})^*$ is then an $\mathfrak{M}$-bimodule homomorphism.

Theorem 6.5.7 *Let $\mathfrak{M}$ be a von Neumann algebra acting on a Hilbert space $\mathfrak{H}$ such that each normal state of $\mathfrak{M}$ is a vector state. Then the map Θ_0 from Lemma 6.5.6 extends uniquely to a w^*-continuous, isometric isomorphism*

$$\Theta : \mathcal{L}^2_{h,w^*}(\mathfrak{M}, \mathbb{C})^* \to {}_{\mathfrak{M}'}\mathcal{L}_{\mathfrak{M}'}(\mathcal{L}(\mathfrak{H}))$$

of Banach $\mathfrak{M}$-bimodules.

Proof Since $\mathfrak{M} \otimes \mathfrak{M}$ is w^*-dense in $\mathcal{L}^2_{h,w^*}(\mathfrak{M}, \mathbb{C})^*$, the uniqueness is clear.

Define $\Theta'_* : \mathcal{L}(\mathfrak{H}) \hat{\otimes} \mathcal{N}(\mathfrak{H}) \to \mathcal{L}^2_{h,w^*}(\mathfrak{M}, \mathbb{C})$ by letting

$$\Theta'_*(T \otimes N)(x \otimes y) := \langle N, xTy \rangle \qquad (x,y \in \mathfrak{M}, T \in \mathcal{L}(\mathfrak{H}), N \in \mathcal{N}(\mathfrak{H})).$$

It is routinely checked (or should at least be) that Θ'_* is a norm decreasing $\mathfrak{M}$-bimodule homomorphism. Let $V \in \mathcal{L}^2_{h,w^*}(\mathfrak{M}, \mathbb{C})$ be such that $\|V\|_h \leq 1$. By Theorem 6.5.5, there are unit vectors $\xi, \eta \in \mathfrak{H}$ and $T \in \mathcal{L}(\mathfrak{H})$ with $\|T\| \leq 1$ such that

$$V(x,y) = \langle yTx\eta, \xi \rangle \qquad (x,y \in \mathfrak{M}), \tag{6.29}$$

so that $V = \Theta'_*(T \otimes (\xi \odot \eta))$. Since $\|T \otimes (\xi \odot \eta)\|_\pi = \|T\|\|(\eta \odot \xi)\|_{\mathcal{N}(\mathfrak{H})} \leq 1$, it follows that Θ'_* is a metric surjection.

A routine calculation (that nevertheless should not be skipped) shows that Θ'_* vanishes on the space F defined in Exercise 6.5.4(i). Consequently, Θ'_* drops to a metric surjection $\Theta_* : (\mathcal{L}(\mathfrak{H})\hat{\otimes}\mathcal{N}(\mathfrak{H}))/F \to \mathcal{L}^2_{h,w^*}(\mathfrak{M}, \mathbb{C})$. Defining $\Theta := (\Theta_*)^*$, we obtain a w^*-continuous, isometric $\mathfrak{M}$-bimodule homomorphism $\Theta : \mathcal{L}^2_{h,w^*}(\mathfrak{M}, \mathbb{C})^* \to {}_{\mathfrak{M}'}\mathcal{L}_{\mathfrak{M}'}(\mathcal{L}(\mathfrak{H}))$ that extends Θ_0.

What remains to be shown is that Θ is surjective. Let $\mathcal{T} \in {}_{\mathfrak{M}'}\mathcal{L}_{\mathfrak{M}'}(\mathcal{L}(\mathfrak{H}))$, and let $V \in \mathcal{L}^2_{h,w^*}(\mathfrak{M}, \mathbb{C})$ be of the form (6.29). Define $\mathbf{T} \in \mathcal{L}^2_{h,w^*}(\mathfrak{M}, \mathbb{C})^*$ through

$$\langle V, \mathbf{T} \rangle := \langle \mathcal{T}(T)\eta, \xi \rangle. \tag{6.30}$$

If this really is a well-defined, bounded linear functional on $\mathcal{L}^2_{h,w^*}(\mathfrak{M}, \mathbb{C})$, it is straigtforward to check that $\Theta(\mathbf{T}) = \mathcal{T}$.

Let $\xi_1, \xi_2, \eta_1, \eta_2$ be unit vectors in $\mathfrak{H}$ and let $T_1, T_2 \in \mathcal{L}(\mathfrak{H})$ be such that

$$V(x,y) = \langle yT_j x\eta_j, \xi_j \rangle \qquad (x,y \in \mathfrak{M}, j = 1,2).$$

For $j = 1,2$, let P_j and Q_j be the orthogonal projections onto $\overline{\mathfrak{M}\xi_j}$ and $\overline{\mathfrak{M}\eta_j}$, respectively. Observe that, for $j = 1,2$, the value of $\langle \mathcal{T}(T_j)\xi_j, \eta_j \rangle$ does not change if we replace T_j by $Q_j T_j P_j$ (since $P_j, Q_j \in \mathfrak{M}'$); we may hence suppose that $T_j = Q_j T_j P_j$.

For $j = 1,2$, let ϕ_j and ψ_j be the vector states given by ξ_j and η_j, respectively. Let $\phi := \frac{1}{2}(\phi_1 + \phi_2)$ and $\psi := \frac{1}{2}(\psi_1 + \psi_2)$. Let $t > 0$, and note that

$$|V(x,y)| \leq \frac{1}{2}\|T_1\| \left(t\langle xx^*, \phi_1 \rangle + \frac{\langle y^*y, \psi_1 \rangle}{t} \right) + \frac{1}{2}\|T_2\| \left(t\langle xx^*, \phi_2 \rangle + \frac{\langle y^*y, \psi_2 \rangle}{t} \right)$$

$$\leq (\|T_1\| + \|T_2\|) \left(t\langle xx^*, \phi \rangle + \frac{\langle y^*y, \psi \rangle}{t} \right) \qquad (x,y \in \mathfrak{M}).$$

Applying Exercise 6.5.1(i), we see that

$$|V(x,y)| \leq (\|T_1\| + \|T_2\|) \sqrt{\langle xx^*, \phi \rangle} \sqrt{\langle y^*y, \psi \rangle} \qquad (x,y \in \mathfrak{M}).$$

Using Theorem 6.5.5, we obtain $\xi_0, \eta_0 \in \mathfrak{H}$ with

$$\langle x, \phi \rangle = \langle x\xi_0, \xi_0 \rangle \quad (x \in \mathfrak{M}) \qquad \text{and} \qquad \langle y, \psi \rangle = \langle y\eta_0, \eta_0 \rangle \quad (y \in \mathfrak{M})$$

as well as $T_0 \in \mathcal{L}(\mathfrak{H})$ such that

$$V(x,y) = \langle yT_0 x\eta_0, \xi_0 \rangle \qquad (x,y \in \mathfrak{M}). \tag{6.31}$$

From [Dixm 3, 2.5.1. Proposition] or [Ped, 3.3.5. Proposition], we obtain $x_j, y_j \in \mathfrak{M}'$ for $j = 1,2$ such that

$$\left\{\begin{array}{ll} \langle \phi_j, x \rangle = \langle xy_j\xi_0, y_j\xi_0 \rangle & \text{and} \quad P_0 y_j P_0 = y_j \\ \langle \psi_j, x \rangle = \langle xz_j\eta_0, z_j\eta_0 \rangle & \text{and} \quad Q_0 z_j Q_0 = z_j \end{array}\right\} \quad (x \in \mathfrak{M}, \; j = 1, 2),$$

where P_0 and Q_0 are the orthogonal projections onto $\overline{\mathfrak{M}\xi_0}$ and $\overline{\mathfrak{M}\eta_0}$, respectively. Define, for $j = 1, 2$, partial isometries $v_j, w_j \in \mathfrak{M}'$ (How?) such that

$$\xi_j = v_j y_j \xi_0 \quad \text{and} \quad \eta_j = w_j z_j \eta_0 \quad (j = 1, 2).$$

For $j = 1, 2$, let $\tilde{T}_j := y_j^* v_j^* T_0 w_j z_j$, and note that $P_0 \tilde{T}_j Q_0 = \tilde{T}_j$. We then have

$$V(x, y) = \langle yT_j x\eta_j, \xi_j \rangle = \langle y\tilde{T}_j x\eta_0, \xi_0 \rangle \quad (x, y \in \mathfrak{M}). \tag{6.32}$$

Since both $P_0 \tilde{T}_j Q_0 = \tilde{T}_j$ for $j = 1, 2$ and $P_0 T Q_0 = T$, it follows from (6.31) and (6.32) that $T_0 = \tilde{T}_j$ for $j = 1, 2$. Consequently, we have:

$$\langle \mathcal{T}(T_j)\eta_j, \xi_j \rangle = \langle y_j^* v_j^* T w_j z_j \eta_0, \xi_0 \rangle = \langle \mathcal{T}(\tilde{T}_j)\eta_0, \xi_0 \rangle = \langle \mathcal{T}(T_0)\eta_0, \xi_0 \rangle \quad (j = 1, 2).$$

Hence, the definition (6.30) of $\mathbf{T}$ is independent of the particular choices for ξ, η, and T.

It remains to be shown that $\mathbf{T}$ does indeed lie in $\mathcal{L}_{w^*}^2(\mathfrak{M}, \mathbb{C})^*$. Let $V_1, V_2 \in \mathcal{L}_{h,w^*}^2(\mathfrak{M}, \mathbb{C})$. Then, for $j = 1, 2$, there are $\xi_j, \eta_j \in \mathfrak{H}$ and $T_j \in \mathcal{L}(\mathfrak{H})$ such that

$$V_j(x, y) = \langle yT_j x\eta_j, \xi_j \rangle \quad (x, y \in \mathfrak{M}, \; j = 1, 2).$$

The argument in the proof that $\mathbf{T}$ is well-defined shows that we may suppose that $\xi_1 = \xi_2 =: \xi$ and $\eta_1 = \eta_2 =: \eta$, so that

$$(V_1 + V_2)(x, y) = \langle y(T_1 + T_2)x\eta, \xi \rangle \quad (x, y \in \mathfrak{M}).$$

It follows that

$$\langle V_1 + V_2, \mathbf{T} \rangle = \langle \mathcal{T}(T_1 + T_2)\eta, \xi \rangle = \langle \mathcal{T}(T_1)\eta, \xi \rangle + \langle \mathcal{T}(T_2)\eta, \xi \rangle = \langle V_1, \mathbf{T} \rangle + \langle V_2, \mathbf{T} \rangle.$$

The homogeneity of $\mathbf{T}$ is immediate. It is also routine to check that $\mathbf{T}$ is bounded (with $\|\mathbf{T}\| \leq \|\mathcal{T}\|$). $\quad\square$

If $\mathfrak{M}$ is a W^*-algebra, it is immediate (How?) that $\Delta_{\mathfrak{M}}^* \mathfrak{M}_* \subset \mathcal{L}_{h,w^*}(\mathfrak{M}, \mathbb{C})$. The bitranspose of $\Delta_{\mathfrak{M}}$ thus drops to an $\mathfrak{M}$-bimodule homomorphism $\Delta_{h,w^*} : \mathcal{L}_{h,w^*}^2(\mathfrak{M}, \mathbb{C}) \to \mathfrak{M}$. We have the following:

Proposition 6.5.8 *Let $\mathfrak{M}$ be an injective W^*-algebra. Then there is $\mathbf{M} \in \mathcal{L}_{h,w^*}^2(\mathfrak{M}, \mathbb{C})$ with $\|\mathbf{M}\| = 1$ such that*

$$x \cdot \mathbf{M} = \mathbf{M} \cdot x \quad \text{and} \quad x\Delta_{h,w^*}\mathbf{M} = x \quad (x \in \mathfrak{M}). \tag{6.33}$$

Proof Using the GNS-construction, we can suppose that $\mathfrak{M}$ acts as a von Neumann algebra on a Hilbert space $\mathfrak{H}$ such that each normal state of $\mathfrak{M}$ is a vector state. Let $\mathcal{E} : \mathcal{L}(\mathfrak{H}) \to \mathfrak{M}'$ be an expectation, i.e. $\|\mathcal{E}\| = 1$ and, by Theorem 6.2.4, $\mathcal{E} \in {}_{\mathfrak{M}'}\mathcal{L}_{\mathfrak{M}}(\mathcal{L}(\mathfrak{H}))$. It is immediate that

$$x \cdot \mathcal{E} = \mathcal{E} \cdot x \quad (x \in \mathfrak{M}) \quad \text{and} \quad \mathcal{E}e_{\mathfrak{M}} = e_{\mathfrak{M}}. \tag{6.34}$$

Let $\Theta : \mathcal{L}^2_{h,w^*}(\mathfrak{M}, \mathbb{C}) \to {}_{\mathfrak{M}'}\mathcal{L}_{\mathfrak{M}'}(\mathcal{L}(\mathfrak{H}))$ be as in Theorem 6.5.7. Then $\mathbf{M} := \Theta^{-1}(\mathcal{E}) \in \mathcal{L}^2_{h,w^*}(\mathfrak{M}, \mathbb{C})$ such that (6.34) translates into (6.33).   □

Does this already mean that every injective W^*-algebra $\mathfrak{M}$ has a normal, virtual diagonal? Not quite yet: A normal, virtual diagonal is, by definition, an element of $\mathcal{L}^2_{w^*}(\mathfrak{M}, \mathbb{C})$, whereas the element $\mathbf{M}$ in Proposition 6.5.8 lies in $\mathcal{L}^2_{h,w^*}(\mathfrak{M}, \mathbb{C})$. We need two more lemmas:

Lemma 6.5.9 *Let $\mathfrak{M}$ be a finite W^*-algebra. Then there is an $\mathfrak{M}$-bimodule homomorphism* $\theta : \mathcal{L}^2_{h,w^*}(\mathfrak{M}, \mathbb{C}) \to \mathcal{L}^2_{w^*}(\mathfrak{M}, \mathbb{C})$ *with* $\|\theta\| \leq 2$ *and* $\Delta_{h,w^*} = \Delta_{w^*} \circ \theta$.

Proof For each function $f : \mathcal{U}(\mathfrak{M}) \to [0, \infty)$ with finite support such that $\sum_{u \in \mathcal{U}(\mathfrak{M})} f(u) = 1$, define $T_f : \mathfrak{M} \to \mathfrak{M}$ through

$$T_f x := \sum_{u \in \mathcal{U}(\mathfrak{M})} f(u) u x u^* \qquad (x \in \mathfrak{M}).$$

Let $\mathcal{F}$ denote the collection of all finite subsets of $\mathfrak{M}$. For each $F = \{x_1, \dots, x_n\} \in \mathcal{F}$ and $n \in \mathbb{N}$, we can find a function $f_{F,n} : \mathcal{U}(\mathfrak{M}) \to [0, \infty)$ with finite support and $\sum_{u \in \mathcal{U}(\mathfrak{M})} f_{F,n}(u) = 1$ along with $z_1, \dots, z_n \in Z(\mathfrak{M})$ such that

$$\|T_{f_{F,n}} x_j - z_j\| < \frac{1}{n} \qquad (j = 1, \dots, n); \tag{6.35}$$

this follows from [K–R, 8.3.5. Theorem]. Let $\mathrm{tr} : \mathfrak{M} \to Z(\mathfrak{M})$ be the $Z(\mathfrak{M})$-valued trace. Since $\mathrm{tr}(T_{f_{F,n}} x_j) = \mathrm{tr}\, x_j$ for $j = 1, \dots, n$, (6.35) implies that

$$\|\mathrm{tr}\, x_j - z_j\| < \frac{1}{n} \qquad (j = 1, \dots, n)$$

and thus

$$\|T_{f_{F,n}} x_j - \mathrm{tr}\, x_j\| < \frac{2}{n} \qquad (j = 1, \dots, n). \tag{6.36}$$

Let $\mathcal{V}$ be an ultrafilter on $F \times \mathbb{N}$ that dominates the canonical order filter. By (6.36), we have

$$\mathrm{tr}\, x = \lim_{\mathcal{V}} T_{f_{F,n}} x \qquad (x \in \mathfrak{M}). \tag{6.37}$$

For any pair $(F, n) \in \mathcal{F} \times \mathbb{N}$, define $\theta_{F,n} : \mathcal{L}^2_{w^*}(\mathfrak{M}, \mathbb{C}) \to \mathcal{L}^2_{w^*}(\mathfrak{M}, \mathbb{C})$ through

$$\theta_{F,n}(V)(x,y) := \sum_{u \in \mathcal{U}(\mathfrak{M})} f_{F,n}(u) V(xu, u^*y) \qquad (V \in \mathcal{L}^2_{w^*}(\mathfrak{M}, \mathbb{C}),\ x, y \in \mathfrak{M}),$$

and define $\theta_* : \mathcal{L}^2_{w^*}(\mathfrak{M}, \mathbb{C}) \to \mathcal{L}^2(\mathfrak{M}, \mathbb{C})$ by letting

$$\theta_*(V)(x,y) := \lim_{\mathcal{V}} \theta_{F,n}(V)(x,y) \qquad (V \in \mathcal{L}^2_{w^*}(\mathfrak{M}, \mathbb{C}),\ x, y \in \mathfrak{M}).$$

It is easy to see that θ_* is a contractive $\mathfrak{M}$-bimodule homomorphism.

We claim that θ_* attains its values in $\mathcal{L}^2_{h,w^*}(\mathfrak{M}, \mathbb{C})$ and that $\|\theta_*\| \leq 2$ when $\mathcal{L}^2_{h,w^*}(\mathfrak{M}, \mathbb{C})$ is equipped with $\|\cdot\|_{h^*}$. To see this, let $V \in \mathcal{L}^2_{w^*}(\mathfrak{M}, \mathbb{C})$. By Theorem 4.2.5, there are $\phi_1, \phi_2, \psi_1, \psi_2 \in S(\mathfrak{M}) \cap \mathfrak{M}_*$ such that

$$|V(x,y)| \leq \|V\| \sqrt{\langle x^*x, \phi_1 \rangle + \langle xx^*, \phi_2 \rangle} \sqrt{\langle y^*y, \psi_1 \rangle + \langle yy^*, \psi_2 \rangle} \qquad (x, y \in \mathfrak{M}).$$

Fix $(F, n) \in \mathcal{F} \times \mathbb{N}$, and note that

$$|\theta_{F,n}(V)(x, y)|$$

$$= \left| \sum_{u \in \mathcal{U}(\mathfrak{M})} f_{F,n}(u) V(xu, u^*y) \right|$$

$$\leq \|V\| \sum_{u \in \mathcal{U}(\mathfrak{M})} f_{F,n}(u) \sqrt{\langle ux^*xu^*, \phi_1 \rangle + \langle xx^*, \phi_2 \rangle} \sqrt{\langle y^*y, \psi_1 \rangle + \langle uyy^*u^*, \psi_2 \rangle}$$

$$\leq \|V\| \sqrt{\sum_{u \in \mathcal{U}(\mathfrak{M})} f_{F,n}(u) \langle ux^*xu^*, \phi_1 \rangle + \langle xx^*, \phi_2 \rangle}$$

$$\sqrt{\langle y^*y, \psi_1 \rangle + \sum_{u \in \mathcal{U}(\mathfrak{M})} f_{F,n}(u) \langle uyy^*u^*, \psi_2 \rangle}$$

$$= \|V\| \sqrt{\langle T_{f_{F,n}}(x^*x), \phi_1 \rangle + \langle xx^*, \phi_2 \rangle} \sqrt{\langle y^*y, \psi_1 \rangle + \langle T_{f_{F,n}}(yy^*), \psi_2 \rangle} \qquad (x, y \in \mathfrak{M}).$$

Passing to the limit along $\mathcal{V}$ and taking (6.37) into account, we obtain:

$$|\theta_*(V)(x, y)| \leq \|V\| \sqrt{\langle \operatorname{tr} x^*x, \phi_1 \rangle + \langle xx^*, \phi_2 \rangle} \sqrt{\langle y^*y, \psi_1 \rangle + \langle \operatorname{tr} yy^*, \psi_2 \rangle}$$

$$= \|V\| \sqrt{\langle \operatorname{tr} xx^*, \phi_1 \rangle + \langle xx^*, \phi_2 \rangle} \sqrt{\langle y^*y, \psi_1 \rangle + \langle \operatorname{tr} y^*y, \psi_2 \rangle} \qquad (x, y \in \mathfrak{M}).$$

Letting $\phi := \frac{1}{2}(\phi_1 \circ \operatorname{tr} + \phi_2)$ and $\psi := \frac{1}{2}(\psi_1 \circ \operatorname{tr} + \psi_2)$, we see that

$$|\theta_*(V)(x, y)| \leq 2\|V\| \sqrt{\langle xx^*, \phi \rangle} \sqrt{\langle y^*y, \psi \rangle} \qquad (x, y \in \mathfrak{M}).$$

It follows that $\theta_*(V) \in \mathcal{L}^2_{w^*}(\mathfrak{M}, \mathbb{C})$ (Why?), and we see from Theorem 6.5.5 that $\theta_*(V) \in \mathcal{L}^2_{h,w^*}(\mathfrak{M}, \mathbb{C})$ with $\|\theta_*(V)\|_{h^*} \leq 2\|V\|$. Letting $\theta := (\theta_*)^*$, we obtain an $\mathfrak{M}$-bimodule homomorphism from $\mathcal{L}^2_{h,w^*}(\mathfrak{M}, \mathbb{C})^*$ into $\mathcal{L}^2_{w^*}(\mathfrak{M}, \mathbb{C})^*$ with $\|\theta\| \leq 2$.

Let $\mathbf{x} = \sum_{j=1}^n x_j \otimes y_j \in \mathfrak{M} \otimes \mathfrak{M}$. For $(F, n) \in \mathcal{F} \times \mathbb{N}$, let

$$\mathbf{x}_{F,n} = \sum_{j=1}^n \sum_{u \in \mathcal{U}(\mathfrak{M})} f_{F,n}(y) x_j u^* u y_j.$$

For $V \in \mathcal{L}^2_{w^*}(\mathfrak{M}, \mathbb{C})$, we have then:

$$\langle V, \theta(\mathbf{x}) \rangle = \langle \mathbf{x}, \theta_*(V) \rangle$$

$$= \lim_{\mathcal{V}} \sum_{j=1}^n \theta_{F,n}(V)(x_j, y_j)$$

$$= \lim_{\mathcal{V}} \sum_{j=1}^n \sum_{u \in \mathcal{U}(\mathfrak{M})} f_{F,n}(u) V(x_j u^*, u y_j)$$

$$= \lim_{\mathcal{V}} \langle V, \mathbf{x}_{F,n} \rangle,$$

so that

$$\mathbf{x} := w^*\text{-}\lim_{\mathcal{V}} \mathbf{x}_{F,n}.$$

By w^*-continuity, it follows that

$$(\varDelta_{w^*} \circ \theta)(\mathbf{x}) = w^*\text{-}\lim_{\mathcal{V}} \varDelta \mathbf{x}_{F,n} = w^*\text{-}\lim_{\mathcal{V}} \sum_{j=1}^{n} \sum_{u \in \mathcal{U}(\mathfrak{M})} f_{F,n}(u) x_j y_j = \varDelta \mathbf{x}.$$

Again by w^*-continuity, this implies $\varDelta_{h,w^*} = \varDelta_{w^*} \circ \theta$. $\square$

Lemma 6.5.10 *Let $\mathfrak{M}$ be a properly finite W^*-algebra. Then there is an $\mathfrak{M}$-bimodule homomorphism $\theta \colon \mathcal{L}_{h,w^*}^2(\mathfrak{M}, \mathbb{C}) \to \mathcal{L}_{w^*}^2(\mathfrak{M}, \mathbb{C})$ with $\|\theta\| \leq 1$ and $\varDelta_{h,w^*} = \varDelta_{w^*} \circ \theta$.*

Proof As in the proof of Lemma 4.2.8, we find a sequence $(p_n)_{n=1}^{\infty}$ of pairwise orthogonal projections in $\mathfrak{M}$ such that $p_n \sim e_{\mathfrak{M}}$ for all $n \in \mathbb{N}$. It follows that $w^*\text{-}\lim_{n \to \infty} p_n = 0$ (to see this just think of $\mathfrak{M}$ acting as a von Neumann algebra on a Hilbert space, and observe that $p_n \to 0$ in the strong operator topology). By the definition of $\sim$, there is a sequence $(v_n)_{n=1}^{\infty}$ in $\mathcal{I}(\mathfrak{M})$ such that $v_n v_n^* = p_n$ for $n \in \mathbb{N}$. Let $\mathcal{U}$ be a free ultrafilter on $\mathbb{N}$, and define an $\mathfrak{M}$-bimodule homomorphism $\theta_* \colon \mathcal{L}_{w^*}^2(\mathfrak{M}, \mathbb{C}) \to \mathcal{L}^2(\mathfrak{M}, \mathbb{C})$ by letting

$$\theta_*(V)(x,y) := \lim_{\mathcal{U}} V(x v_n^*, v_n y) \qquad (V \in \mathcal{L}_{w^*}^2(\mathfrak{M}, \mathbb{C}),\ x,y \in \mathfrak{M}).$$

Fix $V \in \mathcal{L}_{w^*}^2(\mathfrak{M}, \mathbb{C})$, and let $\phi_1, \phi_2, \psi_1, \psi_2 \in S(\mathfrak{M}) \cap \mathfrak{M}_*$ such that

$$|V(x,y)| \leq \|V\| \sqrt{\langle x^* x, \phi_1 \rangle + \langle x x^*, \phi_2 \rangle} \sqrt{\langle y^* y, \psi_1 \rangle + \langle y y^*, \psi_2 \rangle} \qquad (x,y \in \mathfrak{M}).$$

Let $x, y \in B_1[0, \mathfrak{M}]$. Then we have:

$$|V(x v_n^*, v_n y)| \leq \|V\| \sqrt{\langle v_n x^* x v_n^*, \phi_1 \rangle + \langle x x^*, \phi_2 \rangle} \sqrt{\langle y^* y, \psi_1 \rangle + \langle v_n y y^* v_n^*, \psi_2 \rangle}$$
$$\leq \|V\| \sqrt{\langle v_n v_n^*, \phi_1 \rangle + \langle x x^*, \phi_2 \rangle} \sqrt{\langle y^* y, \psi_1 \rangle + \langle v_n v_n^*, \psi_2 \rangle}.$$

Taking limits along $\mathcal{U}$, we see that

$$|\theta_*(V)(x,y)| \leq \|V\| \sqrt{\langle x x^*, \phi_2 \rangle} \sqrt{\langle y^* y, \psi_1 \rangle} \qquad (x,y \in \mathfrak{M}).$$

The rest of the proof is just like that of Lemma 6.5.9. $\square$

We are now ready for the crowning finale of this chapter:

Theorem 6.5.11 *For a W^*-algebra $\mathfrak{M}$, the following are equivalent:*

(i) $\mathfrak{M}$ *is Connes-amenable.*

(ii) $\mathfrak{M}$ *is injective.*

(iii) $\mathfrak{M}$ *is semidiscrete.*

(iv) $\mathfrak{M}$ *has a normal, virtual diagonal.*

(v) $\mathfrak{M}$ *has a normal, virtual diagonal of norm at most 2.*

Proof (i) $\Longrightarrow$ (ii) is Corollary 6.2.8.

(ii) $\Longleftrightarrow$ (iii) are Corollary 6.4.14 and Theorem 6.4.25.

(iv) $\Longrightarrow$ (i) follows from Theorem 4.4.15, and (v) $\Longrightarrow$ (iv) is trivial.

(ii) $\Longrightarrow$ (v): If $\mathfrak{M}$ is finite or properly infinite, the claim follows from a combination of Proposition 6.5.8 with either Lemma 6.5.9 or Lemma 6.5.10. Let $\mathfrak{M}$ be arbitrary. Then there is $p \in Z(\mathfrak{M})$ such that $\mathfrak{M}_1 := p\mathfrak{M}$ is finite and $\mathfrak{M}_2 := (e_{\mathfrak{M}} - p)\mathfrak{M}$ is properly infinite. By the foregoing $\mathfrak{M}_j$ has a normal, virtual diagonal $\mathbf{M}_j$ of norm at most 2 for $j = 1, 2$. Define $\mathbf{M} \in \mathcal{L}_{w^*}^2(\mathfrak{M}, \mathbb{C})^*$ through

$$\langle V, \mathbf{M} \rangle := \langle V|_{\mathfrak{M}_1 \times \mathfrak{M}_1}, \mathbf{M}_1 \rangle + \langle V|_{\mathfrak{M}_2 \times \mathfrak{M}_2}, \mathbf{M}_2 \rangle \qquad (V \in \mathcal{L}^2_{w^*}(\mathfrak{M}, \mathbb{C}).$$

It is routinely checked (by you) that $\mathbf{M}$ is a normal, virtual diagonal with $\|\mathbf{M}\| \leq 2$. □

And for general C^*-algebras, the circle also closes:

Corollary 6.5.12 *For a C^*-algebra $\mathfrak{A}$, the following are equivalent:*

(i) $\mathfrak{A}^{**}$ *is Connes-amenable, etc.*

(ii) $\mathfrak{A}$ *is nuclear.*

(iii) $\mathfrak{A}$ *is amenable.*

(iv) $\mathfrak{A}$ *is 2-amenable.*

Proof Only (i) $\Longrightarrow$ (iv) still needs proof.

While proving Theorem 4.2.4, we already noted that each $V \in \mathcal{L}^2(\mathfrak{A}, \mathbb{C})$ has a unique extension $\tilde{V} \in \mathcal{L}^2_{w^*}(\mathfrak{A}^{**}, \mathbb{C})$. Let $\tilde{\mathbf{M}} \in \mathcal{L}^2_{w^*}(\mathfrak{A}^{**}, \mathbb{C})^*$ be a normal, virtual diagonal with $\|\tilde{\mathbf{M}}\| \leq 2$. Define $\mathbf{M} \in (\mathfrak{A} \hat{\otimes} \mathfrak{A})^{**} \cong \mathcal{L}^2(\mathfrak{A}, \mathbb{C})^*$ by letting

$$\langle V, \mathbf{M} \rangle := \langle \tilde{V}, \tilde{\mathbf{M}} \rangle \qquad (V \in \mathcal{L}^2(\mathfrak{A}, \mathbb{C})).$$

It is routinely checked (by you, again) that $\mathbf{M}$ is a virtual diagonal for $\mathfrak{A}$. □

We conclude this chapter with more examples of amenable C^*-algebras:

Example 6.5.13 Let $\mathfrak{A}$ be a C^*-algebra. A *composition series* for $\mathfrak{A}$ is an increasing family $(I_\beta)_{0 \leq \beta \leq \alpha}$, where α is an ordinal number, of closed ideals of $\mathfrak{A}$ such that $I_0 = \{0\}$, $I_\alpha = \mathfrak{A}$, and

$$I_\beta = \overline{\bigcup_{\gamma < \beta} I_\gamma} \qquad (6.38)$$

for each limit ordinal $\beta \leq \alpha$. Suppose that $\mathfrak{A}$ is *postliminal* (or GCR or smooth or of type I, whatever you prefer ...), i.e. for each irreducible $*$-representation π of $\mathfrak{A}$ on a Hilbert space $\mathfrak{H}$, we have $\pi(\mathfrak{A}) \supset \mathcal{K}(\mathfrak{H})$ (see, e.g. [Dixm 3] or [Ped] for further information); this includes all C^*-algebras of type I_0 ([Ped, 6.1.6. Corollary]). By [Ped, 6.2.6. Theorem], there is a composition series $(I_\beta)_{0 \leq \beta \leq \alpha}$ such that $I_{\beta+1}/I_\beta$ is of type I_0 for all $\beta < \alpha$. We claim that I_β is amenable for each $\beta \leq \alpha$. Assume towards a contradiction that I_β is not amenable for some $\beta \leq \alpha$; by the well ordering principle, there is a minimal β with this property. This means that I_γ is amenable — and thus 2-amenable by Corollary 6.5.12 — for all $\gamma < \beta$. If β is a limit ordinal, (6.38) and Proposition 2.3.17 imply that I_β is also amenable and thus contradict the assumption. If β is a successor ordinal, i.e. $\beta = \gamma + 1$ for some $\gamma < \beta$, the quotient C^*-algebra I_β/I_γ is of type I_0. By Example 6.4.30, it is nuclear and therefore amenable by Corollary 6.5.12. Since I_γ is amenable due to the choice of β, Theorem 2.3.10 yields the amenability of I_β and thus — again — a contradiction. Since $\mathfrak{A} = I_\alpha$, the C^*-algebra $\mathfrak{A}$ is amenable.

6.6 Notes and comments

Theorem 6.1.7 is essentially [Was 1, (1.9) Corollary], but our approach is a bit different. In [Was 1], S. Wassermann shows that a nuclear W^*-algebra has to be subhomogeneous. Our proof of Theorem 6.1.7, which avoids the nuclearity-amenability nexus, is from [Run 4]. The hardest part in the proof of Theorem 6.1.7 is certainly Example 6.1.5, i.e. the proof that ℓ^∞-$\bigoplus_{n=1}^\infty \mathbb{M}_n$ is not of type (QE) and thus not amenable. Disturbingly enough, this is still the easiest proof available for the non-amenability of ℓ^∞-$\bigoplus_{n=1}^\infty \mathbb{M}_n$. Another drawback of this proof is that it is unforgivingly W^*-algebraic: There seems to be no way to adapt it to yield the non-amenability of ℓ^∞-$\bigoplus_{n=1}^\infty \mathcal{L}(\ell_n^p)$ for $p \in [1, \infty] \setminus \{2\}$. In his very recent preprint [Rea 2], C. J. Read establishes the non-amenability of ℓ^∞-$\bigoplus_{n=1}^\infty \mathcal{L}(\ell_n^p)$ for $p \in [1, \infty] \setminus \{2\}$; interestingly, his argument breaks down for $p = 2$. For W^*-algebras which are biduals of C^*-algebras, Remark 6.1.9 can already be found in [L–L–W].

The notion of an injective von Neumann algebra seems to originate in [Conn 1]. Theorem 6.2.4 is [Tom, Theorem 3.1]. The central result of Section 6.2 is Corollary 6.2.8, which is commonly attributed to A. Connes: In [Conn 2], he proves it under the additional hypothesis that $\mathfrak{M}_*$ is separable; his proof makes use of his deep automorphism group machinery developed in [Conn 1]. The more elementary proof we present — and which does not require any separability assumption — is due to J. W. Bunce and W. L. Paschke ([Bu–P]).

Tensor products of C^*- and W^*-algebras are treated extensively in the literature ([Sak], [Tak 2], [Mur], [W–O]). Our exposition only provides the bare essentials. The notion of nuclearity for C^*-algebras was first introduced (albeit under a different name) by M. Takesaki in [Tak 1]; in particular, that paper contains Theorem 6.3.8. The adjective "nuclear" to describe the C^*-algebras characterized in Definition 6.3.4 is used for the first time in [Lanc 1].

That every semidiscrete W^*-algebra is injective was established by E. G. Effros and E. C. Lance in [E–L]; in the same paper Corollaries 6.4.8 and 6.4.15 are proved. The converse implication has a more involved history: It was first proved by A. Connes for the case of a factor acting on a separable Hilbert space ([Conn 1, Theorem 6]). Using some complicated direct integral theory, M. D. Choi and E. G. Effros extended this result to every W^*-algebra that can be represented as a von Neumann algebra acting on a separable Hilbert space ([Ch–E 1]); this establishes Corollary 6.4.27 for separable $\mathfrak{A}$. Later, Choi and Effros proved Corollary 6.4.27 for arbitrary $\mathfrak{A}$ by reducing it to the separable case ([Ch–E 2]). In [Was 2], S. Wassermann modified Connes' arguments and obtained a direct proof Theorem 6.4.25, which not only worked for arbitrary W^*-algebras, but also avoided the use of Connes' theory of automorphism groups. Subsequently, A. Connes indicated how Wassermann's proof could be simplified even further ([Conn 3]). The proof we present is Wassermann's, with Connes' simplifications incorporated into it.

The Powers–Størmer inequality is from [P–St].

The statement made in Corollary 6.4.5 that every nuclear C^*-algebra has the metric approximation property allows for considerable improvement ([Ch–E 3]). The C^*-algebra $C_{\lambda_2}^*(\mathbb{F}_2)$ is not amenable, but nevertheless has the metric approximation property ([Haa 1]).

Normal, virtual diagonals occur for the first time in [Haa 2], albeit only implicitly. Although it is not formulated as a theorem, Haagerup shows in this paper that every injective W^*-algebra has a normal, virtual diagonal of norm one; from there, he obtains that every nuclear C^*-algebra is 1-amenable. In particular, he obtains better estimates than we do in Theorem 6.5.11 and Corollary 6.5.12. His proof, however, relies on the deep fact that every injective W^*-algebra is already approximately finite-dimensional (see below). Our more elementary approach to normal, virtual diagonals is due to E. G. Effros and A. Kishimoto ([E–K]). In [Eff], Effros gives a direct proof for the implication from Connes-amenability to the existence of a normal, virtual diagonal; that proof is similar to that of Theorem 2.2.4.

There are further significant properties of W^*-algebras that are equivalent to Connes-amenability.

A von Neumann algebra $\mathfrak{M}$ acting on a Hilbert space $\mathfrak{H}$ is said to have *Schwartz' property* (P) if, for each $T \in \mathcal{L}(\mathfrak{H})$, the weak operator closure of the convex hull of $\{uTu^* : u \in \mathcal{U}(\mathfrak{M})\}$ has non-empty intersection with $\mathfrak{M}'$ ([Schwa]). A proof for the implication that Schwartz' property (P) implies injectivity is already contained in [Schwa] (5. Lemma). Although defined for a von Neumann algebra acting on a particular Hilbert space, Schwartz' property (P) is in fact Hilbert space independent ([Hak]).

A W^*-algebra $\mathfrak{M}$ is called *AFD* (short for: *approximately finite-dimensional*) if there is a directed family $(\mathfrak{M}_\alpha)_\alpha$ of finite-dimensional W^*-subalgebras of $\mathfrak{M}$ such that $\bigcup_\alpha \mathfrak{M}_\alpha$ is w^*-dense in $\mathfrak{M}$. It is relatively easy to see that an AFD W^*-algebra is Connes-amenable ([J–K–R]). May we suggest the proof as an exercise? The converse is — in my opinion — one of the most baffling results in operator algebra theory. It was originally proved for factors acting on separable Hilbert spaces by A. Connes ([Conn 1]). Later, an alternative proof — still for factors on separable Hilbert spaces — was published by U. Haagerup ([Haa 4]). Haagerup's proof is easier than Connes' in the sense that it avoids Connes' automorphism group machinery; it is still very impressive. A case in which Haagerup's argument is particularly easy is that of a properly infinite factor; in fact, as Haagerup remarks ([Haa 4, p. 200]), it carries over to general properly infinite W^*-algebras. A shorter argument for finite W^*-algebras was subsequently given by S. Popa ([Pop]). With the usual structure theory, this establishes that every Connes-amenable W^*-algebra is indeed AFD. Since finite-dimensional W^*-algebras trivially have Schwartz' property (P), this and [Schwa, 2. Lemma] imply that every Connes-amenable W^*-algebra has Schwartz' property (P) as well.

There is also a characterization of the Connes-amenable von Neumann algebras in the spirit of Chapter 5: A W^*-algebra is Connes-amenable if and only if its predual is injective ([Hel 4]).

In [Joh 1], B. E. Johnson defined a unital C^*-algebra $\mathfrak{A}$ to be *strongly amenable* if, for every Banach $\mathfrak{A}$-bimodule and for every $D \in \mathcal{Z}^1(\mathfrak{A}, E^*)$, there is ϕ in the w^*-closed, convex hull of $\{-(Du) \cdot u^* : u \in \mathcal{U}(\mathfrak{A})\}$ with $D = \mathrm{ad}_\phi$ (a non-unital C^*-algebra is defined as strongly amenable if its unitization is strongly amenable). All postliminal C^*-algebras are strongly amenable ([Joh 1, Theorem 7.9]). Strong amenability implies symmetric amenability ([Joh 8]). The C^*-algebras $\mathcal{O}_n$ with $2 \leq n < \infty$, as introduced by J. Cuntz ([Cun]), however, are nuclear (and thus amenable), but fail to be symmetrically amenable ([Joh 8]).

In [Choi], M. D. Choi constructed a non-nuclear C^*-subalgebra of $\mathcal{O}_2$ thus showing that amenability for C^*-algebras does not always carry over to C^*-subalgebras.

A C^*-algebra $\mathfrak{A}$ is called *exact* if, for each short exact sequence

$$\{0\} \to I \to \mathfrak{B} \to \mathfrak{B}/I \to \{0\}$$

of C^*-algebras, the sequence

$$\{0\} \to \mathfrak{A} \tilde{\otimes}_{\min} I \to \mathfrak{A} \tilde{\otimes}_{\min} \mathfrak{B} \to \mathfrak{A} \tilde{\otimes}_{\min} (\mathfrak{B}/I) \to \{0\} \tag{6.39}$$

is also exact ([Kir 1]). Every quotient of a C^*-subalgebra of a nuclear C^*-algebra is exact ([Ar–B]). More generally, the class of exact C^*-algebras is closed under taking quotients and subalgebras ([Kir 2]). In [K–Ph], E. Kirchberg and N. C. Phillips show that every unital, separable, exact C^*-algebra has a unital embedding into $\mathcal{O}_2$, so that the class of separable, exact C^*-algebras consists precisely of the C^*-subalgebras of separable, nuclear C^*-algebras.

The notions of nuclearity, injectivity, semidiscreteness, and exactness all extend to the context of operator spaces ([E–R]).

7 Operator amenability

Our discussion of amenability of von Neumann algebras showed that Definition 2.1.9 is not well-suited to deal with every class of Banach algebras: Since it ignores the dual space structure of W^*-algebras, it is too strong to encompass sufficiently many interesting examples. The "right" notion of amenability for W^*-algebras is Connes-amenability, which takes the additional structure into account.

There is an analoguous situation for Fourier algebras of locally compact groups: For all locally compact abelian groups G, the Fourier algebra $A(G)$ is amenable, but there are compact groups G, for which $A(G)$ fails to be amenable. Hence, already for very well behaved groups G, the Fourier algebra $A(G)$ behaves badly (if we define good behavior as being amenable).

The reason why Definition 2.1.9 is inappropriate to characterize the Fourier algebras of amenable, locally compact groups is that — as for von Neumann algebras — it ignores an important additional structure of Fourier algebras: They are not only Banach algebras, but operator Banach algebras (Definition D.4.1) in a canonical way (Examples D.4.3(c)). In order to capture the amenability of a locally compact group G through some sort of amenability of $A(G)$, we have to develop a notion of amenability that recognizes the operator space structure which $A(G)$ has as the predual of $VN(G)$. This notion is called *operator amenability*, and as we shall see, G is amenable precisely when $A(G)$ is operator amenable.

Of course, a discussion of operator amenability requires a certain background in the theory of operator spaces. The definitions and results needed in this chapter are collected in Appendix D.

7.1 Bounded approximate identities for Fourier algebras

If G is a locally compact, abelian group with dual group Γ, then $A(G) \cong L^1(\Gamma)$ is certainly amenable. So, which are the locally compact groups G for which $A(G)$ is amenable?

We will first deal with a seemingly more modest question: For which locally compact groups G does $A(G)$ have a bounded approximate identity?

We start with a lemma that improves Exercise 1.1.6(ii):

Lemma 7.1.1 *Let G be an amenable, locally compact group. Then, for each compact subset K of G and each $\epsilon > 0$, there is $f \in P(G)$ such that*

$$\|\delta_g * f - f\|_1 < \epsilon \qquad (g \in K).$$

Proof Fix a non-negative function $f_0 \in P(G)$. Since λ_1 is a representation of G on $L^1(G)$, there is a neighborhood U of e in G such that

$$\|\delta_g * f_0 - f_0\|_1 = \|\lambda_1(g)f_0 - f_0\|_1 < \frac{\epsilon}{2} \qquad (g \in U). \qquad (7.1)$$

Since G is amenable, there is a left invariant mean on $L^\infty(G)$; in fact, as an inspection of the proof of Theorem 1.1.9 shows, this mean can be chosen to be even topologically left invariant. Proceeding as in Exercise 1.1.6, we obtain a net $(f_\alpha)_\alpha$ in $P(G)$ such that

$$\|\tilde{f} * f_\alpha - f_\alpha\|_1 \to 0 \qquad (\tilde{f} \in P(G))$$

and thus, in particular,

$$\|\delta_g * f_0 * f_\alpha - f_\alpha\|_1 \to 0 \qquad (g \in G). \qquad (7.2)$$

Since K is compact, there are $g_1, \ldots, g_n \in K$ such that $K \subset \bigcup_{j=1}^n Ug_j$. By (7.2), there is $f_{\alpha_0} \in P(G)$ such that

$$\|\delta_{g_j} * f_0 * f_{\alpha_0} - f_{\alpha_0}\|_1 < \frac{\epsilon}{4} \quad (j = 1, \ldots, n) \qquad \text{and} \qquad \|f_0 * f_{\alpha_0} - f_{\alpha_0}\|_1 < \frac{\epsilon}{4},$$

so that

$$\|\delta_{g_j} * f_0 * f_{\alpha_0} - f_0 * f_{\alpha_0}\|_1$$
$$\leq \|\delta_{g_j} * f_0 * f_{\alpha_0} - f_{\alpha_0}\|_1 + \|f_0 * f_{\alpha_0} - f_{\alpha_0}\|_1 < \frac{\epsilon}{2} \qquad (j = 1, \ldots, n). \qquad (7.3)$$

Let $f := f_0 * f_{\alpha_0}$, and let $h \in K$. Then there is $j \in \{1, \ldots, n\}$ such that $hg_j^{-1} \in U$. We then obtain:

$$\|\delta_h * f - f\|_1$$
$$= \|\delta_{hg_j^{-1}} * \delta_{g_j} * f_0 * f_{\alpha_0} - f_0 * f_{\alpha_0}\|_1$$
$$\leq \|\delta_{hg_j^{-1}} * \delta_{g_j} * f_0 * f_{\alpha_0} - \delta_{hg_j^{-1}} * f_0 * f_{\alpha_0}\|_1 + \|\delta_{hg_j^{-1}} * f_0 * f_{\alpha_0} - f_0 * f_{\alpha_0}\|_1$$
$$= \|\delta_{g_j} * f_0 * f_{\alpha_0} - f_0 * f_{\alpha_0}\|_1 + \|\delta_{hg_j^{-1}} * f_0 - f_0\|_1$$

(Why do we have equality here?)

$$< \frac{\epsilon}{2} + \|\delta_{hg_j^{-1}} * f_0 - f_0\|_1, \qquad \text{by (7.3),}$$
$$< \frac{\epsilon}{2} + \frac{\epsilon}{2}, \qquad \text{by (7.1), since } hg_j^{-1} \in U,$$
$$= \epsilon.$$

This proves the claim.   □

Let G be a locally compact group. We denote the norm closure of $\tilde{\lambda}_2(L^1(G))$ by $C^*_{\lambda_2}(G)$.

Exercise 7.1.1 Let G be a locally compact group.

(i) Show that the module action of $A(G)$ on $VN(G) = A(G)^*$ restricted to $\tilde{\lambda}_2(L^1(G))$ is given by pointwise multiplication:

$$f \cdot \tilde{\lambda}_2(\tilde{f}) = \tilde{\lambda}_2(\tilde{f}) \cdot f = \tilde{\lambda}_2(f\tilde{f}) \qquad (f \in A(G), \tilde{f} \in L^1(G)).$$

(ii) Show that $C^*_{\lambda_2}(G)$ is a closed $A(G)$-submodule of $VN(G)$.

Exercise 7.1.2 Let G and H be locally compact groups. Show that

$$C^*_{\lambda_2}(G \times H) \cong C^*_{\lambda_2}(G) \tilde{\otimes}_{\min} C^*_{\lambda_2}(H).$$

Lemma 7.1.2 *Let G be a locally compact group. Then the following are equivalent:*

(i) *There is $\phi \in C^*_{\lambda_2}(G)^*$ such that*

$$f \cdot \phi = \phi \cdot f = f \qquad (f \in A(G)).$$

(ii) *There is $\phi \in VN(G)^*$ such that*

$$f \cdot \phi = \phi \cdot f = f \qquad (f \in A(G)).$$

Morover, if (i) *and* (ii) *hold, then ϕ in both* (i) *and* (ii) *can be chosen as a state.*

Proof (i) $\Longrightarrow$ (ii): We may view ϕ as an element of $C^*_{\lambda_2}(G)^{***}$. The universal property of $C^*_{\lambda_2}(G)^{**}$ implies that $\tilde{\lambda}_2$ has an extension $\bar{\lambda}_2$ to $C^*_{\lambda_2}(G)^{**}$ as a W^*-representation with range $VN(G)$. Let $p \in Z(C^*_{\lambda_2}(G)^{**})$ be such that $\bar{\lambda}_2$ restricted to $pC^*_{\lambda_2}(G)^{**}$ is an isomorphism onto $VN(G)$. We may thus view $p \cdot \phi$ as an element of $VN(G)^*$. We obtain:

$$\langle T, f \cdot \phi - f \cdot (p \cdot \phi) \rangle = \langle T \cdot f, \phi - p \cdot \phi \rangle = 0 \qquad (f \in A(G), T \in VN(G)).$$

It follows that

$$f \cdot (p \cdot \phi) = (p \cdot \phi) \cdot f = f \qquad (f \in A(G)).$$

(ii) $\Longrightarrow$ (i): Restrict ϕ to $C^*_{\lambda_2}(G)$.

Suppose that (i) holds. It follows from Exercise 7.1.1(i) that the module action of $A(G)$ on $C^*_{\lambda_2}(G)^* \subset L^\infty(G)$ is given by pointwise multiplication. Hence, ϕ as in (i) is equal to the constant function 1. This implies that ϕ is multiplicative and thus a state. Our argument for (i) $\Longrightarrow$ (ii) shows that if ϕ in (i) is a state, then ϕ in (ii) can also be chosen as a state. $\square$

Theorem 7.1.3 (Leptin's theorem) *The following are equivalent for a locally compact group G:*

(i) *G is amenable.*
(ii) *$A(G)$ has an approximate identity bounded by 1.*
(iii) *$A(G)$ has a bounded approximate identity.*

Proof (i) $\Longrightarrow$ (ii): Let $(f_\alpha)_\alpha$ be a net in $P(G)$ such that

$$\|\delta_g * f_\alpha - f_\alpha\|_1 \to 0$$

uniformly on all compact subsets of G (the existence of such a net is guaranteed by Lemma 7.1.1). Let $(\xi_\alpha)_\alpha$ be defined by $\xi_\alpha := f_\alpha^{\frac{1}{2}}$. Then $(\xi_\alpha)_\alpha$ is a net in $L^2(G)$ with $\|\xi_\alpha\|_2 = 1$ such that — as a consequence of Exercise 4.4.5 —

$$\|\lambda_2(g)\xi_\alpha - \xi_\alpha\|_2 \to 0$$

uniformly on all compact subsets of G. It follows (How?) that

$$\|\tilde{\lambda}_2(f)\xi_\alpha - \langle f, 1\rangle\xi_\alpha\|_2 \to 0 \qquad (f \in L^1(G)).$$

Define $(e_\alpha)_\alpha$ by letting $e_\alpha := \xi_\alpha * \check{\bar{\xi}}_\alpha$; then $(e_\alpha)_\alpha$ is a net in $A(G)$ bounded by one such that

$$|\langle e_\alpha, \tilde{\lambda}_2(f)\rangle - \langle f, 1\rangle| = |\langle \tilde{\lambda}_2(f)\xi_\alpha, \xi_\alpha\rangle - \langle f, 1\rangle\langle \xi_\alpha, \xi_\alpha\rangle|$$
$$\leq \|\tilde{\lambda}_2(f)\xi_\alpha - \langle f, 1\rangle\xi_\alpha\|_2$$
$$\to 0 \qquad (f \in L^1(G)).$$

It follows that the L^∞-function $\phi \equiv 1$ has a continuous extension ϕ to $C_{\lambda_2}^*(G)$, namely the w^*-limit of $(e_\alpha)_\alpha$ in $C_{\lambda_2}^*(G)^*$. It is clear that

$$f \cdot \phi = \phi \cdot f = f \qquad (f \in A(G)).$$

By Lemma 7.1.2, there is $E \in VN(G)^* = A(G)^{**}$ such that

$$f \cdot E = E \cdot f = f \qquad (f \in A(G)).$$

Also by Lemma 7.1.2, we may suppose that $\|E\| = 1$. Let $(\tilde{e}_\alpha)_\alpha$ be a net in $B_1[0, A(G)]$ such that $\tilde{e}_\alpha \to E$ in the w^*-topology. It follows that

$$f = w\text{-}\lim_\alpha f\tilde{e}_\alpha = w\text{-}\lim_\alpha \tilde{e}_\alpha f \qquad (f \in A(G)).$$

Passing to convex combinations, we obtain an approximate identity for $A(G)$ bounded by 1.

(ii) $\Longrightarrow$ (iii) is trivial.

(iii) $\Longrightarrow$ (i): Let $(e_\alpha)_\alpha$ be a bounded approximate identity for $A(G)$. Without loss of generality suppose that $(e_\alpha)_\alpha$ has a w^*-limit $\phi \in VN(G)^*$. It is clear that

$$f \cdot \phi = \phi \cdot f = f \qquad (f \in A(G)).$$

By Lemma 7.1.2, we may suppose that $\phi \in S(VN(G))$. Since $S(VN(G)) \cap A(G)$ is w^*-dense in $S(VN(G))$, we can find a net $(f_\beta)_\beta$ in $S(VN(G)) \cap A(G)$ with $\phi = w^*\text{-}\lim_\beta f_\beta$. Note that $f_\beta \to 1$ pointwise on G. By Theorem A.3.8, there is a net $(\xi_\beta)_\beta$ in $L^2(G)$ with $\|\xi_\beta\| = 1$ such that $f_\beta = \xi_\beta * \check{\bar{\xi}}_\beta$ (Why precisely?). We then have:

$$\|\lambda_2(g)\xi_\beta - \xi_\beta\|_2^2 = \langle \lambda_2(g)\xi_\beta - \xi_\beta, \lambda_2(g)\xi_\beta - \xi_\beta\rangle$$
$$= \|\lambda_2(g)\xi_\beta\|_2^2 + \|\xi_\beta\|_2^2 - 2\operatorname{Re}\langle \lambda_2(g)\xi_\beta, \xi_\beta\rangle$$
$$= 2(1 - \operatorname{Re} f_\beta(g))$$
$$\to 0 \qquad (g \in G).$$

Let $m_\beta := \xi_\beta\bar{\xi}_\beta$ (pointwise product). Then $m_\beta \in P(G)$, and

$$\|\delta_g * m_\beta - m_\beta\|_1 \leq \|(\lambda_2(g)\xi_\beta)\lambda_2(g)\bar{\xi}_\beta - (\lambda_2(g)\xi_\beta)\bar{\xi}_\beta\|_1 + \|(\lambda_2(g)\xi_\beta)\bar{\xi}_\beta - \xi_\beta\bar{\xi}_\beta\|_1$$
$$\leq \|\lambda_2(g)\xi_\beta\|_2\|\lambda_2(g)\bar{\xi}_\beta - \bar{\xi}_\beta\|_2 + \|\lambda_2(g)\xi_\beta - \xi_\beta\|_2\|\bar{\xi}_\beta\|_2$$
$$= 2\|\lambda_2(g)\xi_\beta - \xi_\beta\|_2$$
$$\to 0.$$

Any w^*-accumulation point of $(m_\beta)_\beta$ is then a left invariant mean on $L^\infty(G)$. $\square$

Corollary 7.1.4 *Let G be a locally compact group such that $A(G)$ is amenable. Then G is amenable.*

7.2 (Non-)amenability of Fourier algebras

In view of how nicely the notions of amenability for Banach algebras and for locally compact groups have been fitting together so far, one might be tempted to conjecture that the converse of Corollary 7.1.4 is true as well. As we shall see, this is not the case: There are compact groups G such that $A(G)$ fails to be amenable.

Definition 7.2.1 Let G be a locally compact group. A unitary representation π of G on a Hilbert space $\mathfrak{H}$ is called *irreducible* if $\mathfrak{H}$ and $\{0\}$ are the only closed subspaces of $\mathfrak{H}$ that are invariant under $\pi(G)$. The dimension $\dim \pi$ of π is defined as the Hilbert space dimension of the space $\mathfrak{H}$.

Definition 7.2.2 Let G be a locally compact group, and let π and ρ be unitary representations of G on Hilbert spaces $\mathfrak{H}$ and $\mathfrak{K}$, respectively. Then π and ρ are called *unitarily equivalent*, if there is a unitary $U \in \mathcal{L}(\mathfrak{H}, \mathfrak{K})$ such that

$$\rho(g) = U\pi(g)U^* \qquad (g \in G).$$

It is easy to see that if two unitary representations of a locally compact group are unitarily equivalent and one of them is irreducible, then so is the other. Also, any two unitarily equivalent unitary representations have the same dimension.

Definition 7.2.3 Let G be a locally compact group. Then $\hat{G}$ denotes the collection of equivalence classes of irreducible unitary representations of G with respect to unitary equivalence.

Remark 7.2.4 If G is abelian, then the irreducible, unitary representations of G are one-dimensional. Hence, two irreducible, unitary representations of G are unitarily equivalent if and only if they coincide. It follows that $\hat{G}$ can be identified with the dual group of G.

For the sake of notational simplicity, we will use the same symbol for an irreducible, unitary representation of a locally compact group G and for its equivalence class in $\hat{G}$.

Theorem 7.2.5 *Let G be a compact group. Then $\dim \pi < \infty$ for each $\pi \in \hat{G}$.*

Exercise 7.2.1 Read and understand the proof of the preceding theorem ([H–R, (22.13) Theorem]) in [H–R].

Let G be a locally compact group, and let π be a unitary representation of G on a Hilbert space $\mathfrak{H}$. Then we obtain a *-representation $\tilde{\pi}$ of $L^1(G)$ on $\mathfrak{H}$, i.e. a *-homomorphism $\tilde{\pi} \colon L^1(G) \to \mathcal{L}(\mathfrak{H})$ by letting

$$\langle \tilde{\pi}(f)\xi, \eta \rangle := \int_G f(g)\langle \pi(g)\xi, \eta \rangle \, dm_G(g) \qquad (f \in L^1(G)). \tag{7.4}$$

If G is compact and π is irreducible, $\tilde{\pi}$ is a *-homomorphism into $\mathbb{M}_{\dim \pi}$, so that $\operatorname{Tr} \tilde{\pi}(f)$ is well-defined for $f \in L^1(G)$ where Tr is the canonical trace on $\mathbb{M}_{\dim \pi}$. Note that, if π_1 and π_2 have the same equivalence class in $\hat{G}$, then $\operatorname{Tr} \tilde{\pi}_1(f) = \operatorname{Tr} \tilde{\pi}_2(f)$ for $f \in L^1(G)$ (Why?).

In the sequel, we denote the nuclear norm on $\mathbb{M}_n$ by $\|\cdot\|_{\mathcal{N}}$.

Proposition 7.2.6 *Let G be a compact group. Then:*

(i) *An element $f \in L^1(G)$ has a representative in $A(G)$ if and only if*

$$\sum_{\pi \in \hat{G}} (\dim \pi)\|\tilde{\pi}(f)\|_{\mathcal{N}} < \infty;$$

in this case, we have

$$\|f\|_{A(G)} = \sum_{\pi \in \hat{G}} (\dim \pi)\|\tilde{\pi}(f)\|_{\mathcal{N}}.$$

(ii) *If $\phi: \hat{G} \to \bigcup_{n \in \mathbb{N}} \mathbb{M}_n$ is a function such that $\phi(\pi) \in \mathbb{M}_{\dim \pi}$ for each $\pi \in \hat{G}$ and*

$$\sum_{\pi \in \hat{G}} (\dim \pi)\|\phi(\pi)\|_{\mathcal{N}} < \infty,$$

then there is $f \in A(G)$ such that $\tilde{\pi}(f) = \phi(\pi)$ for all $\pi \in \hat{G}$.

(iii) *If $f \in A(G)$, then*

$$f(e_G) = \sum_{\pi \in \hat{G}} (\dim \pi)\mathrm{Tr}\,\tilde{\pi}(f).$$

Exercise 7.2.2 Identify the statements of Proposition 7.2.6 in [H–R, §34] and understand their proofs (the notation of [H–R] is different from ours; for example, $\hat{G}$ is denoted by Σ and $A(G)$ is by $\mathfrak{R}(G)$).

Exercise 7.2.3 Let G be a compact group, and suppose that $m_G(G) = 1$. Show that $L^2(G)$ and $A(G)$ are Banach algebras with respect to convolution.

Definition 7.2.7 Let G be a compact group. Then $A_\gamma(G)$ denotes the range of the map

$$\gamma: A(G)\hat{\otimes}A(G) \to A(G), \quad f \otimes g \mapsto f * g$$

equipped with the quotient norm $\|\cdot\|_\gamma$ induced by γ.

There is an analogue of Proposition 7.2.6 for $A_\gamma(G)$:

Proposition 7.2.8 *Let G be a compact group.*

(i) *An element $f \in L^1(G)$ has a representative in $A_\gamma(G)$ if and only if*

$$\sum_{\pi \in \hat{G}} (\dim \pi)^2\|\tilde{\pi}(f)\|_{\mathcal{N}} < \infty;$$

in this case, we have

$$\|f\|_\gamma = \sum_{\pi \in \hat{G}} (\dim \pi)^2\|\tilde{\pi}(f)\|_{\mathcal{N}}.$$

(ii) *If $\phi: \hat{G} \to \bigcup_{n \in \mathbb{N}} \mathbb{M}_n$ is a function such that $\phi(\pi) \in \mathbb{M}_{\dim \pi}$ for each $\pi \in \hat{G}$ and*

$$\sum_{\pi \in \hat{G}} (\dim \pi)^2\|\phi(\pi)\|_{\mathcal{N}} < \infty,$$

then there is $f \in A_\gamma(G)$ such that $\tilde{\pi}(f) = \phi(\pi)$ for all $\pi \in \hat{G}$.

Proof First observe that, for each $\pi \in \hat{G}$, the operator γ drops to a metric surjection $\gamma_\pi : \mathbb{M}_{\dim \pi} \hat{\otimes} \mathbb{M}_{\dim \pi} \to \mathbb{M}_{\dim \pi}$, where $\mathbb{M}_{\dim \pi}$ is equipped with $\| \cdot \|_\mathcal{N}$ (Why is this true?). The result then follows from Proposition 7.2.6. $\square$

Exercise 7.2.4 Work out the proof of Proposition 7.2.8 in detail.

We now establish, for a compact group G, a link between the amenability of $A(G)$ and the existence of certain nets in $A_\gamma(G)$:

Lemma 7.2.9 *Let G be a compact group such that $A(G)$ is amenable. Then there is a bounded net $(f_\alpha)_\alpha$ in $A_\gamma(G)$ such that:*

(i) $\|f_\alpha f\|_\gamma \to 0$ *for each $f \in A_\gamma(G)$ with $f(e_G) = 0$;*
(ii) $f_\alpha(e_G) \to 1$;
(iii) $f_\alpha \in Z(L^1(G))$.

Proof Let $(\mathbf{m}_\alpha)_\alpha$ be an approximate diagonal for $A(G)$. Define

$$\check{\gamma} : A(G) \hat{\otimes} A(G) \to A(G), \quad f \otimes g \mapsto \gamma(f \otimes \check{g}).$$

It is easy to see that $\check{\gamma}$ is a continuous map onto $A_\gamma(G)$. Let $f_\alpha := \check{\gamma}(\mathbf{m}_\alpha)$, so that the net $(f_\alpha)_\alpha$ is bounded in $A_\gamma(G)$.

Let $\mathbf{f} \in A(G) \hat{\otimes} A(G)$ belong to $\ker \Delta_{A(G)}$. Then it follows easily (Provide the details!) that $\mathbf{m}_\alpha \bullet \mathbf{f} \to 0$. Define $\Delta_* : \mathcal{C}(G) \to \mathcal{C}(G \times G)$ by letting

$$(\Delta_* f)(g, h) := f(gh^{-1}) \qquad (g, h \in G).$$

It is not hard to see (again: provide the details) that $(\Delta_*(\check{\gamma}(\mathbf{m}_\alpha)))_\alpha$ is also an approximate diagonal for $A(G)$, so that $\Delta_*(\check{\gamma}(\mathbf{m}_\alpha)) \bullet \mathbf{f} \to 0$ holds. Let $f \in A_\gamma(G)$ be such that $f(e_G) = 0$. By Exercise 7.2.5 below, it follows that $\Delta_* f \in A(G) \hat{\otimes} A(G)$, and $f(e_G) = 0$ implies that $\Delta_* f \in \ker \Delta_{A(G)}$. We then have:

$$\Delta_*(f_\alpha f) = (\Delta_* f_\alpha) \bullet (\Delta_* f) = \Delta_*(\check{\gamma}(\mathbf{m}_\alpha)) \bullet \Delta_* f \to 0.$$

Since, again by Exercise 7.2.5 below, $\Delta_* : A_\gamma(G) \to A(G) \hat{\otimes} A(G)$ is an isometry, it follows that $\|f_\alpha f\|_\gamma \to 0$.

Since $A(G)$ is unital (Why?), $\Delta_{A(G)} \mathbf{m}_\alpha \to e_{A(G)}$. Viewing each $\mathbf{m}_\alpha$ as a function on $G \times G$, we thus obtain:

$$f_\alpha(e_G) = \check{\gamma}(\mathbf{m}_\alpha)(e_G) = \int_G \mathbf{m}_\alpha(g, g) \, dm_G(g) \to 1.$$

This proves (ii).

For (iii), replace the net $(f_\alpha)_\alpha$ be the net $(f'_\alpha)_\alpha$ defined through

$$f'_\alpha(h) := \int_G f_\alpha(ghg^{-1}) \, dm_G(g) \qquad (h \in G),$$

where $m_G(G) = 1$. It is not hard to check that, if $(f_\alpha)_\alpha$ satisfies (i) and (ii), so does $(f'_\alpha)_\alpha$, and from the definition it is immediate that $(f'_\alpha)_\alpha$ lies in $Z(L^1(G))$. $\square$

Exercise 7.2.5 With notation as in the proof of Lemma 7.2.9, show that Δ_* maps $A_\gamma(G)$ into $A(G) \hat{\otimes} A(G)$ and that $\Delta_*|_{A_\gamma(G)}$ is an isometric right inverse of $\check{\gamma}$.

We can now show that $A(G)$ is *not* amenable for certain compact groups G:

Theorem 7.2.10 *Let G be an infinite compact group such that, for each $n \in \mathbb{N}$, the set $\{\pi \in \hat{G} : \dim \pi = n\}$ is finite. Then $A(G)$ is not amenable.*

Proof Assume that $A(G)$ is amenable, and let $(f_\alpha)_\alpha$ be a net in $A_\gamma(G)$ as specified in Lemma 7.2.9. Let $C \geq 0$ be such that $\sup_\alpha \|f_\alpha\|_\gamma \leq C$. By Proposition 7.2.8(i), we have

$$\sum_{\pi \in \hat{G}} (\dim \pi)^2 \|\tilde{\pi}(f_\alpha)\|_{\mathcal{N}} \leq C.$$

Since $\| \cdot \|_\gamma \geq \| \cdot \|_\infty$, Lemma 7.2.9(i) implies that $f_\alpha \to 0$ uniformly on compact subsets of $G \setminus \{e_G\}$. From (7.4), it is then follows that $\tilde{\pi}(f_\alpha) \to 0$ for all $\pi \in \hat{G}$ (here we need the fact that G is infinite, so that $m_G(\{e_G\}) = 0$). Fix $N \in \mathbb{N}$. Then we have:

$$\sum_{\pi \in \hat{G}} (\dim \pi) \|\tilde{\pi}(f_\alpha)\|_{\mathcal{N}} = \sum_{\substack{\pi \in \hat{G} \\ \dim \pi < N}} (\dim \pi) \|\tilde{\pi}(f_\alpha)\|_{\mathcal{N}} + \sum_{\substack{\pi \in \hat{G} \\ \dim \pi \geq N}} (\dim \pi) \|\tilde{\pi}(f_\alpha)\|_{\mathcal{N}}$$

$$\leq \sum_{\substack{\pi \in \hat{G} \\ \dim \pi < N}} (\dim \pi) \|\tilde{\pi}(f_\alpha)\|_{\mathcal{N}} + \frac{1}{N} \sum_{\substack{\pi \in \hat{G} \\ \dim \pi \geq N}} (\dim \pi)^2 \|\tilde{\pi}(f_\alpha)\|_{\mathcal{N}}$$

$$\leq \sum_{\substack{\pi \in \hat{G} \\ \dim \pi < N}} (\dim \pi) \|\tilde{\pi}(f_\alpha)\|_{\mathcal{N}} + \frac{C}{N}.$$

Let $\epsilon > 0$, and choose $N \in \mathbb{N}$ so large that $\frac{C}{N} \leq \frac{\epsilon}{2}$. We obtain:

$$\sum_{\pi \in \hat{G}} (\dim \pi) \|\tilde{\pi}(f_\alpha)\|_{\mathcal{N}} \leq \sum_{\substack{\pi \in \hat{G} \\ \dim \pi < N}} (\dim \pi) \|\tilde{\pi}(f_\alpha)\|_{\mathcal{N}} + \frac{\epsilon}{2}. \tag{7.5}$$

The hypothesis on G implies that the second sum in (7.5) is finite, so that

$$\sum_{\substack{\pi \in \hat{G} \\ \dim \pi < N}} (\dim \pi) \|\tilde{\pi}(f_\alpha)\|_{\mathcal{N}} < \frac{\epsilon}{2}$$

for large α; it follows that

$$\sum_{\pi \in \hat{G}} (\dim \pi) \|\tilde{\pi}(f_\alpha)\|_{\mathcal{N}} < \epsilon$$

for large α. Hence, we have

$$\sum_{\pi \in \hat{G}} (\dim \pi) \|\tilde{\pi}(f_\alpha)\|_{\mathcal{N}} \to 0. \tag{7.6}$$

By Lemma 7.2.9(iii), the net $(f_\alpha)_\alpha$ belongs to $Z(L^1(G))$. Consequently, for each $\pi \in \hat{G}$, the net $(\tilde{\pi}(f_\alpha))_\alpha$ lies in $Z(\mathbb{M}_{\dim \pi}) = \mathbb{C}E_{\dim \pi}$, so that $\|\tilde{\pi}(f_\alpha)\|_{\mathcal{N}} = |\mathrm{Tr}\, \tilde{\pi}(f_\alpha)|$. It follows that

$$|f_\alpha(e_G)| = \left| \sum_{\pi \in \hat{G}} (\dim \pi) \mathrm{Tr}\, \tilde{\pi}(f_\alpha) \right|, \qquad \text{by Proposition 7.2.6(iii),}$$

$$\leq \sum_{\pi \in \hat{G}} (\dim \pi) |\mathrm{Tr}\, \tilde{\pi}(f_\alpha)|$$

$$= \sum_{\pi \in \hat{G}} (\dim \pi) \|\tilde{\pi}(f_\alpha)\|_{\mathcal{N}}$$

$$\to 0, \qquad \text{by (7.6).}$$

But this contradicts Lemma 7.2.9(ii)   □

Example 7.2.11 By [H–R, (29.37) Theorem], the special orthogonal group $SO(3)$ satisfies the hypothesis of Theorem 7.2.10. Hence, $A(SO(3))$ is not amenable.

7.3 Operator amenable operator Banach algebras

The Fourier algebra of a locally compact group G is not only a Banach algebra, but — equipped with its operator space structure as the predual of $VN(G)$ — an operator Banach algebra (Examples D.4.3(c)). As will become apparent in the next section, we have to take this operator space structure into account if we want to obtain a satisfactory notion of amenability for $A(G)$.

In this section, we develop a notion of amenability for arbitrary operator Banach algebras. Most of it parallels our discussion in Sections 2.1 and 2.2, so that we leave the proofs as exercises to the reader. The necessary background information on operator spaces is assembled in Appendix D.

Definition 7.3.1 Let $\mathfrak{A}$ be an operator Banach algebra. A an operator space E which is also a left $\mathfrak{A}$-module is called *left operator $\mathfrak{A}$-module* if the bilinear map

$$\mathfrak{A} \times E \to E, \quad (a, x) \mapsto a \cdot x$$

is completely bounded.

Right operator $\mathfrak{A}$-modules and operator $\mathfrak{A}$-bimodules are defined analoguously.

Exercise 7.3.1 Let $\mathfrak{A}$ be a Banach algebra, and let E be a Banach module (left, right or bi-). Show that $MAX(E)$ is an operator $MAX(\mathfrak{A})$-module.

Exercise 7.3.2 Let $\mathfrak{A}$ be an operator Banach algebra, let E be a left operator $\mathfrak{A}$-module, and let F be a right operator $\mathfrak{A}$-module. Show that $E \hat{\otimes} F$ is an operator $\mathfrak{A}$-bimodule in a canonical fashion.

Exercise 7.3.3 Let $\mathfrak{A}$ be an operator Banach algebra, and let E be an operator $\mathfrak{A}$-module (left, right, or bi-). Show that E^* with the corresponding dual module action is again an operator $\mathfrak{A}$-module (right, left, or -bi, respectively).

We can now define operator amenability:

Definition 7.3.2 Let $\mathfrak{A}$ be an operator Banach algebra. Then $\mathfrak{A}$ is called *operator amenable* if, for every operator $\mathfrak{A}$-bimodule E, every completely bounded derivation $D : \mathfrak{A} \to E^*$ is inner.

It is clear that an operator Banach algebra which is amenable as a Banach algebra is operator amenable. As the example of certain Fourier algebras will show, the converse is not true.

Exercise 7.3.4 Let $\mathfrak{A}$ be an operator amenable operator Banach algebra. Show that $\mathfrak{A}$ (as a Banach algebra) has a bounded approximate identity.

Approximate and virtual diagonals have their analogues for operator Banach algebras:

Definition 7.3.3 Let $\mathfrak{A}$ be an operator Banach algebra.

(i) An element $\mathbf{M} \in (\mathfrak{A}\hat{\otimes}\mathfrak{A})^{**}$ is called a *virtual operator diagonal* for $\mathfrak{A}$ if

$$a \cdot \mathbf{M} = \mathbf{M} \cdot a \quad \text{and} \quad a \cdot \Delta_{\mathfrak{A}}^{**}\mathbf{M} = a \quad (a \in \mathfrak{A}).$$

(ii) A bounded net $(\mathbf{m}_\alpha)_\alpha$ in $\mathfrak{A}\hat{\otimes}\mathfrak{A}$ is called an *approximate operator diagonal* for $\mathfrak{A}$ if

$$a \cdot \mathbf{m}_\alpha - \mathbf{m}_\alpha \cdot a \to 0 \quad \text{and} \quad a\Delta_{\mathfrak{A}}\mathbf{m}_\alpha \to a \quad (a \in \mathfrak{A}).$$

There is a perfect analogue of Theorem 2.2.4 for operator amenability:

Theorem 7.3.4 *For an operator Banach algebra $\mathfrak{A}$ the following are equivalent:*

(i) *$\mathfrak{A}$ is operator amenable.*

(ii) *There is an approximate operator diagonal for $\mathfrak{A}$.*

(iii) *There is a virtual operator diagonal for $\mathfrak{A}$.*

Exercise 7.3.5 Prove Theorem 7.3.4. (*Hint*: For (ii) $\Longrightarrow$ (i), first show as in Proposition 2.1.5 that is is sufficient to consider the duals of pseudo-unital operator modules.)

Exercise 7.3.6 Let $\mathfrak{A}$ be a Banach algebra. Show that $\mathfrak{A}$ is amenable if and only if $MAX(\mathfrak{A})$ is operator amenable.

7.4 Operator amenability of Fourier algebras

In this section, we shall prove that a locally compact group G is amenable if and only if its Fourier algebra $A(G)$ is operator amenable. The difficult direction is

$$G \text{ amenable} \quad \Longrightarrow \quad A(G) \text{ operator amenable.}$$

We shall use the characterization of operator amenability through approximate diagonals given in Theorem 7.3.4.

For our first lemma, note that, for any locally compact group, the canonical $A(G)$-bimodule action on $A(G)\hat{\otimes}A(G) \cong A(G \times G)$ extends canonically to $B(G \times G)$ (How?). Also note that $\Delta: A(G)\hat{\otimes}A(G) \to A(G)$ is given by restricting a function in $A(G \times G)$ to $\{(g,g) : g \in G\}$; thus Δ extends canonically to an operator from $B(G \times G)$ to $B(G)$.

Lemma 7.4.1 *Let G be an amenable, locally compact group, and suppose that there is a bounded net $(m_\alpha)_\alpha$ in $B(G \times G)$ such that*

$$\|f \cdot m_\alpha - m_\alpha \cdot f\|_{B(G \times G)} \to 0 \quad (f \in A(G)) \tag{7.7}$$

and

$$\|f\Delta m_\alpha - f\|_{B(G)} \to 0 \quad (f \in A(G)). \tag{7.8}$$

Then $A(G)$ is operator amenable.

Proof By Theorem 7.1.3, $A(G)$ has a bounded approximate identity, say $(e_\beta)_\beta$. Since $A(G \times G)$ is a closed ideal in $B(G \times G)$ (Theorem A.3.13), the net $(e_\beta \cdot m_\alpha \cdot e_\beta)_{\alpha,\beta}$ is an approximate operator diagonal for $A(G)$. $\square$

Hence, in order to show that $A(G)$ is operator amenable for an amenable, locally compact group G, it is sufficient to find a net in $B(G \times G)$ as described in Lemma 7.4.1. This net will eventually consist of elements of the form

$$m_\xi \colon G \times G \to \mathbb{C}, \quad (g, h) \mapsto \langle \lambda_2(g)\rho_2(h)\xi, \xi \rangle,$$

where $\xi \in L^2(G)$ is a unit vector; it is immediate from Theorem A.3.11(i) that $m_\xi \in B(G \times G)$.

Let G be a locally compact group, and define $V \in \mathcal{L}(L^2(G))$ through

$$(V\xi)(g) := \Delta(g)^{-\frac{1}{2}}\xi(g^{-1}) \quad (\xi \in L^2(G), g \in G).$$

It is easily seen (Check it!) that

$$\rho_2(g) = V^*\lambda_2(g)V \qquad (g \in G), \tag{7.9}$$

Also define $W \in \mathcal{L}(L^2(G \times G))$ by letting

$$(W\xi)(g, h) := \xi(g, gh) \quad (\xi \in L^2(G \times G), g, h \in G).$$

Exercise 7.4.1 Let G be a locally compact group. Show that

$$(V^* \otimes \mathrm{id}_{L^2(G)})W^* = W(V^* \otimes \mathrm{id}_{L^2(G)})$$

Lemma 7.4.2 *Let G be a locally compact group, and suppose that there is a net $(\xi_\alpha)_\alpha$ of unit vectors in $L^2(G)$ such that*

$$\|W(\xi_\alpha \otimes \eta) - (\xi_\alpha \otimes \eta)\| \to 0 \qquad (\eta \in L^2(G))$$

and

$$\|\lambda_2(g)\rho_2(g)\xi_\alpha - \xi_\alpha\| \to 0$$

uniformly on all compact subsets of G. Then the net $(m_{\xi_\alpha})_\alpha$ in $B(G \times G)$ satisfies the hypotheses of Lemma 7.4.1.

Proof Let $f \in A(G)$. We may suppose (Why?) that

$$f(g) = \langle \lambda_2(g)\eta, \eta \rangle \qquad (g \in G) \tag{7.10}$$

for some $\eta \in L^2(G)$. From Remarks A.3.12(a), we conclude that

$$(f \cdot m_{\xi_\alpha})(g, h) = \langle \lambda_2(g)\eta, \eta \rangle \langle \lambda_2(g)V^*\lambda_2(h)V\xi_\alpha, \xi_\alpha \rangle, \qquad \text{by (7.9)},$$
$$= \langle (\lambda_2(g)V^*\lambda_2(h)V \otimes \lambda_2(g))(\xi_\alpha \otimes \eta), \xi_\alpha \otimes \eta \rangle \qquad (g, h \in G). \tag{7.11}$$

By Exercise 7.4.1, we have as well:

$$(m_{\xi_\alpha} \cdot f)(g,h)$$
$$= \langle \lambda_2(g) V^* \lambda_2(h) V \xi_\alpha, \xi_\alpha \rangle \langle \lambda_2(h) \eta, \eta \rangle$$
$$= \langle (\lambda_2(g) \otimes \mathrm{id}_{L^2(G)})(V^* \lambda_2(h) V \otimes \lambda_2(h))(\xi_\alpha \otimes \eta), \xi_\alpha \otimes \eta \rangle$$
$$= \langle (\lambda_2(g) \otimes \mathrm{id}_{L^2(G)})(V^* \otimes \mathrm{id}_{L^2(G)})W^*(\lambda_2(h) \otimes \mathrm{id}_{L^2(G)})W(V \otimes \mathrm{id}_{L^2(G)})(\xi_\alpha \otimes \eta),$$
$$\xi_\alpha \otimes \eta \rangle$$
$$= \langle W^*(\lambda_2(g) \otimes \mathrm{id}_{L^2(G)})W(V^* \otimes \mathrm{id}_{L^2(G)})(\lambda_2(h) \otimes \mathrm{id}_{L^2(G)})(V \otimes \mathrm{id}_{L^2(G)})W^*(\xi_\alpha \otimes \eta),$$
$$W^*(\xi_\alpha \otimes \eta) \rangle$$
$$= \langle (\lambda_2(g) \otimes \lambda_2(g))(V^* \lambda_2(h) V \otimes \mathrm{id}_{L^2(G)})W^*(\xi_\alpha \otimes \eta), W^*(\xi_\alpha \otimes \eta) \rangle$$
$$= \langle (\lambda_2(g) V^* \lambda_2(h) V \otimes \lambda_2(g))W^*(\xi_\alpha \otimes \eta), W^*(\xi_\alpha \otimes \eta) \rangle \qquad (g,h \in G). \tag{7.12}$$

Combining (7.11) and (7.12), we see that

$$(f \cdot m_{\xi_\alpha})(g,h) - (m_{\xi_\alpha} \cdot f)(g,h)$$
$$= \langle (\lambda_2(g) V^* \lambda_2(h) V \otimes \lambda_2(g))(\xi_\alpha \otimes \eta), \xi_\alpha \otimes \eta \rangle$$
$$\quad - \langle (\lambda_2(g) V^* \lambda_2(h) V \otimes \lambda_2(g))W^*(\xi_\alpha \otimes \eta), W^*(\xi_\alpha \otimes \eta) \rangle$$
$$= \langle (\lambda_2(g) V^* \lambda_2(h) V \otimes \lambda_2(g))(\xi_\alpha \otimes \eta - W^*(\xi_\alpha \otimes \eta)), \xi_\alpha \otimes \eta \rangle$$
$$\quad + \langle (\lambda_2(g) V^* \lambda_2(h) V \otimes \lambda_2(g))W^*(\xi_\alpha \otimes \eta), (\xi_\alpha \otimes \eta - W^*(\xi_\alpha \otimes \eta)) \rangle \qquad (g,h \in G).$$

From Theorem A.3.11(i), it follows that

$$\|f \cdot m_{\xi_\alpha} - m_{\xi_\alpha} \cdot f\|_{B(G \times G)} \leq 2\|\eta\| \|\xi_\alpha \otimes \eta - W^*(\xi_\alpha \otimes \eta)\|$$
$$= 2\|\eta\| \|W(\xi_\alpha \otimes \eta) - \xi_\alpha \otimes \eta\|$$
$$\to 0.$$

This establishes (7.7).

To prove that $(m_{\xi_\alpha})_\alpha$ satisfies (7.8), we may suppose that η in (7.10) belongs to $C_{00}(G)$. We have:

$$(f \Delta m_{\xi_\alpha})(g)$$
$$= f(g) m_{\xi_\alpha}(g,g)$$
$$= \langle \lambda_2(g) V^* \lambda_2(g) V \xi_\alpha, \xi_\alpha \rangle \langle \lambda_2(g) \eta, \eta \rangle$$
$$= \langle (\lambda_2(g) \otimes \lambda_2(g))(\mathrm{id}_{L^2(G)} \otimes V^* \lambda_2(g) V)(\eta \otimes \xi_\alpha), \eta \otimes \xi_\alpha \rangle$$
$$= \langle W^*(\lambda_2(g) \otimes \mathrm{id}_{L^2(G)})W(\mathrm{id}_{L^2(G)} \otimes V^* \lambda_2(g) V)(\eta \otimes \xi_\alpha), \eta \otimes \xi_\alpha \rangle$$
$$= \langle W^*(\lambda_2(g) \otimes \mathrm{id}_{L^2(G)})(\mathrm{id}_{L^2(G)} \otimes V^* \lambda_2(g) V)W(\eta \otimes \xi_\alpha), \eta \otimes \xi_\alpha \rangle \qquad \text{(Why?)}$$
$$= \langle W^*(\mathrm{id}_{L^2(G)} \otimes V^*)(\lambda_2(g) \otimes \lambda_2(g))(\mathrm{id}_{L^2(G)} \otimes V)W(\eta \otimes \xi_\alpha), \eta \otimes \xi_\alpha \rangle$$
$$= \langle W^*(\mathrm{id}_{L^2(G)} \otimes V^*)W^*(\lambda_2(g) \otimes \mathrm{id}_{L^2(G)})W(\mathrm{id}_{L^2(G)} \otimes V)W(\eta \otimes \xi_\alpha), \eta \otimes \xi_\alpha \rangle$$
$$= \langle (\lambda_2(g) \otimes \mathrm{id}_{L^2(G)})(\mathrm{id}_{L^2(G)} \otimes V)W(\mathrm{id}_{L^2(G)} \otimes V)W(\eta \otimes \xi_\alpha),$$
$$W(\mathrm{id}_{L^2(G)} \otimes V)W(\eta \otimes \xi_\alpha) \rangle \qquad (g \in G).$$

It follows that:

$(f\Delta m_{\xi_\alpha})(g) - f(g)$

$= \langle(\lambda_2(g) \otimes \mathrm{id}_{L^2(G)})(\mathrm{id}_{L^2(G)} \otimes V)W(\mathrm{id}_{L^2(G)} \otimes V)W(\eta \otimes \xi_\alpha),$

$\qquad W(\mathrm{id}_{L^2(G)} \otimes V)W(\eta \otimes \xi_\alpha)\rangle$

$\quad - \langle(\lambda_2(g) \otimes \mathrm{id}_{L^2(G)})(\eta \otimes \xi_\alpha), \eta \otimes \xi_\alpha\rangle$

$= \langle(\lambda_2(g) \otimes \mathrm{id}_{L^2(G)})(W(\mathrm{id}_{L^2(G)} \otimes V)W(\eta \otimes \xi_\alpha) - \eta \otimes \xi_\alpha), W(\mathrm{id}_{L^2(G)} \otimes V)W(\eta \otimes \xi_\alpha)\rangle$

$\quad + \langle(\lambda_2(g) \otimes \mathrm{id}_{L^2(G)})(\eta \otimes \xi_\alpha), W(\mathrm{id}_{L^2(G)} \otimes V)W(\eta \otimes \xi_\alpha) - \eta \otimes \xi_\alpha\rangle \qquad (g \in G).$

Hence, we obtain:

$$\|f\Delta m_{\xi_\alpha} - f\|_{B(G)} \leq 2\|\eta\|\|W(\mathrm{id}_{L^2(G)} \otimes V)W(\eta \otimes \xi_\alpha) - \eta \otimes \xi_\alpha\|. \qquad (7.13)$$

Let $K := \mathrm{supp}(\eta)$. Since $\|\lambda_2(g)\rho_2(g)\xi_\alpha - \xi_\alpha\| \to 0$ uniformly on K, we obtain:

$\|W(\mathrm{id}_{L^2(G)} \otimes V)W(\eta \otimes \xi_\alpha) - \eta \otimes \xi_\alpha\|^2$

$= \int_G \int_G |(W(\mathrm{id}_{L^2(G)} \otimes V)W)(\eta \otimes \xi_\alpha)(g,h) - (\eta \otimes \xi_\alpha)(g,h)|^2 \, dm_G(h) \, dm_G(g)$

$= \int_G \int_G \left|\Delta(g)^{-\frac{1}{2}}\eta(g)\xi_\alpha(ghg^{-1}) - \eta(g)\xi_\alpha(h)\right|^2 \, dm_G(h) \, dm_G(g)$

$= \int_K |\eta(g)|^2 \int_G \left|\Delta(g)^{-\frac{1}{2}}\xi_\alpha(ghg^{-1}) - \xi_\alpha(h)\right|^2 \, dm_G(h) \, dm_G(g)$

$= \int_K |\eta(g)|^2 \|\lambda_2(g)\rho_2(g)\xi_\alpha - \xi_\alpha\|^2 \, dm_G(g)$

$\to 0.$

Together with (7.13), this yields (7.8). $\quad\square$

For an amenable, locally compact group G, we are therefore left with the task of finding a net $(\xi_\alpha)_\alpha$ of unit vectors in $L^2(G)$ that satisfies the conditions of Lemma 7.4.2 ...

Theorem 7.4.3 *Let G be a locally compact group. Then the following are equivalent:*

(i) *G is amenable.*

(ii) *$A(G)$ is operator amenable.*

Proof (ii) $\implies$ (i): If $A(G)$ is operator amenable, it has a bounded approximate identity by Exercise 7.3.4. By Theorem 7.1.3, G is amenable.

(i) $\implies$ (ii): Let $K \subset G$ be compact and let $\epsilon > 0$. By Lemma 7.1.1, there is $f_{K,\epsilon} \in P(G)$ such that

$$\|\delta_g * f_{K,\epsilon} - f_{K,\epsilon}\|_1 < \epsilon \qquad (g \in K).$$

We may suppose that $f_{K,\epsilon} \in \mathcal{C}_{00}(G)$. Let $\mathfrak{U}$ be the basis of all compact neighborhoods of e_G. For $U \in \mathfrak{U}$, define:

$$f_{U,K,\epsilon}(g) := \frac{1}{m_G(U)} \int_G f_{K,\epsilon}(h)\chi_U(h^{-1}gh)\Delta(h) \, dm_G(h) \qquad (g \in G).$$

It is routinely checked (by you) that $f_{U,K,\epsilon} \in P(G)$.

We have:

$$\|\delta_g * f_{U,K,\epsilon} * \delta_{g^{-1}} - f_{U,K,\epsilon}\|_1$$

$$= \frac{1}{m_G(U)} \int_G \left| \int_G f_{K,\epsilon}(k)\chi_U(k^{-1}g^{-1}hgk)\Delta(k)\Delta(g)\,dm_G(k) \right.$$

$$\left. - \int_G f_{K,\epsilon}(k)\chi_U(k^{-1}hk)\Delta(k)\,dm_G(k) \right| dm_G(h)$$

$$= \frac{1}{m_G(U)} \int_G \left| \int_G f_{K,\epsilon}(g^{-1}k)\chi_U(k^{-1}hk)\Delta(k)\,dm_G(k) \right.$$

$$\left. - \int_G f_{K,\epsilon}(k)\chi_U(k^{-1}hk)\Delta(k)\,dm_G(k) \right| dm_G(h)$$

$$\leq \frac{1}{m_G(U)} \int_G \int_G |f_{K,\epsilon}(g^{-1}k) - f_{K,\epsilon}(k)|\chi_U(k^{-1}hk)\Delta(k)\,dm_G(h)\,dm_G(k)$$

$$= \|\delta_g * f_{K,\epsilon} - f_{K,\epsilon}\|_1$$

$$< \epsilon \qquad (g \in K).$$

It follows that

$$\|\delta_g * f_{U,K,\epsilon} * \delta_{g^{-1}} - f_{U,K,\epsilon}\|_1 \to 0$$

uniformly on all compact subsets of G. Let $\xi_{U,K,\epsilon} := f_{U,K,\epsilon}^{\frac{1}{2}}$. As in the proof of Theorem 4.4.13, we see that

$$\|\lambda_2(g)\rho_2(g)\xi_{U,K,\epsilon} - \xi_{U,K,\epsilon}\| \to 0$$

uniformly on all compact subsets of G.

Fix $\eta_1, \ldots, \eta_n \in L^2(G)$. Since λ_2 is continuous with respect to the strong operator topology on $L^2(G)$, there is $U_0 \in \mathfrak{U}$ such that

$$\int_G |\eta_j(gh) - \eta_j(h)|^2 \, dm_G(h) = \|\lambda_2(g^{-1})\eta_j - \eta_j\|_2^2 < \epsilon \qquad (g \in U_0, j = 1, \ldots, n).$$

For any $K \subset G$ compact and $\epsilon > 0$, let $C := \mathrm{supp}(f_{K,\epsilon})$. By [H–R, (4.9) Theorem], we may suppose without loss of generality that

$$gU_0 g^{-1} \subset U_0 \qquad (g \in C).$$

Let $U \in \mathfrak{U}$ be contained in U_0. For $j = 1, \ldots, n$, we obtain:

$$\|W(\xi_{U,K,\epsilon} \otimes \eta_j) - \xi_{U,K,\epsilon} \otimes \eta_j\|^2$$

$$= \int_G \int_G |\xi_{U,K,\epsilon}(g)|^2 |\eta_j(gh) - \eta_j(h)|^2 \, dm_G(g)\,dm_G(h)$$

$$= \int_G \int_G f_{U,K,\epsilon}(g)|\eta_j(gh) - \eta_j(h)|^2 \, dm_G(g)\,dm_G(h)$$

$$= \int_G \int_G \left(\frac{1}{m_G(U)} \int_G f_{K,\epsilon}(k)\chi_U(k^{-1}gk^{-1})\Delta(k)\,dm_G(k) \right)$$

$$|\eta_j(gh) - \eta_j(h)|^2 dm_G(g)\,dm_G(h)$$

$$= \int_C f_{K,\epsilon}(k) \left(\frac{1}{m_G(U)}\chi_U(g) \left(\int_G |\eta(kgk^{-1}h) - \eta(h)|^2 dm_G(h) \right) dm_G(g) \right) dm_G(k)$$

$$\leq \epsilon \qquad \text{(Why?)}.$$

It follows that $(\xi_{U,K,\epsilon})_{U,K,\epsilon}$ has a subnet $(\xi_\alpha)_\alpha$ such that

$$\|W(\xi_\alpha \otimes \eta) - \xi_\alpha \otimes \eta\| \to 0 \qquad (\eta \in L^2(G)).$$

From Lemmas 7.4.2 and 7.4.1, we conclude that $A(G)$ is operator amenable.   □

Example 7.4.4 As we saw immediately after Theorem 7.2.10, $A(SO(3))$ is not amenable. Nevertheless, it is operator amenable by Theorem 7.4.3.

Remark 7.4.5 The previous example also shows that operator amenability depends very much on the particular operator space structure on a Banach algebra: $A(SO(3))$ is operator amenable with respect to its operator space structure as the predual of $VN(SO(3))$. On the other hand, $MAX(A(SO(3)))$ is not operator amenable by Exercise 7.3.6.

Exercise 7.4.2 In [Los], V. Losert shows that, for any locally compact groups G and H, that $A(G)\hat{\otimes}A(G) \cong A(G \times H)$ (not necessarily isometrically) if and only if G or H has a closed, abelian subgroup of finite index. Use this result to show that $A(G)$ is amenable for any locally compact group G with a closed, abelian subgroup of finite index.

7.5 Operator amenability of C^*-algebras

By Examples D.4.3(b), each C^*-algebra is canonically an operator Banach algebra. So, what is the deeper meaning of operator amenability for C^*-algebras?

As it turns out, for C^*-algebras, amenability and operator amenability coincide:

Theorem 7.5.1 *Let $\mathfrak{A}$ be a C^*-algebra. Then the following are equivalent:*

(i) $\mathfrak{A}$ *is amenable.*

(ii) $\mathfrak{A}$ *is operator amenable.*

Proof (i) $\Longrightarrow$ (ii) is true for any operator Banach algebra.

(ii) $\Longrightarrow$ (i): Without loss of generality suppose that $\mathfrak{A}$ is unital (Why?). By Theorem 7.3.4, there is a virtual operator diagonal $\mathbf{M} \in (\mathfrak{A}\hat{\otimes}\mathfrak{A})^{**}$. By Theorem D.3.9, the identity on $\mathfrak{A} \otimes \mathfrak{A}$ extends to a complete contraction $\theta : \mathfrak{A}\hat{\otimes}\mathfrak{A} \to \mathfrak{A}\tilde{\otimes}_h\mathfrak{A}$. Let $\tilde{\mathbf{M}} := \theta^{**}(\mathbf{M})$. Then $\tilde{\mathbf{M}} \in (\mathfrak{A}\tilde{\otimes}_h\mathfrak{A})^{**} \cong \mathcal{L}^2_{w^*,h}(\mathfrak{A}^{**}, \mathbb{C})^*$ satisfies

$$a \cdot \tilde{\mathbf{M}} = \tilde{\mathbf{M}} \cdot a \quad \text{and} \quad a\Delta_{h,w^*}\cdot\tilde{\mathbf{M}} = a \qquad (a \in \mathfrak{A}).$$

Arguing as in the proof of Corollary 6.5.12, we conclude that $\mathfrak{A}$ has a virtual diagonal and thus is amenable by Theorem 2.2.4.   □

7.6 Notes and comments

Theorem 7.1.3 is due to H. Leptin ([Lep]); an alternative proof was later given by A. Derighetti ([Der]): Our proof of (iii) $\Longrightarrow$ (i) is patterned after his approach. Both in [Lep] and in [Der], (i) $\Longrightarrow$ (ii) is proved with a Følner type condition. The statement of Theorem 7.1.3 remains true if we replace the Fourier algebra by any Figà-Talamanca–Herz algebra (see [Pie]). The free group in two generators is not amenable, but, nevertheless, $A(\mathbb{F}_2)$ has a (necessarily unbounded) approximate identity ([Haa 1]). For any locally compact group and $f \in A(G)$, the multiplication operator L_f is completely bounded by Examples D.4.3(c) and Remarks D.4.2(a). If $A(G)$ has an approximate identity $(e_\alpha)_\alpha$ such that $(L_{e_\alpha})_\alpha$ is bounded in $\mathcal{L}_{cb}(A(G))$, the group G is called *weakly amenable* ([C–H]): Amenable and free groups are weakly amenable whereas certain Lie groups aren't. By sheer coincidence, we thus have weakly amenable Banach algebras and weakly amenable, locally compact groups, although these notions are essentially unrelated.

Theorem 7.2.10 is due to B. E. Johnson ([Joh 7]). If G is a compact group such that $\{\dim \pi : \pi \in \hat{G}\}$ is bounded, then $A(G)$ is amenable ([Joh 7, Theorem 5.3]); this is a particular case of Exercise 7.4.2 ([L–L–W, Corollary 4.2]). On the other hand, $A(SO(3))$ is not even weakly amenable ([Joh 7, Corollary 7.3]). Amenability and weak amenability of the Fourier algebra are further investigated in [L–L–W] and [For].

Operator amenability for operator Banach algebras was introduced in [Rua 2]; Theorems 7.4.3 and 7.5.1 are also from that paper.

For a locally compact group G, the pair $(A(G), VN(G))$ is an example of a *Hopf–von Neumann algebra*. In [Rua 3], the operator amenability of the preduals of general Hopf–von Neumann algebras is discussed. In fact, $(A(G), VN(G))$ is even a *Kac algebra*, i.e. a Hopf–von Neumann algebra with additional structure (see [E–S] for the definition). Operator amenability seems to be particularly well-suited to the study of Kac algebras ([Rua 3] and [R–X]).

Hochschild cohomology with coefficients in (particular) operator bimodules and with completely bounded cochains has been studied for quite some time for von Neumann algebras (see [S–S]). The approach in [S–S], however, does not carry over to arbitrary operator Banach algebras: The underlying homological algebra is based on the Haagerup tensor product, so that it works only for closed subalgebras of $\mathcal{L}(\mathfrak{H})$ for some Hilbert space $\mathfrak{H}$; for general operator Banach algebras, however, we need a homological algebra based on the operator projective tensor product. The development of a homological algebra with respect to the Haagerup tensor product — analoguous to Banach homology —, still seems to be in its initial stages (see [Hel 7] for some remarks). There are some partial results: In [Pau 2], for instance, Hochschild cohomology with completely bounded cochains for such algebras is described in terms of Ext-groups (analoguously to Theorem 5.2.28); see [Ari 1] and [Hel 8] for further results in this direction.

In [R–X], it is shown for general operator Banach algebras that operator amenability has a characterization in terms of the splitting of certain short, exact sequences (similar to the one given for amenability in Exercise 2.3.8). In fact, the basics of Banach homology — such as Theorem 5.2.28 — carry over to the operator space/Banach algebra/Banach module world relatively painlessly (see [Ari 2] and [Woo 1]).

Of course, one can define operator variants of amenability-like properties such as superamenability, weak amenability, biprojectivity, etc. In [Ari 2], O. Yu. Aristov systematically develops a theory of operator biprojective operator Banach algebras which pararllels the theory of biprojective Banach algebras; in particular, there is an operator analogue of Selivanov's structure theory for biprojective Banach algebras. The results existing so far further confirm that the operator space structure of $A(G)$ for a locally compact group G is crucial when it comes to amenability, biprojectivity and related properties: N. Spronk showed that $A(G)$ is always operator weakly amenable, i.e. that every derivation from $A(G)$ into $VN(G)$ is inner ([Spr 1]), and P. Wood and O. Yu. Aristov, independently, proved that $A(G)$ is operator biprojective if and only if G is discrete ([Woo 2] and [Ari 2, Theorem 7.30]).

8 Geometry of spaces of homomorphisms

Let $\mathfrak{A}$ be a Banach algebra, and let $\phi \in \Phi_{\mathfrak{A}}$. A *point derivation* at ϕ is a functional $d \in \mathfrak{A}^*$ such that

$$\langle ab, d \rangle = \langle a, \phi \rangle \langle b, d \rangle + \langle b, \phi \rangle \langle a, d \rangle \qquad (a, b \in \mathfrak{A}).$$

It is clear that, whenever there are non-zero point derivations on a Banach algebra $\mathfrak{A}$, then $\mathfrak{A}$ cannot be amenable (not even weakly amenable). Non-zero point derivations, however, arise naturally whenever we have some sort of differentiable structure on $\Phi_{\mathfrak{A}}$, e.g. in the case $\mathfrak{A} = A(\mathbb{D})$ or $\mathfrak{A} = \mathcal{C}^1([0,1])$. It thus seems that amenable Banach algebras and differential geometry are essentially disjoint topics.

As it turns out, however, if $\mathfrak{A}$ is an amenable Banach algebra and if $\mathfrak{B}$ is a dual Banach algebra, then certain sets of homomorphisms from $\mathfrak{A}$ into $\mathfrak{B}$ carry a natural, albeit non-trivial differential geometric structure.

In this chapter, we prove only one result in this direction. In order to provide the background for this, most of the chapter consists of a (necessarily superficial) introduction to differential geometry over Banach spaces. Familiarity with finite-dimensional differential geometry will help, but is not necessary.

8.1 Infinite-dimensional differential geometry

The aim of this section is to provide a quick, self-contained introduction to (complex) infinite-dimensional differential geometry. It is far from being comprehensive, and its purpose is solely to prepare the ground for our investigation of spaces of homomorphisms in the next section.

We start by defining what we mean when saying that a map between Banach spaces is differentiable:

Definition 8.1.1 Let E and F be Banach spaces, and let $U \subset E$ be open. A map $f : U \to F$ is called (complex) *Fréchet differentiable* at $x \in U$, if there is $T \in \mathcal{L}(E, F)$ such that

$$\lim_{\substack{y \neq 0 \\ \|y\| \to 0}} \frac{\|f(x+y) - f(x) - Ty\|}{\|y\|} = 0.$$

In this case, the operator T is called the (complex) *Fréchet derivative* or the *differential* of f at x and denoted by df_x.

Exercise 8.1.1 Let E_1, E_2, and E_3 be Banach spaces, let $U \subset E_1$ and $V \subset E_2$ be open, let $x \in U$, and let $f : U \to E_2$ be Fréchet differentiable at x such that $f(U) \subset V$. Show that, if $g : V \to E_3$ is Fréchet differentiable at $f(x)$, then $g \circ f$ is Fréchet differentiable at x with $d(g \circ f)_x = dg_{f(x)} df_x$.

Definition 8.1.2 Let E and F be Banach spaces, and let $U \subset E$ be open. A map $f : U \to F$ is called *holomorphic*, if it is Fréchet differentiable at each point of U, and if the map

$$U \to \mathcal{L}(E, F), \quad x \mapsto df_x \tag{8.1}$$

is continuous.

Remark 8.1.3 Let E and F be Banach spaces, and let let $U \subset E$ be open. If $f : U \to F$ is Fréchet differentiable at every point of U, then f is clearly continuous and G-holomorphic in the sense of [Muj, 8.1 Definition]. By [Muj, 8.7 Theorem], f is thus holomorphic in the sense of [Muj, 5.1 Definition]. On the other hand, if f is holomorphic according to [Muj, 5.1 Definition], it is obvious that its Fréchet derivative exists everywhere, and that (8.1) is continuous. Hence, the continuity of (8.1) is redundant in Definition 8.1.2.

Examples 8.1.4   (a) Linear maps between Banach spaces are trivially holomorphic. (What is the Fréchet derivative in this case?)

(b) Let $E_1, \dots, E_n$, and F be Banach spaces. Then each $T \in \mathcal{L}(E_1, \dots, E_n; F)$ is holomorphic.

(c) Let $\mathfrak{A}$ be a unital Banach algebra. Then

$$\operatorname{Inv} \mathfrak{A} \to \operatorname{Inv} \mathfrak{A}, \quad a \mapsto a^{-1}$$

is holomorphic.

Exercise 8.1.2 Work out Examples 8.1.4(b) and (c) in detail.

Some results about differentiable maps in finitely many dimensions carry over to the infinite-dimensional context:

Theorem 8.1.5 (mean value theorem) *Let E and F be Banach spaces, let $U \subset E$ be open and convex, and let $f : U \to F$ be holomorphic. Then we have:*

$$\|f(x) - f(y)\| \le \sup\{\|df_z\| : z \in U\}\|x - y\| \qquad (x, y \in U).$$

Proof Let $\phi \in F^*$ be such that $\|\phi\| \le 1$. Then $\phi \circ f : U \to \mathbb{C}$ is holomorphic (Why?). Fix $x, y \in U$. The function

$$g : [0, 1] \to \mathbb{C}, \quad t \mapsto \langle f(tx + (1 - t)y), \phi \rangle$$

is continuously differentiable (Why?) with

$$g'(t) = d(\phi \circ f)_{tx+(1-t)y}(x - y) = \langle df_{tx+(1-t)y}(x - y), \phi \rangle \qquad (t \in [0, 1]).$$

In particular, we have

$$\langle f(x) - f(y), \phi \rangle = g(1) - g(0) = \int_0^1 g'(t)\, dt = \int_0^1 \langle df_{tx+(1-t)y}(x - y), \phi \rangle\, dt.$$

It follows that

$$|\langle f(x) - f(y), \phi \rangle| \le \sup\{\|df_z\| : z \in U\}\|x - y\|.$$

The Hahn–Banach theorem then yields the claim. $\square$

Theorem 8.1.6 (inverse mapping theorem) *Let E and F be Banach spaces, let $U \subset E$ be open, and let $f : U \to F$ be holomorphic. Suppose that $x_0 \in U$ is such that df_{x_0} is an isomorphism of E and F. Then there are open neighborhoods $V \subset U$ of x_0 and W of $f(x_0)$ such that $f|_V : V \to W$ is bijective and has a holomorphic inverse.*

Proof Without loss of generality suppose that $E = F$, $df_{x_0} = \mathrm{id}_E$, and $x_0 = f(x_0) = 0$ (Why is this possible?)

Letting $g := \mathrm{id}_E - f$, we see that g is holomorphic with $dg_{x_0} = 0$. From Definition 8.1.2, it follows that there is $r > 0$ such that

$$\|dg_x\| \le \frac{1}{2} \qquad (x \in B_r[0, E]). \tag{8.2}$$

Together, (8.2) and Theorem 8.1.5 yield that

$$\|g(x)\| \le \frac{1}{2}\|x\| \qquad (x \in B_r[0, E]). \tag{8.3}$$

In particular, g maps $B_r[0, E]$ into $B_{\frac{r}{2}}[0, E]$.

Let $y \in B_{\frac{r}{2}}[0, E]$. We claim that there is a unique $x \in B_r[0, E]$ such that $f(x) = y$. The map

$$B_r[E] \to E, \quad x \mapsto y + g(x) \tag{8.4}$$

maps $B_r[0, E]$ into itself (Why?). Theorem 8.1.5 and (8.3) yield

$$\|(y + g(x_1)) - (y + g(x_2))\| = \|g(x_1) - g(x_2)\| \le \frac{1}{2}\|x_1 - x_2\| \qquad (x_1, x_2 \in B_r[0, E]),$$

so that (8.4) has a unique fixed point in $B_r[0, E]$ by Banach's fixed point theorem. From the definition of (8.4) is is obvious that $x \in B_r[0, E]$ is a fixed point for (8.4) if and only if $f(x) = y$. This proves the claim.

It follows that $f^{-1} : B_r[0, E] \to B_{\frac{r}{2}}[0, E]$ exists. Note that

$$\|x_1 - x_2\| \le \|f(x_1) - f(x_2)\| + \|g(x_1) - g(x_2)\|$$
$$\le \|f(x_1) - f(x_2)\| + \frac{1}{2}\|x_1 - x_2\| \qquad (x_1, x_2 \in B_r[0, E]),$$

where the inequality follows from (8.2) and Theorem 8.1.5. Hence,

$$\|x_1 - x_2\| \le 2\|f(x_1) - f(x_2)\| \qquad (x_1, x_2 \in B_r[0, E])$$

holds, and f^{-1} is continuous.

It can then be shown that f^{-1} is even holomorphic (Exercise 8.1.3 below). $\square$

Exercise 8.1.3 Complete the proof of Theorem 8.1.6 by showing that $f^{-1}|_W : W \to V$ is holomorphic. What is df_y^{-1} for $y \in W$? (*Hint*: Proceed just as in the finite-dimensional situation ...)

With a notion of infinite-dimensional holomorphy at hand, we can introduce manifolds over arbitrary Banach spaces:

Definition 8.1.7 Let M be a topological space, and let E be a Banach space. A *holomorphic atlas* on M over E is a family $((U_\alpha, \phi_\alpha))_\alpha$ of pairs with the following properties:

(i) $(U_\alpha)_\alpha$ is an open cover of M.

(ii) Each ϕ_α is a homeomorphism from U_α onto an open subset of E.

(iii) For any two indices α and β such that $U_\alpha \cap U_\beta \neq \varnothing$, the map $\phi_\beta \circ \phi_\alpha^{-1} : \phi_\alpha(U_\alpha \cap U_\beta) \to \phi_\beta(U_\alpha \cap U_\beta)$ is biholomorphic, i.e. holomorphic and bijective with a holomorphic inverse.

Definition 8.1.8 Let E be a Banach space. A topological space M together with a holomorphic atlas on M over E is called a *Banach manifold* over E.

By the way in which Banach manifolds are defined, they are the natural habitats for holomorphic mappings:

Definition 8.1.9 Let M and N be Banach manifolds — not necessarily over the same Banach space —, and let $((U_\alpha, \phi_\alpha))_\alpha$ and $((V_\beta, \psi_\beta))_\beta$ be the corresponding holomorphic atlases. A map $f : M \to N$ is called *holomorphic* if all the maps $\psi_\beta \circ f \circ \phi_\alpha^{-1}$ are holomorphic. If f is bijective and f^{-1} is also holomoprhic, then f is called *biholomorphic*.

With this definition, we can express when we want to essentially identify two atlasses:

Definition 8.1.10 Two holomorphic atlases $((U_\alpha, \phi_\alpha))_\alpha$ and $((V_\beta, \psi_\beta))_\beta$ on a Banach manifold M are defined to be *equivalent* if the identity map from $(M, ((U_\alpha, \phi_\alpha))_\alpha)$ to $(M, ((V_\beta, \psi_\beta))_\beta)$ is biholomorphic. An equivalence class of holomorphic atlases on M is called a *holomorphic structure* on M.

Of course, the holomorphy of a map between Banach manifolds depends only on the holomorphic structures involved and is independent of the particular atlases.

Another important definition is that of a chart:

Definition 8.1.11 Let M be a Banach manifold over a Banach space E, and let $p \in M$. A *chart* of M at p is a pair (U, ϕ), where U is an open neighborhood of p, $\phi(U)$ is an open subset of E, and $\phi : U \to \phi(U)$ is biholomorphic.

Next, we introduce the concept of the tangent space of a Banach manifold.

Definition 8.1.12 Let M be a Banach manifold over a Banach space E, and let $p \in M$. Two triples (U_1, ϕ_1, x_1) and (U_2, ϕ_2, x_2) — where (U_j, ϕ_j) are charts at p and $x_j \in E$ for $j = 1, 2$ — are defined to be equivalent if $d(\phi_2 \circ \phi_1^{-1})_p x_1 = x_2$. An equivalence class of such triples is called a *tangent vector* of M at p. The collection of all tangent vectors of M at p is called the *tangent space* of M at p and is denoted by $T_p M$.

Exercise 8.1.4 Let M be a Banach manifold over a Banach space E, let $p \in M$ and let (U, ϕ) be a chart at p.

(i) Show that $E \ni x \mapsto (U, \phi, x)$ induces a bijection between E and $T_p M$.

(ii) Show that this bijection is independent of the particular chart (U, ϕ).

The bijection between E and $T_p M$ enables us to identify $T_p M$ with E, so that $T_p M$ is a Banach space in a canonical manner. What happens in the particular case, where M is an open subset of some Banach space?

Exercise 8.1.5 Let M and N be Banach manifolds, let $p \in M$, and let (U, ϕ) and (V, ψ) be charts of M and N at p and $f(p)$, respectively. Show that, for any holomorphic map $f : M \to N$, the differential $d(\psi \circ f \circ \phi^{-1})_{\phi(p)}$ canonically induces a bounded linear map $df_p : T_p M \to T_{f(p)} N$, which is independent of the particular charts (U, ϕ) and (V, ψ).

Extending Definition 8.1.1, the map df_p is also called the *differential* of f at p.

Exercise 8.1.6 Prove a chain rule as in Exercise 8.1.1 for the differentials of holomorphic functions between Banach manifolds.

With these notions, we can extend Theorem 8.1.6 to holomorphic maps between manifolds:

Theorem 8.1.13 *Let M and N be Banach manifolds, let $f : M \to N$ be holomorphic, and let $p \in M$ be such that $df_p : T_pM \to T_{f(p)}N$ is an isomorphism. Then there are open neighborhoods U of p in M and V of $f(p)$ in N such that $f|_U : U \to V$ is bijective and has a holomorphic inverse.*

Exercise 8.1.7 Use Theorem 8.1.6 to prove Theorem 8.1.13.

The notion of a submanifold is relatively intricate — even in the finite-dimensional context. When dealing with Banach manifolds, we also have to take into account that closed subspaces of Banach spaces are rarely complemented:

Definition 8.1.14 Let M be a Banach manifold over a Banach space E. A non-empty subset N of M is called a *submanifold* of M if there is a complemented subspace F of E such that, for each chart (U, ϕ) of M, there is a projection P from E onto F such that $P \circ \phi : N \cap U \to F$ is open and bijective.

Exercise 8.1.8 Show that a submanifold of a Banach manifold is again a Banach manifold. (Over which space?)

The next proposition identifies certain submanifolds of a given Banach manifold:

Proposition 8.1.15 *Let M and N be Banach manifolds, and let $f : M \to N$ be holomorphic such that, for each $p \in M$, the bounded linear map $df_p : T_pM \to T_{f(p)}N$ is surjective with complemented kernel. Then:*

(i) *For each $p \in M$, the set $f^{-1}(f(p))$ is a submanifold of M.*

(ii) *For each $p \in M$, there are open neighborhoods U of p in M and V of $f(p)$ in N such that $f|_U : U \to V$ has a holomorphic right inverse.*

Proof We first make a few preliminary considerations.

Fix $q \in M$, and choose a chart (U, ϕ) of M at q. We may suppose (Why?) that $\phi(q) = 0$ and $d\phi_q : T_qM \to E$ is the identity on E. Let $P : T_qM \to \ker df_q$ be the projection onto $\ker df_q$, and define

$$\tilde{f} : U \to N \times \ker df_q, \quad p \mapsto (f(p), P\phi(p)). \tag{8.5}$$

Then $\tilde{f}$ is holomorphic, and

$$d\tilde{f}_q = (df_q, P) : T_qM \to T_{f(q)}N \times \ker df_q$$

is an isomorphism of Banach spaces (Why?). By Theorem 8.1.13, there are open neighborhoods V of q in M and W of $(f, 0)$ in $N \times \ker df_q$ such that $\tilde{f}|_V : V \to W$ is biholomorphic.

Making W smaller, if necessary, we may suppose that $W = W_1 \times W_2$, where W_1 is an open neighborhood of $f(q)$ in N, and W_2 is an open neighborhood of 0 in $\ker df_q$.

For the proof of (i), let $p \in M$, let $q \in f^{-1}(f(p))$, and choose a chart (U, ϕ) of M at q. By the foregoing, there are open neighborhoods V of q in M and $W_1 \times W_2$ of $(f(q), 0)$ in $N \times \ker df_q$ such that $\tilde{f} : V \to W_1 \times W_2$ — defined as in (8.5) — is biholomorphic. Then $\tilde{f}(f^{-1}(f(p))) \subset f(p) \times \ker df_q$ with $\tilde{f}(U \cap f^{-1}(f(p))) = f(p) \times W_2$ (Why?). In particular, $P\phi : U \cap f^{-1}(f(p)) \to W_2$ is a homeomorphism. This shows that $f^{-1}(f(p))$ is indeed a submanifold.

For (ii), let $p \in M$, and choose (U, ϕ), V, W_1, and W_2 as in our preliminary considerations, but with p in the rôle of q. Define

$$g : W_1 \to V, \quad q \mapsto \tilde{f}^{-1}(q, 0).$$

It is then clear that g is holomorphic, and a routine calculation shows that it is also a right inverse of f. □

Definition 8.1.16 A group G which is also a Banach manifold such that the maps

$$G \to G, \quad g \mapsto g^{-1} \quad \text{and} \quad G \times G \to G, \quad (g_1, g_2) \mapsto g_1 g_2$$

are holomorphic is called a *Banach–Lie group*. A subgroup H of a Banach–Lie group G is called a *Banach–Lie subgroup* of G if it is also a submanifold of G.

Examples 8.1.17   (a) Let $\mathfrak{A}$ be a unital Banach algebra. Then, by Examples 8.1.4(b) and (c), $\mathrm{Inv}\,\mathfrak{A}$ is a Banach–Lie group.

(b) Let $\mathfrak{A}$ be any Banach algebra, and let $\mathrm{Aut}(\mathfrak{A})$ denote the automorphism group of $\mathfrak{A}$, i.e. the invertible homomorphisms in $\mathcal{L}(\mathfrak{A})$. Let

$$U := \{\theta \in \mathrm{Aut}(\mathfrak{A}) : \|\mathrm{id}_{\mathfrak{A}} - \theta\| < 1\},$$

so that U is open in $\mathrm{Aut}(\mathfrak{A})$, and define

$$\log : U \to \mathcal{Z}^1(\mathfrak{A}, \mathfrak{A}), \quad \theta \mapsto \sum_{n=1}^{\infty} (-1)^{n-1} \frac{(\theta - \mathrm{id}_{\mathfrak{A}})^n}{n}$$

(Why does that land us in $\mathcal{Z}^1(\mathfrak{A}, \mathfrak{A})$?). Then $\log$ is a homeomorphism onto its image (with continuous inverse $\exp|_{\log U}$). For each $\sigma \in \mathrm{Aut}(\mathfrak{A})$, define $\phi_\sigma(\cdot) := \log(\sigma^{-1} \cdot)$. Then $(\sigma U, \phi_u)_{\sigma \in \mathrm{Aut}(\mathfrak{A})}$ is a holomorphic atlas on $\mathrm{Aut}(\mathfrak{A})$ over $\mathcal{Z}^1(\mathfrak{A}, \mathfrak{A})$. It is not hard to see (it is done the same way as in the following exercise) that this turns $\mathrm{Aut}(\mathfrak{A})$ into a Banach–Lie group.

Exercise 8.1.9 Let $\mathfrak{A}$ be a unital Banach algebra.

(i) Define a holomorphic atlas on $\mathrm{Inv}\,\mathfrak{A}$ as in Examples 8.1.17(b) and show that this turns $\mathrm{Inv}\,\mathfrak{A}$ into a Banach–Lie group.

(ii) Show that this atlas is equivalent to the canonical one on $\mathrm{Inv}\,\mathfrak{A}$, i.e. as an open subset of $\mathfrak{A}$.

Definition 8.1.18 Let G be a Banach–Lie group acting on a Banach manifold M. We say that the action of G on M is *holomorphic* if the map

$$G \times M \to M, \quad (g, p) \mapsto g \cdot p$$

is holomorphic.

Let G be a group acting on a non-empty set X, and let $x \in X$. We denote the *orbit* of x by $O(x)$ and define

$$\tau^x : G \to X, \quad g \mapsto g \cdot x. \tag{8.6}$$

We can finally introduce the notion of a Banach homogeneous space:

Definition 8.1.19 Let G be a Banach–Lie group acting holomorphically and transitively on a Banach manifold M. We say that M is a *Banach homogeneous space* under the action of G if there is $p \in M$ such that

(i) $\ker d\tau^p_{e_G}$ is a complemented subspace of $T_{e_G}G$, and
(ii) $d\tau^p_{e_G} : T_{e_G}G \to T_pM$ is surjective.

The conditions in Definition 8.1.19 are reminiscent of the hypotheses of Proposition 8.1.15, except that in Definition 8.1.19 only one particular point of the manifold is considered. This turns out to not be a restriction:

Lemma 8.1.20 *Let G be a Banach–Lie group acting on a Banach manifold M such that M is a Banach homogeneous space under the action of G. Then, for each $q \in M$ and $g \in G$, we have:*

(i) *$\ker d\tau^q_g$ is a complemented subspace of T_gG;*
(ii) *$d\tau^q_g : T_gG \to T_{g \cdot q}M$ is surjective.*

Proof Let $p \in M$ be as in Definition 8.1.19. For $g \in G$, define biholomorphic maps

$$l^g : G \to G, \quad h \mapsto gh \quad \text{and} \quad L^g : M \to M, \quad q \mapsto g \cdot q.$$

Note that $\tau^p = L^g \circ \tau^p \circ l^{g^{-1}}$ for each $g \in G$, so that

$$d\tau^p_g = (dL^q_p)(d\tau^p_{e_G})(dl^{g^{-1}}_g)$$

by Exercise 8.1.6. Since L_g and $l^{g^{-1}}$ are biholomorphic, dL^q_p and $dl^{g^{-1}}_g$ are isomorphisms of Banach spaces. It follows that $d\tau^p_g$ is surjective with complemented kernel for each $g \in G$.

Let now $g \in G$ and $q \in M$ be arbitrary. Since the action of G on M is transitive, there is $h \in G$ such that $q = h \cdot p$. Let

$$r^h : G \to G, \quad k \mapsto kh,$$

so that $\tau^q = \tau^p \circ r^h$. It follows that

$$d\tau^p_g = d\tau^p_{gh} dr^h_g,$$

so that $d\tau^p_g$ is surjective with complemented kernel. $\square$

Combining Lemma 8.1.20 with Proposition 8.1.15, we obtain:

Corollary 8.1.21 *Let G be a Banach–Lie group acting on a Banach manifold M such that M is a Banach homogeneous space under the action of G. Then:*

(i) *For each $p \in M$, the set $\{g \in G : g \cdot p = p\}$ is a Lie subgroup of G.*

(ii) *For each $p \in M$ and each $g \in G$, there are open neighborhoods U of g in G and V of $g \cdot p$ in M such that $\tau^p|_U : U \to V$ has a holomorphic right inverse.*

Our next theorem will be very important in the next section, when it comes to identifying certain sets of homomorphisms as Banach homogeneous spaces:

Theorem 8.1.22 *Let $\mathfrak{A}$ be a unital Banach algebra such that $G = \mathrm{Inv}\,\mathfrak{A}$ acts holomorphically on a Banach space E, let $x \in E$, and suppose that there is $p \in O(x)$ such that:*

(i) $\ker d\tau_{e_G}^p$ *is complemented in $T_{e_G}G$;*

(ii) $\mathrm{ran}\,d\tau_{e_G}^p$ *is a complemented subspace of E;*

(iii) $\tau^p : G \to O(x)$ *is open.*

Then $O(x)$ is a submanifold of E which is a Banach homogeneous space such that we have $T_pO(x) \cong \mathrm{ran}\,d\tau_{e_G}^p$.

Proof Let F_1 be a closed subspace of $T_{e_G}G$ such that $T_{e_G}G = F_1 \oplus \ker d\tau_{e_G}^p$. Define

$$\phi : T_{e_G}G = F_1 \oplus \ker d\tau_{e_G}^p \to G, \quad (y, z) \mapsto \exp(y)\exp(z).$$

Using Theorem 8.1.13 (Work out the details if you're in doubt!), we can find an open neighborhood U of 0 in $T_{e_G}G$ and V of e_G in G such that $\phi|_U : U \to V$ is biholomorphic. Define

$$\psi : T_{e_G} \to O(x), \quad y \mapsto \phi(y) \cdot p.$$

Then ψ is an open map from U onto a neighborhood of p in $O(x)$. Since

$$\frac{d}{dt}\exp(ty) \cdot p\bigg|_{t=0} = 0 \quad (y \in \ker d\tau_{e_G}^p)$$

(Why is this the case?), it follows that $\exp(y) \cdot p = p$ for all $y \in \ker d\tau_{e_G}^p$. It follows that

$$\psi(y, z) = \psi(y, 0) \quad (y \in F_1, z \in \ker d\tau_{e_G}^p),$$

so that ψ maps $F_1 \cap U$ onto a neighborhood of p in $O(x)$.

Let F_2 be a closed subspace of E such that $E = \mathrm{ran}\,d\tau_{e_G}^p \oplus F_2$. Another application of Theorem 8.1.13 shows that

$$F_1 \oplus F_2 \mapsto E, \quad (y, z) \mapsto \psi(y) + z$$

maps an open neighborhood of 0 biholomorphically onto an open neighborhood of p in E. In particular, there is an open neighborhood of p in $O(x)$ which is a submanifold of E. A translation argument as in the proof of Lemma 8.1.20 eventually yields that $O(x)$ is a submanifold of E (please, work out the details, in particular the bit on the tangent space). $\square$

8.2 Spaces of homomorphisms

If $\mathfrak{A}$ and $\mathfrak{B}$ are Banach algebras, we write $\mathrm{Hom}(\mathfrak{A}, \mathfrak{B})$ to denote the set of algebra homomorphisms from $\mathfrak{A}$ into $\mathfrak{B}$, i.e. those maps in $\mathcal{L}(\mathfrak{A}, \mathfrak{B})$ which are multiplicative. Every $\theta \in \mathrm{Hom}(\mathfrak{A}, \mathfrak{B})$ turns $\mathfrak{B}$ into a Banach $\mathfrak{A}$-bimodule. It is not at all clear why any subset of $\mathrm{Hom}(\mathfrak{A}, \mathfrak{B})$ should be a Banach manifold (over which space?).

The following theorem is the technical heart of this section:

Theorem 8.2.1 *Let $\mathfrak{A}$ be an amenable Banach algebra with virtual diagonal* $\mathbf{M}$*, let $\mathfrak{B}$ be a dual Banach algebra with identity, and let $\theta \in \mathrm{Hom}(\mathfrak{A}, \mathfrak{B})$ be such that $\theta^{**}(\Delta_{\mathfrak{A}}^{**}\mathbf{M}) = e_{\mathfrak{B}}$. Then $\mathcal{K}_\theta \colon \mathcal{L}(\mathfrak{A}, \mathfrak{B}) \to \mathfrak{B}$, defined through*

$$\langle \phi, \mathcal{K}_\theta(T) \rangle := \langle \phi, \Delta_{\mathfrak{B}}^{**}(T \otimes \theta)^{**}(\mathbf{M}) \rangle \qquad (\phi \in \mathfrak{B}_*, \, T \in \mathcal{L}(\mathfrak{A}, \mathfrak{B})),$$

is a bounded linear map such that:

(i) $\|\mathcal{K}_\theta\| \leq \|\mathbf{M}\| \|\theta\|$.

(ii) *For every $b \in \mathfrak{B}$ and $T \in \mathcal{L}(\mathfrak{A}, \mathfrak{B})$, we have:*

$$\mathcal{K}_\theta(L_b T) = b \mathcal{K}_\theta(T).$$

(iii) *For every $a \in \mathfrak{A}$ and $T \in \mathcal{L}(\mathfrak{A}, \mathfrak{B})$, we have:*

$$\mathcal{K}_\theta(T L_a) = \mathcal{K}_\theta(T) \theta(a).$$

(iv) $\mathcal{K}_\theta(\theta) = e_{\mathfrak{B}}$.

Proof (i), (ii), and (iv) are straightforward to verify.

For (iii), let $a \in \mathfrak{A}$, $T \in \mathcal{L}(\mathfrak{A}, \mathfrak{B})$, and $\phi \in \mathfrak{B}$. We obtain:

$$
\begin{aligned}
\langle \phi, \mathcal{K}_\theta(T L_a) \rangle - \langle \phi, \mathcal{K}_\theta(T)\theta(a) \rangle &= \langle \phi, \Delta_{\mathfrak{B}}^{**}(T L_a \otimes \theta)^{**}(\mathbf{M}) \rangle - \langle \phi, \Delta_{\mathfrak{B}}^{**}(T \otimes \theta)^{**}(\mathbf{M})\theta(a) \rangle \\
&= \langle \phi, \Delta_{\mathfrak{B}}^{**}(T \otimes \theta)^{**}(a \cdot \mathbf{M}) \rangle - \langle \phi, \Delta_{\mathfrak{B}}^{**}(T \otimes \theta)^{**}(\mathbf{M})\theta(a) \rangle \\
&= \langle \phi, \Delta_{\mathfrak{B}}^{**}(T \otimes \theta)^{**}(\mathbf{M} \cdot a) \rangle - \langle \phi, \Delta_{\mathfrak{B}}^{**}(T \otimes \theta)^{**}(\mathbf{M})\theta(a) \rangle \\
&= \langle \phi, \Delta_{\mathfrak{B}}^{**}(T \otimes \theta)^{**}(\mathbf{M})\theta(a) \rangle - \langle \phi, \Delta_{\mathfrak{B}}^{**}(T \otimes \theta)^{**}(\mathbf{M})\theta(a) \rangle \\
&= 0,
\end{aligned}
$$

as claimed. $\quad\square$

Theorem 8.2.1 allows for an alternative approach to Theorem 4.4.11(i):

Exercise 8.2.1 With $\mathfrak{A}$, $\mathfrak{B}$, and θ as in Theorem 8.2.1, let

$$\mathcal{Q}_\theta \colon \mathfrak{B} \to \mathfrak{B}, \quad b \mapsto \mathcal{K}_\theta(\theta R_b).$$

Show that $\mathcal{Q}_\theta$ is a quasi-expectation onto $Z_{\mathfrak{B}}(\theta(\mathfrak{A}))$.

Exercise 8.2.2 With $\mathfrak{A}$, $\mathfrak{B}$, and θ as in Theorem 8.2.1, define

$$\mathrm{ad} \colon \mathfrak{B} \to \mathcal{L}(\mathfrak{A}, \mathfrak{B}), \quad b \mapsto \mathrm{ad}_b \circ \theta,$$

and let $\mathcal{P}_\theta := \mathrm{ad} \circ \mathcal{K}_\theta$.

(i) Show that $\operatorname{ran} \mathcal{Q}_\theta = \ker \operatorname{ad}$ and $\mathcal{K}_\theta \circ \operatorname{ad} = \operatorname{id}_\mathfrak{B} - \mathcal{Q}_\theta$.

(ii) Show that $\mathcal{P}_\theta$ is a projection onto $\mathcal{Z}^1(\mathfrak{A}, \mathfrak{B})$.

(iii) Show that we have a canonical exact sequence

$$\{0\} \to \ker \mathcal{Q}_\theta \to \mathfrak{B} \overset{\mathcal{Q}_\theta}{\to} \mathfrak{B} \overset{\operatorname{ad}}{\to} \mathcal{L}(\mathfrak{A}, \mathfrak{B}) \overset{\operatorname{id}-\mathcal{P}_\theta}{\to} \mathcal{L}(\mathfrak{A}, \mathfrak{B})/\mathcal{Z}^1(\mathfrak{A}, \mathfrak{B}) \cong \mathcal{K}_\theta^{-1}(Z_\mathfrak{B}(\theta(\mathfrak{A}))) \to \{0\}.$$

Let $\mathfrak{A}$ and $\mathfrak{B}$ be Banach algebras such that $\mathfrak{B}$ is unital. Define a group action of $\operatorname{Inv}\mathfrak{A}$ on $\mathcal{L}(\mathfrak{A}, \mathfrak{B})$ through

$$(b \cdot T)(a) := bT(a)b^{-1} \qquad (b \in \operatorname{Inv} B,\, T \in \mathcal{L}(\mathfrak{A}, \mathfrak{B}),\, a \in \mathfrak{A}). \tag{8.7}$$

Note that $O(\theta) \subset \operatorname{Hom}(\mathfrak{A}, \mathfrak{B})$ for each $\theta \in \operatorname{Hom}(\mathfrak{A}, \mathfrak{B})$.

Exercise 8.2.3 Let $\mathfrak{A}$ and $\mathfrak{B}$ be Banach algebras such that $\mathfrak{B}$ is unital.

(i) Show that the action of $\operatorname{Inv}\mathfrak{B}$ on $\mathcal{L}(\mathfrak{A}, \mathfrak{B})$ as defined in (8.7) is holomorphic.

(ii) Let $T \in \mathfrak{B}$, and let $\tau^T \colon \operatorname{Inv}\mathfrak{B} \to \mathcal{L}(\mathfrak{A}, \mathfrak{B})$ be defined as in (8.6). Show that $d\tau^T_{e_\mathfrak{B}} = \operatorname{ad}$.

Proposition 8.2.2 *Let $\mathfrak{A}$ be an amenable Banach algebra, let $\mathfrak{B}$ be a dual Banach algebra with identity, and suppose that $\mathfrak{A}$ has a virtual diagonal $\mathbf{M}$ such that $\theta^{**}(\Delta_\mathfrak{A}^{**}\mathbf{M}) = e_\mathfrak{B}$. Then, for each $\theta \in \operatorname{Hom}(\mathfrak{A}, \mathfrak{B})$, there is an open neighborhood U of θ in $\operatorname{Hom}(\mathfrak{A}, \mathfrak{B})$ such that*

$$\mathcal{K}_\theta(\sigma) \in \operatorname{Inv}\mathfrak{B} \quad \text{and} \quad \mathcal{K}_\theta(\sigma) \cdot \theta = \sigma \qquad (\sigma \in U).$$

Proof Let

$$U := \left\{ \sigma \in \operatorname{Hom}(\mathfrak{A}, \mathfrak{B}) : \|\sigma - \theta\| < \frac{1}{\|\mathbf{M}\|\|\theta\|} \right\}.$$

Let $\sigma \in U$, and note that

$$\|e_\mathfrak{B} - \mathcal{K}_\theta(\sigma)\| = \|\mathcal{K}_\theta(\theta - \sigma)\| \leq \|\mathbf{M}\|\|\theta\|\|\theta - \sigma\| < 1,$$

so that $\mathcal{K}_\theta(\sigma)$ is invertible. By Theorem 8.2.1(ii) and (iii), we also have:

$$\sigma(a)\mathcal{K}_\theta(\sigma) = \mathcal{K}_\theta(L_\sigma(a)\sigma) = \mathcal{K}_\theta(\sigma L_a) = \mathcal{K}_\theta(\sigma)\theta(a) \qquad (a \in \mathfrak{A}).$$

But this means that

$$(\mathcal{K}_\theta(\sigma) \cdot \theta)(a) = \mathcal{K}_\theta(\sigma)\theta(a)\mathcal{K}_\theta(\sigma)^{-1} = \sigma(a) \qquad (a \in \mathfrak{A}),$$

as claimed.   □

Exercise 8.2.4 Let $\mathfrak{A}$ and $\mathfrak{B}$ be as in Proposition 8.2.2.

(i) Let $b \in \mathfrak{B}$ be such that $\|e_\mathfrak{B} - b\| < 1$. Show that there is $c \in \mathfrak{B}$ such that $b = \exp c$.

(ii) Let $\exp\mathfrak{B}$ denote the subgroup of $\operatorname{Inv}\mathfrak{B}$ generated by the set $\{\exp b : b \in \mathfrak{B}\}$. Show that $\exp\mathfrak{B}$ is the component of $\operatorname{Inv}\mathfrak{B}$ containing $e_\mathfrak{B}$.

(iii) Let $\theta \in \operatorname{Hom}(\mathfrak{A}, \mathfrak{B})$. Show that $\sigma \in \operatorname{Hom}(\mathfrak{A}, \mathfrak{B})$ lies in the component of $\operatorname{Hom}(\mathfrak{A}, \mathfrak{B})$ containing θ if and only if there are $b_1, \ldots, b_n \in \mathfrak{B}$ such that

$$\sigma(a) = (\exp b_1) \cdots (\exp b_n)\theta(a)(\exp b_n)^{-1} \cdots (\exp b_1)^{-1} \qquad (a \in \mathfrak{A}).$$

We can now use the information provided by Proposition 8.2.2 to identify a differential structure on orbits of homomorphism:

Theorem 8.2.3 *Let $\mathfrak{A}$ be an amenable Banach algebra, let $\mathfrak{B}$ be a dual Banach algebra with identity, and suppose that $\mathfrak{A}$ has a virtual diagonal $\mathbf{M}$ such that $\theta^{**}(\Delta_{\mathfrak{A}}^{**}\mathbf{M}) = e_{\mathfrak{B}}$. Then for each $\theta \in \mathrm{Hom}(\mathfrak{A}, \mathfrak{B})$ the following hold:*

(i) *The orbit $O(\theta)$ is a Banach homogeneous space, and thus, in particular, a Banach submanifold of $\mathcal{L}(\mathfrak{A}, \mathfrak{B})$.*

(ii) *For every $\sigma \in O(\theta)$, the tangent space $T_\sigma O(\pi)$ is isomorphic to $\mathcal{Z}^1(\mathfrak{A}, \mathfrak{B})$.*

Proof By Exercise 8.2.3(ii), we have $d\tau_{e_{\mathfrak{B}}}^\theta = \mathrm{ad}$. Since $\ker \mathrm{ad} = \mathrm{ran}\, \mathcal{Q}_\theta$, and since $\mathcal{Q}_\theta$ is a projection, it follows that $\ker d\tau_{e_{\mathfrak{B}}}^\theta$ is complemented in $\mathfrak{B} \cong T_{e_{\mathfrak{B}}}(\mathrm{Inv}\,\mathfrak{B})$. The same is true for $\mathrm{ran}\, d\tau_{e_{\mathfrak{B}}}^\theta = \ker(\mathrm{id} - \mathcal{P}_\theta)$ (by Exercise 8.2.2(iii)). Also, by Proposition 8.2.2, $\tau^\theta: \mathrm{Inv}\,\mathfrak{B} \to O(\theta)$ is open, so that the hypotheses of Theorem 8.1.22 are satisfied. This settles (i).

For (ii), note that we may suppose without loss of generality that $\sigma = \theta$ (since $O(\theta) = O(\sigma)$). By Theorem 8.1.22, we have

$$T_\theta O(\theta) \cong \mathrm{ran}\,\mathrm{ad} = \mathcal{Z}^1(\mathfrak{A}, \mathfrak{B})$$

(here, as in the solution to Exercise 8.2.2, we need that $\mathfrak{A}$ is amenable and $\mathfrak{B}$ is dual, so that $\mathcal{Z}^1(\mathfrak{A}, \mathfrak{B}) = \mathcal{B}^1(\mathfrak{A}, \mathfrak{B})$). $\square$

8.3 Notes and comments

Infinite-dimensional holomorphy is discussed in great detail in the monographs [Muj] and [Din]. As in the finite-dimensional case, there are numerous equivalent characterizations of holomorphic functions in the infinite-dimensional setting.

A standard source on differential geometry (over $\mathbb{R}^n$) is [K-N], whereas [Lang 3] treats manifolds over arbitrary Banach spaces. Our exposition is patterned after [Lang 3] and [Rae]. In particular, the discussion of Banach homogeneous spaces is taken from [Rae].

Our discussion of the differential geometry of spaces of homomorphisms is from the work of G. Corach and J. E. Galé ([C–G 2]). In order to keep the necessary background from differential geometry to a minimum, we have not stated Theorem 8.2.3 in its full strength; in addition to what we prove, one can show that $\tau^\theta: \mathrm{Inv}\,\mathfrak{B} \to O(\theta)$ is a so-called *Banach principal bundle* whose structure group is easily identified (see [K-N] for the relevant definitions). The results from [C–G 2] are generalized in [C–G 1], where a class of Banach algebras is considered which includes all amenable Banach algebras, but also all dual Banach algebras with a normal, virtual diagonal.

An apparently very important concept in the differential geometry of spaces of homomorphisms is that of a *reductive structure* (see [C–G 2] for the definition). In [A–C–St], nuclear C^*-algebras and Connes-amenable von Neumann algebras are characterized through the existence of certain reductive structures. A similar result for group algebras is proved in [C–G 2].

Open problems

The following problems are — to my knowledge — still open: Some of them have been around for quite a while, sometimes for decades; others are new. Some, in particular the older problems, are quite hard, and their solution would be a major breakthrough in the field. Others are hopefully easier, and just the fact that I don't know (yet) how to solve them shouldn't discourage anyone from trying.

Amenable, locally compact groups

Since the main focus of these notes are amenable Banach algebras, we only present one problem on amenable, locally compact groups. It's a classic, due to J. Dixmier:

Problem 1 Let G be a locally compact group such that each uniformly bounded representation of G on a Hilbert space is similar to a unitary representation. Does this mean that G is amenable?

This problem falls into the category of so-called similarity problems ([Pis 2]). In [Pis 4], G. Pisier gives a new approach to this problem (in the discrete case) and related ones, along with partial solutions.

Amenable Banach algebras

In [Joh 8], B. E. Johnson introduced the notion of a symmetrically amenable Banach algebra. His definition is in terms of approximate diagonals.

Problem 2 Is there a characterization of symmetrically amenable Banach algebras in terms of derivations?

Some results from [Joh 8] suggest that one might have to consider Lie or Jordan derivations.

In Theorem 4.4.11, we used (Connes-)amenability to establish the existence of certain quasi-expectations. Does a converse hold?

Problem 3 Let $\mathfrak{A}$ be a Banach algebra with the following property: For every dual Banach algebra $\mathfrak{B}$ and for every homomorphism $\theta : \mathfrak{A} \to \mathfrak{B}$ there is a quasi-expectation $\mathcal{Q} : \mathfrak{B} \to Z_{\mathfrak{B}}(\theta(\mathfrak{A}))$. Is $\mathfrak{A}$ necessarily amenable?

The answer is affirmative for C^*-algebras: If $\mathfrak{A}$ is a C^*-algebra acting via the universal *-representation on a Hilbert space $\mathfrak{H}$, then the property in Problem 3 implies that there is a quasi-expecation from $\mathcal{L}(\mathfrak{H})$ onto the von Neumann algebra $\mathfrak{A}'$. Hence, $\mathfrak{A}'$ and thus $\mathfrak{A}'' = \mathfrak{A}^{**}$ is injective by Theorem 6.2.6. By Corollary 6.5.12, this means that $\mathfrak{A}$ is amenable.

Sometimes, amenability forces members of a certain class of Banach algebras to have a particular structure (think of Theorems 2.3.14 and 6.1.7). The following problems, the first of which saw the light of day in [G–R–W], ask for further results of this kind:

Problem 4 Let $\mathfrak{A}$ be an amenable Banach algebra, let $\mathfrak{B}$ be any Banach algebra, and let $\theta : \mathfrak{A} \to \mathfrak{B}$ be a weakly compact homomorphism. Is $\theta(\mathfrak{A})$ necessarily finite-dimensional and semi-simple?

Partial answers are given in [G–R–W] and also in [Joh 6], but the general case seems to be open even when $\mathfrak{A} = L^1(G)$ for a locally compact group G. A special case of Problem 4 is when $\mathfrak{A} = \mathfrak{B}$ and θ is the identity:

Problem 5 Let $\mathfrak{A}$ be an amenable Banach algebra whose underlying Banach space is reflexive. Is $\mathfrak{A}$ finite-dimensional?

Essentially the same argument as in the proof of Proposition 4.1.2 shows that, if the algebra in Problem 5 is finite-dimensional, it must be of the form $\mathbb{M}_{n_1} \oplus \cdots \oplus \mathbb{M}_{n_k}$ with $n_1, \ldots, n_k \in \mathbb{N}$. In the Hilbert space case, the problem has an affirmative answer ([Gh–L–W]; see also [Joh 6]). Other partial results are given in [Zha], [Run 1], and [Run 3].

Another problem including Problem 5 as a special case is the following:

Problem 6 Let $\mathfrak{A}$ be a Banach algebra such that $\mathfrak{A}^{**}$ — equipped with either Arens product — is amenable. Does this imply that $\mathfrak{A}$ is isomorphic to a subhomogeneous C^*-algebra?

All we positively know about $\mathfrak{A}$ is that it must also be amenable ([Gh–L–W] or [Gou 2]). However, the only known examples of Banach algebras $\mathfrak{A}$ such that $\mathfrak{A}^{**}$ is amenable are the subhomogeneous C^*-algebras. Even demanding that $\mathfrak{A}$ be Arens regular doesn't seem to bring us closer to a solution of Problem 6. The following is a particular case of Problem 6 which one should perhaps try to tackle first:

Problem 7 Let $\mathfrak{A}$ be a Banach algebra such that $\mathfrak{A}^{**}$ is commutative and amenable. Does this imply that $\mathfrak{A} \cong C_0(\Omega)$ for some locally compact Hausdorff space Ω?

Examples of amenable Banach algebras

There are a surprising number of Banach algebras around for which it it still unknown if they are amenable or not.

The following was already asked by Johnson ([Joh 1]):

Problem 8 Is there an infinite-dimensional Banach space E such that $\mathcal{L}(E)$ is amenable?

Intuitively, there should be no such space: Algebras of the form $\mathcal{L}(E)$ for infinite-dimensional E are simply too "large" to be amenable. It follows from Theorem 6.1.7 that the solution to Problem 8 is negative for Hilbert spaces; other Banach spaces E for which

$\mathcal{L}(E)$ is known to be non-amenable are ℓ^1 ([Rea 2]) and spaces of the form $\ell^p \oplus \ell^q$ for certain values of p and q (Example 4.4.9). But still, the following rather special case of Problem 8 seems to be open:

Problem 9 Is $\mathcal{L}(\ell^p)$ amenable for some $p \in (1, \infty) \setminus \{2\}$?

Probably not, but where is a proof?

The following problem is related to Problem 8:

Problem 10 Let $(E_n)_{n=1}^\infty$ be a sequence of finite-dimensional Banach spaces such that $\lim_{n \to \infty} \dim E_n = \infty$. Can $\ell^\infty\text{-}\bigoplus_{n=1}^\infty \mathcal{L}(E_n)$ be amenable?

As a consequence of Theorem 6.1.7, Problem 10 has a negative answer if $E_n = \ell_n^2$. Another case in which a negative answer to Problem 10 was given only very recently is where $E_n = \ell_n^p$ for some $p \in [1, \infty] \setminus \{2\}$ independent of n ([Rea 1]). Interestingly, the methods to prove the non-amenability of $\ell^\infty\text{-}\bigoplus_{n=1}^\infty \mathcal{L}(\ell_n^p)$ are completely different depending on whether $p = 2$ or $p \in [1, \infty] \setminus \{2\}$. Of course, a negative answer to Problem 6 would entail a negative answer to Problem 10 in its full generality.

We call a Banach operator algebra on some Banach space E a *Banach operator ideal* if it is an ideal in $\mathcal{L}(E)$. The only known amenable Banach operator ideals are of the form $\mathcal{A}(E)$ for a suitable Banach space E. This naturally leads to the following question, which generalizes Problem 6:

Problem 11 Let $\mathfrak{A}$ be an amenable Banach operator ideal on some Banach space E. Is $\mathfrak{A} = \mathcal{A}(E)$?

The central problem for Banach algebras of approximable operators is the following:

Problem 12 Characterize in Banach space theoretic terms those Banach spaces E for which $\mathcal{A}(E)$ is amenable.

The long open question of whether a commutative, radical, amenable Banach algebra existed was eventually answerd by C. J. Read (Corollary 3.2.8). The following question was raised (and discussed) in [Loy *et al.*]:

Problem 13 Is there a commutative, radical, amenable Banach algebra with compact multiplication?

In [Loy *et al.*], a singly-generated, radical, *weakly* amenable Banach algebra was constructed, which is also an integral domain and has a bounded approximate identity of a rather peculiar form. That algebra, however, cannot be amenable. If Problem 13 has a positive answer, Exercise 4.3.4 implies that there is a commutative, radical, biprojective Banach algebra (which would positively answer Problem 20 below).

Theorem 7.2.10 and related results from [Joh 7] and [For] lead to the following challenging problem:

Problem 14 Which intrinsic condition characterizes those locally compact groups G for which $A(G)$ is amenable?

In view of Theorem 7.2.10 and Exercise 7.4.2, it is a good guess that these are precisely those locally compact groups with a closed, abelian subgroup of finite index. If one wishes to increase the challenge, one can replace $A(G)$ in Problem 14 with an arbitrary Figà-Talamanca–Herz algebra.

Amenability-like properties

Obviously, the following is the central problem regarding super-amenable Banach algebras:

Problem 15 Is there an infinite-dimensional, super-amenable Banach algebra?

If there is such an algebra, Theorem 4.1.5 implies that the algebra itself as well as many modules on which it acts lack the approximation property. Surprisingly, not even the following, rather special case of Problem 15 has been solved yet:

Problem 16 Is there an infinite-dimensional Banach space E such that $\mathcal{L}(E)$ is super-amenable?

By Lemma 4.1.4, no such Banach space can have the approximation property. As shown in [Joh 8], super-amenability of $\mathcal{L}(E)$ forces E to be even more bizarre ...

In [D–Gh–G], H. G. Dales, F. Ghahramani, and N. Grœnbæk generalized Theorem 4.2.3 by showing that $\mathcal{H}^1(L^1(G), L^1(G)^{n*}) = \{0\}$ for every locally compact group G and for all *odd* $n \in \mathbb{N}$. The following is still open:

Problem 17 Is $\mathcal{H}^1(L^1(G), L^1(G)^{n*}) = \{0\}$ for every locally compact group G and for all $n \in \mathbb{N}$?

Adopting the terminology of [D–Gh–G], Problem 17 is the question of whether $L^1(G)$ is *permanently weakly amenable* for every locally compact group G. The answer is obviously positive for amenable groups, but also for every free group ([Joh 9]). The relevance of Problem 17 lies in the fact that $\mathcal{H}^1(L^1(G), L^1(G)^{**}) = \{0\}$ implies $\mathcal{H}^1(L^1(G), M(G)) = \{0\}$, so that a positive answer to Problem 17 would solve a classical problem posed by Johnson:

Problem 18 Is $\mathcal{H}^1(L^1(G), M(G)) = \{0\}$ for every locally compact group G?

For partial results, see [Joh 1], [Gh–R–W], and [Joh 10].

In view of how difficult the proof of Theorem 4.2.4 is relative to that of Theorem 4.2.3, we ask:

Problem 19 Is there a "simple" way to prove that all C^*-algebras are weakly amenable?

Here, "simple" means that the proof should at least avoid the Grothendieck inequality. We only have one question on biprojective and biflat Banach algebras:

Problem 20 Is there a radical, biprojective Banach algebra?

As we already mentioned, a positive answer to Problem 13 would automatically solve Problem 20.

Connes-amenability has so far received relatively little attention outside the context of von Neumann algebras ...

Problem 21 Let $\mathfrak{A}$ be an Arens regular Banach algebra such that $\mathfrak{A}^{**}$ is Connes-amenable. Is $\mathfrak{A}$ amenable?

For C^*-algebras, the answer is "yes" by Corollary 6.4.27; a partial result in the general situation is Theorem 4.4.8.

There are some results on Connes-amenable W^*-algebras, which can be formulated for arbitrary Connes-amenable, dual Banach algebras. But are they still true?

For a von Neumann algebra, Connes-amenability is the same as injectivity, so that the following problem has a positive answer for W^*-algebras by Theorem 6.2.6:

Problem 22 Let $\mathfrak{A}$ be a dual Banach algebra with the following property: For every dual Banach algebra $\mathfrak{B}$ and for every w^*-continuous homomorphism $\theta \colon \mathfrak{A} \to \mathfrak{B}$ there is a quasi-expectation $\mathcal{Q} \colon \mathfrak{B} \to Z_{\mathfrak{B}}(\theta(\mathfrak{A}))$. Is $\mathfrak{A}$ necessarily Connes-amenable?

In view of Theorems 2.2.4, 4.4.15 and 6.5.11, the following is natural:

Problem 23 Let $\mathfrak{A}$ be a Connes-amenable, dual Banach algebra. Is there a normal, virtual diagonal for $\mathfrak{A}$?

The answer is positive if $\mathfrak{A}$ is a W^*-algebra ([Eff]) or the measure algebra of a locally compact group ([Run 5]). The main difficulty when trying to mimic the proof of Theorem 2.2.4 is that, for a dual Banach algebra $\mathfrak{A}$, the dual Banach $\mathfrak{A}$-bimodule $\mathcal{L}^2_{w^*}(\mathfrak{A}, \mathbb{C})$ need not be normal. The methods used in [Eff], [E–K], and [Run 5] to overcome this obstacle for particular dual Banach algebras do not carry over to the general situation. My guess is that Problem 23 has a negative answer for arbitrary $\mathfrak{A}$.

In [Run 4], the notion of *strong Connes-amenability* was introduced, and it was shown that strong Connes-amenability is equivalent to the existence of a normal virtual diagonal. A positive answer to Problem 23 would thus entail the equivalence of Connes-amenability and strong Connes-amenability.

In view of [Hel 4], the following question — posed to the author by Helemskiǐ — is canonical to ask:

Problem 24 Let $\mathfrak{A}$ be a Connes-amenable, dual Banach algebra with predual $\mathfrak{A}_*$. Is $\mathfrak{A}_*$ injective?

Banach homology

There is an article by Helemskiǐ which — as stated in its title — contains 31 open problems from topological homology ([Hel 6]); most of them are still open.

Let $\mathfrak{A}$ be a Banach algebra, let E be a Banach $\mathfrak{A}$-bimodule, and let, for $n \in \mathbb{N}$, the n-th algebraic Hochschild cohomology group be denoted by $H^n(\mathfrak{A}, E)$. For each $n \in \mathbb{N}_0$, there is a canonical group homomorphism $c^n \colon \mathcal{H}^n(\mathfrak{A}, E) \to H^n(\mathfrak{A}, E)$ (see the notes and remarks section of Chapter 2).

Problem 25 Let $\mathfrak{A}$ be a Banach algebra, and let E be a Banach $\mathfrak{A}$-bimodule. Can we say anything meaningful about $c^3 \colon \mathcal{H}^3(\mathfrak{A}, E) \to H^3(\mathfrak{A}, E)$ for certain $\mathfrak{A}$ and E?

Maybe the methods of [Sol], [D–V], or [Wod] can be adapted to yield partial answers to this problem. Maybe . . .

C^*- and W^*-algebras

The results on the (Connes-)amenability of C^*- and W^*-algebras we have presented in these notes have one thing in common: Their proofs are much deeper and more complicated than those of all the results we had previously encountered. In particular, the proofs for the non-amenability of the W^*-algebras $\ell^\infty\text{-}\bigoplus_{n=1}^\infty \mathbb{M}_n$ and $\mathcal{L}(\mathfrak{H})$, where $\mathfrak{H}$ is an infinite-dimensional Hilbert space, seem to be inappropriately deep for an answer to an (apparently) simple question about simple algebras.

We thus ask:

Problem 26 Is there a more elementary proof for the non-amenability of the W^*-algebra $\ell^\infty\text{-}\bigoplus_{n=1}^\infty \mathbb{M}_n$, preferably one that establishes the non-amenability of $\ell^\infty\text{-}\bigoplus_{n=1}^\infty \mathcal{L}(\ell_n^p)$ for all $p \in [1,\infty]$ simultaneously?

As we already mentioned, the methods from [Rea 2] that establish the non-amenability of $\ell^\infty\text{-}\bigoplus_{n=1}^\infty \mathcal{L}(\ell_n^p)$ for $p \in [1,\infty] \setminus \{2\}$ do not work for $p = 2$.

Problem 27 Let $\mathfrak{H}$ be an infinite-dimensional Hilbert space. Is there a more elementary proof for the non-amenability of $\mathcal{L}(\mathfrak{H})$, perhaps one that could be modified to solve Problem 9?

Among the properties of W^*-algebras equivalent to Connes-amenability, being approximately finite-dimensional is the formally strongest — injectivity, semidiscreteness, and the existence of normal, virtual diagonals are easy to establish for AFD W^*-algebras. We have not given a proof for the implication from Connes-amenable/injective/semidiscrete to AFD because a reasonably self-contained proof of this implication — even with the simplications due to Haagerup and Popa — would probably have been longer than the remainder of these notes. We thus ask (perhaps naively):

Problem 28 Is there a "reasonably" simple proof for

$$\text{Connes-amenable/injective/semidiscrete} \quad \Longrightarrow \quad \text{AFD?}$$

Let $\mathfrak{H}$ be a Hilbert space, and let $\mathfrak{A}$ and $\mathfrak{B}$ be closed subalgebras of $\mathcal{L}(\mathfrak{H})$. Then $\mathfrak{A}$ and $\mathfrak{B}$ are called *similar* if there is an invertible operator $T \in \mathcal{L}(\mathfrak{H})$ such that $\mathfrak{A} = T\mathfrak{B}T^{-1}$. An intriguing open problem — another similiarity problem — asks for a characterization of those closed subalgebras of $\mathcal{L}(\mathfrak{H})$ that are similar to C^*-algebras. An important rôle here seems to be played by the following property: If $\mathcal{H}^1(\mathfrak{A}, \mathcal{L}(\mathfrak{K})) = \{0\}$ for every Hilbert space $\mathfrak{K}$ and for every bounded representation $\pi\colon \mathfrak{A} \to \mathcal{L}(\mathfrak{K})$, then the algebra $\mathfrak{A}$ is said to have the *total reduction property* (this is not quite the definition, but equivalent to it by [Gif, 9 Theorem]). It is clear that, if $\mathfrak{A}$ is amenable, it has the total reduction property. On the other hand, $\mathcal{L}(\mathfrak{H})$ has the total reduction property for every Hilbert space $\mathfrak{H}$ ([Pis 2, Corollary 7.14]). The question of whether every C^*-algebra has the total reduction property is equivalent to a famous similarity problem due to R. V. Kadison ([Pis 2, Problem 0.2] and [Gif, 16 Corollary]).

The following question is raised in [Gif]:

Problem 29 Let $\mathfrak{H}$ be a Hilbert space, and let $\mathfrak{A}$ be a closed subalgebra of $\mathcal{L}(\mathfrak{H})$ with the total reduction property. Is $\mathfrak{A}$ then similar to a C^*-algebra?

A partial answer is due to J. Gifford: It is positive if $\mathfrak{A} \subset \mathcal{K}(\mathfrak{H})$ ([Gif, 37 Theorem]; see also [Wil]).

A perhaps easier variant of Problem 29 is:

Problem 30 Let $\mathfrak{H}$ be a Hilbert space, and let $\mathfrak{A}$ be a closed, amenable subalgebra of $\mathcal{L}(\mathfrak{H})$. Is $\mathfrak{A}$ then similar to a C^*-algebra?

This seems to be open even for commutative $\mathfrak{A}$.

Operator amenability

The notion of operator amenability is still very young, so that one should expect a lot of tractable problems in that field.

The results existing so far (see [Rua 2], [R–X], and [Ari 2]) suggest that the theory of operator amenable operator Banach algebras very much parallels the theory of amenable Banach algebras. Nevertheless, the existence of natural operator Banach algebra structures over a Banach algebra $\mathfrak{A}$ which may be different from $MAX(\mathfrak{A})$ allows for interesting new twists (as the example of the Fourier algebra shows).

If $\mathfrak{A}$ is a uniform algebra, then $MIN(\mathfrak{A})$ is an operator Banach algebra. Theorem 2.3.14 suggests the following:

Problem 31 Let $\mathfrak{A}$ be a uniform algebra such that $MIN(\mathfrak{A})$ is operator amenable. Is $\mathfrak{A}$ already a C^*-algebra?

Let G be a locally compact group, and let $B_{\lambda_2}(G) := C^*_{\lambda_2}(G)^*$. Since $C^*_{\lambda_2}(G)^*$ is a quotient of $C^*(G)$, the space $B_{\lambda_2}(G)$ is a closed subspace of $B(G)$; in fact, $B_{\lambda_2}(G)$ is a closed ideal of $B(G)$ ([Eym, (2.16) Proposition]). The unitary representation

$$G \to \mathcal{L}(L^2(G \times G)), \quad g \mapsto \lambda_2(g) \otimes \lambda_2(g)$$

extends to a $*$-homomorphism from $C_{\lambda_2}(G)$ into $\mathcal{M}(C^*_{\lambda_2}(G))$ ([Lanc 2, pp. 83–86]) such that the adjoint map induces the algebra multiplication on $B_{\lambda_2}(G)$ ([Lanc 2, Proposition 8.5]). It follows that $B_{\lambda_2}(G)$ is an operator Banach algebra.

Problem 32 Which are the locally compact groups G such that $B_{\lambda_2}(G)$ is operator amenable?

Certainly, if G is compact, then $B_{\lambda_2}(G) = B(G) = A(G)$ is operator amenable by Theorem 7.4.3. Suppose conversely that $B_{\lambda_2}(G)$ is operator amenable. Since $A(G)$ is a closed $C^*_{\lambda_2}(G)$ submodule of $B_{\lambda_2}(G)$, there is a central projection $p \in C^*_{\lambda_2}(G)^{**}$ such that $A(G) = p \cdot B_{\lambda_2}(G)$. In particular, $A(G)$ is complemented in $B_{\lambda_2}(G)$ with a completely bounded projection onto $A(G)$. As in the proof of Theorem 2.3.7, we conclude that $A(G)$ has a bounded approximate identity, so that G is amenable by Theorem 7.1.3. My guess is that the operator amenability of $B_{\lambda_2}(G)$ not only implies the amenability, but the compactness of G.

Recently, N. Spronk has shown that, for a locally compact group G, its Fourier–Stieltjes algebra $B(G)$ is also an operator Banach algebra ([Spr 2]). One can then, of course, pose the analogue of Problem 32 for $B(G)$ (with probably the same answer). It is very likely that, for algebras such as $B_{\lambda_2}(G)$ and $B(G)$, the "right" notion of amenability, i.e. one that precisely captures the amenability of G, is a hybrid of operator amenability and Connes-amenability.

Apparently the notion of biflatness has not yet been studied in the operator Banach algebras context. By [R–X], all operator amenable operator Banach algebras are biflat, so that $A(G)$ is operator biflat for all amenable locally compact groups. Trivially, operator biprojective operator Banach algebras are operator biflat, so that $A(G)$ is also operator biflat by [Woo 2] or [Ari 2, Theorem 7.30], respetively. It is highly plausible that a property of locally compact groups which all amenable and all discrete groups have is, in fact, ejoyed by all locally compact groups. Hence, the following question is likely to have a positive answer:

Problem 33 Let G be a locally compact group. Is $A(G)$ operator biflat?

By [R–X, Corollary 4.5], the answer to Problem 33 is indeed positive if G is a so-called $[SIN]$-group, i.e. if $L^1(G)$ has a a bounded approximate identity belonging to $Z(L^1(G))$ (a somewhat weaker condition is, in fact, sufficient ([R–X, Theorem 4.4])). Another piece of circumstantial evidence that supports the belief that Problem 33 has an affirmative answer is the main result of [Spr 1] because it is very likely that there is an operator analogue of Theorem 5.3.13.

Let G be a locally compact group. Theorem 7.1.3 remains valid if we replace $A(G)$ by $A_p(G)$ for some $p \in (1, \infty)$. The question of whether Theorem 7.4.3 also remains true for arbitrary Figà-Talamanca–Herz algebras $A_p(G)$ with $p \in (0, \infty) \setminus \{2\}$ faces an immediate obstacle: There is no canonical operator space structure (except $MAX(A_p(G))$, of course) that turns $A_p(G)$ into an operator Banach algebra.

We thus ask:

Problem 34 Let G be a locally compact group. What is the "right" notion of amenability for the Figà-Talamanca–Herz algebras $A_p(G)$ for $p \in (1, \infty) \setminus \{2\}$?

In [Pis 5], G. Pisier uses his interpolation theory of operator spaces ([Pis 3]) to introduce a canonical operator space structure on all L^p-spaces. In [Run 6], this is used to introduce, for a locally compact group G and $p \in (1, \infty)$, an operator space analogue $OA_p(G)$ of $A_p(G)$. These algebras $OA_p(G)$ are operator Banach algebras, but for $p \neq 2$, we only have a contractive inclusion $A_p(G) \subset OA_p(G)$ in general and not identity. Also, $A(G)$ and $OA_2(G)$ are identical only as Banach spaces, but generally not as operator spaces.

Geometry of spaces of homomorphisms

If $\mathfrak{A}$ is an amenable Banach algebra, then there are differential geometric structures on certain spaces of homomorphisms from $\mathfrak{A}$. The big general question is, of course, if the presence of such structures characterizes the amenable Banach algebras:

Problem 35 Let $\mathfrak{A}$ be a Banach algebra such that for each dual Banach algebra $\mathfrak{B}$ with identity and for each $\theta \in \mathrm{Hom}(\mathfrak{A}, \mathfrak{B})$, there is an open neighborhood U of θ and a continuous map $\mathcal{K} : U \to \mathrm{Inv}\, \mathfrak{B}$ such that

$$\mathcal{K}_\theta(\sigma)\theta(a)\mathcal{K}(\sigma)^{-1} = \sigma(a) \qquad (\sigma \in U,\, a \in \mathfrak{A}).$$

Does this mean that $\mathfrak{A}$ is amenable?

A closely related condition does indeed imply amenability ([Gou 1, Theorem 3]).

Problem ...

A Abstract harmonic analysis

A *locally compact group* is a group G which is also a locally compact Hausdorff space such that the maps

$$G \times G \to G, \quad (g,h) \mapsto gh \quad \text{and} \quad G \to G, \quad g \mapsto g^{-1}$$

are continuous. Abstract harmonic analysis is the study of locally compact groups and of the spaces and algebras associated with them.

This appendix serves as a storage space for the notions and facts from abstract harmonic analysis required for these notes. In order not to bloat it, we more or less refrain from giving proofs and refer to the literature instead.

A.1 Convolution of measures and functions

Let G be a locally compact group. We write $M(G)$ for the space of all (finite) complex, regular Borel measures on G. By Riesz' representation theorem (e.g. [Coh, Theorem 7.3.5]), $M(G)$ can be identified with the dual space of $\mathcal{C}_0(G)$. We want to make $M(G)$ into a Banach algebra.

Definition A.1.1 Let G be a locally compact group, and let $\mu, \nu \in M(G)$. The *convolution product* $\mu * \nu \in M(G)$ of $\mu, \nu \in M(G)$ is defined through

$$\langle f, \mu * \nu \rangle := \int_G \left(\int_G f(gh)\, d\mu(g) \right) d\nu(h) \qquad (f \in \mathcal{C}_0(G)). \tag{A.1}$$

By Fubini's theorem, the order of integration in (A.1) is irrelevant. It is routine (albeit dull) to check that $(M(G), *)$ is a unital Banach algebra — the *measure algebra* of G —, which is commutative if and only if G is abelian. By the way, what is the identity of $M(G)$?

Exercise A.1.1 Let G be a locally compact group, and let H be closed subgroup of G. Show that the adjoint of the restriction map from $\mathcal{C}_0(G)$ into $\mathcal{C}_0(H)$ is an isometric algebra isomorphism from $M(H)$ into $M(G)$ whose range is the set of those measures in $M(G)$ with support in H.

Next, we consider two particular subalgebras of the measure algebra.

Exercise A.1.2 Let G be a group. For $\mu, \nu \in \ell^1(G)$ define their convolution product $\mu * \nu \in \ell^1(G)$ through

$$(\mu * \nu)(g) := \sum_{h \in G} \mu(gh)\nu(h^{-1}) \qquad (h \in G).$$

(i) Show that $(\ell^1(G), *)$ is a unital Banach algebra.

(ii) Let G be equipped with the discrete topology, so that $\ell^1(G) \cong M(G)$ isometrically as Banach spaces. Show that this isomorphism is an isomorphism of Banach algebras if both $M(G)$ and $\ell^1(G)$ are equipped with their respective convolution products.

We now relate $\ell^1(G)$ and $M(G)$ for an arbitrary locally compact group G.

Definition A.1.2 Let G be a locally compact group.

(i) For $g \in G$, the *point mass* $\delta_g \in M(G)$ at g is defined through

$$\langle f, \delta_g \rangle := f(g) \qquad (f \in \mathcal{C}_0(G)).$$

(ii) A measure in $M(G)$ is called *discrete* if it is in the closed linear span of $\{\delta_g : g \in G\}$.

Exercise A.1.3 Let G be a locally compact group. Show that

$$\ell^1(G) \to M(G), \quad \mu \mapsto \sum_{g \in G} \mu(g)\delta_g \tag{A.2}$$

is a unital, isometric homomorphism of Banach algebras whose range is the set of discrete measures in $M(G)$.

We shall thus identify $\ell^1(G)$ with the subalgebra of $M(G)$ consisting of all discrete measures.

For the second subalgebra of $M(G)$ we are interested in, we require the following result ([H–R, (15.5) Theorem] and [H–R, (15.8) Remarks]) which is one of the most fundamental results in abstract harmonic analysis:

Theorem A.1.3 *Let G be a locally compact group. Then there is a non-zero, regular (positive) Borel measure m_G on G — left Haar measure — which is left invariant, i.e. $m_G(gE) = m_G(E)$ for all $g \in G$ and all Borel subsets E of G. If m_G' is another non-zero, regular, left invariant Borel measure on G, then there is a constant $c > 0$ such that $m_G' = c\,m_G$.*

Exercise A.1.4 Let G be a locally compact group, and let $\varnothing \neq U \subset G$ be open. Show that $m_G(U) > 0$.

Remark A.1.4 Left Haar measure on a locally compact group G is unique only up to a mutliplicative constant. For general G, there is no canonical way to normalize Haar measure. However, both for discrete and for compact groups a normalization is possible. In the discrete case, we suppose that $m_G(\{g\}) = 1$ for all $g \in G$, i.e. m_G is counting measure; for compact G, regularity of m_G forces m_G to be a finite measure, so that we can suppose without loss of generality that m_G is a probability measure (there is a conflict between these normalizations in case $G \neq \{e_G\}$ is finite ...).

Sometimes we might omit (or simply forget) the adjective "left" in these notes when speaking of Haar measure. We thus make the tacit convention that "Haar measure" always means "left Haar measure".

Examples A.1.5   (a) On $\mathbb{R}^n$, Haar measure is n-dimensional Lebesgue measure.

(b) If G is discrete, counting measure is Haar measure.

Exercise A.1.5 Let G be a non-discrete, locally compact group. Show that $m_G(\{g\}) = 0$ for $g \in G$. Conclude that the homomorphism (A.2) is an isomorphism only if G is discrete.

As is customary, for a locally compact group, we write $L^1(G)$ for those (equivalence classes of) Borel functions on G which are integrable with respect to m_G.

Definition A.1.6 Let G be a locally compact group, and let $f_1, f_2 \in L^1(G)$. Their *convolution product* $f_1 * f_2 \in L^1(G)$ is defined through

$$(f_1 * f_2)(g) := \int_G f_1(h) f_2(h^{-1}g) \, dm_G(h) \qquad (g \in G). \tag{A.3}$$

Exercise A.1.6 Convince yourself that the convolution product on $L^1(G)$ as introduced in Definition A.1.6 is well defined, i.e. the integral in (A.3) exists for all $g \in G$ outside a set of Haar measure zero and $f_1 * f_2$ as defined there is again in $L^1(G)$. If you have difficulties, look at [Coh], [Fol], or [H–R].

Remark A.1.7 Alternative formulas for the convolution product of two L^1-functions are given in [H–R, (20.10) Theorem]; for some of them, the modular function (introduced below) is needed.

As in the case of the measure algebra it is routinely, though somewhat tediously checked that $(L^1(G), *)$ is a Banach algebra. Does that deter you from actually checking it?

If G is discrete, $L^1(G) = \ell^1(G)$, so that $L^1(G)$ is unital in this case. For arbitrary locally compact G, this is no longer true, but at least, we have a bounded approximate identity (see [B–D, §11] for a definition):

Theorem A.1.8 *Let G be a locally compact group, let $\mathfrak{U}$ be a basis of neighborhoods of e_G, and let $(e_U)_{U \in \mathfrak{U}}$ be a net in $L^1(G)$ with the following properties:*

(i) $e_U \geq 0$ *for all $U \in \mathfrak{U}$;*
(ii) $\mathrm{supp}(e_U) \subset U$ *for all $U \in \mathfrak{U}$;*
(iii) $\|e_U\|_1 = 1$ *for all $U \in \mathfrak{U}$.*

Then $(e_U)_{U \in \mathfrak{U}}$ is a bounded approximate identity for $L^1(G)$.

Corollary A.1.9 *Let G be a locally compact group. Then $L^1(G)$ has a bounded approximate identity in $P(G)$.*

Theorem A.1.8 is an immediate consequence of the slightly more general [H–R, (28.52) Theorem].

There is an element of asymmetry in the definition of $L^1(G)$, since we have chosen to define it via *left* Haar measure. Although left Haar measure need not be right invariant, it behaves "relatively nicely" under right translations ([H–R, (15.11) Theorem]):

Theorem A.1.10 *Let G be a locally compact group. Then there is a unique, continuous function $\Delta_G : G \to (0, \infty)$ — the modular function of G — such that*

(i) $\Delta_G(gh) = \Delta_G(g)\Delta_G(h)$ *for all $g, h \in G$, and*
(ii) $m_G(Eg) = \Delta_G(g)m_G(E)$ *for each $g \in G$ and for each Borel subset E of G.*

Remark A.1.11 If $\Delta_G \equiv 1$, the group G is called *unimodular*. Trivially, abelian and discrete groups are unimodular, but so are all compact groups: In this case, Theorem A.1.10(i) implies that $\Delta_G(G)$ is a compact subgroup of $(0, \infty)$, so that $\Delta_G(G) = \{1\}$ (Why exactly?). Examples of locally compact groups which are not unimodular can be found in [H–R, (15.17)].

If it is clear to which locally compact group G we are referring, we simply write Δ instead of Δ_G. We use the symbol Δ also for the diagonal operator (introduced in Section 2.2) of a Banach algebra as well as for the symmetric difference of sets, but no confusion should arise ...

Exercise A.1.7 Let G be a locally compact group, and let $f \in L^1(G)$. Define $\tilde{f}$ through

$$\tilde{f}(g) := \Delta(g^{-1})f(g^{-1}) \qquad (g \in G).$$

(i) Show that $\tilde{f}$ is well defined, i.e. does not depend on the choice of a particular representative for f.

(ii) Show that $\tilde{f} \in L^1(G)$ with

$$\int_G \tilde{f}(g)\, dm_G(g) = \int_G f(g)\, dm_G(g).$$

For any locally compact group G, the map

$$L^1(G) \to M(G), \quad f \mapsto f\, dm_G \tag{A.4}$$

is an isometric isomorphism whose range consists of those measures that are absolutely continuous with respect to Haar measure. Although Haar measure need not be σ-finite (just think of counting measure on an uncountable group), there is a version of the Radon–Nikodým theorem for measures on locally compact spaces ([Coh, Proposition 7.3.8]) which ascertains that every absolutely continuous measure in $M(G)$ lies in the range of (A.4). In fact, more is true ([H–R, (19.18) Theorem]):

Theorem A.1.12 *Let G be a locally compact group. Then the set of all measures in $M(G)$ which are absolutely continuous with respect to m_G is a closed ideal of $M(G)$ and, via (A.4), is isometrically isomorphic to $L^1(G)$ as a Banach algebra.*

We shall thus, for any locally compact group G, identify $L^1(G)$ with its image under (A.4) in $M(G)$.

Exercise A.1.8 Let G be a locally compact group. Show that $L^1(G)$ is w^*-dense in $M(G)$.

Exercise A.1.9 Let G be a locally compact group such that $L^1(G)$ is unital. Show that G is discrete.

The next proposition shows how the convolution product of an L^1-function with an arbitrary element of $M(G)$ is computed ([H–R, (20.9) Theorem]):

Proposition A.1.13 *Let G be a locally compact group, let $\mu \in M(G)$, and let $f \in L^1(G)$. Then*

$$(\mu * f)(g) = \int_G f(h^{-1}g)\, d\mu(h) \qquad and \qquad (f * \mu)(g) = \int_G f(gh^{-1})\Delta(h^{-1})\, d\mu(h)$$

hold for m_G-almost all $g \in G$.

For any locally compact group G, the dual space of $L^1(G)$ can be canonically identified with $L^\infty(G)$, the space of all (equivalence classes of) essentially bounded, m_G-measurable functions on G, where two such functions are identified if their difference is locally m_G-zero ([H–R, (12.18) Theorem]). Surprisingly, every element of $L^\infty(G)$ has a representative which is a Borel function ([Coh, Theorem 9.4.8]). We conclude this section with the definition of the convolution of a measure with an L^∞-function:

Definition A.1.14 Let G be a locally compact group, let $\mu \in M(G)$, and let $\phi \in L^\infty(G)$. Then $\mu * \phi, \phi * \mu \in L^\infty(G)$ are defined through

$$(\mu * \phi)(g) = \int_G \phi(h^{-1}g)\, d\mu(h) \quad \text{and} \quad (\phi * \mu)(g) = \int_G \phi(gh^{-1})\, d\mu(h) \qquad \text{(A.5)}$$

for locally m_G-almost all $g \in G$.

Remarks A.1.15   (a) It can be shown that $\mu * \phi$ and $\phi * \mu$ as defined in (A.5) are well defined and satisfy

$$\|\mu * \phi\|_\infty \le \|\mu\|\|\phi\|_\infty \quad \text{and} \quad \|\phi * \mu\|_\infty \le \|\phi\|_\infty \|\mu\|.$$

For convolution from the left see [H–R, (20.12) Theorems]; the corresponding claim for convolution from the right can be proved analoguously.

(b) Let $\mu \in M(G)$, and let $\phi \in L^1(G) \cap L^\infty(G)$. Then the meanings of $\mu * \phi$ according to Proposition A.1.13 and Definition A.1.14 are consistent. However, unless G is unimodular, the formulae for $\phi * \mu$ in Proposition A.1.13 and Definition A.1.14 may define different objects.

Our definition of the convolution of a measure from the right with an L^∞-function has the drawback that we use one symbol to denote possibly different objects: We have to be careful if we consider a certain function as an element of L^1 or of L^∞. Its advantage becomes apparent in the following exercise:

Exercise A.1.10 Let G be a locally compact group.

(i) For $f \in \mathcal{C}_0(G)$ and $\mu \in M(G)$, define $\check{f} \in \mathcal{C}_0(G)$ through $\check{f}(g) := f(g^{-1})$ for $g \in G$ and $\tilde{\mu} \in M(G)$ through $\langle f, \tilde{\mu} \rangle := \langle \check{f}, \mu \rangle$. Furthermore, let $\cdot$ denote the dual module action of $M(G)$ on $L^\infty(G)$ (Exercise 2.1.1). Show that

$$\mu * \phi = \phi \cdot \tilde{\mu} \quad \text{and} \quad \phi * \mu = \tilde{\mu} \cdot \phi \qquad (\mu \in M(G),\ \phi \in L^\infty(G)).$$

(ii) Conclude that Definition A.1.14 defines a Banach bimodule action of $M(G)$ on $L^\infty(G)$.

(iii) Let $\mu = f\, dm_G$ for some $f \in L^1(G)$. Show that $\tilde{\mu}$ is also absolutely continuous with respect to m_G and compute its Radon–Nikodým derivative.

A.2 Invariant subspaces of $L^\infty(G)$

Let G be a locally compact group. In Definition 1.1.3(ii), we defined what we mean by a left invariant subspace of $L^\infty(G)$. Although we haven't given a formal definition, it should be obvious what a right invariant subspace of $L^\infty(G)$ is supposed to mean; a subspace of

$L^\infty(G)$ which is both left and right invariant is simply called *invariant*. It is clear from these definitions that $L^\infty(G)$ along with $\mathcal{C}_0(G)$ are invariant subspaces of $L^\infty(G)$; the space $\mathcal{C}_b(G)$ of all bounded, continuous functions on G is also invariant.

In this section, we discuss three more invariant subspaces of $L^\infty(G)$.

Definition A.2.1 Let G be a locally compact group. A function $f \in \mathcal{C}_b(G)$ is called

(i) *left uniformly continuous* if the map

$$G \to \mathcal{C}_b(G), \quad g \mapsto \delta_g * f$$

is continuous (with respect to the norm topology on $\mathcal{C}_b(G)$),
(ii) *right uniformly continuous* if the map

$$G \to \mathcal{C}_b(G), \quad g \mapsto f * \delta_g$$

is continuous, and
(iii) *uniformly continuous* if it is both left and right uniformly continuous.

Remark A.2.2 Our definitions of left and right uniform continuity are consistent with those in [Fol]. Readers of [H–R] should be aware that a function that is left uniformly continuous in the sense of Definition A.2.1 is called right uniformly continuous in [H–R] (and vice versa).

We use the following notation:

$$LUC(G) := \{f \in \mathcal{C}_b(G) : f \text{ is left uniformly continuous}\};$$
$$RUC(G) := \{f \in \mathcal{C}_b(G) : f \text{ is right uniformly continuous}\};$$
$$UC(G) := \{f \in \mathcal{C}_b(G) : f \text{ is uniformly continuous}\}.$$

Our first proposition lists the basic properties of $LUC(G)$, $RUC(G)$, and $UC(G)$ for a locally compact group G; note that $L^\infty(G)$ equipped with "pointwise" multiplication is a C^*-algebra (in fact, a W^*-algebra).

Proposition A.2.3 *Let G be a locally compact group. Then $LUC(G)$, $RUC(G)$, and $UC(G)$ are closed invariant subspaces and unital C^*-subalgebras of $L^\infty(G)$.*

Proposition A.2.3 follows more or less directly from Definition A.2.1, so that we leave its proof to the reader:

Exercise A.2.1 Prove Proposition A.2.3.

Remark A.2.4 By [Fol, (2.6) Proposition], every function in $\mathcal{C}_{00}(G)$ with compact support belongs to $UC(G)$. It follows that $\mathcal{C}_0(G) \subset UC(G)$.

Exercise A.2.2 Let G be a locally compact group, let $\mu \in M(G)$, and let $\phi \in \mathcal{C}_0(G) \subset L^\infty(G)$. Use Remarks A.2.4(b) to show that $\mu * \phi, \phi * \mu \in \mathcal{C}_0(G)$ as well.

Our second proposition is a combination of [H–R, (20.16) Theorem] and Cohen's factorization theorem (see [H–R, (32.45)]) for (i); (ii) and (iii) follow analoguously):

Proposition A.2.5 *For a locally compact group G the following hold:*

(i) $LUC(G) = \{f * \phi : f \in L^1(G),\ \phi \in L^\infty(G)\}$;

(ii) $RUC(G) = \{\phi * f : f \in L^1(G),\ \phi \in L^\infty(G)\}$;

(iii) $UC(G) = \{f_1 * \phi * f_2 : f_1, f_2 \in L^1(G),\ \phi \in L^\infty(G)\}$.

Exercise A.2.3 Show that, for a locally compact group G, the following hold:

(i) $LUC(G) = \{f * \phi : f \in L^1(G),\ \phi \in LUC(G)\}$;

(ii) $RUC(G) = \{\phi * f : f \in L^1(G),\ \phi \in RUC(G)\}$;

(iii) $UC(G) = \{f_1 * \phi * f_2 : f_1, f_2 \in L^1(G),\ \phi \in UC(G)\}$.

Although left and right uniformly continuous functions do not show up explicitly in the following exercise (whose third part is needed for the proof of Theorem 2.1.8), you should keep Proposition A.2.5 in mind when doing it. Note that, for any locally compact group G, the measure algebra (and thus all of its subalgebras) can be isometrically embedded into $\mathcal{C}_b(G)^*$ via

$$\langle f, \mu \rangle := \int_G f(g)\, d\mu(g) \qquad (f \in \mathcal{C}_b(G),\ \mu \in M(G)). \tag{A.6}$$

Exercise A.2.4 Let G be a locally compact group.

(i) View $M(G)$ as a subspace of $\mathcal{C}_b(G)^*$ via (A.6). Show that $\ell^1(G)$ is dense in $M(G)$ with respect to the w^*-topology on $\mathcal{C}_b(G)^*$.

(ii) The *weakly strict topology* on $M(G)$ is defined through the seminorms $(p_{f,\phi})_{f \in L^1(G),\, \phi \in L^\infty(G)}$, where

$$p_{f,\phi}(\mu) := |\langle \mu * f, \phi \rangle| + |\langle f * \mu, \phi \rangle| \qquad (\mu \in M(G),\ f \in L^1(G),\ \phi \in L^\infty(G)).$$

Show that $\ell^1(G)$ is weakly strictly dense in $M(G)$. (*Hint*: Proposition A.2.5 and (i).)

(iii) Show that $\ell^1(G)$ is strictly dense in $M(G)$. (*Hint*: Show that the strict and the weakly strict topology on $M(G)$ have the same bounded linear functionals.)

A.3 Regular representations on $L^p(G)$ and associated algebras

For any $p \in [1, \infty)$ we define two representations of G on $L^p(G)$:

Definition A.3.1 Let G be a locally compact group, and let $p \in [1, \infty)$.

(i) The *left regular representation* $\lambda_p \colon G \to \mathcal{L}(L^p(G))$ is defined through

$$(\lambda_p(g)\xi)(h) := \xi(g^{-1}h) \qquad (g, h \in G,\ \xi \in L^p(G)).$$

(ii) The *right regular representation* $\rho_p \colon G \to \mathcal{L}(L^p(G))$ is defined through

$$(\rho_p(g)\xi)(h) := \Delta(g)^{\frac{1}{p}}\xi(hg) \qquad (g, h \in G,\ \xi \in L^p(G)).$$

Remarks A.3.2   (a) It is is immediate from Definition A.3.1 that both λ_p and ρ_p map G into the invertible isometries on $L^p(G)$, i.e. into the unitaries if $p = 2$.

(b) By [H–R, (20.4) Theorem], both λ_p and ρ_p are indeed representations of G on $L^p(G)$ in the sense of Definition 1.4.1 (this follows from the fact that $\mathcal{C}_{00}(G) \subset UC(G)$ is norm dense in $L^p(G)$).

(c) For $p = 1$, we have the following formulas:

$$\lambda_1(g)\xi = \delta_g * \xi \quad \text{and} \quad \rho_1(g)\xi = \xi * \delta_{g^{-1}} \qquad (g \in G, \, \xi \in L^1(G)).$$

For each $p \in [1, \infty)$, the left and the right regular representation, respectively, of a locally compact group G canonically induce representations of the measure algebra $M(G)$ — and thus of $L^1(G)$ — on $L^p(G)$; they are denoted by $\tilde{\lambda}_p$ and $\tilde{\rho}_p$, respectively, and are given through:

$$\langle \tilde{\lambda}_p(\mu)\xi, \eta \rangle := \int_G \langle \lambda_p(g)\xi, \eta \rangle \, d\mu(g)$$

$$\text{and} \quad \langle \tilde{\rho}_p(\mu)\xi, \eta \rangle := \int_G \langle \rho_p(g)\xi, \eta \rangle \, d\mu(g) \qquad (\mu \in M(G), \, \xi \in L^p(G), \, \eta \in L^q(G)),$$

where $q \in (1, \infty]$ is such that $\frac{1}{p} + \frac{1}{q} = 1$ (so that $L^q(G) \cong L^p(G)^*$).

Exercise A.3.1 Let G be a locally compact group, and let $p \in [1, \infty)$. Show that $\tilde{\lambda}_p$ and $\tilde{\rho}_p$ are faithful, i.e. injective.

Definition A.3.3 Let G be a locally compact group, and let $p \in (1, \infty)$. The Banach algebra $PM_p(G)$ of p-*pseudomeasures* on G is the closure of $\tilde{\lambda}_p(L^1(G))$ in $\mathcal{L}(L^p(G))$ with respect to the weak operator topology.

It is an immediate consequence of the Kreĭn–Šmulian theorem that $PM_p(G)$ is w^*-closed in $\mathcal{L}(L^p(G)) \cong (L^p(G) \hat{\otimes} L^q(G))^*$ (Exercise B.2.10) for each locally compact group G and for all $p, q \in (1, \infty)$ with $\frac{1}{p} + \frac{1}{q} = 1$. It follows that $PM_p(G)$ is itself a dual space. So, what's its predual?

For any function f on a locally compact group G, we define $\check{f}$ through $\check{f}(g) := f(g^{-1})$ for $g \in G$ (as we already did in Exercise A.1.10 for $f \in \mathcal{C}_0(G)$). Let $q \in (1, \infty)$, and let $\eta \in L^q(G)$; then $\check{\eta}$ does not depend on the particular representative of η, so that we may speak unambiguously of $\check{\eta} \in L^q(G)$ (compare Exercise A.1.7(a)).

Exercise A.3.2 Let G be a locally compact group, let $p, q \in (1, \infty)$ be such that $\frac{1}{p} + \frac{1}{q} = 1$, let $\xi \in L^p(G)$, and let $\eta \in L^q(G)$. Show that

$$(\xi * \check{\eta})(g) := \int_G \xi(h)\eta(g^{-1}h) \, dm_G(h) \qquad (g \in G)$$

defines a function in $\mathcal{C}_0(G)$ such that $\|\xi * \check{\eta}\|_\infty \leq \|\xi\|_p \|\eta\|_q$. (*Hint*: Use the fact that $\mathcal{C}_{00}(G)$ is dense in both $L^p(G)$ and $L^q(G)$ along with Remarks A.2.4(b).)

Yet another use of the convolution symbol $*$, but this time it's at least consistent with Definition A.1.6 ...

Definition A.3.4 Let G be a locally compact group, and let $p, q \in (1, \infty)$ be such that $\frac{1}{p} + \frac{1}{q} = 1$.

(i) The space $A_p(G)$ consists of those $f \in \mathcal{C}_0(G)$ such that there are sequences $(\xi_n)_{n=1}^\infty$ in $L^p(G)$ and $(\eta_n)_{n=1}^\infty$ in $L^q(G)$ with

$$\sum_{n=1}^\infty \|\xi_n\|_p \|\eta_n\|_q < \infty \tag{A.7}$$

and

$$f = \sum_{n=1}^{\infty} \xi_n * \check{\eta}. \tag{A.8}$$

(ii) For $f \in A_p(G)$, let $\|f\|_{A_p(G)}$ be the infimum over all numbers (A.7), where $(\xi_n)_{n=1}^{\infty}$ and $(\eta_n)_{n=1}^{\infty}$ are sequences in $L^p(G)$ and $L^q(G)$, respectively, such that (A.8) holds.

For a locally compact group G, and $p, q \in (1, \infty)$ with $\frac{1}{p} + \frac{1}{q} = 1$, the linear space $(A_p(G), \|\cdot\|_{A_p(G)})$ is a quotient space of $L^p(G) \hat{\otimes} L^q(G)$ and thus a Banach space. But more is true ([Her, Corollary to Theorem A]):

Theorem A.3.5 *Let G be a locally compact group, and let $p \in (1, \infty)$. Then the Banach space $(A_p(G), \|\cdot\|_{A_p(G)})$ is a Banach algebra with respect to pointwise operations.*

Exercise A.3.3 Let G be a locally compact group.

(i) Show that $C_{00}(G) \cap A_p(G)$ is dense in $A_p(G)$ with respect to $\|\cdot\|_{A_p(G)}$.
(ii) Show that $A_p(G)$ is norm dense in $C_0(G)$.

Banach algebras of this type are called *Figà-Talamanca–Herz algebras*. The connection between Figà-Talamanca–Herz algebras and pseudomeasures is a straightforward duality ([Pie, Proposition 10.3]):

Theorem A.3.6 *Let G be a locally compact group, and let $p, q \in (1, \infty)$ be such that $\frac{1}{p} + \frac{1}{q} = 1$. Then $A_p(G)^* \cong PM_q(G)$, where the duality is given through*

$$\langle \xi * \check{\eta}, T \rangle := \langle T\eta, \xi \rangle \qquad (T \in PM_q(G),\ \xi \in L^p(G),\ \eta \in L^q(G)). \tag{A.9}$$

Exercise A.3.4 Let G be a locally compact group, and let $p, q \in (1, \infty)$ be such that $\frac{1}{p} + \frac{1}{q} = 1$.

(i) Show that $\tilde{\lambda}_p(M(G)) \subset PM_p(G)$.
(ii) Show that

$$\langle f, \tilde{\lambda}_p(\mu) \rangle = \int_G f(g)\, d\mu(g) \qquad (f \in A_p(G),\ \mu \in M(G)).$$

Conclude that $\tilde{\lambda}_p \colon M(G) \to PM_p(G)$ is w^*-continuous with w^*-dense range.

We now consider to the special case where $p = 2$. Traditionally, terminology is a bit different in this situation:

Definition A.3.7 Let G be a locally compact group. Then:

(i) $VN(G) := PM_2(G)$ is the *group von Neumann algebra* of G.
(ii) $A(G) := A_2(G)$ is the *Fourier algebra* of G.

The space $L^2(G)$ is a Hilbert space for any locally compact group G, so that it is more convenient to work with conjugate duality, rather than with duality. One thus considers $A(G)$ as consisting of those $f \in C_0(G)$ for which there are sequences $(\xi_n)_{n=1}^{\infty}$ and $(\eta_n)_{n=1}^{\infty}$ in $L^2(G)$ with $\sum_{n=1}^{\infty} \|\xi_n\|_2 \|\eta_n\|_2 < \infty$ and $f = \sum_{n=1}^{\infty} \xi_n * \bar{\check{\eta}}_n$, where $^-$ denotes complex conjugation (What's the minor difference between this description and Definition A.3.4?). The duality (A.9) between $A(G)$ and $VN(G)$ then becomes

$$\langle \xi * \bar{\check{\eta}}, T \rangle := \langle T\bar{\eta}, \xi \rangle \qquad (T \in VN(G),\ \xi, \eta \in L^2(G)).$$

Exercise A.3.5 Let G be a locally compact group.

(i) For $\mu \in M(G)$ define $\mu^* := \tilde{\bar{\mu}}$ (recall the definition of $\tilde{\mu}$ from Exercise A.1.10). Show that * is an involution on $M(G)$ which leaves $L^1(G)$ invariant.

(ii) Show that $\tilde{\lambda}_2 : M(G) \to \mathcal{L}(L^2(G))$ is a *-homomorphism.

(iii) Conclude that $VN(G)$ is not only called a von Neumann algebra, but really is one.

Exercise A.3.6 Let G be a locally compact group. Show that $VN(G) = \lambda_2(G)''$.

The elements of the Fourier algebra allow for another description ([Eym, Théorème, p. 218]):

Theorem A.3.8 *Let G be a locally compact group, and let $f \in A(G)$. Then there are $\xi, \eta \in L^2(G)$ with $f = \xi * \bar{\tilde{\eta}}$ and $\|f\|_{A(G)} = \|\xi\|_2 \|\eta\|_2$.*

Remarks A.3.9   (a) As a consequence of Theorem A.3.8, $A(G)$ consists of all functions of the form

$$G \to \mathbb{C}, \quad g \mapsto \langle \lambda_2(g)\xi, \eta \rangle$$

with $\xi, \eta \in L^2(G)$.

(b) Also as a consequence of Theorem A.3.8, the weak operator topology and the ultra-weak topology on $VN(G)$ coincide — not only on bounded sets, but on all of $VN(G)$.

(c) Let G be a locally compact abelian group with dual group Γ. Then the image of $L^1(G)$ under the Fourier transform is also denoted by $A(\Gamma)$ and called the Fourier algebra of Γ. By Plancherel's theorem ([Rud 1, 1.6.1 Theorem]) and Theorem A.3.8, both definitions characterize the same object ([Rud 1, 1.6.3 Theorem]).

Let G be a locally compact group. Then λ_2 induces a C^*-norm on $L^1(G)$. By [B–D, Lemma 39.2], the supremum

$$|f| := \sup\{\|f\| : \|\cdot\| \text{ is a } C^*\text{-norm on } L^1(G)\} \qquad (f \in L^1(G))$$

defines a C^*-norm on $L^1(G)$ with

$$\|\lambda_2(f)\| \le |f| \le \|f\|_1 \qquad (f \in L^1(G)).$$

The completion $C^*(G)$ of $L^1(G)$ with respect to $|\cdot|$ is a C^*-algebra, the *group C^*-algebra* of G (not to be confused with the reduced group C^*-algebra $C^*_{\lambda_2}(G)$, which is the norm closure of $\tilde{\lambda}_2(L^1(G))$ in $\mathcal{L}(L^2(G))$).

Definition A.3.10 Let G be a locally compact group. Then $B(G) := C^*(G)^*$ is called the *Fourier–Stieltjes algebra* of G.

You probably wonder why this dual space is called an algebra . . .

Since $|\cdot| \le \|\cdot\|_1$, the space $B(G)$ embeds continuously into $L^1(G)^* = L^\infty(G)$ for each locally compact group G; we can thus identify $B(G)$ with a subspace of $L^\infty(G)$. The following theorem collects the basic facts about $B(G)$ ([Eym, (2.1) Proposition, (2.4) Lemma, and (2.16) Proposition]):

Theorem A.3.11 *Let G be a locally compact group. Then the following hold:*

(i) *A function $f \in L^\infty(G)$ belongs to $B(G)$ if and only if there is a unitary representation π of G on some Hilbert space $\mathfrak{H}$ and $\xi, \eta \in \mathfrak{H}$ such that*

$$f(g) = \langle \pi(g)\xi, \eta \rangle \qquad (g \in G) \tag{A.10}$$

and (automatically) $\|f\| \le \|\xi\|\|\eta\|$. The vectors ξ and η can be chosen such that $\|f\| = \|\xi\|\|\eta\|$.

(ii) *With pointwise multiplication, $B(G)$ is a Banach algebra with identity.*

Remarks A.3.12   (a) There is a (fairly straightforward) link between multiplication in $B(G)$ and the representation (A.10) of functions in $B(G)$: Let $f_1, f_2 \in B(G)$; for $j = 1, 2$, let π_j be a unitary representation of G on some Hilbert space $\mathfrak{H}_j$, and let $\xi_j, \eta_j \in \mathfrak{H}_j$ be such that

$$f_j(g) = \langle \pi_j(g)\xi_j, \eta_j \rangle \qquad (g \in G).$$

Then we have:

$$\begin{aligned}
f_1(g)f_2(g) &= \langle \pi_1(g)\xi_1, \eta_1 \rangle \langle \pi_2(g)\xi_2, \eta_2 \rangle \\
&= \langle (\pi_1(g) \otimes \pi_2(g))(\xi_1 \otimes \xi_2), \eta_1 \otimes \eta_2 \rangle \qquad (g \in G).
\end{aligned}$$

(b) Let G be a locally compact abelian group with dual group Γ. Then $C^*(G) \cong \mathcal{C}_0(\Gamma)$, so that $B(G) \cong M(\Gamma)$ via the Fourier–Stieltjes transform (see [Rud 1]). As in the case of the Fourier algebra, Definitions A.3.10 extends the notion of the Fourier–Stieltjes algebra of a locally compact abelian group to arbitrary locally compact groups.

Exercise A.3.7 Conclude from Theorem A.3.11 that, for any locally compact group G, we have $B(G) \subset UC(G)$.

From Theorem A.3.8 and Theorem A.3.11(ii), it follows that $A(G)$ embeds isometrically into $B(G)$ for any locally compact group G. More is true ([Eym, (3.4) Proposition]):

Theorem A.3.13 *Let G be a locally compact group. Then $A(G)$ is a closed ideal of $B(G)$.*

A.4 Notes and comments

Our principal source on abstract harmonic analysis is [H–R]: an encyclopedic treatise which contains virtually everything there was to know about abstract harmonic analysis up to 1970. Nevertheless, its sheer size along with its scope and generality make reading it a somewhat daunting experience for the beginner. An alternative source is the much slimmer book [R-St]. A more recent text is [Fol]: it really deserves the word "course" in its title. Although [Coh] is a textbook on measure theory, it also contains an excellent exposition on the basics (Haar measure, convolution, etc.) of abstract harmonic analysis. The slender monograph [Rud 1] is a classic that confines itself to the abelian case.

The Fourier algebra as well as the Fourier–Stieltjes algebra of an arbitrary locally compact group were introduced by P. Eymard in [Eym]. General Figà-Talamanca–Herz algebras were studied by C. Herz ([Her]).

B Tensor products

Since Banach space tensor products are rarely covered in an introductory course on functional analysis, this appendix gives an essentially self-contained crash course on tensor products: We start with the algebraic tensor product, then proceed to the injective and projective tensor products, and finish with the Hilbert space tensor product.

Unlike in Appendix A, we do not refer to the literature for proofs, but either give them in reasonable detail or leave them to the reader as exercises (with hints, of course, where necessary).

B.1 The algebraic tensor product

The tensor product $E_1 \otimes E_2$ of two linear spaces E_1 and E_2 is a universal linearizer: Every bilinear map from $E_1 \times E_2$ factors uniquely through $E_1 \otimes E_2$.

We give the formal definition of a — we still have to justify that we may speak of *the* — tensor product of linear spaces for an arbitrary finite number of spaces:

Definition B.1.1 Let $E_1, \dots, E_n$ be linear spaces. A *tensor product* of $E_1, \dots, E_n$ is a pair $(\mathcal{T}, \tau)$, where $\mathcal{T}$ is a linear space and $\tau : E_1 \times \cdots \times E_n \to \mathcal{T}$ is an n-linear map with the following (universal) property: For each linear space F, and for each n-linear map $V : E_1 \times \cdots \times E_n \to F$, there is a unique linear map $\tilde{V} : \mathcal{T} \to F$ such that $V = \tilde{V} \circ \tau$.

It is clear that a tensor product $(\mathcal{T}, \tau)$ of $E_1, \dots, E_n$ is not unique: If $\mathcal{T}'$ is another linear space, and if $\theta : \mathcal{T}' \to \mathcal{T}$ is an isomorphism of linear spaces, then $(\mathcal{T}', \theta \circ \tau)$ is another tensor product of $E_1, \dots, E_n$. We call tensor products arising from one another in this fashion *isomorphic*: Given two tensor products $(\mathcal{T}_1, \tau_1)$ and $(\mathcal{T}_2, \tau_2)$, an isomorphism of $(\mathcal{T}_1, \tau_1)$ and $(\mathcal{T}_2, \tau_2)$ is an isomorpism $\theta : \mathcal{T}_1 \to \mathcal{T}_2$ of linear spaces such that $\tau_2 = \theta \circ \tau_1$. The best we can thus hope for is uniqueness up to isomorphism, which is indeed what we have:

Exercise B.1.1 Let $E_1, \dots, E_n$ be linear spaces, and let $(\mathcal{T}_1, \tau_1)$ and $(\mathcal{T}_2, \tau_2)$ be tensor products of $E_1, \dots, E_n$. Show that there is a unique isomorphism of $(\mathcal{T}_1, \tau_1)$ and $(\mathcal{T}_2, \tau_2)$.

We may thus speak of *the* tensor product of n — in most situations: two — linear spaces. From now on, we shall also use the standard notation for tensor products: Given linear spaces $E_1, \dots, E_n$ and their tensor product $(\mathcal{T}, \tau)$, we write $E_1 \otimes \cdots \otimes E_n$ for $\mathcal{T}$.

Furthermore, we define

$$x_1 \otimes \cdots \otimes x_n := \tau(x_1, \dots, x_n) \qquad (x_1 \in E_1, \dots, x_n \in E_n). \tag{B.1}$$

Elements of $E_1 \otimes \cdots \otimes E_n$ are called *tensors*, and elements of the form (B.1) are called *elementary tensors*. Not every tensor is an elementary tensor: The set of all elementary tensors is (except in trivial cases) not a linear space. Hence, the tensor product must contain at least all linear combinations of elementary tensors. As the following proposition shows, these are all there is:

Proposition B.1.2 *Let $E_1, \ldots, E_n$ be linear spaces, and let $x \in E_1 \otimes \cdots E_n$. Then there is $m \in \mathbb{N}$, and for each $j = 1, \ldots, n$ there are $x_j^{(1)}, \ldots, x_j^{(m)} \in E_j$ such that*

$$\mathbf{x} = \sum_{k=1}^{m} x_1^{(k)} \otimes \cdots \otimes x_n^{(k)}. \tag{B.2}$$

Proof Let F be the set of all tensors of the form (B.2), and define

$$V : E_1 \times \cdots \times E_n \to F, \quad (x_1, \ldots, x_n) \mapsto x_1 \otimes \cdots \otimes x_n.$$

It is routinely seen that F is a linear space, and that V is n-linear. From the defining property of $E_1 \otimes \cdots \otimes E_n$, there is a unique linear map $\tilde{V} : E_1 \otimes \cdots \otimes E_n \to F$ such that

$$\tilde{V}(x_1 \otimes \cdots \otimes x_n) = V(x_1, \ldots, x_n) \quad (x_1 \in E_1, \ldots, x_n \in E_n).$$

It follows that $\tilde{V}$ is the identity on $E_1 \otimes \cdots \otimes E_n$ (Why not just on F?).   □

Exercise B.1.2 Let E be a linear space. Show that $E \times \mathbb{C} \ni (x, \lambda) \mapsto \lambda x$ induces an isomorphism of $E \otimes \mathbb{C}$ and E.

Exercise B.1.3 Show that the tensor product is associative: If E_1, E_2, and E_3 are linear spaces, then

$$E_1 \otimes E_2 \otimes E_3 \cong E_1 \otimes (E_2 \otimes E_3) \cong (E_1 \otimes E_2) \otimes E_3$$

through canonical isomorphisms.

Exercise B.1.4 Let $E_1, F_1, \ldots, E_n, F_n$ be linear spaces and let $T_j : E_j \to F_j$ be linear for $j = 1, \ldots, n$. Show that there is a unique linear map $T_1 \otimes \cdots \otimes T_n : E_1 \otimes \cdots \otimes E_n \to F_1 \otimes \cdots \otimes F_n$ such that

$$(T_1 \otimes \cdots \otimes T_n)(x_1 \otimes \cdots \otimes x_n) = T_1 x_1 \otimes \cdots \otimes T_n x_n \quad (x_1 \in E_1, \ldots, x_n \in E_n).$$

Exercise B.1.5 Let $\mathfrak{A}$ be an algebra, let E be a left $\mathfrak{A}$-module, and let F be a right $\mathfrak{A}$-module. Show that there is a unique bimodule action of $\mathfrak{A}$ on $E \otimes F$ such that

$$a \cdot (x \otimes y) = a \cdot x \otimes y \quad \text{and} \quad (x \otimes y) \cdot a = x \otimes y \cdot a \quad (a \in \mathfrak{A}, x \in E, y \in F). \tag{B.3}$$

Exercise B.1.6 Let $\mathfrak{A}$ and $\mathfrak{B}$ be algebras. Show that there is a unique product $\bullet$ on $\mathfrak{A} \otimes \mathfrak{B}$ turning $\mathfrak{A} \otimes \mathfrak{B}$ into an algebra such that

$$(a_1 \otimes b_1) \bullet (a_2 \otimes b_2) = (a_1 a_2 \otimes b_1 b_2) \quad (a_1, a_2 \in \mathfrak{A}, b_1, b_2 \in \mathfrak{B}). \tag{B.4}$$

If $\mathfrak{A}$ and $\mathfrak{B}$ have identity elements $e_{\mathfrak{A}}$ and $e_{\mathfrak{B}}$, respectively, show that $e_{\mathfrak{A}} \otimes e_{\mathfrak{B}}$ is an identity for $\mathfrak{A} \otimes \mathfrak{B}$.

Exercise B.1.7 Let $n \in \mathbb{N}$.

(i) Let E be a linear space. Show that

$$\mathbb{M}_n \otimes E \to \mathbb{M}_n(E), \quad [\lambda_{j,k}]_{\substack{j=1,\ldots,n \\ k=1,\ldots,n}} \otimes x \mapsto [\lambda_{j,k} x]_{\substack{j=1,\ldots,n \\ k=1,\ldots,n}} \tag{B.5}$$

is an isomorphism of $\mathbb{M}_n \otimes E$ and $\mathbb{M}_n(E)$ as linear spaces.

(ii) Let $\mathfrak{A}$ be an algebra. Show that the linear isomorphism (B.5) between $\mathbb{M}_n \otimes \mathfrak{A}$ and $\mathbb{M}_n(\mathfrak{A})$ is an algebra isomorphism.

(iii) Let $m \in \mathbb{N}$. Show that canonically

$$\mathbb{M}_n \otimes \mathbb{M}_m \cong \mathbb{M}_n(\mathbb{M}_m) \cong \mathbb{M}_{nm}$$

as algebras.

So far we have dealt with tensor products without bothering to ask if they exist at all. Luckily, they do:

Theorem B.1.3 *Let $E_1, \ldots, E_n$ be linear spaces. Then their tensor product $E_1 \otimes \cdots \otimes E_n$ exists.*

Proof For convenience, we switch back (in this proof only!) to the notation of Definition B.1.1.

Let $\tilde{\mathcal{T}}$ be the linear space of all maps from the set $E_1 \times \cdots \times E_n$ to $\mathbb{C}$ with finite support. For $(x_1, \ldots, x_n) \in E_1 \times \cdots \times E_n$ define $\delta_{(x_1,\ldots,x_n)}$, the *point mass* at $(x_1, \ldots, x_n)$, through

$$\delta_{(x_1,\ldots,x_n)}(y_1, \ldots, y_n) := \begin{cases} 1, \text{ if } (y_1, \ldots, y_n) = (x_1, \ldots, x_n), \\ 0, \text{ otherwise.} \end{cases}$$

Also define

$$\tilde{\tau} \colon E_1 \times \cdots \times E_n \to \tilde{\mathcal{T}}, \quad (x_1, \ldots, x_n) \to \delta_{(x_1,\ldots,x_n)}.$$

Note that $\tilde{\tau}$ is not an n-linear map: In order to "make it linear", we have to factor out a certain subspace. Let $\tilde{\mathcal{T}}_0$ be the subspace of $\tilde{\mathcal{T}}$ spanned by all elements of the form

$$\lambda \delta_{(x_1,\ldots,x_j,\ldots,x_n)} + \mu \delta_{(x_1,\ldots,y_j,\ldots,y_n)} - \delta_{(x_1,\ldots,\lambda x_j + \mu y_j,\ldots,x_n)}$$
$$(\lambda, \mu \in \mathbb{C}, \ x_1, y_1 \in E_1, \ldots, x_j, y_j \in E_j, \ldots, x_n, y_n \in E_n), \tag{B.6}$$

where j ranges from 1 to n. Define $\mathcal{T} := \tilde{\mathcal{T}}/\tilde{\mathcal{T}}_0$, let $\pi \colon \tilde{\mathcal{T}} \to \mathcal{T}$ be the quotient map, and let $\tau := \pi \circ \tilde{\tau}$. From the definition of $\tilde{\mathcal{T}}_0$, it follows that τ is n-linear (work out the details for yourself).

We claim that $(\mathcal{T}, \tau)$ is a tensor product. To prove this, let F be another linear space, and let $V \colon E_1 \times \cdots \times E_n \to F$ be an n-linear map. Define a linear map $\bar{V} \colon \tilde{\mathcal{T}} \to F$ through

$$\bar{V}(\delta_{(x_1,\ldots,x_n)}) = V(x_1, \ldots, x_n) \qquad (x_1 \in E_1, \ldots, x_n \in E_n).$$

Then V satisfies $V = \bar{V} \circ \tilde{\tau}$; since every element of $\tilde{\mathcal{T}}$ is a finite linear combination of point masses, V is uniquely determined by this property. Define

$$\tilde{V}(x + \tilde{\mathcal{T}}_0) := \bar{V}(x) \qquad (x \in \tilde{\mathcal{T}}). \tag{B.7}$$

It is clear that $V = \tilde{V} \circ \tau$ — once we have established that $\tilde{V}$ is well defined. However, since $V = \bar{V} \circ \tilde{\tau}$, and since V is n-linear, it is easily, albeit tediously verified (nevertheless, do it) that $\bar{V}$ vanishes on tensors of the form (B.6) and thus on all of $\tilde{\mathcal{T}}_0$, which shows that $\tilde{V}$ is indeed well defined.

Finally, suppose that $\tilde{W}: \mathcal{T} \to F$ is another linear map satisfying $\phi = \tilde{W} \circ \tau$. Then $\tilde{W}$ and $\bar{V}$ are necessarily related in the same way as $\tilde{V}$ and $\bar{V}$ are in (B.7) (Why?). Since $\bar{V}$, however, is uniquely determined through $V = \bar{V} \circ \tilde{\tau}$, this establishes $\tilde{V} = \tilde{W}$. $\square$

The construction of $E_1 \otimes \cdots \otimes E_n$ is not very illuminating, and we won't come back to it anymore; actually, it is needed only to establish once and for all that tensor products of linear spaces exist. Whenever we work with tensor products, however, we shall either use their defining property or Proposition B.1.2.

For use in the next section, we require the following lemma:

Lemma B.1.4 *Let* $m \in \mathbb{N}$, *let* $E_1, \ldots, E_n$ *be linear spaces, and, for* $j = 1, \ldots, n$, *let* $x_j^{(1)}, \ldots, x_j^{(m)} \in E_j$ *be such that*

$$\sum_{k=1}^{m} x_1^{(k)} \otimes \cdots \otimes x_n^{(k)} = 0. \tag{B.8}$$

Then, if $x_n^{(1)}, \ldots, x_n^{(m)}$ *are linearly independent, we have*

$$x_1^{(k)} \otimes \cdots \otimes x_{n-1}^{(k)} = 0 \qquad (k = 1, \ldots, m).$$

Proof Suppose that there is $k_0 \in \{1, \ldots, m\}$ such that $x_1^{(k_0)} \otimes \cdots \otimes x_{n-1}^{(k_0)} \neq 0$. Define a linear map $\phi: E_n \to \mathbb{C}$ via a Hamel basis argument such that

$$\left\langle x_n^{(k_0)}, \phi \right\rangle = 1 \qquad \text{and} \qquad \left\langle x_n^{(k)}, \phi \right\rangle = 0 \quad (k \neq k_0).$$

Then

$$E_1 \times \cdots \times E_{n-1} \times E_n \to E_1 \otimes \cdots \otimes E_{n-1} \quad (x_1, \cdots, x_{n-1}, x_n) \mapsto \langle x_n, \phi \rangle (x_1 \otimes \cdots \otimes x_{n-1})$$

is an n-linear map and thus induces a linear map $\psi: E_1 \otimes \cdots \otimes E_{n-1} \otimes E_n \to E_1 \otimes \cdots \otimes E_{n-1}$ such that

$$\left\langle \sum_{k=1}^{m} x_1^{(k)} \otimes \cdots \otimes x_n^{(k)}, \psi \right\rangle = x_1^{(k_0)} \otimes \cdots \otimes x_{n-1}^{(k_0)} \neq 0$$

contradicting (B.8). $\square$

B.2 Banach space tensor products

Suppose that $E_1, \ldots, E_n$ are Banach spaces. Then their tensor product $E_1 \otimes \cdots \otimes E_n$ exists by Theorem B.1.3. Except in rather trivial cases, however, there is no need for $E_1 \otimes \cdots \otimes E_n$ to be a Banach space.

The first problem is to define a norm on $E_1 \otimes \cdots \otimes E_n$. Once we have found a suitable norm, we can obtain a Banach space tensor product of $E_1, \ldots, E_n$ by completing $E_1 \otimes \cdots \otimes E_n$ with respect to this norm. Of course, one can define norms on $E_1 \otimes \cdots \otimes E_n$ as one pleases via elementary Hamel basis arguments. However, very few norms obtained in this fashion will yield a useful notion of a Banach space tensor product. In order to obtain Banach space tensor products we can work with, we have to require at least one property for the norm on $E_1 \otimes \cdots \otimes E_n$:

Definition B.2.1 Let $E_1, \ldots, E_n$ be Banach spaces. A norm $\|\cdot\|$ on $E_1 \otimes \cdots \otimes E_n$ is called a *cross norm* if

$$\|x_1 \otimes \cdots \otimes x_n\| = \|x_1\| \cdots \|x_n\| \qquad (x_1 \in E_1, \ldots, x_n \in E_n).$$

Two questions arise naturally in connection with Definition B.2.1:

1. Is there a cross norm on $E_1 \otimes \cdots \otimes E_n$?
2. Is there more than one cross norm on $E_1 \otimes \cdots \otimes E_n$?

We deal with the first question first.

B.2.1 The injective tensor product

Let $E_1, \ldots, E_n$ be Banach spaces with dual spaces $E_1^*, \ldots, E_n^*$, and let $\phi_j \in E_j^*$ for $j = 1, \ldots, n$. Since $\mathbb{C} \otimes \cdots \otimes \mathbb{C} \cong \mathbb{C}$ by Exercise B.1.2, $\phi_1 \otimes \cdots \otimes \phi_n$ is a linear functional on $E_1 \otimes \cdots \otimes E_n$.

Definition B.2.2 Let $E_1, \ldots, E_n$ be Banach spaces with dual spaces $E_1^*, \ldots, E_n^*$. Then we define for $\mathbf{x} \in E_1 \otimes \cdots \otimes E_n$

$$\|\mathbf{x}\|_\epsilon := \sup\{|\langle \mathbf{x}, \phi_1 \otimes \cdots \otimes \phi_n \rangle| : \phi_j \in B_1[0, E_j^*] \text{ for } j = 1, \ldots, n\}.$$

We call $\|\cdot\|_\epsilon$ the *injective norm* on $E_1 \otimes \cdots \otimes E_n$.

Although we have just called $\|\cdot\|_\epsilon$ the injective norm, this is just a name tag: We don't know yet if it is a norm at all. It is not difficult, however, to show this (and even more):

Proposition B.2.3 *Let $E_1, \ldots, E_n$ be Banach spaces. Then $\|\cdot\|_\epsilon$ is a cross norm on $E_1 \otimes \cdots \otimes E_n$.*

Proof It is clear from the definition that

$$\|x_1 \otimes \cdots \otimes x_n\|_\epsilon \leq \|x_1\| \cdots \|x_n\| \qquad (x_1 \in E_1, \ldots, x_n \in E_n).$$

This also implies, by Proposition B.1.2, that the supremum in Definition B.2.2 is always finite. Let $x_j \in E_j$ for $j = 1, \ldots, n$. By the Hahn–Banach theorem, there are $\phi_j \in E_j^*$ with $\|\phi_j\| = 1$ and $\langle x_j, \phi_j \rangle = \|x_j\|$ for $j = 1, \ldots, n$, so that

$$\|x_1 \otimes \cdots \otimes x_n\|_\epsilon \geq |\langle x_1 \otimes \cdots \otimes x_n, \phi_1 \otimes \cdots \otimes \phi_n \rangle| = \|x_1\| \cdots \|x_n\|.$$

All that remains to be shown, is thus that $\|\cdot\|_\epsilon$ is a norm.

It is easy to see (but please check it, nevertheless) that $\|\cdot\|_\epsilon$ is a seminorm on $E_1 \otimes \cdots \otimes E_n$. What has to be shown is therefore that $\|\mathbf{x}\|_\epsilon = 0$ implies $\mathbf{x} = 0$ for all $\mathbf{x} \in E_1 \otimes \cdots \otimes E_n$. $\square$

We postpone the remainder of the proof for an instant:

Exercise B.2.1 Let $E_1, \ldots, E_{n-1}, E_n$ be Banach spaces and let $\|\cdot\|_\epsilon^{(n-1)}$ and $\|\cdot\|_\epsilon^{(n)}$ denote the injective norms on $E_1 \otimes \cdots \otimes E_{n-1}$ and $E_1 \otimes \cdots \otimes E_n$, respectively (if $n = 2$, $\|\cdot\|_\epsilon^{(1)}$ is just supposed to be the norm on E_1). Show that, for $\mathbf{x} \in E_1 \otimes \cdots \otimes E_n$,

$$\|\mathbf{x}\|_\epsilon^{(n)} := \sup\left\{\|(\mathrm{id} \otimes \phi)(\mathbf{x})\|_\epsilon^{(n-1)} : \phi \in B_1[0, E_n^*]\right\}.$$

Proof of Proposition B.2.3 (continued) We prove that $\| \cdot \|_\epsilon$ is indeed a norm by induction on n. Let $\mathbf{x} \in E_1 \otimes \cdots \otimes E_n \setminus \{0\}$. By Proposition B.1.2, there is $m \in \mathbb{N}$, and for each $j = 1, \ldots, n$ there are $x_j^{(1)}, \ldots, x_j^{(m)} \in E_j$ such that

$$\mathbf{x} = \sum_{k=1}^m x_1^{(k)} \otimes \cdots \otimes x_n^{(k)}.$$

We may suppose that $x_n^{(1)}, \ldots, x_n^{(m)}$ are linearly independent (Why?). By Lemma B.1.4, this means that there is $k_0 \in \{1, \ldots, m\}$ such that $x_1^{(k_0)} \otimes \cdots \otimes x_{n-1}^{(k_0)} \neq 0$. By the induction hypothesis, we have

$$\left\| x_1^{(k_0)} \otimes \cdots \otimes x_{n-1}^{(k_0)} \right\|_\epsilon^{(n-1)} \neq 0$$

(with notation as in Exercise B.2.1). Use the Hahn–Banach theorem to find $\phi \in B_1[0, E_n^*]$ such that

$$\left\langle x_n^{(k_0)}, \phi \right\rangle > 0 \quad \text{and} \quad \left\langle x_n^{(k)}, \phi \right\rangle = 0 \quad (k \neq k_0).$$

From Exercise B.2.1, we conclude that

$$\|\mathbf{x}\|_\epsilon \geq \|(\mathrm{id} \otimes \phi)(\mathbf{x})\|_\epsilon^{(n-1)} = \left\langle x_n^{(k_0)}, \phi \right\rangle \left\| x_1^{(k_0)} \otimes \cdots \otimes x_{n-1}^{(k_0)} \right\|_\epsilon^{(n-1)} > 0,$$

which completes the proof. $\square$

Definition B.2.4 Let $E_1, \ldots, E_n$ be Banach spaces. Then their *injective tensor product* $E_1 \check{\otimes} \cdots \check{\otimes} E_n$ is the completion of $E_1 \otimes \cdots \otimes E_n$ with respect to $\| \cdot \|_\epsilon$.

Exercise B.2.2 Prove an associative law, as in Exercise B.1.3, for the injective tensor product (with isometric isomorphisms).

Exercise B.2.3 Let $E_1, F_1, \ldots, E_n, F_n$ be Banach spaces, and let $T_j \in \mathcal{L}(E_j, F_j)$. Show that $T_1 \otimes \cdots \otimes T_n$ is continuous with respect to the injective norms on $E_1 \otimes \cdots \otimes E_n$ and $F_1 \otimes \cdots \otimes F_n$, respectively, and satisfies

$$\|T_1 \otimes \cdots \otimes T_n\| = \|T_1\| \cdots \|T_n\|.$$

Exercise B.2.4 Let $\mathfrak{A}$ be a Banach algebra, let E be a left Banach $\mathfrak{A}$-module, and let F be a right Banach $\mathfrak{A}$-module. Show that the bimodule action (B.3) of $\mathfrak{A}$ on $E \otimes F$ turns $E \check{\otimes} F$ into a Banach $\mathfrak{A}$-bimodule.

Towards the end of our discussion of the injective tensor product, we give a concrete description of the injective tensor product in a particular case.

Let Ω be a set, and let E be a linear space. For $f \in \mathbb{C}^\Omega$ and $x \in E$, we define $fx \in E^\Omega$ through

$$(fx)(\omega) := f(\omega)x \quad (\omega \in \Omega).$$

Theorem B.2.5 *Let Ω be a locally compact Hausdorff space, and let E be a Banach space. Then the bilinear map*

$$\mathcal{C}_0(\Omega) \times E \to \mathcal{C}_0(\Omega, E), \quad (f, x) \mapsto fx \tag{B.9}$$

induces an isometric isomorphism $\mathcal{C}_0(\Omega) \check{\otimes} E \cong \mathcal{C}_0(\Omega, E)$.

Proof We only treat the case where Ω is compact.

From the defining property of the algebraic tensor product $\mathcal{C}(\Omega) \otimes E$, it follows that (B.9) extends to a linear map from $\mathcal{C}(\Omega) \otimes E$ into $\mathcal{C}(\Omega, E)$. With an argument similar to that used to solve Exercise B.2.1 (how is it done precisely?), we see that this linear map is an isometry and thus injective with closed range. It remains to be shown that it has dense range as well.

Let $f \in \mathcal{C}(\Omega, E)$, and $\epsilon > 0$. Being the continuous image of a compact space, $K := f(\Omega) \subset E$ is compact. We may thus find $x_1, \ldots, x_n \in E$ such that

$$K \subset B_\epsilon(x_1, E) \cup \cdots \cup B_\epsilon(x_n, E).$$

Let $U_j := f^{-1}(B_\epsilon(x_j, E))$ for $j = 1, \ldots, n$. Choose $f_1, \ldots, f_n \in \mathcal{C}(\Omega)_+$ such that

$$f_1 + \cdots + f_n \equiv 1 \quad \text{and} \quad \operatorname{supp}(f_j) \subset U_j \quad (j = 1, \ldots, n).$$

For $\omega \in \Omega$, we then have

$$\|f(\omega) - (f_1 x_1 + \cdots + f_n x_n)(\omega)\| \leq \sum_{j=1}^n f_j(\omega)\|f(\omega) - x_j\|. \tag{B.10}$$

It easy to see (just check how easy) that the right hand side of (B.10) is less than ϵ. This completes the proof. $\square$

Exercise B.2.5 Prove Theorem B.2.5 for general locally compact spaces.

In view of Theorem B.2.5, we feel justified in writing $f \otimes x$ for the map $fx \in \mathcal{C}_0(\Omega, E)$.

Exercise B.2.6 Let Ω be a locally compact Hausdorff space, and let $\mathfrak{A}$ be a Banach algebra. Show that the multiplication $\bullet$ on $\mathcal{C}_0(\Omega) \otimes \mathfrak{A}$ defined in Exercise B.1.6 extends to $\mathcal{C}_0(\Omega) \check{\otimes} \mathfrak{A}$ turning the isomorphism $\mathcal{C}_0(\Omega) \check{\otimes} \mathfrak{A} \cong \mathcal{C}_0(\Omega, \mathfrak{A})$ from Theorem B.2.5 into an algebra isomorphism, where $\mathcal{C}_0(\Omega, \mathfrak{A})$ is equipped with pointwise multiplication.

Exercise B.2.7 Let Ω_1 and Ω_2 be locally compact Hausdorff spaces. Show that there is a canonical isometric algebra isomorphism $\mathcal{C}_0(\Omega_1) \check{\otimes} \mathcal{C}_0(\Omega_2) \cong \mathcal{C}_0(\Omega_1 \times \Omega_2)$.

There is a connection between the injective tensor product and approximable operators on Banach spaces:

Definition B.2.6 Let E be a Banach space, let $x_1, \ldots, x_n \in E$, and let $\phi_1, \ldots, \phi_n \in E^*$. Then $T := \sum_{j=1}^n x_j \odot \phi_j \in \mathcal{F}(E)$ is defined through

$$Tx = \sum_{j=1}^n \langle x, \phi_j \rangle x_j \quad (x \in E).$$

It is clear that every finite rank operator arises in this fashion.

Proposition B.2.7 *Let E be a Banach space. Then the linear map*

$$E \otimes E^* \to \mathcal{L}(E), \quad x \otimes \phi \mapsto x \odot \phi$$

is an isometry with respect to the injective norm on $E \otimes E^*$, and thus extends to an isometric isomorphism of $E \check{\otimes} E^*$ and $\mathcal{A}(E)$.

Proof For $x_1, \ldots, x_n \in E$ and $\phi_1, \ldots, \phi_n \in E^*$, let $T := \sum_{j=1}^n x_j \odot \phi_j$ and $\mathbf{x} := \sum_{j=1}^n x_j \otimes \phi_j$. Then

$$
\begin{aligned}
\|T\| &= \|T^{**}\| \\
&= \sup\{\|T^{**}X\| : X \in E^{**}, \|X\| \le 1\} \\
&= \sup\{|\langle \phi, T^{**}X \rangle| : \phi \in E^*, \|\phi\| \le 1, X \in E^{**}, \|X\| \le 1\} \\
&= \sup\left\{\left|\sum_{j=1}^n \langle \phi_j, X \rangle \langle x_j, \phi \rangle\right| : \phi \in E^*, \|\phi\| \le 1, X \in E^{**}, \|X\| \le 1\right\} \\
&= \sup\{|\langle \mathbf{x}, \phi \otimes X \rangle| : \phi \in E^*, \|\phi\| \le 1, X \in E^{**}, \|X\| \le 1\} \\
&= \|\mathbf{x}\|_\epsilon.
\end{aligned}
$$

This completes the proof. $\square$

B.2.2 The projective tensor product

The injective tensor product of Banach spaces has its drawbacks:

1. It is not characterized by the obvious functional analytic analogue of the definining (universal) property of the algebraic tensor product.
2. For two Banach algebras $\mathfrak{A}$ and $\mathfrak{B}$, the injective norm $\|\cdot\|_\epsilon$ on $\mathfrak{A} \otimes \mathfrak{B}$ is, in general, not submultiplicative with respect to the product defined in Exercise B.1.6, so that $\bullet$ does not extend to $\mathfrak{A} \check{\otimes} \mathfrak{B}$ (except in special cases such as the one discussed in Exercise B.2.6).

As we shall now see, there is a cross norm such that the corresponding tensor product of Banach spaces has the desired properties.

Definition B.2.8 Let $E_1, \ldots, E_n$ be Banach spaces. Then we define for $\mathbf{x} \in E_1 \otimes \cdots \otimes E_n$:

$$
\|\mathbf{x}\|_\pi := \inf\left\{\sum_{k=1}^m \left\|x_1^{(k)}\right\| \cdots \left\|x_n^{(k)}\right\| : \mathbf{x} = \sum_{k=1}^m x_1^{(k)} \otimes \cdots \otimes x_n^{(k)}\right\}.
$$

We call $\|\cdot\|_\pi$ the *projective norm* on $E_1 \otimes \cdots \otimes E_n$.

Proposition B.2.9 *Let $E_1, \ldots, E_n$ be Banach spaces. Then $\|\cdot\|_\pi$ is a cross norm on $E_1 \otimes \cdots \otimes E_n$ such that*

$$
\|\mathbf{x}\| \le \|\mathbf{x}\|_\pi \qquad (\mathbf{x} \in E_1 \otimes \cdots \otimes E_n),
$$

for any cross norm $\|\cdot\|$ on $E_1 \otimes \cdots \otimes E_n$.

Proof It is immediate that $\|\cdot\|$ is a seminorm on $E_1 \otimes \cdots \otimes E_n$ satisfying

$$
\|x_1 \otimes \cdots \otimes x_n\|_\pi \le \|x_1\| \cdots \|x_n\| \qquad (x_1 \in E_1, \ldots, x_n \in E_n).
$$

Let $\mathbf{x} \in E_1 \otimes \cdots \otimes E_n$. For $j = 1, \ldots, n$ choose $x_j^{(1)}, \ldots, x_j^{(m)} \in E_j$ such that $\mathbf{x} = \sum_{k=1}^m x_1^{(k)} \otimes \cdots \otimes x_n^{(k)}$. For any cross norm $\|\cdot\|$, we have

$$\|\mathbf{x}\| \le \sum_{k=1}^{m} \left\|x_1^{(k)}\right\| \cdots \left\|x_n^{(k)}\right\|.$$ (B.11)

Since $\|\mathbf{x}\|_\pi$ is the infimum over all possible expressions occurring as right hand sides of (B.11), we obtain $\|\mathbf{x}\| \le \|\mathbf{x}\|_\pi$. In the special case where $\|\cdot\| = \|\cdot\|_\epsilon$, we obtain $\|\cdot\|_\epsilon \le \|\cdot\|_\pi$, which establishes that $\|\cdot\|_\pi$ is a norm, and

$$\|x_1\| \cdots \|x_n\| = \|x_1 \otimes \cdots \otimes x_n\|_\epsilon \le \|x_1 \otimes \cdots \otimes x_n\|_\pi \qquad (x_1 \in E_1, \dots, x_n \in E_n),$$

so that $\|\cdot\|_\pi$ is a cross norm. $\square$

Definition B.2.10 Let $E_1, \dots, E_n$ be Banach spaces. Then their *projective tensor product* $E_1 \hat{\otimes} \cdots \hat{\otimes} E_n$ is the completion of $E_1 \otimes \cdots \otimes E_n$ with respect to $\|\cdot\|_\pi$.

Exercise B.2.8 Show that the projective tensor product of Banach spaces satisfies the following universal property: Let $E_1, \dots, E_n$ be Banach spaces. Then for every Banach space F, and for every $V \in \mathcal{L}^n(E_1, \dots, E_n; F)$, there is a unique $\tilde{V} \in \mathcal{L}(E_1 \hat{\otimes} \cdots \hat{\otimes} E_n; F)$ with $\|\tilde{V}\| = \|V\|$ such that

$$V(x_1, \dots, x_n) = \tilde{V}(x_1 \otimes \cdots \otimes x_n) \qquad (x_1 \in E_1, \dots, x_n \in E_n).$$

Exercise B.2.9 Let $E_1, \dots, E_n$ be Banach spaces, and let $E_1 \tilde{\otimes} \cdots \tilde{\otimes} E_n$ be the completion of $E_1 \otimes \cdots \otimes E_n$ with respect to some cross norm. Show that the identity on $E_1 \otimes \cdots \otimes E_n$ extends to a contraction from $E_1 \hat{\otimes} \cdots \hat{\otimes} E_n$ to $E_1 \tilde{\otimes} \cdots \tilde{\otimes} E_n$.

Exercise B.2.10 Let $E_1, \dots, E_n, F$ be Banach spaces. Show that

$$\langle x_1 \otimes \cdots \otimes x_n \otimes y, V \rangle := \langle y, V(x_1, \dots, x_n) \rangle$$

$$(x_1 \in E_1, \dots, x_n \in E_n, y \in F, V \in \mathcal{L}^n(E_1, \dots, E_n; F^*))$$

induces an isometric isomorphism of $(E_1 \hat{\otimes} \cdots \hat{\otimes} E_n \hat{\otimes} F)^*$ and $\mathcal{L}^n(E_1, \dots, E_n; F^*)$.

Exercise B.2.11 Prove an associative law, as in Exercise B.1.3, for the projective tensor product (with isometric isomorphisms).

Exercise B.2.12 Let $E_1, F_1, \dots, E_n, F_n$ be Banach spaces, and let $T_j \in \mathcal{L}(E_j, F_j)$.

(i) Prove an analogue of Exercise B.2.3 for the projective tensor product.
(ii) Show that, if $T_j \in \mathcal{K}(E_j, F_j)$ for $j = 1, \dots, n$, then $T_1 \otimes \cdots \otimes T_n$ is also compact.

Exercise B.2.13 Let $\mathfrak{A}$ be a Banach algebra, let E be a left Banach $\mathfrak{A}$-module, and let F be a right Banach $\mathfrak{A}$-module. Show that the bimodule action (B.3) of $\mathfrak{A}$ on $E \otimes F$ turns $E \hat{\otimes} F$ into a Banach $\mathfrak{A}$-bimodule.

Exercise B.2.14 Let $\mathfrak{A}$ and $\mathfrak{B}$ be Banach algebras. Show that $\|\cdot\|_\pi$ is submultiplicative with respect to the product $\bullet$ on $\mathfrak{A} \otimes \mathfrak{B}$ defined in Exercise B.1.6, so that $\bullet$ extends to $\mathfrak{A} \hat{\otimes} \mathfrak{B}$ turning it into a Banach algebra. If $\mathfrak{A}$ and $\mathfrak{B}$ have bounded approximate identities $(e_\alpha)_\alpha$ and $(e_\beta)_\beta$, respectively, show that $(e_\alpha \otimes e_\beta)_{\alpha,\beta}$ is a bounded approximate identity for $\mathfrak{A} \hat{\otimes} \mathfrak{B}$.

There is a useful analogue of Proposition B.1.2 for the projective tensor product:

Proposition B.2.11 Let $E_1, \dots, E_n$ be Banach spaces, and let $\mathbf{x} \in E_1 \hat{\otimes} \cdots \hat{\otimes} E_n$. Then there are sequences $\left(x_j^{(k)}\right)_{k=1}^{\infty}$ in E_j for $j = 1, \dots, n$ such that

$$\sum_{k=1}^{\infty} \left\| x_1^{(k)} \right\| \cdots \left\| x_n^{(k)} \right\| < \infty \tag{B.12}$$

and

$$\mathbf{x} = \sum_{k=1}^{\infty} x_1^{(k)} \otimes \cdots \otimes x_n^{(k)} \tag{B.13}$$

Moreover, $\|\mathbf{x}\|_\pi$ is the infimum over all infinite series (B.12) such that (B.13) is satisfied.

Proof For each $\mathbf{x} \in E_1 \hat{\otimes} \cdots \hat{\otimes} E_n$ having a series representation as in (B.13), let $\|x\|_{\tilde{\pi}}$ be the infimum over all infinite series (B.12) such that (B.13) holds. It is easy to see that $\|\cdot\|_{\tilde{\pi}}|_{E_1 \otimes \cdots \otimes E_n}$ is a cross norm such that the resulting completion of $E_1 \otimes \cdots \otimes E_n$ enjoys the same universal property as the projective tensor product (Exercise B.2.8). It follows that $\|\cdot\|_{\tilde{\pi}} = \|\cdot\|_\pi$ (Why?).

Let F be the subspace of $E_1 \hat{\otimes} \cdots \hat{\otimes} E_n$ consisting of all x having a series representation as in (B.13). It is not difficult to see that $(F, \|\cdot\|_{\tilde{\pi}})$ is a Banach space (*Hint*: Show that every absolutely converging series in $(F, \|\cdot\|_{\tilde{\pi}})$ converges.). It follows that $F = E_1 \hat{\otimes} \cdots \hat{\otimes} E_n$. □

Exercise B.2.15 Work out the proof of Proposition B.2.11 in detail.

There is an analogue of Theorem B.2.5 for the projective tensor product. For a measure space $(\Omega, \mathcal{S}, \mu)$ and a Banach space E, let $L^1(\Omega, \mathcal{S}, \mu; E)$ denote the space of all (equivalence classes of) μ-integrable functions on Ω with values in E. If you don't feel comfortable about Banach space valued integration, have a look at [Lang 1]: there, integration theory is developed from scratch for Banach space valued functions.

Theorem B.2.12 *Let $(\Omega, \mathcal{S}, \mu)$ be a measure space, and let E be a Banach space. Then the bilinear map*

$$L^1(\Omega, \mathcal{S}, \mu) \times E \to L^1(\Omega, \mathcal{S}, \mu; E), \quad (f, x) \mapsto fx \tag{B.14}$$

induces an isometric isomorphism of $L^1(\Omega, \mathcal{S}, \mu) \hat{\otimes} E$ and $L^1(\Omega, \mathcal{S}, \mu; E)$.

Proof It follows immediately from the universal property of the projective tensor product, that (B.14) induces a contraction from $L^1(\Omega, \mathcal{S}, \mu) \hat{\otimes} E$ into $L^1(\Omega, \mathcal{S}, \mu; E)$. From the definition of $L^1(\Omega, \mathcal{S}, \mu; E)$, it is clear that this contraction has dense range; it remains to be shown that it is also an isometry.

For $f_1, \ldots, f_n \in L^1(\Omega, \mathcal{S}, \mu)$ and $x_1, \ldots, x_n \in E$, let $f = \sum_{j=1}^n f_j x_j$ and $\mathbf{x} = \sum_{j=1}^n f_j \otimes x_j$. We claim that $\|f\|_1 = \|\mathbf{x}\|_\pi$. A simple density argument shows that we may confine ourselves to the case where $f_1, \ldots, f_n$ are step functions. Furthermore (Why?), we may suppose without loss of generality that there are mutually disjoint $\Omega_1, \ldots, \Omega_n \in \mathcal{S}$ with $\mu(\Omega_j) < \infty$ for $j = 1, \ldots, n$ such that $f_j = \chi_{\Omega_j}$ for $j = 1, \ldots, n$. It follows that

$$\|f\|_1 = \int_\Omega \|f\|\, d\mu = \sum_{j=1}^n \|f_j x_j\|_1 = \sum_{j=1}^n \mu(\Omega_j)\|x_j\| = \sum_{j=1}^n \|f_j\|_1 \|x\| \geq \|\mathbf{x}\|_\pi,$$

which completes the proof. □

Exercise B.2.16 Let G be a locally compact group, and let $\mathfrak{A}$ be a Banach algebra.

(i) Show that $L^1(G, \mathfrak{A})$ becomes a Banach algebra through the convolution formula

$$(f_1 * f_2)(g) = \int_G f_1(gh^{-1})f_2(h)\, dm_G(h) \qquad (f_1, f_2 \in L^1(G, \mathfrak{A})).$$

(ii) Show that, if $L^1(G) \hat{\otimes} \mathfrak{A}$ is equipped with the product $\bullet$ from Exercise B.2.14, then the isomorphism of $L^1(G) \hat{\otimes} \mathfrak{A}$ and $L^1(G, \mathfrak{A})$ from Theorem B.2.12 is an algebra isomorphism, where $L^1(G, \mathfrak{A})$ is equipped with the convolution product from (i).

Exercise B.2.17 Let G_1 and G_2 be locally compact groups. Show that there is a canonical isometric algebra isomorphism $L^1(G_1) \hat{\otimes} L^1(G_2) \cong L^1(G_1 \times G_2)$.

We conclude this section on Banach space tensor products with an easy exercise that relates the injective and the projective tensor product:

Exercise B.2.18 Let E be a Banach space, and let F be a finite-dimensional Banach space. Show that $E \otimes F$ is complete both in the injective and in the projective norm. Conclude that $E \check{\otimes} F \cong E \hat{\otimes} F$.

B.3 The Hilbert space tensor product

Let $\mathfrak{H}_1$ and $\mathfrak{H}_2$ be Hilbert spaces. When is $\mathfrak{H}_1 \hat{\otimes} \mathfrak{H}_2$ a Hilbert space (or $\mathfrak{H}_1 \check{\otimes} \mathfrak{H}_2$)? The answer is: hardly ever (even if we are willing to put up with merely topological, but not isometric isomorphism). If we want a tensor product of Hilbert spaces to again be a Hilbert space, we need a different construction.

Exercise B.3.1 Let $\mathfrak{H}_1, \ldots, \mathfrak{H}_n$ be Hilbert spaces.

(i) Show that there is a unique positive definite, sesquilinear form on $\mathfrak{H}_1 \otimes \cdots \otimes \mathfrak{H}_n$ such that

$$\langle \xi_1 \otimes \cdots \otimes \xi_n, \eta_1 \otimes \cdots \otimes \eta_n \rangle := \langle \xi_1, \eta_1 \rangle \cdots \langle \xi_n, \eta_n \rangle \qquad (\xi_1, \eta_1 \in \mathfrak{H}_1, \ldots, \xi_n, \eta_n \in \mathfrak{H}_n).$$

(ii) Show that the norm on $\mathfrak{H}_1 \otimes \cdots \otimes \mathfrak{H}_n$ induced by the scalar product from (i) is a cross norm.

Definition B.3.1 Let $\mathfrak{H}_1, \ldots, \mathfrak{H}_n$ be Hilbert spaces, and let $\langle \cdot, \cdot \rangle$ be as in Exercise B.3.1. Then the Hilbert space tensor product $\mathfrak{H}_1 \bar{\otimes} \cdots \bar{\otimes} \mathfrak{H}_n$ of $\mathfrak{H}_1, \ldots, \mathfrak{H}_n$ is the completion of $\mathfrak{H}_1 \otimes \cdots \otimes \mathfrak{H}_n$ with respect to the norm induced by $\langle \cdot, \cdot \rangle$.

Exercise B.3.2 Let $\mathfrak{H}_1$ and $\mathfrak{H}_2$ be Hilbert spaces with orthonormal bases $\left(e_\alpha^{(1)}\right)_\alpha$ and $\left(e_\beta^{(2)}\right)_\beta$, respectively.

(i) Show that $\left(e_\alpha^{(1)} \otimes e_\beta^{(2)}\right)_{\alpha, \beta}$ is an orthonormal basis for $\mathfrak{H}_1 \bar{\otimes} \mathfrak{H}_2$.

(ii) Show that each element of $\mathfrak{H}_1 \bar{\otimes} \mathfrak{H}_2$ has the form $\sum_{n=1}^\infty \xi_n^{(1)} \otimes \xi_n^{(2)}$, where $\left(\xi_n^{(1)}\right)_{n=1}^\infty$ and $\left(\xi_n^{(2)}\right)_{n=1}^\infty$ are sequences in $\mathfrak{H}_1$ and $\mathfrak{H}_2$, respectively, such that $\sum_{n=1}^\infty \left\|\xi_n^{(1)}\right\|^2 \left\|\xi_n^{(2)}\right\|^2 < \infty$.

Exercise B.3.3 Prove an analogue of Exercise B.2.3 for the Hilbert space tensor product.

There is an analogue of Theorem B.2.12:

Exercise B.3.4 Let $(\Omega, \mathcal{S}, \mu)$ be a measure space, and let $\mathfrak{H}$ be a Hilbert space. Then the bilinear map

$$L^2(\Omega, \mathcal{S}, \mu) \times \mathfrak{H} \to L^2(\Omega, \mathcal{S}, \mu; \mathfrak{H}), \quad (f, \xi) \mapsto f\xi$$

induces an isometric isomorphism of $L^2(\Omega, \mathcal{S}, \mu) \bar{\otimes} \mathfrak{H}$ and $L^2(\Omega, \mathcal{S}, \mu; \mathfrak{H})$.

Exercise B.3.5 Let G_1 and G_2 be locally compact groups. Show that there is a canonical isometric isomorphism $L^2(G_1)\bar{\otimes}L^2(G_2) \cong L^2(G_1 \times G_2)$ such that

$$\lambda_2((g_1, g_2)) = \lambda_2(g_1) \otimes \lambda_2(g_2) \qquad (g_1 \in G_1,\, g_2 \in G_2)$$

with the appropriate identifications made.

That was it, already!

B.4 Notes and comments

The algebraic tensor product of linear spaces (or, more generally, of modules over commutative rings), is a standard construction in abstract algebra; see, e.g., [Lang 2].

The theory of Banach space tensor products, initiated by A. Grothendieck, is a vast and deep area. Of course, the injective and the projective norms are not the only cross norms on tensor products of Banach spaces — and by far not the only interesting ones. We have limited ourselves to these two norms because they are the only ones we need in these notes. If you want to dig much deeper, look into [D–F]. If E and F are any two Banach spaces such that $E\check{\otimes}F \cong E\hat{\otimes}F$, then, in most cases, E or F is finite-dimensional ([D–F, Exercise 4.12]). In 1981, G. Pisier constructed an infinite-dimensional Banach space E such that $E\check{\otimes}E \cong E\hat{\otimes}E$ ([Pis 1]) although the contrary had been conjectured by Grothendieck.

C Banach space properties

In these notes, we need some background from the theory of Banach spaces for several reasons:

- If E is a Banach space, the amenability (or non-amenability) of $\mathcal{A}(E)$ corresponds to certain Banach space properties of E.
- When dealing with super-amenable or — more generally — biprojective Banach algebras, we require the approximation property for the proofs of some results.
- For C^*-algebras, amenability implies the approximation property.

In this appendix, we collect the required notions and results. As far as the extent to which we present proofs is concerned, this appendix lies somewhere between Appendices A and B. The fact that some proof is not included does not necessarily imply that it is too deep or too technical to be presented here, but just that it would have taken too much space.

C.1 Approximation properties

Definition C.1.1 Let E be a Banach space. Then:

(i) E has the *approximation property* if there is a net $(S_\alpha)_\alpha$ in $\mathcal{F}(E)$ such that $S_\alpha \to \mathrm{id}_E$ uniformly on compact subsets of E.

(ii) E has the *C-approximation property* with $C \geq 1$ if there is a net $(S_\alpha)_\alpha$ in $\mathcal{F}(E)$ with $\sup_\alpha \|S_\alpha\| \leq C$ such that $S_\alpha \to \mathrm{id}_E$ uniformly on compact subsets of E.

(iii) E has the *bounded approximation property* if it has the C-approximation property for some $C \geq 1$.

(iv) E has the *metric approximation property* if it has the 1-approximation property.

Examples C.1.2   (a) A *basis* for a Banach space E is a sequence $(x_k)_{k=1}^\infty$ in E with the property that, for each $x \in E$, there is a unique sequence $(\lambda_k)_{k=1}^\infty$ in $\mathbb{C}$ such that

$$x = \sum_{k=1}^\infty \lambda_k x_k; \tag{C.1}$$

for example, the spaces c_0 and ℓ^p for $p \in [1, \infty)$ have a (canonical) basis (What is it?). Let E be a Banach space with a basis $(x_k)_{k=1}^\infty$. For each $n \in \mathbb{N}$, let $P_n : E \to E$ be the unique projection which, for each $x \in E$ with a representation (C.1), is defined as

$$P_n x = \sum_{k=1}^n \lambda_k x_k.$$

Certainly, $P_n \to \mathrm{id}_E$ uniformly on all finite subsets of E. In order to establish the (bounded) approximation property for E, it suffices (Why?) to show that each P_n is continuous and that $\sup_{n\in\mathbb{N}} \|P_n\| < \infty$. Define a new norm $\|\|\cdot\|\|$ on E through $\|\|x\|\| := \sup_{n\in\mathbb{N}} \|P_n x\|$ for $x \in E$. It is immediate from this definition that each P_n is bounded and has norm one with respect to $\|\|\cdot\|\|$. Then $\|\cdot\| \leq \|\|\cdot\|\|$, and it is routinely checked (Do it!) that $(E, \|\|\cdot\|\|)$ is complete. The open mapping theorem implies that $\|\cdot\|$ and $\|\|\cdot\|\|$ are equivalent.

(b) For any Hilbert space $\mathfrak{H}$, let $\mathcal{F}$ be the collection of all finite-dimensional subspaces of $\mathfrak{H}$. For each $F \in \mathcal{F}$, let P_F be the orthogonal projection onto F. Then $P_F \to \mathrm{id}_\mathfrak{H}$ in the strong operator topology and thus uniformly on compact subsets of $\mathfrak{H}$. Hence, $\mathfrak{H}$ has the metric approximation property.

(c) All the examples of Banach spaces with property (A) discussed in Examples 3.1.12 have the bounded approximation property.

Remarks C.1.3   (a) There are Banach spaces without the approximation property; in fact, for any infinite-dimensional Hilbert space $\mathfrak{H}$, the space $\mathcal{L}(\mathfrak{H})$ lacks the approximation property ([Sza]).

(b) A reflexive Banach space has the approximation property if and only if it has the metric approximation property; the same is true for separable dual Banach spaces ([D–F, 16.4, Corollary 4]).

(c) There are (separable) Banach spaces which have the approximation property, but not the bounded approximation property ([L-Tz, I, Example 1.e.20]).

We have a brief (and rather superficial) look at the hereditary properties of the approximation property:

Exercise C.1.1 Let E be a Banach space with the (bounded) approximation property, and let F be a complemented subspace of E. Show that F has the (bounded) approximation property as well.

Remark C.1.4 Both c_0 and ℓ^p for $p \in [1, \infty) \setminus \{2\}$ contain subspaces lacking the approximation property ([L-Tz, I, Theorem 2.d.6 and II, Theorem 1.g.4]). Hence, the approximation property is not inherited by arbitrary subspaces. Dualizing, we see that also quotients need not inherit the approximation property.

Exercise C.1.2 For $j = 1, 2$, let E_j be a Banach space with the C_j-approximation property for some $C_j \geq 1$. Show that both $E_1 \hat\otimes E_2$ and $E_1 \check\otimes E_2$ have the $C_1 C_2$-approximation property.

For any Banach spaces $E_1, \ldots, E_n$, the identity on $E_1 \otimes \cdots \otimes E_n$ extends to a contraction from $E_1 \hat\otimes \cdots \hat\otimes E_n$ into $E_1 \check\otimes \cdots \check\otimes E_n$ by Exercise B.2.9. This map can be used to characterize the approximation property:

Theorem C.1.5 *For a Banach space E the following are equivalent:*

(i) *E has the approximation property.*

(ii) *For any Banach space F, the canonical map from $E\hat\otimes F$ into $E\check\otimes F$ is injective.*

(iii) *The canonical map from $E\hat\otimes E^*$ into $E\check\otimes E^*$ is injective.*

(iv) *The map*

$$E \hat{\otimes} E^* \to \mathcal{N}(E), \quad x \otimes \phi \mapsto x \odot \phi \qquad (C.2)$$

is an isometric isomorphism.

Proof (i) $\Longrightarrow$ (ii): Let $\mathbf{x} \in E \hat{\otimes} F \setminus \{0\}$. Then there are sequences $(x_n)_{n=1}^{\infty}$ in E and $(y_n)_{n=1}^{\infty}$ in F such that $\sum_{n=1}^{\infty} \|x_n\| \|y_n\| < \infty$ and $\mathbf{x} = \sum_{n=1}^{\infty} x_n \otimes y_n$ (Proposition B.2.11). Without loss of generality, we may suppose that $\|x_n\| \to 0$ and $\sum_{n=1}^{\infty} \|y_n\| < \infty$ (Why?). Since $\{x_n : n \in \mathbb{N}\} \cup \{0\}$ is compact, the approximation property for E implies that there is $S \in \mathcal{F}(E)$ such that

$$\|Sx_n - x_n\| < \frac{\|\mathbf{x}\|_\pi}{2 \sum_{n=1}^{\infty} \|y_n\|} \qquad (n \in \mathbb{N}). \qquad (C.3)$$

Let $\mathbf{y} := \sum_{n=1}^{\infty} Sx_n \otimes y_n$. Then (C.3) implies that $\|\mathbf{x} - \mathbf{y}\|_\pi < \frac{1}{2}\|\mathbf{x}\|_\pi$, so that $\mathbf{y} \neq 0$. Since there are $z_1, \dots, z_m \in E$ and $\phi_1, \dots, \phi_m$ with $S = \sum_{k=1}^{m} z_k \odot \phi_k$, this means that there is $\phi \in E^*$ with $\sum_{n=1}^{\infty} \langle x_n, \phi \rangle y_n \neq 0$. The Hahn–Banach theorem yields $\psi \in F^*$ such that $\sum_{n=1}^{\infty} \langle x_n, \phi \rangle \langle y_n, \psi \rangle \neq 0$. From the definition of $\| \cdot \|_\epsilon$, it is now immediate that the image of $\mathbf{x}$ in $E \check{\otimes} F$ is non-zero.

(ii) $\Longrightarrow$ (iii) is trivial.

(iii) $\Longrightarrow$ (iv): Clearly, (C.2) is the composition of the canonical map from $E \hat{\otimes} E^*$ into $E \check{\otimes} E^*$ with the canonical isomorphism between $E \check{\otimes} E^*$ and $\mathcal{A}(E)$ (Proposition B.2.7). Since both maps are injective, so is (C.2). By definition, (C.2) is onto, and the definition of the nuclear norm implies that (C.2) is even an isometric isomorphism.

(iv) $\Longrightarrow$ (i): Let τ denote the topology on $\mathcal{L}(E)$ of uniform convergence on all compact subsets of E. Assume that E lacks the approximation property; this means that id_E does not lie in the τ-closure of $\mathcal{F}(E)$. The Hahn–Banach theorem then implies that there is a τ-continuous linear functional on $\mathcal{L}(E)$ that vanishes on $\mathcal{F}(E)$, but not at id_E. By [L-Tz, I, Proposition 1.e.3], each τ-continuous functional is of the form

$$\mathcal{L}(E) \to \mathbb{C}, \quad T \mapsto \sum_{n=1}^{\infty} \langle Tx_n, \phi_n \rangle,$$

where $(x_n)_{n=1}^{\infty}$ and $(\phi_n)_{n=1}^{\infty}$ are sequences in E and E^*, respectively, with the property that $\sum_{n=1}^{\infty} \|x_n\| \|\phi_n\| < \infty$. Hence, there exist such sequences $(x_n)_{n=1}^{\infty}$ and $(\phi_n)_{n=1}^{\infty}$ with

$$\sum_{n=1}^{\infty} \langle Sx_n, \phi_n \rangle = 0 \qquad (S \in \mathcal{F}(E)), \qquad (C.4)$$

but

$$\sum_{n=1}^{\infty} \langle x_n, \phi_n \rangle \neq 0. \qquad (C.5)$$

From (C.4), it follows (How exactly?) that $\sum_{n=1}^{\infty} \langle x, \phi_n \rangle x_n = 0$ for all $x \in E$. It follows that $\mathbf{x} := \sum_{n=1}^{\infty} x_n \otimes \phi_n \in E \hat{\otimes} E^*$ lies in the kernel of (C.2). Since (C.2) is injective, this means $\mathbf{x} = 0$, but this contradicts (C.5). $\square$

With Theorem C.1.5 at hand, we can establish yet another hereditary property for the approximation property:

Corollary C.1.6 *Let E be a Banach space such that E^* has the approximation property. Then E has the approximation property.*

Proof Since E^* has the approximation property, the canonical map from $E^* \hat{\otimes} E^{**}$ into $E^* \check{\otimes} E^{**}$ is injective. From the definition of the injective norm, it is immediate (Do you believe that?) that the inclusion of $E \otimes E^*$ into $E^{**} \otimes E^*$ is an isometry with respect to the injective norms on both spaces. Hence, the canonical map from $E \hat{\otimes} E^*$ into $E \check{\otimes} E^*$ is injective as well, so that E has the approximation property by Theorem C.1.5.   □

C.2 The Radon–Nikodým property

Let E and F be Banach spaces, and define a bilinear map $\kappa_{E,F} \colon E^* \times F^* \to (E \check{\otimes} F)^*$ through

$$\kappa_{E,F}(\phi, \psi)(x, y) := \langle x, \phi \rangle \langle y, \psi \rangle \qquad (x \in E,\ \phi \in E^*,\ y \in F,\ \psi \in F^*).$$

It is immediate from the definition of the injective norm that $\kappa_{E,F}$ has norm one. The universal property of the projective tensor product implies that $\kappa_{E,F}$ induces a linear map of norm one from $E^* \hat{\otimes} F^*$ into $(E \check{\otimes} F)^*$. We are interested in the question of when $\kappa_{E,F}$ is an isomorphism.

Definition C.2.1 A Banach space E has the *Radon–Nikodým property* if, for each finite measure space $(\Omega, \mathcal{S}, \mu)$ and each bounded linear operator $T \colon L^1(\Omega, \mathcal{S}, \mu) \to E$, there is a bounded μ-measurable function $\phi \colon \Omega \to E$ such that

$$Tf = \int f\phi \, d\mu \qquad (f \in L^1(\Omega, \mathcal{S}, \mu)). \tag{C.6}$$

Exercise C.2.1   (i) Use the classical Radon–Nikodým theorem to conclude that $\mathbb{C}$ has the Radon–Nikodým property.

(ii) Let $(E_\alpha)_\alpha$ be a family of Banach spaces each of which has the Radon–Nikodým property. Show that their ℓ^1-direct sum $\ell^1\text{-}\bigoplus_\alpha E_\alpha$ has the Radon–Nikodým property as well.

(iii) Conclude that $\ell^1(\mathbb{I})$ has the Radon–Nikodým property for any index set $\mathbb{I}$.

Exercise C.2.2 Let E be a Banach space with the Radon–Nikodým property, and let F be a closed subspace of F. Show that F has the Radon–Nikodým property.

Examples C.2.2   (a) Any reflexive Banach space has the Radon–Nikodým property: This follows immediately from the strong Dunford–Pettis theorem ([D–F, C4, Theorem]).

(b) Any separable, dual Banach space has the Radon–Nikodým property (this follows from [D–F, C5, Proposition]).

(c) Let $T \colon L^1([0,1]) \to c_0$ be the operator that assigns to each function $f \in L^1(\mathbb{T})$ the sequence of its Fourier coefficients (with positive index):

$$Tf := \left(\int_0^1 f(t) e^{2\pi i n t} \, dt \right)_{n \in \mathbb{N}}.$$

Assume that there is a bounded Lebesgue-measurable function $\phi \colon [0,1] \to c_0 \subset \ell^\infty$ such that (C.6) holds. It follows (How exactly?) that

$$\phi(t) = (e^{2\pi i n t})_{n \in \mathbb{N}}$$

for almost all $t \in [0,1]$. For all $t \in [0,1]$, however, $|e^{2\pi i n t}| = 1$ holds, so that $\phi(t) \notin c_0$ for almost all $t \in [0,1]$. Hence, c_0 lacks the Radon–Nikodým property.

Exercise C.2.3 Let Ω be a locally compact Hausdorff space with infinitely many points. Use Exercise C.2.2 and Examples C.2.2(c) to show that $C_0(\Omega)$ lacks the Radon–Nikodým property.

For the problem we are interested in, the Radon–Nikodým property is relevant due to the following theorem ([D–F, 16.6, Theorem(1)]):

Theorem C.2.3 *Let E and F be Banach spaces such that*

(i) *E^* or F^* has the approximation property, and*
(ii) *E^* or F^* has the Radon–Nikodým property.*

Then $\kappa_{E,F}: E^* \hat{\otimes} F^* \to (E \check{\otimes} F)^*$ is an isometric isomorphism.

Remark C.2.4 Suppose that E is a Banach space such that $\kappa_{E,F}: E^* \hat{\otimes} F^* \to (E \check{\otimes} F)^*$ is a topological isomorphism for every Banach space F. Then E^* has both the Radon–Nikodým property and the metric approximation property ([D–F, 16.6, Theorem(2)]).

C.3 Local theory of Banach spaces

By the local theory of Banach spaces we mean the study of those properties of a Banach space which are accessible through the investigation of its finite-dimensional subspaces.

It is elementary that every finite-dimensional subspace F of a Banach space E is complemented, i.e. there is a bounded projection P from E onto F. It is a highly non-trivial problem to determine what the best, i.e. smallest, value for $\|P\|$ is. For arbitrary finite-dimensional subspaces of arbitrary Banach spaces the following estimate is the best one available ([Woj, II.B.7., 10 Theorem]):

Theorem C.3.1 *Let E be a Banach spaces, let $n \in \mathbb{N}$, and let F be an n-dimensional subspace of E. Then there is a projection P of E onto F with $\|P\| \leq \sqrt{n}$.*

The following notion is of fundamental importance in the local theory of Banach spaces ([Woj, II.E.15]):

Definition C.3.2 Let E_1 and E_2 be Banach spaces. Then E_1 is *finitely representable* in E_2 if there is $C \geq 1$ such that, for each finite-dimensional subspace F of E_1, there is $\tau: F \to E_2$ such that $\|\tau\| \| \tau^{-1}|_{\tau(F)}\| \leq C$.

Sloppily speaking, a Banach space E_1 is finitely representable in the Banach space E_2 if each finite-dimensional subspace of E_1 "is" a subspace of E_2.

For every Banach space E, its bidual E^{**} is finitely representable in E. This follows immediately from the local reflexivity principle ([Woj, II.E.14, Theorem]):

Theorem C.3.3 (local reflexivity principle) *Let E be a Banach space, let F be a finite-dimensional subspace of E^{**}, let $\phi_1, \dots, \phi_n \in E^*$, and let $\epsilon > 0$. Then there is $\tau: F \to E$ such that $\tau|_{F \cap E} = \mathrm{id}_{F \cap E}$, $\|\tau\|\|\tau^{-1}|_{\tau(F)}\| < 1 + \epsilon$, and*

$$\langle \tau(X), \phi_j \rangle = \langle \phi_j, X \rangle \qquad (X \in F, j = 1, \dots, n).$$

Exercise C.3.1 Use Theorem C.3.3 to prove an analogue of Corollary C.1.6 for the bounded approximation property: If E is a Banach space such that E^* has the C-approximation property for some $C \geq 1$, then E has the C-approximation property as well.

Exercise C.3.2 Use the previous exercise along with Exercise C.1.2 to exhibit a separable Banach space with the metric approximation property whose dual lacks the approximation property.

To pass from local to global statements about Banach spaces, the following construction is useful:

Definition C.3.4 Let E be a Banach space, let $\mathbb{I}$ be an index set, and let $\mathcal{U}$ be an ultrafilter on $\mathbb{I}$. Then:

(i) $\ell^\infty(\mathbb{I}, E)$ is the space of all bounded families $(x_\alpha)_{\alpha \in \mathbb{I}}$ in E.
(ii) $\mathcal{N}_\mathcal{U}$ is the subspace of $\ell^\infty(\mathbb{I}, E)$ consisting of those $(x_\alpha)_{\alpha \in \mathbb{I}}$ for which $\lim_\mathcal{U} \|x_\alpha\| = 0$.
(iii) The quotient space $(E)_\mathcal{U} := \ell^\infty(\mathbb{I}, E)/\mathcal{N}_\mathcal{U}$ is called an *ultrapower* of E.

Exercise C.3.3 Let E be a Banach space, let $\mathbb{I}$ be an index set, and let $\mathcal{U}$ be an ultrafilter on $\mathbb{I}$.

(i) Let $\ell^\infty(\mathbb{I}, E)$ be equipped with its canonical norm. Show that $\mathcal{N}_\mathcal{U}$ is a closed subspace of $\ell^\infty(\mathbb{I}, E)$.
(ii) Show that, for any $x = (x_\alpha)_{\alpha \in \mathbb{I}} \in \ell^\infty(\mathbb{I}, E)$, the equality

$$\|x\|_\mathcal{U} := \lim_\mathcal{U} \|x_\alpha\| = \|x + \mathcal{N}_\mathcal{U}\|$$

holds.
(iii) Conclude that $((E)_\mathcal{U}, \|\cdot\|_\mathcal{U})$ is a Banach space.

Every Banach space is canonically contained in all of its ultrapowers (How?).

Exercise C.3.4 Let E be a finite-dimensional Banach space, and let $(E)_\mathcal{U}$ be an ultrapower of E. Show that the canonical embedding of E into $(E)_\mathcal{U}$ is an isomorphism.

Unless the ultrafilter is fixed, i.e. consists of all sets containing a given point, the corresponding ultrapower of an infinite-dimensional Banach space is much larger than the original space. Nevertheless, we have the following:

Proposition C.3.5 *Let E be a Banach space, and let $(E)_\mathcal{U}$ be an ultrapower of E. Then $(E)_\mathcal{U}$ is finitely representable in E.*

Proof Let F be a finite-dimensional subspace of $(E)_\mathcal{U}$ with Hamel basis $x_1, \dots, x_n$. Let $\mathbb{I}$ be the index set over which $\mathcal{U}$ is defined. For each $j = 1, \dots, n$, let $(x_{j,\alpha})_{\alpha \in \mathbb{I}} \in \ell^\infty(\mathbb{I}, E)$ be such that $x_j = (x_{j,\alpha})_{\alpha \in \mathbb{I}} + \mathcal{N}_\mathcal{U}$. For $\alpha \in \mathbb{I}$, define $\tau_\alpha : F \to E$ through

$$\tau_\alpha(x_j) := x_{j,\alpha} \qquad (j = 1, \dots, n).$$

Clearly, $C := \sup_{\alpha \in \mathbb{I}} \|\tau_\alpha\| < \infty$. Let $x \in F$. Then there are $\lambda_1, \ldots, \lambda_n$ such that $x = \sum_{j=1}^n \lambda_j x_j$. It follows that

$$\lim_{\mathcal{U}} \|\tau_\alpha(x)\| = \lim_{\mathcal{U}} \left\|\sum_{j=1}^n \lambda_j x_{j,\alpha}\right\| = \left\|\sum_{j=1}^n \lambda_j x_j\right\|_{\mathcal{U}} = \|x\|_{\mathcal{U}}. \tag{C.7}$$

For any $x \in F$, (C.7) yields $I_x \in \mathcal{U}$, such that

$$\frac{1}{2}\|x\| \leq \|\tau_\alpha(x)\| \leq 2\|x\| \qquad (\alpha \in I_x). \tag{C.8}$$

For $\delta > 0$ choose a δ-net, say $x_1', \ldots, x_m'$ in $B_1(0, F)$. Choosing δ sufficiently small (How exactly is this done?), we obtain from (C.8) for all $\alpha \in I_{x_1'} \cap \cdots \cap I_{x_m'}$ that

$$\frac{1}{3}\|x\| \leq \|\tau_\alpha(x)\| \leq 3\|x\| \qquad (x \in F).$$

This completes the proof. $\square$

Remark C.3.6 It is not hard to show even more: The map τ_α in the proof of Proposition C.3.5 can be chosen such that $\|\tau_\alpha\|\|\tau_\alpha^{-1}|_{\tau_\alpha(F)}\| < 1 + \epsilon$, where $\epsilon > 0$ is arbitrarily chosen. Try it for yourself, and if you don't succeed, have a look at [Hei, Proposition 6.1] and its proof.

Theorem C.3.7 *For two Banach spaces E_1 and E_2, the following are equivalent:*

(i) *E_1 is finitely representable in E_2.*
(ii) *There are $C \geq 1$ and a directed family $\mathcal{F}$ of finite-dimensional subspaces of E_1 such that*
    (a) *$\bigcup\{F : F \in \mathcal{F}\}$ is dense in E_1, and*
    (b) *for each $F \in \mathcal{F}$, there is $\tau: F \to E_2$ such that $\|\tau\|\|\tau^{-1}|_{\tau(F)}\| \leq C$.*
(iii) *E_1 is topologically isomorphic to a closed subspace of some ultrapower of E_2.*

Proof (i) $\implies$ (ii) is obvious.

(ii) $\implies$ (iii): Let $\mathcal{U}$ be an ultrafilter on $\mathcal{F}$ that dominates the order filter. For each $F \in \mathcal{F}$, there is $\tau_F: F \to E_2$ such that $\|\tau_F\|\|\tau_F^{-1}|_{\tau(F)}\| \leq C$. Multiplying τ_F by an appropriate scalar, we can suppose that $\|\tau_F\| = 1$. Let $E_0 := \bigcup\{F : F \in \mathcal{F}\}$, and define $\tau: E_0 \to \ell^\infty(\mathbb{I}, E_2)$ through

$$\tau(x) := \begin{cases} \tau_F(x), & \text{if } x \in F, \\ 0 & \text{otherwise.} \end{cases}$$

Note that τ need not be linear. Nevertheless, the map

$$\tilde{\tau}: E_0 \to (E_2)_{\mathcal{U}}, \quad x \mapsto \tau(x) + \mathcal{N}_{\mathcal{U}}.$$

is linear (Why?). Let $x \in E_0$. Then we have

$$\|\tau_F(x)\| \geq \frac{1}{\|\tau^{-1}|_{\tau(F)}\|}\|x\| \geq \frac{1}{C}\|x\| \qquad (F \in \mathcal{F}, \, x \in F),$$

so that $\tilde{\tau}$ is bounded from below by the definition of $\|\cdot\|_{\mathcal{U}}$. Since E_0 is dense in E_1, the map $\tilde{\tau}$ has a unique extension to all of E_1, which is again bounded from below. It follows that (the unique extension of) $\tilde{\tau}$ is a topological isomorphism of E_1 and a closed subspace of $(E_2)_{\mathcal{U}}$.

(iii) $\implies$ (i): This follows from Proposition C.3.5 (How exactly?) $\square$

Exercise C.3.5 Let E be a Banach space. Show that E^{**} is isometrically isomorphic to a complemented subspace of some ultrapower of E. (*Hint*: To show that E^{**} embeds isometrically into an ultrapower of E, use Theorem C.3.3 and slightly vary the idea of the proof of Theorem C.3.7. To show that the resulting subspace of the ultrapower is complemented, think of how you (hopefully) solved Exercise C.3.4.)

Corollary C.3.8 *Let* $p \in [1, \infty)$*, let* $(\Omega, \mathcal{S}, \mu)$ *be a measure space, and let* E *be a Banach space which is finitely representable in* $L^p(\Omega, \mathcal{S}, \mu)$*. Then there is another measure space* $(\tilde{\Omega}, \tilde{\mathcal{S}}, \tilde{\mu})$ *such that* E *is topologically isomorphic to a subspace of* $L^p(\tilde{\Omega}, \tilde{\mathcal{S}}, \tilde{\mu})$*.*

Proof By Theorem C.3.7, E is topologically isomorphic to a closed subspace of some ultrapower $(L^p(\Omega, \mathcal{S}, \mu))_{\mathcal{U}}$. As a consequence of the Kakutani–Bohnenblust–Nakano theorem ([Al–B, Theorem 12.26]), there is a measure space $(\tilde{\Omega}, \tilde{\mathcal{S}}, \tilde{\mu})$ such that $(L^p(\Omega, \mathcal{S}, \mu))_{\mathcal{U}} \cong L^q(\tilde{\Omega}, \tilde{\mathcal{S}}, \tilde{\mu})$ ([Hei, Theorem 3.3(ii)]). $\square$

We shall now use Corollary C.3.8 to show that ℓ^p is not finitely representable in ℓ^q for certain values of p and q. Our main tool is the following theorem due to W. Orlicz ([Woj, II.D.6., Theorem]):

Theorem C.3.9 *Let* $p \in [1, 2]$*, let* $(\Omega, \mathcal{S}, \mu)$ *be a probability space, and let* $(f_n)_{n=1}^{\infty}$ *be a sequence in* $L^p(\Omega, \mathcal{S}, \mu)$*. Then* $\sum_{n=1}^{\infty} \|f_n\|_p^2 < \infty$*.*

Corollary C.3.10 *Let* $p \in (2, \infty)$*, let* $q \in [1, 2]$*, and let* $(\Omega, \mathcal{S}, \mu)$ *be a measure space. Then* ℓ^p *is not finitely representable in* $L^q(\Omega, \mathcal{S}, \mu)$*.*

Proof Assume towards a contradiction that ℓ^p is finitely represented in $L^q(\Omega, \mathcal{S}, \mu)$. By Corollary C.3.8, we may suppose that ℓ^p is topologically isomorphic to a closed linear subspace of $L^q(\Omega, \mathcal{S}, \mu)$. Since this subspace of $L^q(\Omega, \mathcal{S}, \mu)$ is obviously separable, we may use [D–S, III.8.5 Lemma] to suppose that $L^q(\Omega, \mathcal{S}, \mu)$ itself is separable. As a consequence of [Hal, Theorem 41.C] (work out the details in Exercise C.3.6 below), $L^q(\Omega, \mathcal{S}, \mu)$ then is isometrically isomorphic to a subspace of $L^q([0, 1])$. We may thus suppose without loss of generality that there is an embedding $\tau \colon \ell^p \to L^q([0, 1])$ with closed range.

Since $p > 2$, there is a sequence $(\lambda_n)_{n=1}^{\infty} \in \ell^p \setminus \ell^2$. Let $(e_n)_{n=1}^{\infty}$ be the standard basis of ℓ^p. Then $\sum_{n=1}^{\infty} \lambda_n e_n$ converges conditionally and so does $\sum_{n=1}^{\infty} \lambda_n \tau(e_n)$. By Theorem C.3.9 this means that $\sum_{n=1}^{\infty} |\lambda_n|^2 \|\tau(e_n)\|_q^2 < \infty$. Since the sequence $(\|\tau(e_n)\|_p)_{n=1}^{\infty}$ is bounded below, this implies that $(\lambda_n)_{n=1}^{\infty} \in \ell^2$, which is a contradiction. $\square$

Exercise C.3.6 Let $p \in [1, \infty)$, and let $(\Omega, \mathcal{S}, \mu)$ be a measure space such that $L^p(\Omega, \mathcal{S}, \mu)$ is separable. Use [Hal, Thoerem 41.C] to show that $L^p(\Omega, \mathcal{S}, \mu)$ is isometrically isomorphic to a subspace of $L^p([0, 1])$.

C.4 Notes and comments

It was a long standing open problem in Banach space theory as to whether every Banach space has the approximation property. Eventually, P. Enflo constructed a Banach space without the approximation property ([Enf]). Subsequently, his construction was simplified by A. M. Davie; his construction is presented in [L-Tz]. All those spaces are "Frankenstein spaces", i.e. they are interesting for no other reason, but for their lack of the approximation property. The first space "occurring in nature" which was shown to lack the approximation property was $\mathcal{L}(\mathfrak{H})$ for an infinite-dimensional Hilbert space $\mathfrak{H}$ ([Sza]). More generally, every W^*-algebra which is not amenable does not have the approximation property; see Remark 6.1.9. It is still an open problem whether or not $H^\infty(\mathbb{D})$, the Banach algebra of all bounded holomorphic functions on the open unit disc, has the approximation property ([D–F, 5.2 (4)]).

The Radon–Nikodým property is discussed in [D–F] and [D-U]. In [D-U] various equivalent formulations of the Radon–Nikodým property are given along with a list of spaces having or lacking that property ([D-U, p. 217–219]).

For more on the local theory of Banach spaces, we refer to [Woj]. Ultrapowers and, more generally, ultraproducts of Banach spaces are treated in the survey article [Hei].

D Operator spaces

The theory of operator spaces — sometimes also termed *quantized functional analysis* (see the preface of [E–R]) — is a fairly new and rapidly evolving part of functional analysis. Until quite recently, no account in book form was available.

This appendix collects the necessary background material for our discussion of operator amenability in Chapter 7. We we confine ourselves to the bare essentials and don't even attempt to give proofs. For most results we give references to the recent monograph [E–R] by E. G. Effros and Z.-J. Ruan (also highly recommended for further reading).

D.1 Abstract and concrete operator spaces

The idea behind operator space theory is simple: Instead of only looking at a normed space E, we simultaneously consider all spaces of matrices $\mathbb{M}_n(E)$ for $n \in \mathbb{N}$.

We need a few conventions about matrices.

For a linear space E and $n, m \in \mathbb{N}$, let $\mathbb{M}_{m,n}(E)$ denote the $m \times n$ matrices, i.e. with m rows and n columns, with entries in E; if $E = \mathbb{C}$, we simply write $\mathbb{M}_{n,m}$. We identify $\mathbb{M}_{m,n}$ with $\mathcal{L}(\ell_n^2, \ell_m^2)$, and denote the resulting norm on $\mathbb{M}_{m,n}$ by $|\cdot|_{m,n}$; if $m = n$, we write $|\cdot|_n$ for $|\cdot|_{n,n}$. If E is a linear space, $n, m \in \mathbb{N}$, $\lambda \in \mathbb{M}_{m,n}$, $x \in \mathbb{M}_m(E)$, and $\mu \in \mathbb{M}_{n,m}$, the expression $\lambda x \mu \in \mathbb{M}_n(E)$ is defined by formal matrix multiplication. Furthermore, for a linear space E, for $n, m \in \mathbb{N}$, $x \in \mathbb{M}_n(E)$ and $y \in \mathbb{M}_m(E)$, let

$$x \oplus y := \left[\begin{array}{c|c} x & 0 \\ \hline 0 & y \end{array}\right] \in \mathbb{M}_{n+m}(E).$$

Definition D.1.1 Let E be a linear space. A *matricial norm* on E is a family $(\|\cdot\|_n)_{n=1}^{\infty}$ such that $\|\cdot\|_n$ is a norm on $\mathbb{M}_n(E)$ for $n \in \mathbb{N}$ with the following properties:

$$\|\lambda x \mu\|_n \leq |\lambda|_{m,n} \|x\|_m |\mu|_{n,m} \qquad (\lambda \in \mathbb{M}_{m,n}, \, x \in \mathbb{M}_m(E), \, \mu \in \mathbb{M}_{n,m})$$

and

$$\|x \oplus y\|_{n+m} = \max\{\|x\|_n, \|y\|_m\} \qquad (x \in \mathbb{M}_n(E), \, y \in \mathbb{M}_m(E)). \tag{D.1}$$

Definition D.1.2 A linear space E equipped with a matricial norm $(\|\cdot\|_n)_{n=1}^{\infty}$ is called a *matricially normed space*. If each space $(\mathbb{M}_n(E), \|\cdot\|_n)$ is a Banach space, E is called an (abstract) *operator space*.

Exercise D.1.1 Let E be a linear space, and let $(\|\cdot\|_n)_{n=1}^{\infty}$ be a family such that $\|\cdot\|_n$ is a norm on $\mathbb{M}_n(E)$ for $n \in \mathbb{N}$ with the following properties:

$$\|\lambda x \mu\|_n \leq |\lambda|_n \|x\|_n |\mu|_n \qquad (\lambda, \mu \in \mathbb{M}_n,\, x \in \mathbb{M}_n(E))$$

and

$$\|x \oplus y\|_{n+m} = \max\{\|x\|_n, \|y\|_m\} \qquad (x \in \mathbb{M}_n(E),\, y \in \mathbb{M}_m(E)).$$

Show that E is a matricially normed space.

Exercise D.1.2 Let E be a matricially normed space and let $n \in \mathbb{N}$.

(i) Let $x \in \mathbb{M}_n(E)$, and let $\lambda, \mu \in \mathbb{M}_n$ be unitary. Show that $\|\lambda x \mu\|_n = \|x\|_n$.

(ii) Let $x, y \in \mathbb{M}_n(E)$ be such that y is obtained from x by permuting the rows or columns of x. Show that $\|x\|_n = \|y\|_n$.

Exercise D.1.3 Let E be a matricially normed space.

(i) Show that E is an operator space if and only if $(E, \|\cdot\|_1)$ is a Banach space.

(ii) Show that the completion of $(E, \|\cdot\|_1)$ is an operator space in a canonical fashion.

Examples D.1.3   (a) Let $\mathfrak{A}$ be a C^*-algebra. By Exercise 6.1.5, there is a unique C^*-norm $\|\cdot\|_n$ on $\mathbb{M}_n(\mathfrak{A})$ for each $n \in \mathbb{N}$. It is routinely checked that $(\|\cdot\|_n)_{n=1}^{\infty}$ is a matricial norm on $\mathfrak{A}$. Consequently, if E is any closed subspace of $\mathfrak{A}$, the restrictions of the norms $\|\cdot\|_n$ to $\mathbb{M}_n(E)$ for $n \in \mathbb{N}$ turn E into an operator space. Such operator spaces are called *concrete operator spaces*.

(b) Let E be any Banach space, and let Ω be the closed unit ball of E^* equipped with the w^*-topology. Then Ω is a compact Hausdorff space, and E embeds isometrically into the commutative C^*-algebra $\mathcal{C}(\Omega)$. By (a), we thus obtain a matricial norm $(\|\cdot\|_n)_{n=1}^{\infty}$ on E such that $\|\cdot\|_1$ coincides with the given norm. The resulting operator space is called the *minimal operator space* of E and denoted by $MIN(E)$.

(c) Let E be any Banach space. Let $n \in \mathbb{N}$; for $[x_{j,k}]_{\substack{j=1,\dots,n \\ k=1,\dots,n}} \in \mathbb{M}_n(E)$, let

$$\left\| [x_{j,k}]_{\substack{j=1,\dots,n \\ k=1,\dots,n}} \right\|_n := \sup \left\{ \left\| [Tx_{j,k}]_{\substack{j=1,\dots,n \\ k=1,\dots,n}} \right\|_n : T \in \mathcal{L}(E, \mathcal{L}(\ell^2)) \text{ is a contraction} \right\}. \tag{D.2}$$

Then $(\|\cdot\|_n)_{n=1}^{\infty}$ is a matricial norm on E such that $\|\cdot\|_1$ coincides with the given norm. The resulting operator space is called the *maximal operator space* of E and is denoted by $MAX(E)$.

So, if every Banach space "is" already an operator space, why bother defining operator spaces at all? The answer is that the same Banach space may carry very different operator space structures, i.e. given a Banach space E there are different matricial norms $(\|\cdot\|_n)_{n=1}^{\infty}$ and $(\|\|\cdot\|\|_n)_{n=1}^{\infty}$ on E such that nevertheless $\|\cdot\|_1 = \|\|\cdot\|\|_1$ is the given norm on E. This leads us immediately to the problem of when we can identify two operator spaces, i.e. of finding the right notion of isomorphism for operator spaces.

Exercise D.1.4 Let E be a matricially normed space, let F be a subspace of E, and let $n \in \mathbb{N}$. Check that F, E/F (if F is closed), and $\mathbb{M}_n(E)$ are matricially normed spaces in a canonical fashion.

D.2 Completely bounded maps

Let E and F be linear spaces, and let $T\colon E \to F$ be linear. At the beginning of Section 6.4, we defined, for $n \in \mathbb{N}$, the n-th *amplification* $T^{(n)}\colon \mathbb{M}_n(E) \to \mathbb{M}_n(F)$ of T: apply T to each matrix entry. Clearly (Why?), if each E and F are matricially normed spaces, $T \in \mathcal{L}(E,F)$ implies that $T^{(n)} \in \mathcal{L}(\mathbb{M}_n(E), \mathbb{M}_n(F))$ for each $n \in \mathbb{N}$.

Definition D.2.1 Let E and F be matricially normed spaces. An operator $T \in \mathcal{L}(E,F)$ is called *completely bounded* if

$$\|T\|_{cb} := \sup\left\{ \left\| T^{(n)} \right\| : n \in \mathbb{N} \right\} < \infty.$$

The collection of all completely bounded operators in $\mathcal{L}(E,F)$ is denoted by $\mathcal{L}_{cb}(E,F)$. If $T^{(n)}$ is a contraction for each $n \in \mathbb{N}$, we call T a *complete contraction*; if $T^{(n)}$ is an isometry for each $n \in \mathbb{N}$, we call T a *complete isometry*.

Clearly, for any two matricially normed spaces E and F, $\| \cdot \|_{cb}$ is a norm on the linear space $\mathcal{L}_{cb}(E,F)$.

Exercise D.2.1 Let E be a matricially normed space, and let F be an operator space. Show that $(\mathcal{L}_{cb}(E,F), \| \cdot \|_{cb})$ is a Banach space.

Examples D.2.2   (a) Let $\mathfrak{A}$ be a unital C^*-algebra, and let E be a closed subspace of $\mathfrak{A}$ containing $e_{\mathfrak{A}}$. Then, for any C^*-algebra $\mathfrak{B}$, we have

$$\|T\|_{cb} = \|T\| = \|Te_{\mathfrak{A}}\|$$

for all completely positive $T\colon E \to \mathfrak{B}$ ([E–R, Proposition 2.2.6]).

(b) Let E be a matricially normed space, let $n \in \mathbb{N}$, and let $T \in \mathcal{L}(E, \mathbb{M}_n)$. Then $T \in \mathcal{L}_{cb}(E, \mathbb{M}_n)$ with $\|T\|_{cb} = \left\| T^{(n)} \right\|$ ([E–R, Proposition 2.2.2]);this fact is known as *Smith's lemma*. In particular, E^* and $\mathcal{L}_{cb}(E, \mathbb{C})$ are isometrically isomorphic.

(c) Let E be an operator space, and let $\mathfrak{A}$ be a commutative C^*-algebra. Then $\mathcal{L}(E, \mathfrak{A}) = \mathcal{L}_{cb}(E, \mathfrak{A})$ with

$$\|T\|_{cb} = \|T\| (T \in \mathcal{L}(E, \mathfrak{A})),$$

so that $\mathcal{L}(E, \mathfrak{A})$ and $\mathcal{L}_{cb}(E, \mathfrak{A})$ are isometrically isomorphic ([Pau 1, Theorem 3.8]). It follows that, if F is an arbitrary Banach space, then $\mathcal{L}(E,F)$ and $\mathcal{L}_{cb}(E, MIN(F))$ are isometrically isomorphic.

(d) Let E be a Banach space, and let F be a subspace of $\mathcal{L}(\ell^2)$. From (D.2), it is immediate that $\mathcal{L}(E,F)$ and $\mathcal{L}_{cb}(MAX(E), F)$ are isometrically isomorphic.

(e) For $n \in \mathbb{N}$, the transpose map

$$T\colon \mathbb{M}_n \to \mathbb{M}_n, \quad A \mapsto A^t$$

is completely bounded (by (b)) with $\|T\|_{cb} = n$ ([E–R, Proposition 2.2.7]). It follows that, for any infinite-dimensional Hilbert space $\mathfrak{H}$, taking the Banach space adjoint

$$\mathcal{L}(\mathfrak{H}) \to \mathcal{L}(\mathfrak{H}), \quad S \mapsto S^*$$

is an isometry, but not completely bounded.

We can now formulate one of the central theorems of operator space theory ([Rua 1, Theorem 3.1]):

Theorem D.2.3 (Ruan's theorem) *Let E be an operator space. Then E is completely isometrically isomorphic to a concrete operator space.*

Remark D.2.4 Originally (see [Pau 1], for example), an operator space was defined to be what we call a concrete operator space. Ruan's theorem shows that operator spaces can (up to completely isometric isomorphism) be characterized through a very simple set of axioms. The advantage of this axiomatic approach to operator spaces is that it is much easier to establish hereditary properties and to show that a given space is indeed an operator space. This can perhaps be best illustrated with the analoguous result for Banach spaces: Up to isometric isomorphism, each Banach space is a closed subspace of $\mathcal{C}(\Omega)$ for some compact Hausdorff space Ω, so that we could *define* a Banach space as a closed subspace of $\mathcal{C}(\Omega)$ for a compact Hausdorff space Ω. With that "definition", however, it is extremely hard (if not impossible) to verify even very basic hereditary properties of Banach spaces, for example, that quotients and dual spaces of Banach spaces are again Banach spaces.

Exercise D.2.2 Let E be an operator space, and let F be a Banach space. Show that $\mathcal{L}(E, F)$ and $\mathcal{L}_{cb}(E, MIN(F))$ are isometrically isomorphic.

Exercise D.2.3 Let E be a Banach space, and let F be an operator space. Show that $\mathcal{L}(E, F)$ and $\mathcal{L}_{cb}(MAX(E), F)$ are isometrically isomorphic.

Let E and F be matricially normed spaces, and let $n \in \mathbb{N}$. For any $[T_{j,k}]_{\substack{j=1,\dots,n \\ k=1,\dots,n}} \in \mathbb{M}_n(\mathcal{L}_{cb}(E, F))$, we define a unique element in $\mathcal{L}_{cb}(E, \mathbb{M}_n(F))$ through

$$E \to \mathbb{M}_n(F), \quad x \mapsto [T_{j,k} x]_{\substack{j=1,\dots,n \\ k=1,\dots,n}}.$$

Conversely, every operator in $\mathcal{L}_{cb}(E, \mathbb{M}_n(F))$ yields a matrix in $\mathbb{M}_n(\mathcal{L}_{cb}(E, F))$. We can thus identify $\mathbb{M}_n(\mathcal{L}_{cb}(E, F))$ and $\mathcal{L}_{cb}(E, \mathbb{M}_n(F))$ as linear spaces. Through these identifications, we can equip $\mathcal{L}_{cb}(E, F)$ with a matricial norm:

Proposition D.2.5 *Let E and F be matricially normed spaces. For $n \in \mathbb{N}$, let $\|\cdot\|_n$ on $\mathbb{M}_n(\mathcal{L}_{cb}(E, F))$ be the norm $\|\cdot\|_{cb}$ on $\mathcal{L}_{cb}(E, \mathbb{M}_n(F))$. Then $\mathcal{L}_{cb}(E, F)$ is a matricially normed space which is an operator space if F is one.*

Exercise D.2.4 Prove Proposition D.2.5.

Remarks D.2.6   (a) By Examples D.2.2(b), $E^* = \mathcal{L}_{cb}(E, \mathbb{C})$ isometrically for each matricially normed space E. By Proposition D.2.5, there is thus a matricial norm $(\|\cdot\|_n)_{n=1}^\infty$ on E^* such that $\|\cdot\|_1$ is the dual Banach space norm on E^*. This turns E^* into an operator space — the *dual operator space* of E — and each subspace of E^* into a matricially normed space.

(b) Note that for $n \in \mathbb{N}$, although the Banach spaces $\mathbb{M}_n(E^*)$ and $\mathbb{M}_n(E)^*$ are isomorphic, the norm $\|\cdot\|_n$ on $\mathbb{M}_n(E^*)$ is *not* the dual norm of $\|\cdot\|_n$ on $\mathbb{M}_n(E)$. For $X = [x_{j,k}]_{\substack{j=1,\dots,n \\ k=1,\dots,n}} \in \mathbb{M}_n(E)$ and $\Phi = [\phi_{j,k}]_{\substack{j=1,\dots,n \\ k=1,\dots,n}} \in \mathbb{M}_n(E^*)$, let

$$\langle\langle X, \Phi \rangle\rangle := [\langle x_{j,k}, \phi_{l,m}\rangle]_{\substack{j,l=1,\dots,n \\ k,m=1,\dots,n}} \in \mathbb{M}_{n^2}.$$

We have ([E–R, p. 41]):

$$\|\Phi\|_n = \sup\{|\langle\langle X, \Phi\rangle\rangle|_{n^2} : X \in \mathbb{M}_n(E), \|X\|_n \leq 1\} \qquad (\Phi \in \mathbb{M}_n(E^*)).$$

(c) Let F be a closed subspace of E. Then we have a completely isometric isomorphism $F^\perp \cong (E/F)^*$ ([E–R, Proposition 4.2.1]). The proof requires an operator valued Hahn–Banach theorem.

(d) The dual operator space of E^* is again an operator space which endows its Banach space bidual E^{**} with an operator space structure. The canonical embedding of E into E^{**} is a complete isometry ([E–R, Proposition 3.2.1]).

The following is true ([E–R, Proposition 3.2.1] and [Ble, Lemma 1.1]):

Proposition D.2.7 *Let E and F be matricially normed spaces. Then*

$$\mathcal{L}_{cb}(E, F) \to \mathcal{L}_{cb}(F^*, E^*), \quad T \mapsto T^*$$

is a complete isometry.

We now turn from linear to bilinear maps. Let E_1, E_2 and F be matricially normed spaces, let $n, n_1, n_2 \in \mathbb{N}$ with $n = n_1 n_2$, and let $T \in \mathcal{L}(E_1, E_2; F)$. Then the (n_1, n_2)-amplification $T^{(n_1,n_2)} : \mathbb{M}_{n_1}(E_1) \times \mathbb{M}_{n_2}(E_2) \to \mathbb{M}_{n_1}(\mathbb{M}_{n_2}(F)) = \mathbb{M}_n(F)$ of T is defined through

$$T^{(n_1,n_2)}\left([x_{j,k}]_{\substack{j=1,\ldots,n_1 \\ k=1,\ldots,n_1}}, [y_{l,m}]_{\substack{l=1,\ldots,n_2 \\ m=1,\ldots,n_2}}\right)$$

$$:= \left[[T(x_{j,k}, y_{l,m})]_{\substack{l=1,\ldots,n_2 \\ m=1,\ldots,n_2}}\right]_{\substack{j=1,\ldots,n_1 \\ k=1,\ldots,n_1}}$$

$$\left([x_{j,k}]_{\substack{j=1,\ldots,n_1 \\ k=1,\ldots,n_1}} \in \mathbb{M}_{n_1}(E_1), [y_{l,m}]_{\substack{l=1,\ldots,n_2 \\ m=1,\ldots,n_2}} \in \mathbb{M}_{n_2}(E_2)\right).$$

Definition D.2.8 Let E_1, E_2 and F be matricially normed spaces. A bilinear map $T \in \mathcal{L}(E_1, E_2; F)$ is called *completely bounded* if

$$\|T\|_{cb} := \sup\left\{\left\|T^{(n_1,n_2)}\right\| : n_1, n_2 \in \mathbb{N}\right\} < \infty.$$

The collection of all completely bounded bilinear maps in $\mathcal{L}(E_1, E_2; F)$ is denoted by $\mathcal{L}_{cb}(E_1, E_2; F)$. If $\|T\|_{cb} \leq 1$, we say that T is *completely contractive*.

Exercise D.2.5 Let E_1, E_2 and F be matricially normed spaces, and let $T \in \mathcal{L}_{cb}(E_1, E_2; F)$. Show that

$$\|T\|_{cb} := \sup\left\{\left\|T^{(n,n)}\right\| : n \in \mathbb{N}\right\} < \infty.$$

Remark D.2.9 As in the case of completely bounded operators, $\|\cdot\|_{cb}$ is a norm on $\mathcal{L}_{cb}(E_1, E_2; F)$ which is complete if F is an operator space. Identifying $\mathbb{M}_n(\mathcal{L}_{cb}(E_1, E_2; F))$ and $\mathcal{L}_{cb}(E_1, E_2; \mathbb{M}_n(F))$ for $n \in \mathbb{N}$, we also obtain a matricial norm on $\mathcal{L}_{cb}(E_1, E_2; F)$.

D.3 Tensor products of operator spaces

Let E_1 and E_2 be matricially normed spaces. The canonical embedding of $E_1 \otimes E_2$ into $\mathcal{L}(E_1, E_2^*)^*$ (Exercise B.2.10) yields an identification of $E_1 \otimes E_2$ with a subspace of $\mathcal{L}_{cb}(E_1, E_2^*)^*$. This enables us to define a matricial norm on $E_1 \otimes E_2$:

Definition D.3.1 Let E_1 and E_2 be operator spaces. Then the matricial norm $(\|\cdot\|_{\pi,n})_{n=1}^{\infty}$ obtained through the embedding $E_1 \otimes E_2 \hookrightarrow \mathcal{L}_{cb}(E_1, E_2^*)^*$ is called the *operator projective tensor norm* on $E_1 \otimes E_2$. The completion of $E_1 \otimes E_2$ with respect to this matricial norm is called the *operator projective tensor product* of E_1 and E_2 and is denoted by $E_1 \hat{\otimes} E_2$.

Remarks D.3.2 (a) It is immediate from this definition that $\|\cdot\|_{\pi,1} \leq \|\cdot\|_{\pi}$.

(b) Let E_1 and E_2 be Banach spaces. Then we have a completely isometric isomorphism

$$MAX(E_1) \hat{\otimes} MAX(E_2) \cong MAX(E_1 \hat{\otimes} E_2)$$

(see [E–R, (8.2.9)]).

(c) Let $n, n_1, n_2 \in \mathbb{N}$ be such that $n = n_1 n_2$, and let $x \in \mathbb{M}_{n_1}(E_1)$ and $y \in \mathbb{M}_{n_2}(E_2)$. Then, if we identify $\mathbb{M}_{n_1}(E_1) \otimes \mathbb{M}_{n_2}(E_2)$ and $M_n(E_1 \otimes E_2)$, we have

$$\|x \otimes y\|_{\pi,n} = \|x\|_{\pi,n_1} \|y\|_{\pi,n_2}.$$

In particular, $\|\cdot\|_{\pi,1}$ is a cross norm. In fact, $(\|\cdot\|_{\pi,n})_{n=1}^{\infty}$ is the largest matricial norm $(\|\cdot\|_n)_{n=1}^{\infty}$ on $E_1 \otimes E_2$ such that

$$\|x \otimes y\|_n = \|x\|_{n_1} \|y\|_{n_2} \quad (n, n_1, n_2 \in \mathbb{N}, \ n = n_1 n_2, \ x \in \mathbb{M}_{n_1}(E_1), \ y \in \mathbb{M}_{n_2}(E_2)) \tag{D.3}$$

holds ([Bl–P, Theorem 5.5]).

(d) The operator projective norm on $E_1 \otimes E_2$ can also be defined intrinsically, in a way reminiscent of the definition of $\|\cdot\|_{\pi}$ ([E–R, p. 124]).

Like the projective tensor product of Banach spaces (see Exercise B.2.8), the operator projective tensor product of operator spaces is characterized by a universal property ([E–R, Proposition 7.1.2]):

Proposition D.3.3 *Let E_1, E_2, and F be operator spaces. Then we have canonical completely isometric isomorphisms*

$$\mathcal{L}_{cb}(E_1, E_2; F) \cong \mathcal{L}_{cb}(E_1 \hat{\otimes} E_2, F) \cong \mathcal{L}_{cb}(E_1, \mathcal{L}_{cb}(E_2, F)).$$

In particular, $(E_1 \hat{\otimes} E_2)^* \cong \mathcal{L}_{cb}(E_1, E_2^*)$ as operator spaces.

Exercise D.3.1 Let E_1, F_1, E_2, and F_2 be operator spaces, and let $T_j \in \mathcal{L}_{cb}(E_j, F_j)$ for $j = 1, 2$. Then $T_1 \otimes T_2 : E_1 \otimes E_2 \to F_1 \otimes F_2$ has an extension in $\mathcal{L}_{cb}(E_1 \hat{\otimes} E_2, F_1 \hat{\otimes} F_2)$ such that $\|T_1 \otimes T_2\|_{cb} = \|T_1\|_{cb} \|T_2\|_{cb}$.

Let $\mathfrak{M}$ and $\mathfrak{N}$ be W^*-algebras. Their preduals $\mathfrak{M}_*$ and $\mathfrak{N}_*$ are operator spaces, so that $\mathfrak{M}_* \hat{\otimes} \mathfrak{N}_*$ is well-defined. The following theorem identifies $\mathfrak{M}_* \hat{\otimes} \mathfrak{N}_*$ as the predual of another W^*-algebra ([E–R, Theorem 7.2.4]):

Theorem D.3.4 *Let $\mathfrak{M}$ and $\mathfrak{N}$ be W^*-algebras. Then we have a canonical completely isometric isomorphism*

$$\mathfrak{M}_* \hat{\otimes} \mathfrak{N}_* \cong (\mathfrak{M} \bar{\otimes}_{W^*} \mathfrak{N})_*.$$

Applying Theorem D.3.4 to the case where $\mathfrak{M} = VN(G)$ and $\mathfrak{N} = VN(H)$ for locally compact groups G and H, we obtain (see Theorem A.3.6 and Exercise 6.3.8):

Corollary D.3.5 *Let G and H be locally compact groups. Then we have a canonical completely isometric isomorphism*

$$A(G) \hat{\otimes} A(H) \cong A(G \times H).$$

Let E be a matricially normed space, and let $m, n \in \mathbb{N}$. If $k \geq \max\{n, m\}$, we may embed $\mathbb{M}_{n,m}(E)$ into $\mathbb{M}_k(E)$ by putting $\mathbb{M}_{n,m}(E)$ into the upper left corner of $\mathbb{M}_k(E)$. We thus obtain a norm $\|\cdot\|_{n,m}$ on $\mathbb{M}_{n,m}(E)$ by restricting $\|\cdot\|_k$ to $\mathbb{M}_{n,m}(E)$. It follows from (D.1), that this norm is independent of k.

Definition D.3.6 Let E and F be linear spaces, let $n, m, r \in \mathbb{N}$, let $X = [x_{j,k}]_{\substack{j=1,\ldots,n \\ k=1,\ldots,r}} \in \mathbb{M}_{n,r}(E)$ and let $Y = [y_{j,k}]_{\substack{j=1,\ldots,r \\ k=1,\ldots,m}} \in \mathbb{M}_{r,m}(F)$. Then the *matrix inner product* $X \boxtimes Y \in \mathbb{M}_{n,m}(E \otimes F)$ of X and Y is the matrix $[z_{j,k}]_{\substack{j=1,\ldots,n \\ k=1,\ldots,m}}$ with

$$z_{j,k} := \sum_{l=1}^{r} x_{j,l} \otimes y_{l,k} \qquad (j = 1, \ldots, n, \ k = 1, \ldots, m).$$

Definition D.3.7 Let E and F be matricially normed spaces, and let $n \in \mathbb{N}$. For $z \in \mathbb{M}_n(E \otimes F)$, let

$$\|z\|_{h,n} := \inf\{\|x\|_{n,r}\|y\|_{r,n} : r \in \mathbb{N}, \ x \in \mathbb{M}_{n,r}(E), \ y \in \mathbb{M}_{r,n}(F) \text{ such that } z = x \boxtimes y\}.$$
$$\text{(D.4)}$$

Remark D.3.8 For $z \in \mathbb{M}_n(E \otimes F)$, there are $r \in \mathbb{N}$, $x \in \mathbb{M}_{n,r}(E)$, and $y \in \mathbb{M}_{r,n}(F)$ such that $z = x \boxtimes y$ ([E–R, Lemma 9.1.1]), so that the infimum (D.4) is always finite.

We have the following ([E–R, Theorem 9.2.1]):

Theorem D.3.9 *Let E and F be operator spaces. Then $(\|\cdot\|_{h,n})_{n=1}^{\infty}$ is a matricial norm on $E \otimes F$ such that*

$$\|\cdot\|_{h,n} \leq \|\cdot\|_{\pi,n} \qquad (n \in \mathbb{N}).$$

Definition D.3.10 Let E and F be operator spaces. Then $(\|\cdot\|_{h,n})_{n=1}^{\infty}$ is called the *Haagerup norm* on $E \otimes F$. The completion of $E \otimes F$ with respect to this matricial norm is called the *Haagerup tensor product* of $E \otimes F$ and denoted by $E \tilde{\otimes}_h F$.

Remark D.3.11 It follows from Exercise 6.5.3 and [E–R, Theorem 9.4.1] that for two unital C^*-algebras, the Haagerup norm $\|\cdot\|_h$ on $\mathfrak{A} \otimes \mathfrak{B}$ as defined in Definition 6.5.1 coincides with $\|\cdot\|_{h,1}$.

D.4 Operator Banach algebras

Definition D.4.1 An *operator Banach algebra* is an algebra $\mathfrak{A}$ which is also an operator space such that the product map

$$\mathfrak{A} \times \mathfrak{A} \to \mathfrak{A}, \quad (a, b) \mapsto ab \tag{D.5}$$

is completely contractive.

Remarks D.4.2   (a) Let $\mathfrak{A}$ be an operator Banach algebra. Then $L_a, R_a \in \mathcal{L}_{cb}(\mathfrak{A})$ for all $a \in \mathfrak{A}$.

(b) If $\mathfrak{A}$ is an operator Banach algebra, (D.5) induces a complete contraction from $\mathfrak{A} \hat{\otimes} \mathfrak{A}$ to $\mathfrak{A}$. Like the diagonal operator from $\mathfrak{A} \check{\otimes} \mathfrak{A}$ to $\mathfrak{A}$, we denote this map by $\Delta_{\mathfrak{A}}$.

Note that operator Banach algebras are not to be confused with Banach operator algebras

. . .

Examples D.4.3   (a) Let $\mathfrak{A}$ be a Banach algebra. By Exercise D.2.3, the diagonal map $\Delta_{\mathfrak{A}} : MAX(\mathfrak{A} \check{\otimes} \mathfrak{A}) \to MAX(\mathfrak{A})$ is completely contractive. From Remarks D.3.2(b) and Proposition D.3.3, we conclude that

$$MAX(\mathfrak{A}) \times MAX(\mathfrak{A}) \to MAX(\mathfrak{A}), \quad (a,b) \mapsto ab$$

is completely contractive. Hence, $MAX(\mathfrak{A})$ is an operator Banach algebra.

(b) Let $\mathfrak{H}$ be a Hilbert space, and let $n, n_1, n_2 \in \mathbb{N}$ with $n = n_1 n_2$. Let $T_j \in \mathcal{L}(\mathfrak{H}^{n_j})$ for $j = 1, 2$. Let

$$\tilde{T}_1 := \underbrace{\begin{bmatrix} T_1 & \cdots & ,0 \\ \vdots & \ddots & \vdots \\ 0 & \cdots & T_1 \end{bmatrix}}_{n_2 \text{ blocks}} \in \mathcal{L}((\mathfrak{H}^{n_1})^{n_2})$$

and

$$\tilde{T}_2 := \underbrace{\begin{bmatrix} T_2 & \cdots & 0 \\ \vdots & \ddots & \vdots \\ 0 & \cdots & T_2 \end{bmatrix}}_{n_1 \text{ blocks}} \in \mathcal{L}((\mathfrak{H}^{n_2})^{n_1}),$$

and note that $\|\tilde{T}_j\| = \|T_j\|$ for $j = 1, 2$. Let $U \in \mathcal{L}((\mathfrak{H}^{n_2})^{n_1}, (\mathfrak{H}^{n_1})^{n_2})$ be the unitary operator defined by

$$\left(\xi_1^{(1)}, \ldots, \xi_1^{(n_2)}\right) \oplus \cdots \oplus \left(\xi_{n_1}^{(1)}, \ldots, \xi_{n_1}^{(n_2)}\right)$$
$$\mapsto \left(\xi_1^{(1)}, \ldots, \xi_{n_1}^{(1)}\right) \oplus \cdots \oplus \left(\xi_1^{(n_2)}, \ldots, \xi_{n_1}^{(n_2)}\right)$$

Identifying $(\mathfrak{H}^{n_2})^{n_1}$ and $\mathfrak{H}^n$, we obtain a contraction

$$\mathcal{L}(\mathfrak{H}^{n_1}) \times \mathcal{L}(\mathfrak{H}^{n_1}) \to \mathcal{L}(\mathfrak{H}^n), \quad T_1 \times T_2 \mapsto U^* \tilde{T}_1 U \tilde{T}_2 \tag{D.6}$$

If we make the canonical identifications $\mathcal{L}(\mathfrak{H}^{n_1}) \cong \mathbb{M}_{n_1}(\mathcal{L}(\mathfrak{H}))$, $\mathcal{L}(\mathfrak{H}^{n_2}) \cong \mathbb{M}_{n_2}(\mathcal{L}(\mathfrak{H}))$, and $\mathcal{L}(\mathfrak{H}^n) \cong \mathbb{M}_n(\mathcal{L}(\mathfrak{H}))$, we see that (D.6) is just the (n_1, n_2)-amplification of the product map of $\mathcal{L}(\mathfrak{H})$. Hence, $\mathcal{L}(\mathfrak{H})$ is an operator Banach algebra. Since every C^*-algebra can be viewed as a closed subalgebra of $\mathcal{L}(\mathfrak{H})$ for some Hilbert space $\mathfrak{H}$, it follows that every C^*-algebra is an operator Banach algebra.

(c) Let G be a locally compact group. Identifying $L^2(G) \hat{\otimes} L^2(G)$ and $L^2(G \times G)$, we obtain a unitary representation

$$G \to \mathcal{L}(L^2(G \times G)), \quad g \mapsto \lambda_2(g) \otimes \lambda_2(g). \tag{D.7}$$

Let $W \in \mathcal{L}(L^2(G \times G))$ be defined by

$$(W\xi)(g,h) := \xi(g, gh) \qquad (\xi \in L^2(G \times G), \, g, h \in G).$$

Clearly, W is unitary, and

$$\lambda_2(g) \otimes \lambda_2(g) = W^*(\lambda_2(g) \otimes \mathrm{id}_{L^2(G)})W \qquad (g \in G) \tag{D.8}$$

holds. From (D.8), it is immediate that (D.7) extends to a W^*-representation $\Delta^{VN(G)}$: $VN(G) \to \mathcal{L}(L^2(G \times G))$ which attains its values in $VN(G \times G)$, a so-called *comulti-plication* for $VN(G)$. Since $\Delta^{VN(G)}$ is a *-homomorphism, it is a complete contraction; also, $\Delta^{VN(G)} : VN(G) \to VN(G \times G)$ is w^*-continuous. Hence, $\Delta^{VN(G)}$ is the transpose of a complete contraction $\Delta_*^{VN(G)} : A(G \times G) \to A(G)$. By Remarks A.3.9(a) and with the identification $A(G \times G) \cong A(G) \hat{\otimes} A(G)$ (Corollary D.3.5), it follows that $\Delta_*^{VN(G)} : A(G) \hat{\otimes} A(G) \to A(G)$ is just $\Delta_{A(G)}$.

Remark D.4.4 Let $\mathfrak{A}$ be a closed subalgebra of $\mathcal{L}(\mathfrak{H})$ for some Hilbert space $\mathfrak{H}$. Then $\mathfrak{A}$, with its concrete operator space structure, is an operator Banach algebra. However, not every operator Banach algebra arises in this fashion: Let G be an infinite, locally compact abelian group with dual group Γ. Then $L^1(G) \cong A(\Gamma)$ is an operator Banach algebra by Examples D.4.3(c). However, since $L^1(G)$ is not Arens regular, it cannot occur as a closed subalgebra of a C^*-algebra. Those operator Banach algebras that can be completely isometrically embedded into $\mathcal{L}(\mathfrak{H})$ for some Hilbert space $\mathfrak{H}$ can be characterized in terms of their matricial norm ([E–R, Theorem 17.1.2]).

D.5 Notes and comments

The definitive account of operator space theory is the recent monograph [E–R]; all the material in this appendix can be found there. The notes [Pis 6] provide a somewhat different point of view. The survey article [Wit *et al.*] (available online), which we follow mostly as far as our choice of terminology is concerned, gives an excellent overview of the theory without hiding the fundamental ideas behind a barrage of detailed proofs. For an impression of what operator space theory — or rather: the theory of completely bounded maps between concrete operator spaces — looked like before Ruan's theorem was available, we refer to [Pau 1].

Our operator Banach algebras are generally called *completely contractive Banach algebras* in the literature.

A *quantization* is the act of replacing functions by operators: The observables of classical mechanics are functions, those of quantum mechanics are operators on Hilbert space. A Banach space is (up to isometric isomorphism) a space of continuous functions, and an operator space is (up to completely isometric isomorphism) a space of operators on Hilbert space: This motivates the name *quantized functional analysis* for the theory of operator spaces. Large parts of quantized functional analysis can be developed in analogy with classical functional analysis: For instance, the Hahn–Banach theorem and the Kreĭn–Milman theorem both have a respective counterpart in quantized functional analysis ([E–R, Theorem 4.1.5] and [Wit *et al.*, 9.2]).

An important result from Banach space theory that fails to carry over smoothly into the category of operator spaces is the principle of local reflexivity. Consequently, an operator space is called *locally reflexive* if the (operator space version of the) local reflexivity principle holds for it. The notion of local reflexivity for operator spaces is intimately connected with the notions of nuclearity, injectivity, and semidiscreteness ([E–R, Chapter 14]).

List of symbols

Page numbers (should) refer to the first occurrence of a symbol in the text. If a symbol is defined in the appendices, the first page number refers to the first occurrence in the main text and the second one to its definition.

References

[Al–B]   C. D. ALIPRANTIS and O. BURKINSHAW, *Positive Operators*. Academic Press, 1985.

[A–C–St]   E. ANDRUCHOW, G. CORACH, and D. STOJANOFF, A geometric characterization of nuclearity and injectivity. *J. Funct. Anal.* **133** (1995), 474–494.

[Ar–B]   R. J. ARCHBOLD and C. J. K. BATTY, C^*-tensor norms and slice maps. *J. London Math. Soc.* (2) **22** (1980), 127–138.

[Ari 1]   O. YU. ARISTOV, On the definition of a flat operator module. In: A. YA. HELEMSKIĬ (ed.), *Topological Homology*, pp. 29–38. Nova, 2000.

[Ari 2]   O. YU. ARISTOV, Biprojective algebras and operator spaces. Preprint (2001).

[Aup]   B. AUPETIT, *A Primer on Spectral Theory*. Springer Verlag, 1991.

[B–D–L]   W. G. BADE, H. G. DALES, and Z. A. LYKOVA, Algebraic and strong splittings of extensions of Banach algebras. *Mem. Amer. Math. Soc.* **656**.

[B–C–D]   W. G. BADE, P. C. CURTIS, JR., and H. G. DALES, Amenability and weak amenability for Beurling and Lipschitz algebras. *Proc. London Math. Soc.* (3) **55** (1987), 359–377.

[Ban]   S. BANACH, Sur le problème de la mesure. *Fund. Math.* **4** (1923), 7–33.

[B–T]   S. BANACH and A. TARSKI, Sur la décomposition des ensembles de points en parts respectivement congruents. *Fund. Math.* **6** (1924), 244–277.

[B–P]   E. BERKSON and H. PORTA, Representations of $\mathfrak{B}(X)$. *J. Funct. Anal.* **3** (1969), 1–34.

[Ble]   D. P. BLECHER, Tensor products of operator spaces, II. *Canad. J. Math.* **44** (1992), 75–90.

[Bl–P]   D. P. BLECHER and V. I. PAULSEN, Tensor products of operator spaces. *J. Funct. Anal.* **99** (1991), 262–292.

[B–D]   F. F. BONSALL and J. DUNCAN, *Complete Normed Algebras*. Springer Verlag, 1973.

[B–M]   J. A. BONDAY and U. S. R. MURTY, *Graph Theory with Applications*. North-Holland, 1976.

[Bu–P]   J. W. BUNCE and W. L. PASCHKE, Quasi-expectations and amenable von Neumann algebras. *Proc. Amer. Math. Soc.* **71** (1978), 232–236.

[C–E]   H. CARTAN and S. EILENBERG, *Homological Algebra*. Princeton University Press, 1956.

[Choi]   M. D. CHOI, A simple C^*-algebra generated by two finite-order unitaries. *Canad. J. Math.* **31** (1979), 347–350.

[Ch–E 1]   M. D. CHOI and E. G. EFFROS, Separable nuclear C^*-algebras and injectivity. *Duke. J. Math.* **43** (1976), 309–322.

[Ch–E 2]   M. D. CHOI and E. G. EFFROS, Nuclear C^*-algebras and injectivity: the general case. *Indiana Univ. Math. J.* **26** (1977), 443–446.

[Ch–E 3]   M. D. CHOI and E. G. EFFROS, Nuclear C^*-algebras and the approximation property. *Amer. J. Math.* **100** (1978), 61–79.

[Ch–S]   E. CHRISTENSEN and A. M. SINCLAIR, Completely bounded isomorphisms of injective von Neumann algebras. *Proc. Edinburgh Math. Soc* **32** (1983), 317–327.

[C–L–M]   J. CIGLER, V. LOSERT, and P. MICHOR, *Banach Modules and Functors on Categories of Banach Spaces*. Marcel Dekker, 1979.

[Coh]   D. L. COHN, *Measure Theory*. Birkhäuser, 1980.

[Conn 1]   A. CONNES, Classification of injective factors. *Ann. of Math.* **104** (1976), 73–114.

[Conn 2]   A. CONNES, On the cohomology of operator algebras. *J. Funct. Anal.* **28** (1978), 248–253.

[Conn 3]   A. CONNES, On the equivalence between injectivity and semidiscreteness for operator algebras. In : *Algèbres d'opérateurs et leurs applications en physique mathématique*, pp. 107–112. C.N.R.S., 1979.

[Conw 1]   J. B. CONWAY, *Functions of One Complex Variable*. Springer Verlag, 1978.

[Conw 2]   J. B. CONWAY, *A Course in Functional Analysis*. Springer Verlag, 1978.

[C–G 1]   G. CORACH and J. E. GALÉ, Averaging with virtual diagonals and geometry of representations. In: E. ALBRECHT and M. MATHIEU (ed.s), *Banach Algebras '97*, pp. 87–100. Walter de Grutyer, 1998.

[C–G 2]   G. CORACH and J. E. GALÉ, On amenability and geometry of spaces of bounded representations. *J. London Math. Soc.* (2) **59** (1999), 311–329.

[C–H]   M. COWLING and U. HAAGERUP, Completely bounded multipliers of the Fourier algebra of a simple Lie group of real rank one. *Invent. Math.* **96** (1989), 507–549.

[Cun]   J. CUNTZ, Simple C^*-algebras generated by isometries. *Comm. Math. Phys.* **57** (1977), 173–185.

[C–L]   P. C. CURTIS, JR., and R. J. LOY, The structure of amenable Banach algebras. *J. London Math. Soc.* (2) **40** (1989), 89–104.

[Dae]   A. VAN DAELE, *Continuous Crossed Products and Type III von Neumann Algebras*. Cambridge University Press, 1978.

[Dal]   H. G. DALES, *Banach Algebras and Automatic Continuity*. Oxford University Press, 2001.

[D–Gh–G]   H. G. DALES, F. GHAHRAMANI, and N. GRŒNBÆK, Derivations into iterated duals of Banach algebras. *Studia Math.* **128** (1998), 19–54.

[D–Gh–H]   H. G. DALES, F. GHAHRAMANI, and A. YA. HELEMSKIĬ, The amenability of measure algebras. Preprint (2001).

[D–V]   H. G. DALES and A. R. VILLENA, Continuity of derivations, intertwining maps, and cocycles from Banach algebras. *J. London Math. Soc.* (2) **63** (2001), 215–225.

[Day]   M. M. DAY, Means on semigroups and groups. *Bull. Amer. Math. Soc.* **55** (1949), 1054–1055.

[D–F]   A. DEFANT and K. FLORET, *Tensor Norms and Operator Ideals*. North-Holland, 1993.

[Der]   A. DERIGHETTI, Some results on the Fourier–Stieltjes algebra of a locally compact group. *Comm. Math. Helv.* **46** (1973), 328–339.

[D–Gh]   M. DESPIĆ and F. GHAHRAMANI, Weak amenability of group algebras of locally compact groups. *Canad. Math. Bull.* **37** (1994), 165–167.

[D–U]   J. DIESTEL and J. J. UHL, *Vector Measures*. American Mathematical Society, 1977.

[Din]   S. DINEEN, *Complex Analysis on Infinite Dimensional Spaces*. Springer Verlag, 1999.

[Dixm 1]   J. DIXMIER, Les moyennes invariantes dans les semi-groupes et leurs applications. *Acta Sci. Math.* (Szeged) **12** (1950), 213–227.

[Dixm 2]   J. DIXMIER, *Les algèbres d'opérateurs dans l'espace Hilbertien*. Gauthier–Villars, 1969.

[Dixm 3]   J. DIXMIER, C^*-*Algebras* (translated from the French). North-Holland, 1977.

[Dixo]   P. G. DIXON, Left approximate identities in algebras of compact operators on Banach spaces. *Proc. Royal Soc. Edinburgh* **104A** (1989), 169–175.

[D–S]   N. DUNFORD and J. T. SCHWARTZ, *Linear Operators*, I. Wiley Classics Library, 1988.

[E–L]   E. G. EFFROS and E. C. LANCE, Tensor products of operator algebras. *J. Funct. Anal.* **25** (1977), 1–34.

[Eff]   E. G. EFFROS, Amenability and virtual diagonals for von Neumann algebras. *J. Funct. Anal.* **78** (1988), 137–156.

[E–K]   E. G. EFFROS and A. KISHIMOTO, Module maps and Hochschild–Johnson cohomology. *Indiana Univ. Math. J.* **36** (1987), 257–276.

[E–R]   E. G. EFFROS and Z.-J. RUAN, *Operator Spaces.* Oxford University Press, 2000.

[Enf]   P. ENFLO, A counterexample to the approximation property in Banach spaces. *Acta. Math.* **130** (1973), 309–317.

[E–S]   M. ENOCK and J.-M. SCHWARTZ, *Kac Algebras and Duality of Locally Compact Groups.* Springer Verlag, 1992.

[E–P]   J. ESCHMEIER and M. PUTINAR, *Spectral Decompositions and Analytic Sheaves.* Oxford University Press, 1996.

[Eym]   P. EYMARD, L'algèbre de Fourier d'un groupe localement compact. *Bull. Soc. Math. France* **92** (1964), 181–236.

[Fol]   G. B. FOLLAND, *A Course in Abstract Harmonic Analysis.* CRC Press, 1995.

[Føl]   E. FØLNER, On groups with full Banach mean value. *Math. Scand.* **3** (1955), 243–254.

[For]   B. E. FORREST, Amenability and weak amenability of the Fourier algebra. Preprint (2000).

[Fre]   R. M. FRENCH, The Banach–Tarski theorem. *Mathematical Intelligencer* **10** (1988), 21–28.

[G–R–W]   J. E. GALÉ, T. J. RANSFORD, and M. C. WHITE, Weakly compact homomorphisms. *Trans. Amer. Math. Soc.* **331** (1992), 815–824.

[Gh–L–W]   F. GHAHRAMANI, R. J. LOY, and G. A. WILLIS, Amenability and weak amenability of second conjugate Banach algebras. *Proc. Amer. Math. Soc.* **124** (1996), 1489–1497.

[Gh–S]   F. GHAHRAMANI and YU. V. SELIVANOV, The global dimension theorem for weighted convolution algebras. *Proc. Edinburgh Math. Soc.* (2) **41** (1998), 393–406.

[Gh–R–W]   F. GHAHRAMANI, V. RUNDE, and G. A. WILLIS, Derivations on group algebras. *Proc. London Math. Soc.* (3) **80** (2000), 360–390.

[Gif]   J. GIFFORD, Operator algebras with a reduction property. Preprint (1997).

[G–L]   I. GLICKSBERG and K. DE LEEUW, The decomposition of certain group representations. *J. Anal. Math.* **15** (1965), 135–192.

[God]   R. GODEMENT, Les fonctions de type positif et la théorie des groupes. *Trans. Amer. Math. Soc.* **63** (1948), 1–84.

[Gou 1]   F. GOURDEAU, Amenability of Lipschitz algebras. *Math. Proc. Cambridge Phil. Soc.* **112** (1992), 581–588.

[Gou 2]   F. GOURDEAU, Amenability and the second dual of a Banach algebra. *Studia Math.* **125** (1997), 75–81

[Gre]   F. P. GREENLEAF, *Invariant Means on Locally Compact Groups.* Van Nostrand, 1969.

[Gri]   R. I. GRIGORCHUK, On the Milnor problem of group growth (in Russian). *Dokl. Akad. Nauk SSSR* **271** (1983), 30–33.

[Gro]   M. GROSSER, *Bidualräume und Vervollständigungen von Banachmoduln.* Springer Verlag, 1979.

[Grø 1]   N. GRØNBÆK, Amenability and weak amenability of tensor algebras and algebras of nuclear operators. *J. Austral. Math. Soc. Ser. A* **51** (1991), 483–488.

[Grø 2]   N. GRØNBÆK, Weak and cyclic amenability for noncommutative Banach algebras. *Proc. Edinburgh Math. Soc.* (2) **35** (1992), 315–328.

[G–W]   N. GRØNBÆK and G. A. WILLIS, Approximate identities in Banach algebras of compact operators. *Canad. Math. Bull.* **36** (1993), 45–53.

[G–J–W]   N. GRØNBÆK, B. E. JOHNSON, and G. A. WILLIS, Amenability of Banach algebras of compact operators. *Israel J. Math.* **87** (1994), 289–324.

[Gui]   A. GUICHARDET, Sur l'homologie et la cohomologie des algèbres de Banach. *C. R. Acad. Sci. Paris*, Sér. A **262** (1966), 38–42.

[Haa 1]   U. HAAGERUP, An example of a nonnuclear C^*-algebra, which has the metric approximation property. *Invent. math.* **50** (1978/79), 279–293.

[Haa 2]   U. HAAGERUP, All nuclear C^*-algebras are amenable. *Invent. math.* **74** (1983), 305–319.

[Haa 3]   U. HAAGERUP, The Grothendieck inequality for bilinear forms on C^*-algebras. *Adv. Math.* **56** (1985), 93–116.

[Haa 4]   U. HAAGERUP, A new proof of the equivalence of injectivity and hyperfiniteness for factors on a separable Hilbert space. *J. Funct. Anal.* **62** (1985), 160–201.

[H–L]   U. HAAGERUP and N. J. LAUSTSEN, Weak amenability of C^*-algebras and a theorem of Goldstein. In: E. ALBRECHT and M. MATHIEU (ed.s), *Banach Algebras '97*, pp. 221–243. Walter de Grutyer, 1998.

[Hak]   J. HAKEDA, On property P of von Neumann algebras. *Tôhoku Math. J.* **19** (1967), 238–242.

[Hal]   P. R. HALMOS, *Measure Theory*. Van Nostrand, 1950.

[Hau]   F. HAUSDORFF, *Grundzüge der Mengenlehre*. Veit, 1914.

[Hei]   S. HEINRICH, Ultraproducts in Banach space theory. *J. reine angew. Math.* **313** (1980), 72–104.

[Hel 1]   A. YA. HELEMSKIĬ, Flat Banach modules and amenable algebras. *Trans. Moscow Math. Soc.* **47** (1985), 199–224.

[Hel 2]   A. YA. HELEMSKIĬ, Some remarks about ideas and results of topological homology. In: R. J. LOY (ed.), *Conference on Automatic Continuity and Banach Algebras (Canberra, 1989)*, pp. 203–238. Australian National Universtiy, 1989.

[Hel 3]   A. YA. HELEMSKIĬ, *The Homology of Banach and Topological Algebras* (translated from the Russian). Kluwer Academic Publishers, 1989.

[Hel 4]   A. YA. HELEMSKIĬ, Homological essence of amenability in the sense of A. Connes: the injectivity of the predual bimodule (translated from the Russion). *Math. USSR–Sb* **68** (1991), 555–566.

[Hel 5]   A. YA. HELEMSKIĬ, *Banach and Locally Convex Algebras* (translated from the Russian). Oxford University Press, 1993.

[Hel 6]   A. YA. HELEMSKIĬ, 31 problems of the homology of the algebras of analysis. In: V. P. HAVIN and N. NIKOLSKIĬ (ed.s), *Linear and Complex Analysis, Problem Book 3*, I, pp. 54–78. Springer Verlag, 1994.

[Hel 7]   A. YA. HELEMSKIĬ, Homology for the algebras of analysis. In: *Handbook of Algebra*, pp. 151–274. North-Holland, 2000.

[Hel 8]   A. YA. HELEMSKIĬ, Wedderburn-type theorems for operator algebras and modules: traditional and "quantized" approaches. In: A. YA. HELEMSKIĬ (ed.), *Topological Homology*, pp. 57–92. Nova, 2000.

[Her]   C. HERZ, The theory of p-spaces with applications to convolution operators. *Trans. Amer. Math. Soc.* **154** (1971), 69–82.

[H–R]   E. HEWITT and K. A. ROSS, *Abstract Harmonic Analysis*, I and II. Springer Verlag, 1963 and 1970.

[Hoch 1]   G. HOCHSCHILD, On the cohomology groups of an associative algebra. *Ann. of Math.* (2) **46** (1945), 58–67.

[Hoch 2]   G. HOCHSCHILD, On the cohomology theory for associative algebras. *Ann. of Math.* (2) **47** (1946), 568–579.

[Hul]   A. HULANICKI, Means and Følner conditions on locally compact groups. *Studia Math.* **24** (1964), 87–104.

[Joh 1]   B. E. JOHNSON, Cohomology in Banach algebras. *Mem. Amer. Math. Soc.* **127** (1972).

[Joh 2]   B. E. JOHNSON, Approximate diagonals and cohomology of certain annihilator Banach algebras. *Amer. J. Math.* **94** (1972), 685–698.

[Joh 3]   B. E. JOHNSON, Perturbations of Banach algebras. *Proc. London Math. Soc.* (3) **34** (1977), 439–458.

[Joh 4]   B. E. JOHNSON, Derivations from $L^1(G)$ into $L^1(G)$ and $L^\infty(G)$. In: J. P. PIER (ed.), *Harmonic Analysis (Luxembourg, 1987)*, pp. 191–198. Springer Verlag, 1988.

[Joh 5]   B. E. JOHNSON, Weak amenability of group algebras. *Bull. London Math. Soc.* **23** (1991), 281–284.

[Joh 6]   B. E. JOHNSON, Weakly compact homomorphisms between Banach algebras. *Math. Proc. Cambridge Phil. Soc.* **112** (1992), 157–163.

[Joh 7]   B. E. JOHNSON, Non-amenability of the Fourier algebra of a compact group. *J. London Math. Soc.* (2) **50** (1994), 361–374.

[Joh 8]   B. E. JOHNSON, Symmetric amenability and the nonexistence of Lie and Jordan derivations. *Math. Proc. Cambridge Phil. Soc.* **120** (1996), 455–473.

[Joh 9]   B. E. JOHNSON, Permanent weak amenability of group algebras of free groups. *Bull. London Math. Soc.* **31** (1999), 569–573.

[Joh 10]   B. E. JOHNSON, The derivation problem for group algebras of connected locally compact groups. *J. London Math. Soc.* (2) **63** (2001), 441–452.

[J–K–R]   B. E. JOHNSON, R. V. KADISON, and J. RINGROSE, Cohomology of operator algebras, III. *Bull. Soc. Math. France* **100** (1972), 73–79.

[K–R]   R. V. KADISON and J. RINGROSE, *Fundamentals of the Theory of Operator Algebras*, I and II. American Mathematical Society, 1997.

[Kam]   H. KAMOWITZ, Cohomology groups of commutative Banach algebras. *Trans. Amer. Math. Soc.* **102** (1962), 352–372.

[Kir 1]   E. KIRCHBERG, The Fubini theorem for exact C^*-algebras. *J. Operator Theory* **10** (1983), 3–8.

[Kir 2]   E. KIRCHBERG, Commutants of unitaries in UHF-algebras and functorial properties of exactness. *J. reine angew. Math.* **452** (1994), 39–77.

[K–Ph]   E. KIRCHBERG and N. C. PHILLIPS, Embeddings of exact C^*-algebras in the Cuntz algebra $\mathcal{O}_2$. *J. reine angew. Math.* **525** (2000), 17–54.

[K–N]   SH. KOBAYASHI and K. NOMIZU, *Foundations of Differential Geometry*, I and II. Wiley-Interscience, 1963 and 1969.

[Lanc 1]   E. C. LANCE, On nuclear C^*-algebras. *J. Funct. Anal.* **12** (1973), 157–176.

[Lanc 2]   E. C. LANCE, *Hilbert C^*-modules — a Toolkit for Operator Algebraists*. Cambridge University Press, 1995.

[Lang 1]   S. LANG, *Real Analysis* (Second Edition). Addison–Wesley, 1983.

[Lang 2]   S. LANG, *Algebra* (Second Edition). Addison–Wesley, 1983.

[Lang 3]   S. LANG, *Differential Manifolds*. Springer Verlag, 1985.

[L–L–W]   A. T.-M. LAU, R. J. LOY, and G. A. WILLIS, Amenability of Banach and C^*-algebras on locally compact groups. *Studia Math.* **119** (1996), 161–178.

[L–P]   A. T.-M. LAU and A. L. T. PATERSON, Inner amenable locally compact groups. *Trans. Amer. Math. Soc.* **325** (1991), 155–169.

[Lep]   H. LEPTIN, Sur l'algèbre de Fourier d'un groupe localement compact. *C. R. Acad. Sci. Paris*, Sér. A **266** (1968), 1180–1182.

[L-Tz]   J. LINDENSTRAUSS and L. TZAFRIRI, *Classical Banach Spaces*, I and II. Springer Verlag, Classics in Mathematics, 1996.

[Los]   V. LOSERT, On tensor products of Fourier algebras. *Arch. Math. (Basel)* **43** (1984), 370–372.

[Loy *et al.*] R. J. LOY, C. J. READ, V. RUNDE, and G. A. WILLIS, Amenable and weakly amenable Banach algebras with compact multiplication. *J. Funct. Anal.* **171** (2000), 78–114.

[MacL]   S. MACLANE, *Homology*. Springer Verlag, 1995.

[M-U]   S. MAZUR and S. ULAM, Sur les transformations isométriques d'espaces vectoriels, normés. *C. R. Acad. Sci. Paris* **194** (1932), 946–948.

[Muj]   J. MUJICA, *Complex Analysis in Banach Spaces*. North Holland, 1986.

[Mur]   G. J. MURPHY, *C^*-Algebras and Operator Theory*. Academic Press, 1990.

[Neu]   J. VON NEUMANN, Zur allgemeinen Theorie des Maßes. *Fund. Math.* **13** (1929), 73–116.

[Pal]   T. W. PALMER, *Banach Algebras and the General Theory of *-Algebras*, I. Cambridge University Press, 1994.

[Pau 1]   V. I. PAULSEN, *Completely Bounded Maps and Dilations*. Longman Scientific & Technical, 1986.

[Pau 2]   V. I. PAULSEN, Relative Yoneda cohomology for operator spaces. *J. Funct. Anal.* **157** (1998), 358–393.

[Pat 1]   A. L. T. PATERSON, *Amenability*. American Mathematical Society, 1988.

[Pat 2]   A. L. T. PATERSON, Virtual diagonals and n-amenability for Banach algebras. *Pacific J. Math.* **175** (1996), 161–185.

[Ped]   G. K. PEDERSEN, *C^*-Algebras and their Automorphism Groups*. Academic Press, 1979.

[Pie]   J. P. PIER, *Amenable Locally Compact Groups*. Wiley-Interscience, 1984.

[Pis 1]   G. PISIER, *Factorization of Linear Operators and Geometry of Banach Spaces*. American Mathematical Society, 1986.

[Pis 2]   G. PISIER, *Similarity Problems and Completely Bounded Maps*. Springer Verlag, 1991.

[Pis 3]   G. PISIER, The operator Hilbert space OH, complex interpolation and tensor norms. *Mem. Amer. Math. Soc.* **585** (1996).

[Pis 4]   G. PISIER, The similarity degree of an operator algebra. *Algebra i Analiz* **10** (1998), 132–186.

[Pis 5]   G. PISIER, Non-commutative vector valued L_p-spaces and completely p-summing maps. *Astérisque* **247** (1998).

[Pis 6]   G. PISIER, *An Introduction to the Theory of Operator Spaces*. Notes du Cours du Centre Emile Borel, 2000.

[Pop]   S. POPA, A short proof of "injectivity implies hyperfiniteness" for finite von Neumann algebras. *J. Operator Theory* **16** (1986), 261–272.

[Pot]   S. POTT, An account of the global homological dimension theorem of A. Ya. Helemskiĭ. *Ann. Univ. Sarav. Ser. Math.* **9** (1999), 155–194.

[P-St]   R. T. POWERS and E. STØRMER, Free states of the canonical anticommutation relations. *Comm. Math. Phys.* **16** (1970), 1–33.

[Rae]   I. RAEBURN, The relationship between a commutative Banach algebra and its maximal ideal spaces. *J. Funct. Anal.* **25** (1977), 366–390.

[R-T]   I. RAEBURN and J. L. TAYLOR, Hochschild cohomology and perturbations of Banach algebras. *J. Funct. Anal.* **25** (1977), 258–266.

[Rea 1]   C. J. READ, Commutative, radical amenable Banach algebras. *Studia Math.* **140** (2000), 199–212.

[Rea 2]   C. J. READ, Relative amenability and the non-amenability of $B(l^1)$. Preprint (2001).

[R-St]   H. REITER and J. D. STEGEMAN, *Classical Harmonic Analysis and Locally Compact Groups*. Oxford University Press, 2000.

[Rua 1]   Z.-J. RUAN, Subspaces of C^*-algebras. *J. Funct. Anal.* **76** (1988), 217–230.

[Rua 2]   Z.-J. RUAN, The operator amenability of $A(G)$. *Amer. J. Math.* **117** (1995), 1449–1474.

[Rua 3]   Z.-J. RUAN, Amenability of Hopf–von Neumann algebras and Kac algebras. *J. Funct. Anal.* **139** (1996), 466–499.

[R-X]   Z.-J. RUAN and G. XU, Splitting properties of operator bimodules and operator a-menability of Kac algebras. In: A. GHEONDEA, R. N. GOLOGAN and D. TIMOTIN, *Operator Theory, Operator Algebras and Related Topics*, pp. 193–216. The Theta Foundation, 1997.

[Rud 1]   W. RUDIN, *Fourier Analysis on Groups*. Wiley Classics Library, 1990.

[Rud 2]   W. RUDIN, *Functional Analysis* (Second Edition). McGraw-Hill, 1991.

[Run 1]   V. RUNDE, The structure of contractible and amenable Banach algebras. In: E. ALBRECHT and M. MATHIEU (ed.s), *Banach Algebras '97*, pp. 415–430. Walter de Grutyer, 1998.

[Run 2]   V. RUNDE, The Banach–Tarski paradox or What mathematics and miracles have in common. *π in the Sky* **2** (2000), 13–15.

[Run 3]   V. RUNDE, Banach space properties forcing a reflexive, amenable Banach algebra to be trivial. *Arch. Math. (Basel)* (to appear).

[Run 4]   V. RUNDE, Amenability for dual Banach algebras. *Studia Math.* (to appear).

[Run 5]   V. RUNDE, Connes-amenability and normal, virtual diagonals for measure algebras. *J. London Math. Soc.* (to appear).

[Run 6]   V. RUNDE, Operator Figà-Talamanca–Herz algebras. Preprint (2001).

[Sak]   S. SAKAI, C^*-*Algebras and W^*-Algebras*. Springer Verlag, 1971.

[Sam]   C. SAMUEL, Bounded approximate identities in the algebra of compact operators on a Banach space. *Proc. Amer. Math. Soc.* **117** (1993), 1093–1096.

[Schwa]   J. SCHWARTZ, Two finite, non-hyperfinite, non-isomorphic factors. *Comm. Pure Appl. Math.* **16** (1963), 19–26.

[Sel 1]   YU. V. SELIVANOV, On Banach algebras of small global dimension zero (in Russian). *Uspekhi Mat. Nauk* **31** (1976), 227–228.

[Sel 2]   YU. V. SELIVANOV, Biprojective Banach algebras. *Math. USSR Izvestija* **15** (1980), 387–399.

[Sel 3]   YU. V. SELIVANOV, Homological dimensions of tensor products of Banach algebras. In: E. ALBRECHT and M. MATHIEU (ed.s), *Banach Algebras '97*, pp. 441–459. Walter de Grutyer, 1998.

[Sel 4]   YU. V. SELIVANOV, Cohomological characterizations of biprojective and biflat Banach algebras. *Monatsh. Math.* **128** (1999), 35–60.

[Sel 5]   YU. V. SELIVANOV, Biflat Banach algebras of weak homological bidimension one. Preprint (2001).

[Sheĭ]   M. V. SHEĬNBERG, A characterization of the algebra $C(\Omega)$ in terms of cohomology groups (in Russian). *Uspekhi Mat. Nauk* **32** (1977), 203–204.

[S-S]   A. M. SINCLAIR and R. R. SMITH, *Hochschild Cohomology of von Neumann Algebras*. Cambridge University Press, 1995.

[Sol]   M. SOLOVEJ, The cohomology comparison problem for Banach algebras. *J. London Math. Soc.* (2) **55** (1997), 499–514.

[Spr 1]   N. SPRONK, Operator weak amenability of the Fourier algebra. Preprint (2001).

[Spr 2]   N. SPRONK, Ph.D. thesis, University of Waterloo (in preparation).

[Stra]   Ş. STRĂTILĂ, *Modular Theory in Operator Algebras*. Editura Academiei, 1981.

[Stro]   K. STROMBERG, The Banach–Tarski paradox. *Amer. Math. Monthly* **86** (1979), 151–161

[Sza]   A. SZANKOWSKI, $B(\mathcal{H})$ does not have the approximation property. *Acta. Math.* **147** (1981), 89–108.

[SzN]   B. SZ.-NAGY, On uniformly bounded linear transformations in Hilbert space. *Acta Sci. Math.* (Szeged) **11** (1947), 152–157.

[Tak 1]   M. TAKESAKI, On the crossnorm of the direct product of C^*-algebras. *Tôhoku Math. J.* **16** (1964), 111–122.

[Tak 2]   M. TAKESAKI, *Theory of Operator Algebras*, I. Springer Verlag, 1979.

[Tar]   A. TARSKI, Algebraische Fassung des Maßproblems. *Fund. Math.* **31** (1938), 47–66.

[Tay]   J. A. TAYLOR, Homology and cohomology for topological algebras. *Adv. in Math.* **9** (1970), 137–182.

[Tom]   J. TOMIYAMA, *Tensor Products and Projections of Norm One in von Neumann Algebras*. Lecture notes, University of Copenhagen, 1970.

[Wag]   S. WAGON, *The Banach–Tarski Paradox*. Cambridge University Press, 1985.

[Was 1]   S. WASSERMANN, On Tensor products of certain group C^*-algebras. *J. Funct. Anal.* **23** (1976), 239–254.

[Was 2]   S. WASSERMANN, Injective W^*-algebras. *Math. Proc. Cambridge Phil. Soc.* **82** (1977), 39–47.

[W-O]   N. E. WEGGE-OLSEN, K-*Theory and* C^*-*Algebras*. Oxford University Press, 1993.

[Wei]   C. A. WEIBEL, *An Introduction to Homological Algebra*. Cambridge University Press, 1994.

[Wil]   G. A. WILLIS, When the algebra generated by an operator is amenable. *J. Operator Theory* **34** (1995), 239–249.

[Wit *et al.*]   G. WITTSTOCK *et al.*, Was sind Operatorräume? — Ein Internetlexikon. URL: http://www.math.uni-sb.de/~ag-wittstock/Lexikon99/ (1999).

[Wod]   M. WODZICKI, Resolution of the cohomology comparison problem for amenable Banach algebras. *Invent. Math.* **106** (1991), 541–547.

[Woj]   P. WOJTASZCZYK, *Banach Spaces for Analysts*. Cambridge University Press, 1991.

[Woo 1]   P. J. WOOD, *Homological Algebra in Operator Spaces with Applications to Harmonic Analysis*. Ph.D. thesis, University of Waterloo, 1999.

[Woo 2]   P. J. WOOD, The operator biprojectivity of the Fourier algebra. Preprint (2001).

[Zha]   Y. ZHANG, Maximal ideals and the structure of contractible and amenable Banach algebras. *Bull. Austral. Math. Soc.* **62** (2000), 221–226.

Index

Recent Reprints and New Editions